EARTH REINFORCEMENT
VOLUME 2

PROCEEDINGS OF THE INTERNATIONAL SYMPOSIUM ON EARTH REINFORCEMENT
FUKUOKA / KYUSHU / JAPAN / 12-14 NOVEMBER 1996

Earth Reinforcement

Edited by
HIDETOSHI OCHIAI
NORIYUKI YASUFUKU
KIYOSHI OMINE
Kyushu University, Fukuoka, Japan

VOLUME 2

Under the auspices of the Japanese Geotechnical Society
A.A. BALKEMA / ROTTERDAM / BROOKFIELD / 1997

IS Kyushu '96

The texts of the various papers in this volume were set individually by typists under the supervision of each of the authors concerned.

Published by
A.A. Balkema, P.O. Box 1675, 3000 BR Rotterdam, Netherlands (Fax: +31.10.4135947)
A.A. Balkema Publishers, Old Post Road, Brookfield, VT 05036-9704, USA (Fax: +1.802.2763837)

For the complete set of two volumes, ISBN 90 5410 833 9
For Volume 1, ISBN 90 5410 834 7
For Volume 2, ISBN 90 5410 835 5

Printed in the Netherlands

Earth Reinforcement, Ochiai, Yasufuku & Omine (eds) © 1997 Balkema, Rotterdam, ISBN 90 5410 833 9

Table of contents

Opening and closing ceremonies

Earth Reinforcement, Ochiai, Yasufuku & Omine (eds) © *1997 Balkema, Rotterdam, ISBN 90 5410 833 9*

Opening ceremony

Prof. Hidetoshi Ochiai
(Chairman of IS Kyushu '96)

Thank you, Mr. Chairman. Ladies and Gentlemen. It is indeed a great pleasure and honour to declare the opening of the International Symposium on Earth Reinforcement, IS Kyushu '96.

About 30 years have passed successfully since the modern concept of earth reinforcement was proposed by Henry Vidal in France in the middle of the 1960s. Expansion of experiences and knowledge in earth reinforcement practice has succeeded in developing new techniques and their applications to geotechnical engineering problems.

As a result of that, over the past decade, the earth reinforcing techniques have become a useful, powerful and economical solution to many problems in geotechnical and geoenvironmental engineering practice. So, it is a great advantage to us to get together and talk about our experiences and our knowledge and to share our wisdom in this field. Of course, this wisdom first came to us from our founding fathers like Henry Vidal.

Professor Toyotoshi Yamanouchi, who I am sure most of you here recognize him as one of the founding fathers, first proposed and organized the international symposium on Earth Reinforcement, here in Fukuoka in 1988, which was attended by delegates from 24 different countries. As you know, this first symposium provided successful contributions for the development of earth reinforcement.

The second one was held in 1992 with an attendance of about 400 participants, and was of great help for further promotion of earth reinforcement practice. These symposia are now known as "IS Kyushu".

This third symposium is a continuation and expansion of the two previous symposia. In this symposium, one special and three keynote lectures will be delivered by distinguished scholars in the field. The special reports on the performance of reinforced soil structures under the two great earthquakes, The Northridge Earthquake in 1994 and The Great Hanshin Earthquake in 1995, will also be presented. The summary discussion sessions on "Testing methods" and "Design methods" are organized to summarize the result of research activities developed in the last three decades in these areas. A total of 151 papers submitted from 30 different countries will be presented in technical and poster sessions during the symposium.

It is our hope that this symposium will lead to further understanding of current earth reinforcement practice and provide a base for further research.

The symposium is held under the auspices of the Japanese Geotechnical Society, and is supported by the International Geosynthetics Society and the Japan Society of Civil Engineers. I would like to express my sincere gratitude to these societies for their encouragement and support to the symposium.

Finally, there are 440 participants here attending this symposium, including about 130 overseas participants. On behalf of the symposium committee, I would like to extend a hearty welcome to all of you.

Thank you.

Prof. Kenji Ishihara
(President of the Japanese Geotechnical Society)

Good morning, distinguished quests, my dear colleagues, Ladies and Gentlemen.

It is a great pleasure to welcome all of you, from abroad as well as from within Japan, here to Fukuoka, and to have an opportunity to exchange views and discuss issues of earth reinforcement which has now become a subject of prime importance in our community.

It is my understanding that, while it is evolving rapidly, the discipline of earth reinforcement is still in a chaotic stage and a number of new ideas and techniques are being proposed and discussed. I hope some new perspectives are addressed and some initiatives to synthesize contents are developed in the course of sessions during this conference.

Generally speaking, the development of a discipline takes place stepwise from one conference to another. Thus, I wish this conference will place an important milestone in the further development of earth-reinforcing technology.

As you might know, Japanese Geotechnical Society has been hosting a series of International Symposia called IS's on specific subjects of current interest and of prime importance. One or two such symposia are held each year. Last week, we held IS Osaka, that is, 2nd International Congress on Environmental Geotechnology. Next year in 1997, Japanese National Society is going to host the International Symposium on Deformation and Progressive Failure in Geomechanics which will be held in Nagoya in October. I hope many of you will participate in it and find it useful and rewarding.

Lastly, I would like to express my sincere thanks and appreciation to members of Organizing Committee, and Chairman Professor Ochiai for their enormous efforts in bringing this conference to such a great event. I also wish all of you enjoy the conference and related events such as Poster sessions, Technical session, Exhibition and Sightseeing tours.

Thank you.

Prof. Iichiro Kono
(Representative of Japan Society of Civil Engineers)

Good morning, Ladies and Gentlemen. I am Kono, Professor of Civil Engineering of Okayama University. I am currently Division head of Investigation and Research in the Japan Society of Civil Engineers (JSCE).

Chairman of IS Kyushu, Professor Ochiai, invited me to attend this opening ceremony of International Symposium Kyushu '96. And JSCE President, Dr. Matsuo asked me to send warm and heartfelt welcome greetings to all the participations of the symposium at the occasion.

We, civil engineers, have the strong interest in the recent developments of geotechnical engineering and earth reinforcement. I know that many and various, theoretical and experimental reports will be presented in this symposium. I sincerely hope that the participants would make fruitful discussion, broadly and frankly. And you would make information exchange of knowledge of earth reinforcement.

I have read in a book that the origin of word "symposium" is Greek language. In the old age of Greece, philosophers discussed exhaustively various problems, theoretical and practical, drinking wine. That drinking-together discussion meeting was "symposium" in those days.

Before that the theory-orientated philosophers discussed the engineering problems impractically for a long time, but they could not get good solutions, they could not get much progress, because they did not have enough experimental evidences for practical improvement. They discussed fruitlessly problems only in the theoretical field.

On the other hand, the skill-orientated technicians received their training from their old masters only by following and copying their predecessor's examples without theoretical explanations. They could not get much progress by apprenticeship.

Thus, they needed the "symposium" in the original meaning; meeting together of theoretical philosophers and practical technicians, meeting together of desk-work planners and construction-site workmen. All of these people met together and exchanged their information about their own fields with each other, and discuss frankly and freely common problems. The symposium was needed for both philosophers and technicians to give and take knowledge to and from each other.

This symposium, Kyushu '96, is the place for both types of practitioners of earth reinforcement to study each other's problems, understand each others' situations.

I believe that the International Symposium Kyushu '96 is timely academic meeting, which will offer a lot of hints to solve the geotechnical problems, earth reinforcement. I would like to encourage all of the participants to have enthusiastic and frank discussion, and exchange information.

I hope finally that the study of earth reinforcement will make great advance in Kyushu, Japan and internationally in the world. I also hope again that this international symposium will come to happy end in great success.

Thank you for your kind attention.

Earth Reinforcement, Ochiai, Yasufuku & Omine (eds) © 1997 Balkema, Rotterdam, ISBN 90 5410 833 9

Closing ceremony

Prof. C.J.F.P. Jones
(President of the International Geosynthetic Society)

Thank you, Mr.Chairman. Ladies and Gentlemen.

This is the third conference on Reinforced Soil that has been held in Kyushu. The first in 1988 attracted delegates from 21 countries. In 1992, 25 countries were represented, and at this conference there have been delegates from 33 countries and all the continents. And it is appropriate to say that IS Kyushu series of reinforced soil symposiums are now in due represent the major conference on the subject of earth reinforcement.

Since the last conference in 1992 we have had two major earthquakes, one in California and one in Japan. Although we regret the loss of life and great damage that occurred in these events, these earthquakes provide a dramatic demonstration of the benefits and safety of reinforced soil structures. In addition, we have seen the publication, since 1992, of a number of national design codes and specifications. The result of this, the earthquakes and the codes, is that there has been a marked growth in confidence in reinforced soil structures since the last conference in 1992 and this is reflected in this symposium.

The papers at the symposium do reflect a significant contribution to knowledge and a great deal of information of the highest quality has been presented. It is also apparent that we are no longer exploring the basics of the technology but are interested in the wider aspects of reinforced soil. Much attention is now being given to difficult soils and the solution of difficult technical problems using earth reinforcement. What is apparent is that we still have a lot of research to do.

In the past, the practitioners and manufacturers associated with reinforced soil led the technology, and the academic community were behind the practitioners in their understanding. It may be, when you look at the papers at the conference, that the academic community has caught up and that our theoretical understanding may in future lead practice. In other words, the cross-over point may have occurred at this conference. But that said, we do not fully understand the mechanisms of the system in particular when we start talking about difficult soils, and we have still a lot to learn. The subject, if anything, is growing.

It would be remiss of me not to mention geosynthetic materials. With problem soils, and it will appear that we are going towards the use of reinforced soil in problem soils, it seems that geosynthetic reinforcement and drainage products are the primary elements that we are using and propose to use to effect the necessary improvements and stabilities. It is clear that the testing and properties of these materials is still a subject of great interest and debate. But the information base is clearly growing, a lot of information has been given here and it is very clear that a lot of people know a great deal about these materials now. However, we still don't have and still need agreed methods of testing of these products and there is evidence, though, that this is been provided.

Now, the format of this conference has been a model of how to organize an international symposium. A significant proportion of the conference time has been devoted to structured discussions. This is very refreshing and it also has been very valuable and I believe the high point of this symposium. We have to recognize that to produce a structured discussion an awful lot of work is done by quite a number of people, and I should think we should acknowledge, give our thanks to all those who have worked at this aspect of the symposium.

Conferences have a social side and this symposium is no exception. Much of the benefits of coming to a conference or symposium is to meet colleagues from other countries and to discuss, perhaps not in the conference hall but outside, interests and also to make friends and acquaintances and this is to a certain extent how we grow and how our understanding grows. Perhaps, it is outside the conference hall where the main work is done. IS Kyushu provides us with a very nice venue to do this.

We have had and have a large number of overseas delegates at the conference and on their behalf I should like to thank our Japanese hosts for their hospitality. Kyushu represents, to many visitors to Japan, what we perceive to be the ideal Japan. Everyone is very helpful, everyone is very cheerful, everyone is very friendly, everything works, it's unusual, it doesn't happen in our country anyway. The food is strange and sometimes exotic, the drink sometimes rather powerful, and then we even have Sumo. Your hospitality is overwhelming, and I am surprised so many people actually managed to get here at 8:30 every morning. I found it quite difficult.

Finally, on behalf of all of the delegates I should like to express our heartfelt thanks and gratitude to Professor Ochiai and his organizing committee for providing us with such an enjoyable conference.

Thank you very much indeed.

Prof. Shigenori Hayashi
(Vice Chairman of IS Kyushu '96)

Thank you Mr. Chairman. Good evening, Ladies and Gentlemen.

First of all, I would like to express my sincere appreciation to professor Colin Jones for his excellent Concluding Remarks of the Symposium.

As professor Jones mentioned at his concluding remarks that we have held three times Conference on Earth Reinforcement, IS Kyushu '88, '92 and this Symposium, IS Kyushu '96, so we have held, in this time, two Summary Discussion Sessions, on Testing Methods and Design Methods, and the Special Reports on two earthquakes.

We can share the developments and the state of the arts on earth reinforcement technics, and it make sure that the earth reinforcement is one of the most useful technics in geotechnical field, especially with seismic and environmental concerns.

However, I think that there are still so many problems for research and new development of construction methods, for example, how to evaluate the environmental effects, the deformation, the ductility and durability of the earth reinforcement structures, and how to decide the many partial safety factors, and so on.

In closing this International symposium on Earth Reinforcement, IS Kyushu '96, I wish to express, on behalf of all the members of our Organizing Committee, our sincere gratitude to the Special Lecturer, Special Reporters, Keynote Lecturers, Discussion Leaders, Panelists, each Session Chairmen and Cochairmen, Oral and Poster Presenters, and all of you, who have so actively participated in this Symposium to make it such a success.

Now I have the duty and the honor to declare this International Symposium on Earth Reinforcement, IS Kyushu '96, officially closed. And also I would like to declare the new challenges and new development on Earth Reinforcement Techniques started.

Thank you, and I wish you all a safe trip home. Thank you!

Special lecture

Earth Reinforcement, Ochiai, Yasufuku & Omine (eds) © 1997 Balkema, Rotterdam, ISBN 90 5410 833 9

Geosynthetic materials with improved reinforcement capabilities

C.J.F.P.Jones, A.Fakher, R.Hamir & I.M.Nettleton
Department of Civil Engineering, University of Newcastle upon Tyne, UK

ABSTRACT: The use of geosynthetic materials as reinforcement in soil is established practice. The materials used can take any form, the most common being strips, woven textiles and grids. The reinforcing mechanism is the provision of a tensile component to the soil strength developed through bond, anchorage or mechanical interlock. With the growing maturity of reinforced soil improved forms of reinforcement can be used to resolve specific technical problems.

Although all soils can be reinforced, good quality fill is usually specified and the use of fine grained soils in some codes of practice is not accepted. This can be a restriction of the use of reinforced soil, particularly when it is difficult to obtain good quality cohesionless fill. It has been shown that many soils not accepted by established codes of practice can be used to form serviceable structures. In some cases the explanation for the success of these structures is not apparent if analysed in accordance with recognised analytical procedures. However, incorporating drainage into the reinforced soil construction can improve the reinforcement mechanism and composite geosynthetic materials with the combined functions of drainage and reinforcement can be a logical choice in some reinforced soil applications. The use of electrically conductive geosynthetic materials introduces a new dimension in reinforced soil technology.

1 INTRODUCTION

The modern concept of earth reinforcement and soil structures was proposed by Casagrande who idealised the problem in the form of a weak soil reinforced by high-strength membranes laid horizontally in layers (Westergaard, 1938). The modern form of soil reinforcement was introduced by Vidal in the 1960s. Vidal's concept was for a composite material formed from flat (metal) reinforcing strips laid horizontally in a frictional soil, the interaction between the soil and the reinforcing members being solely by friction generated by gravity.

The introduction of the Vidal structures led to rapid development. Much fundamental work was sponsored by various national bodies, notably at the Laboratoire des Ponts et Chaussées (LCPC) in France (Schlosser, 1977), by the United States Department of Transportation (Walkinshaw, 1975) and by the United Kingdom Department of Transport (Murray, 1977). This work led to a better understanding of the fundamental concepts involved and the introduction of improved forms of reinforcement such as grids, first used by the California Department of Transportation (Forsyth, 1978).

Material development is interrelated with soil structure developments. Whereas historic reinforced soil structures were formed using organic materials such as timber, straw or reed for reinforcement, as early as the 19th century, Pasley (1822) recognised the potential of more advanced forms of reinforcement; his use of canvas as a reinforcing membrane is arguably the first application of geotextile reinforcement. Canvas could only have been expected to have a limited life before deterioration and Pasley's structures would not have been expected to be very durable.

The use of geotextiles for permanent reinforcement could not be contemplated until the development of synthetic polymer based materials. Synthetic fabrics were known prior to 1940 but it was not until the late 1960s and early 1970s that the advances in synthetic fabric and geotextile developments produced materials whose longevity could be assured.

1.1 *Polymeric Reinforcement Properties*

Polymer materials used as soil reinforcement have four main requirements. They must be strong, relatively stiff, durable and bond with the soil. Of critical importance is that the strength of the reinforcement is sufficient to support the force required to achieve stability of the structure. The magnitude of the required force will vary depending upon the application.

In a steep slope strengthened by a geotextile reinforcing layer each reinforcement might have to support a force of 10 - 40 kN/m; alternatively a single geotextile reinforcing layer at the base of an embankment on soft soil may be required to support a tensile force of 100 - 400 kN/m to ensure stability. The tensile strength required to provide support of structures in areas of subsidence can be significantly greater with long term strengths in excess of 1,000 kN/m.

The requirement of the geotextile to be stiff is so that the required force can be mobilised at a tensile strain which is compatible with the deformation of the soil. The concept of strain compatibility between the reinforced soil and the soil is implicit in any reinforced soil structure, Jewell (1992). The allowable tensile strain depends on the application and in the case of a reinforced slope on soft soil the allowable extension can vary from 5 - 10 per cent. In the case of a reinforced soil wall the design *allowable* tensile extension of the reinforcement is unlikely to exceed 2 - 4 per cent, with a limitation of <1 per cent strain occurring after construction. The *actual* extension is likely to be <2 per cent in total.

Durability of the polymeric reinforcement is influenced by time and has to be considered together with the environment conditions. With permanent structures durability is a dominant consideration of designers.

The mechanical requirement for bond between the reinforcement and the soil is important but often a function of the form of the polymer reinforcement. Geogrids and conventional geotextiles in the form of sheets provide good bond with the soil either due to the large surface offered by the geotextile or, in the case of geogrids, by soil/reinforcement interlock. In the case of strip or bar reinforcement, bond can become a critical consideration particularly in the top of a reinforced soil structure, Hassan (1992).

2 MATERIAL DEVELOPMENT

The successful development of many innovative technologies depends upon the active support of influential "champions". In the case of geosynthetic reinforced soil structures in the United Kingdom, the active support of the Department of Transport (DTp), the Transport Research Laboratory (TRL), the Highway Authorities and the Universities has been essential. In addition, the material manufacturers have been a major influence in initiating research and development.

Different forms of high quality geosynthetic reinforcement materials providing the essential properties detailed above have been developed by a number of manufacturers. Brief details of the main geosynthetic materials developed in the United Kingdom during the ten year period 1973-83 are shown in Table 1. Several of these have been the subject of continuous development and improvement, an example being Tensar reinforcement which over a period of 10 years shows an increase in tensile strength/weight of material of 150 per cent, Netlon (1996).

Developments of polymeric reinforcing materials have occurred in other parts of the world, notably in Europe, Japan and the United States. Nearly all are associated with different forms of grid materials or geotextiles in the form of needle-punched, heat-bonded or woven/knitted products.

2.1 *Benefits of PolymericReinforcement*

The initial development of polymeric reinforcement was to address the potential problem of corrosion of steel reinforcing elements and provide economies. Polymeric materials do not corrode, although durability has to be assured as described previously. A second major incentive to the use of polymeric reinforcement is the possibility of using indigenous and waste fills which can provide significant savings in construction costs. The use of pulverised fuel ash as a fill in conjunction with geosynthetic reinforcement is now established in the UK and can provide benefits not only in cost savings but also in the technical benefits arising from the cohesive nature of many of these fills which produce a lighter and inherently stronger structure, Jones *et al* (1990). The use of indigenous (cohesive) fill in steepened slopes

Table 1: Main geosynthetic reinforcing materials developed in the United Kingdom during 1973-83

Material (name)	Form	% extension at working load	Load capacity or characteristic tensile strength kN/m² or kN	Manufacturer (approx. date of introduction to the market)
Glass reinforced plastic (Fibretain)	strip §	0.2	16 - 80	Pilkington plc (1973)
Molecularly orientated grid (Tensar)	grid	< 3 *	varies (typically 97)	Netlon (1978)
Woven fabric (Terran RF12)	fabric	< 3 *	varies #	Imperial Chemical Industry (ICI) plc (1975)
Linear composite strip (paraweb)	tape	< 2	10 - 100	ICI/Linear Composites (1976)
Linear composite grid (paragrid)	grid	2 - 5	varies (300 - 1000+)	ICI/Exxon Chemical Geopolymers Ltd (early 1980s)

* Extension at working load is usually in range 1 - 2 %
\# Depends upon structure and weight per square metre
§ Normal width 40 - 160 mm

is common and this is the application where the majority of polymeric reinforcement is used.

The acute lack of suitable frictional fill in some countries such as parts of Japan has lead to the use of cohesive soils in major reinforced soil structures. In the case of Japan notable uses of cohesive fill include widening of railway embankments supporting the railway network. Extensive research by Japan Railways including the construction of full scale trial walls has established that geosynthetic reinforced soil structures formed using cohesive or cohesive-frictional fill are potentially more stable than structures formed from purely frictional fills, Tatsuoka *et al* (1992), Tatsuoka (1992). It is notable that this form of construction survived the Kobe earthquake, Tatsuoka *et al* (1995).

In France the development of geosynthetic reinforced structures has been continuing for over twenty years. A major argument advanced for the use of geosynthetic reinforcements is that on site material can be used for construction, whatever its nature. This concept has been demonstrated in the Lezat experimental wall where on site silt was used in the works. One of the largest structures built in France to date is a 21 m high structure, formed in three 7 m tiers with two 3 m berms, Gourc and Matichard (1994).

In Italy permanent geosynthetic reinforced soil walls have become an accepted and widely used technique. The low cost of the constituent materials forming the structures, the speed of construction and the possibility of developing alternative solutions are quoted as primary reasons for their success, Cazzuffi *et al* (1994).

3 TRIGGERS FOR ADVANCED GEOSYNTHETICS

Development of new and advanced forms of geosynthetic reinforcement is brought about by a number of factors including market forces, technical challenges, innovation and better understanding of the fundamental mechanisms of reinforced soil applications. Each of these can be shown to have been the trigger for new materials or material developments.

Market Forces As the use of reinforced soil has grown so competition in the material industry has increased. This in turn stimulates development of reinforcement products by manufacturers either using different reinforcing materials or improved manufacturing techniques. Examples of such developments can be found in the improvement in the range and quality of geogrid products offered by a number of manufacturers.

Technical Challenges As experience with reinforced soil has grown so the potential usage of the technique has expanded. Design concepts previously unknown to civil engineering have been possible; an example is the construction on super soft soil. The use of reinforced soil for different applications exposes areas where knowledge is lacking and identifies areas of uncertainty; examples have been the influence of repeated loading or seismic activity on reinforced soil structures and the soil-reinforcement bond with some forms of reinforcement. Consideration of the economics of reinforced soil illustrates that the use of waste fill or poor quality fill could be extremely beneficial.

Each of these technical challenges has been the trigger for the development of new forms of reinforcement or studies leading to a better understanding of the underlying mechanism which in turn leads to better design and economic benefit. Some of these are illustrated below.

3.1 *Reinforcement Bond*

When sheets, bars and strips are used as reinforcement the coefficient of friction between the reinforcement and soil is a critical property: the higher the friction the more efficient the reinforcement. Thus an ideally rough bar, strip or sheet is significantly better than a reinforcement with a smooth surface, Schlosser and Elias (1978).

Reinforcement in the form of grids has been shown to be significantly more effective in bond. The frictional adherence between the longitudinal members of a grid and the soil is a function of the surface properties and the coefficient of friction between the longitudinal members and the soil. However, the influence of the horizontal bearing capacity of the transverse elements is of an order greater, Forsyth (1978).

Reinforcement bond can be increased by the use of reinforcement in the form of anchors. This concept evolved simultaneously in Europe, Japan and the USA in the 1980s. The multi-anchor system was developed by Fukuoka (1980) for the Japanese Ministry of Construction. The anchor is in the form of a rectangular steel plate. The NEW retaining wall system, developed in Austria, is based upon an elemental concrete facing and polymeric ties in the form of a loop, Brandl & Dalmatiner (1986). In the USA and the UK anchors formed from waste automobile tyres illustrate both the economic and the environmental benefits of reinforced soil. Steel anchors formed from a single piece of rebar were developed by the Transport Research Laboratory in the UK, Murray & Irwin (1981). The first polymeric anchor was developed in 1992, Jones & Hassan (1992), Figure 1. The use of a polymeric anchor offers a reduction in the total quantity of reinforcing materials whilst still satisfying the design requirements of existing Codes of Practice. Using a polymeric anchor, material savings of up to 40 per cent are possible, Figure 2.

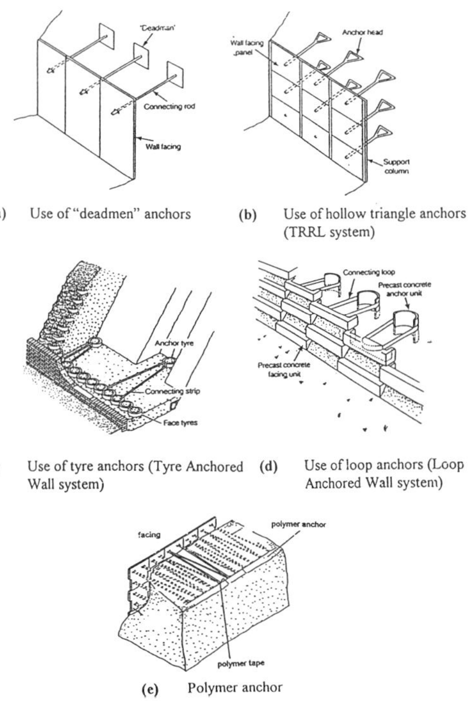

(a) Use of "deadmen" anchors (b) Use of hollow triangle anchors (TRRL system)

(c) Use of tyre anchors (Tyre Anchored Wall system) (d) Use of loop anchors (Loop Anchored Wall system)

(e) Polymer anchor

Figure 1: Anchored soil retaining wall systems

3.2 *Construction on Super Soft Soil*

In the past the construction method used when dealing with very soft soil was excavation and replacement. This is expensive and not always

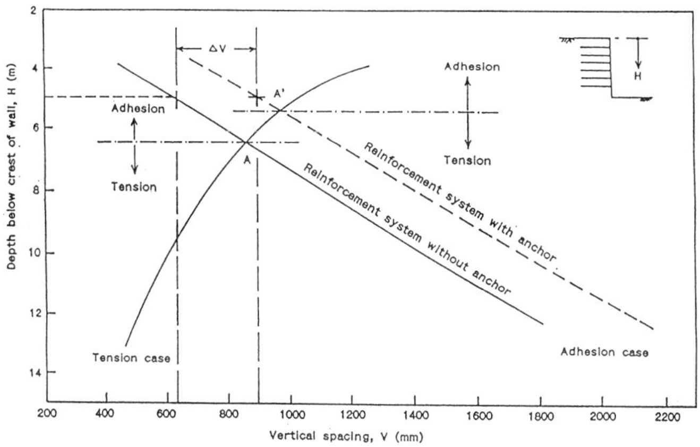

Figure 2: Economic benefits of polymeric anchor (increase in vertical spacing Δv of reinforcement)

practical; as a result treatment is becoming more popular. The most common method to stabilise soft soil is by preloading, frequently used in conjunction with vertical drains. When the soil is extremely soft, commonly referred to as a *super soft* soil, where the moisture content is higher than the liquid limit, preloading is not possible as the foundation has no effective bearing capacity. In this case a primary construction stage may be the creation of a working platform on which remedial treatment can be based, Yamanouchi (1970). The working platform (primary construction) frequently consists of geosynthetic reinforcement laid on the surface of the super soft soil supporting a thin uniform layer of cohesionless fill (sand), Figure 3. A number of methods have been used to form a fill layer over super soft soil using a geosynthetic membrane or reinforcement (Broms, 1987; Tan *et al*, 1994; Toh *et al*, 1994; Yano *et al*, 1985). Although this construction technique is demonstrably successful, there is no general agreement with respect to the reinforcement mechanism or how the reinforcement improves the bearing capacity.

The development of construction procedures for super soft soil was initially centred on South East Asia and Japan in particular. An early development in Japan was the use of low strength geogrids possessing an element of *in-plane* rigidity. Progressively these materials, although successful, have been replaced by more modern materials possessing high tensile strength. Recent research using model tests aimed at identifying the controlling mechanism involved in the primary stage construction concept have shown that the controlling parameters are, Zakaria (1994):

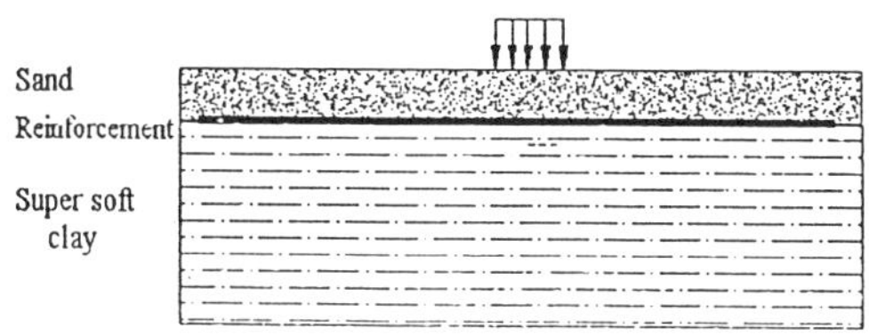

Figure 3: Using geosynthetics and sand layer to improve bearing capacity

$$f(q, B, D, a, S, R, c) = 0 \qquad (1)$$

where f = a function which governs the system; q = bearing capacity (kN/m²); B = diameter of the applied load (footing) (m); D = thickness of fill layer (m); a = grid aperture (m); S = stiffness of the reinforcement in tension (kN/m); R = in-plane bending stiffness (rigidity) of the reinforcement (kN/m); c = shear strength of the clay (kN/m²)

By the use of *dimensional analysis* it is possible to convert data from model tests into design information for a large prototype. Dimensional analysis also reduces parametric variables leading to easier interpretation of the model tests.

Model tests of the mechanism of the primary construction stage technique indicate that geogrid reinforcement offers a higher bearing capacity than geotextile reinforcement, and that the main contributing property from the reinforcement is the in-plane rigidity, R, not the tensile stiffness, S. The size of the grid aperture was shown to be a minor contributor to the improvement of the bearing capacity. Using dimensional analysis a classification system for the bearing capacity based upon the dimensionless parameter ($R/S.B^2$) has been developed, Fakher *et al* (1996).

3.3 *Use of Cohesive Fills*

Many codes of practice do not permit the use of cohesive soil in the construction of reinforced soil structures for permanent work. The reasons given are the potential problems of low strength, high moisture content, creep and low bond strength between the reinforcement and the soil. Early research into reinforced soil associated with the development of internal friction of cohesive fill in undrained conditions showed that the parameter controlling the shear strength of reinforced soil was the relative volume of the fine grained portion of the fill, Schlosser & Long (1974). Murray & Boden (1979) constructed a full scale structure using cohesive soil reinforced with impermeable glass fibre reinforcing strips. They measured substantial rises in pore water pressures which took time to dissipate. Ingold (1979) in triaxial tests on clay reinforced with metallic foil found that the inclusion of impermeable reinforcement led to a reduction in the compressive strength with a strength ratio, F, defined as:

$$F = \frac{(\sigma_1 - \sigma_3 \text{ reinforced}}{(\sigma_1 - \sigma_3)\text{unreinforced}} = 0.62 - 0.79 \qquad (2)$$

This effect was also noticed by Lee (1976) in tests using clay reinforced with mylor strips. The conclusion from these studies is that the inclusion of *impermeable* reinforcements in a clay fill results in induced excess pore water pressure creating possible reductions in the overall strength of the structure in the short term. If the excess pore water pressures could be reduced or removed more stable structures would result. This leads to the concept of the use of a *permeable* reinforcing element to act as a drainage layer.

The use of a permeable reinforcing material enables the build up of pore water pressure to be controlled. Primarily the reinforcing material must be permeable in the normal direction, allowing the passage of water from the soil above the reinforcement to that below. Secondly, the material needs in-plane permeability to speed the consolidation process. Ingold (1979) investigated the initial stability phase (undrained condition) of reinforced clay in a triaxial apparatus using a permeable geosynthetic reinforcement. The use of the permeable reinforcement resulted in an increase in shear strength with measurements of the pore pressure parameter A (as defined by Skempton (1954), B = 1) similar to an unreinforced control sample. However, there were still significant increases in pore water pressure. It was deduced that the permeable reinforcement had an A value lower than that of the soil; as a result one of the problems associated with the use of cohesive fill, the transfer of effective stress to the reinforcing element was effectively eliminated. Even if the reinforcement does not have high in-plane permeability the A value effect relives the pore water pressure developed by the increase in $\Delta\sigma_3$ developed by the presence of the reinforcement, allowing an immediate stress transfer. However, while the structure is approaching drained conditions at the soil-reinforcement interface, the remainder of the soil is still undergoing undrained loading which may lead to the soil failing in this region. This is dramatically illustrated in the full scale trial embankments reported by Tatsuoka & Yamauchi (1986). These tests were undertaken to study the effects of reinforcing clay slopes with non-woven permeable geotextiles. Figure 4 shows the configuration of the reinforcement in the two sides of one trial embankment and Figure 5 the resulting displacements. It is apparent that in the left-hand slope the spacing of the geosynthetic material was too large.

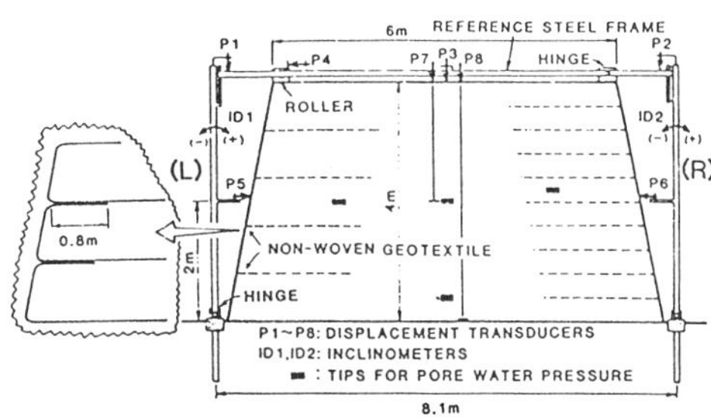

Figure 4: Reinforced clay slope (After Tatsuoka & Yamanuchi, 1986)

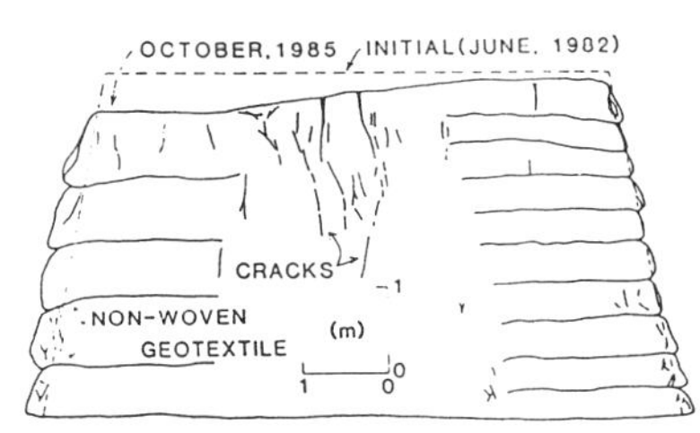

Figure 5: Reinforced clay slope (After Tatsuoka & Yamanuchi, 1986)

A second factor which needs to be addressed is the development of soil-reinforcement bond. Smith *et al* (1979) reported shear box texts investigating the development of bond using both woven and non-woven geotextiles. The results showed that the non-woven fabrics developed resistances to deformation at an earlier stage than the woven materials.

Non-woven geotextiles do not have great strength or in-plane stiffness and are not ideal in respect of the pure reinforcement function. An *ideal* reinforcing material for cohesive soil requires the drainage characteristics of a non-woven geotextile and the strength of a custom made reinforcement such as a geogrid. Alternatively it is possible to combine existing materials, ie using a non-woven drainage geotextile with a geogrid reinforcement. Such a combination is used in the Textomur system, Figure 6. The layout of the different geosynthetic materials in Figure 6 can be criticised as the reinforcement is placed in an area where the highest excess pore water pressure will occur and when dissipation will be slowest. The ideal arrangement would have the drainage function acting with the reinforcement function at the same place.

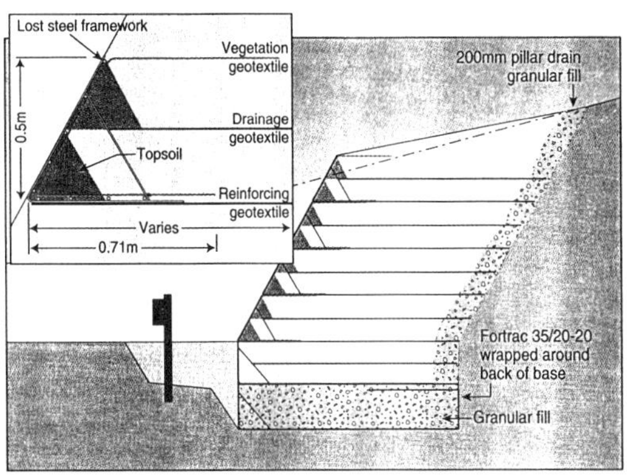

Figure 6: Textomur system

In order to study the effects of combining a drainage material with grid reinforcement in clay soil, a series of tests have been undertaken, Heshmati (1993). The objective of the tests was to identify the separate roles and the contribution of the different materials in improving the shear strength characteristics of clay. The drainage materials considered were non-woven geotextiles and the reinforcements used were grids. Grids are permeable in the plane perpendicular to the plane in which the reinforcement is laid and therefore could be expected to provide an increase in the shear strength.

A number of consolidated undrained (CU) and consolidated drained (CD) triaxial compression tests were undertaken using 100 mm diameter samples containing a single layer of different geotextiles placed at mid-height in the sample. The arrangement of the triaxial apparatus and the geotextile composite material used in the tests is shown in Figure 7. The form and nature of the materials tested are shown in Table 2.

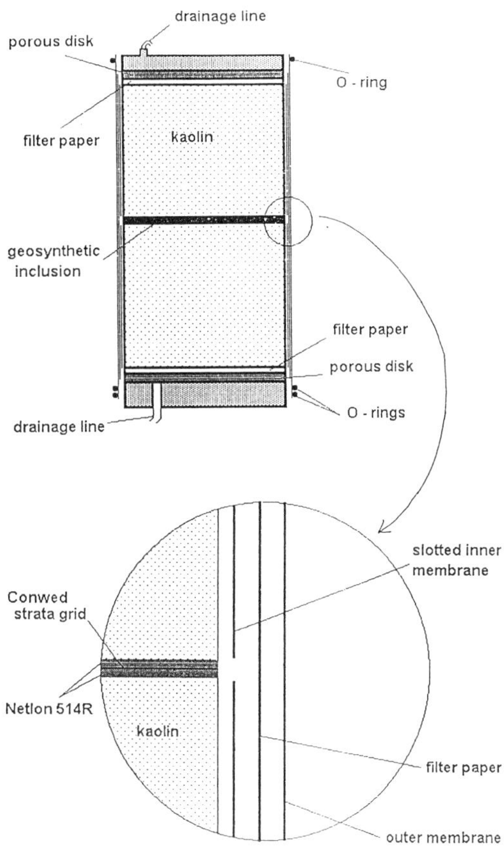

Figure 7: Arrangement of specimen within the cell

A major part of the experimental work was the deliberate separation of the reinforcement function provided by the geosynthetic materials identified as geogrids G2, G3, G4, and geotextile G5, and the drainage function provided by the materials G1 and G6, Table 2. The action of geotextile G1

Table 2: Nature and form of the geotextile materials tested

Samples	Figures	Description
K		Kaolin only, China Clay, 0.002 mm Particle Size
K + G1	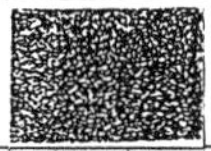	Kaolin with, Netlon 514R, non woven needle punched polypropylene fibres. (providing both drainage and reinforcement)
K + G1CT		Kaolin with, Netlon 514R, non woven needle punched polypropylene fibres cut to equal wedges. (providing drainage)
K + G2	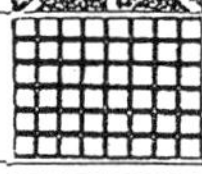	Kaolin with, Conwed Strata grid, geogrid, polyester 10x10 mm. (providing reinforcement)
K + G3	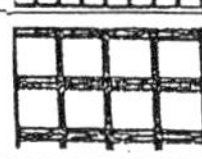	Kaolin with, Conwed 5033 geogrid, polyester, 25x25 mm. (providing reinforcement)
K + G4	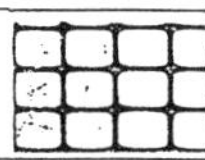	Kaolin with, Tensar SS1 geogrid, polypropylene, 28x38 mm. (providing reinforcement)
K + G5	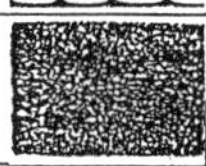	Kaolin with, non woven heat-bonded polypropylene geotextile.(providing reinforcement)
K + G6	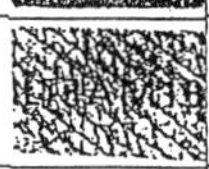	Kaolin with, Filtram 1B1, Netlon, Composite geotextile, mesh core spacer with filter on both side. (providing both drainage and reinforcement)
K + G7	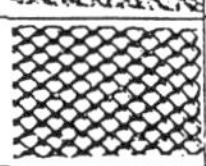	Kaolin with, Filtram 1B1, Netlon, mesh core, without filter on sides. (providing reinforcement)
K + G1G2	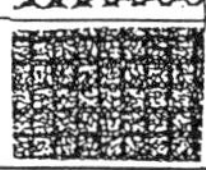	Kaolin with, Combined both G1 and G2 together. (providing both drainage and reinforcement)

was limited to provide a drainage function only by cutting the intact material thereby destroying any reinforcement properties (G1CT). Similarly the material G6 could be made to provide reinforcement properties only by removal of the outer filter membranes (G7). G1 and G6 were used to provide both reinforcement and drainage functions, and G1 and G2 were combined together to provide both reinforced and drainage functions (G1G2).

Unreinforced samples of kaolin were tested to provide base data and samples were tested at cell pressures of 150, 300 and 500 kPa. Samples were prepared in the laboratory in accordance with BS 1377 (1990) by compacting kaolin which had been mechanically mixed with 29 per cent water to obtain an optimum moisture content at 5 per cent voids.

Results of the consolidated undrained tests (CU) are shown in Table 3. The results of the consolidated drained tests (CD) are shown in Table 4.

It can be seen from the results that the inclusion of the drainage geotextiles in the fully consolidated undrained tests produced a *reduction* in the effective angle of internal friction (ϕ') [$K_{23.6}°\rightarrow G1_{20.1}°$], as did the cut geotextile [$K_{23.6}°\rightarrow G1CT_{20.6}°$], but a major *increase* in effective cohesion (c') [$K_{9.8}\rightarrow G1_{25.5}$], as did the cut geotextile [$K_{9.8}\rightarrow G1CT_{25}$].

In both the consolidated undrained and consolidated drained tests the geogrids (G2 and G7) acted as reinforcement which produced a *reduction* in the effective angle of internal friction (ϕ'); for the undrained tests [$K_{23.6}°\rightarrow G2_{20.4}°\rightarrow G7_{21.1}°$]; for the drained tests [$K_{23}°\rightarrow G2_{20.7}°\rightarrow G7_{20.5}°$]. However, the presence of the reinforcement resulted in a major *increase* in effective cohesion (c') for the undrained tests [$K_{9.8}\rightarrow G2_{25}\rightarrow G7_{36}$], for the drained tests [$K_{18.5}\rightarrow G2_{43}\rightarrow G7_{44}$].

The results showed that the geogrids (G3 and G4) acting as reinforcement were not as effective as

Table 3: Summary of Triaxial Test Result
Consolidated - Undrained Triaxial Compression Test (CU Test)

Sample	Effective Cell Pressure kPa 150-300-500	Rate of Strain (mm/min)	Deviator Stress at Failure kPa $(\sigma'_1-\sigma'_3)_f$	$(\sigma'_1)_f$ kPa	Strain at Failure %	Shear Strength Parameter Cohesion - Angle of C'_{cu}	ϕ'_{cu}
K	150	0.035	140	235.5	10.5		
	300	0.028	220	393.4	9.5	9.8	23.6°
	500	0.030	308	566.4	9.0		
K + GI	150	0.074	215	357.7	10.7		
	300	0.028	283	611.5	8.0	25.5	20.1°
	500	0.040	325	795.3	10.7		
K + GICT	150	0.048	170	317.9	8.6		
	300	0.074	340	575.6	10.7	18.3	20.6°
	500	0.077	418	637.4	11.0		
K + G2	150	0.054	130	218.9	7.2		
	300	0.090	170	345.7	9.4	25	20.4°
	500	0.11	350	709.5	7.8		
K + G3	150	0.041	162	281.4	10.5		
	300	0.11	262	453.7	8.2	16	21.7°
	500	0.033	358	634.5	8.8		
K + G4	150	0.041	145	248.2	8.2		
	300	0.053	250	402.7	10.5	16	20.5°
	500	0.059	391	715.8	11.5		
K + G5	150	0.055	175	276.5	9.2		
	300	0.079	240	418.7	8.7	18	19.3°
	500	0.078	390	632.3	11.5		
K + G6	150	0.067	210	306.4	6.5		
	300	0.028	331	221.7	9.5	39.5	18.9°
	500	0.030	392	674.5	9.0		
K + G7	150	0.033	229	364.9	8.2		
	300	0.057	282	465.5	9.8	36.1	21.2°
	500	0.047	388	703.1	10.7		
K + GIG2	150	0.035	160	262.5	8.5		
	300	0.035	270	453.7	8.2	16	20.5°
	500	0.037	407	718.1	9.0		

the materials (G2 and G7) probably due to the large aperture of the grid. G1G2 was thought to have the ideal properties required to provide the dual action of drainage and reinforcement; however the results produced a *reduction* in strength compared to the same drainage and reinforcement materials acting separately, the explanation being the development of a plane of weakness at the interface of the two materials. The implication of this is that it is an essential requirement when combining the functions of drainage and reinforcement for the materials to be made *integral*. This has been confirmed by Cunningham (1995).

4 NEW GEOSYNTHETIC MATERIALS

The increasing demands for construction materials, land and the redevelopment of derelict land are leading to the development of new technologies, and the integration of existing technologies. The development of *electrically conductive geosynthetics* (EKGs) is an example of this. The concept is to develop a range of geosynthetics which in addition to providing filtration, drainage and reinforcement can be enhanced by electrokinetic techniques for the transport of water and chemical species within fine grained low permeability soils, which are otherwise difficult or impossible to deal with. In addition, transivity, sorption, wicking and hydrophobic tendencies may also be incorporated in the geosynthetic to enhance other properties.

The ability of electrokinetic phenomena to move water, charged particles and free ions through fine grained low permeability soil is established. Electrokinetic phenomena will occur in any soil; however, in medium to coarse grained soils

Table 4: Summary of Triaxial Test Result
Consolidated - Drained Triaxial Compression Test (CD Test)

Sample	Effective Cell Pressure kPa 150-300-500	Deviator Stress at Failure kPa $(\sigma'_1-\sigma'_3)_f$	Strain at Failure $\varepsilon_f\%$	Shear Strength Parameter Cohesion - Angle of Internal Friction C'_d	ϕ'_d
K	150	118	11.7		
	300	220	9.8	18.5	23.0°
	500	330	9.2		
K + G1	150	262	10.2		
	300	420	11.0	31.5	22.5°
	500	798	11.9		
K + G1CT	150	260	9.5		
	300	396	11.0	25.0	22.5°
	500	623	9.0		
K + G2	150	377	8.0		
	300	462	10.0	43.0	20.7°
	500	650	10.0		
K + G3	150	239	9.5		
	300	362	12.4	23.0	22.2°
	500	710	10.0		
K + G4	150	220	7.7		
	300	424	8.7	27.0	21.1°
	500	602	8.7		
K + G5	150	230	9.7		
	300	419	11.5	29.5	18.7°
	500	580	12.0		
K + G6	150	245	9.6		
	300	282	8.0	41.5	18.6°
	500	372	7.6		
K + G7	150	280	9.5		
	300	400	11.0	44.0	20.5°
	500	610	11.0		
K + G1G2	150	238	6.5		
	300	418	8.0	30.0	18.9°
	500	579	8.0		

electrokinetic phenomena provide a less effective transport mechanism than hydraulic flows, due to high permeability of the soil. In the case of fine grained soils electrokinetic phenomena provide much more important transport mechanisms than hydraulic flows, due to the low hydraulic permeabilities of these soils.

The first application of electrokinetic phenomena for civil engineering processes was undertaken by Casagrande in 1939 for the dewatering and stabilisation of railway cuttings at Salzgitter, Germany, Casagrande (1952). Since then there have been many other applications of electroosmosis for dewatering and stabilisation of soils, including: (i) The use of electroosmosis for dewatering and strengthening Norwegian quick clays, Bjerrum *et al* (1967); (ii) Electroosmotic stabilisation of West Branch Dam, Fetzer (1967); (iii) Electroosmotic treatment to improve pile friction, Milligan (1994).

4.1 *Electrokinetic Transport Phenomena*

There are five principle electrokinetic phenomena: Streaming Potential, Migration Potential, Electroosmosis, Ion Migration and Electrophoresis. The first two of these phenomena are concerned with the generation of electrical potential due to the movement of charges and charged particles respectively. The remaining three are concerned with the transport mechanisms developed upon application of an electrical field across a soil mass.

Electroosmosis When an electrical field is applied across a fine grained soil mass cations are attracted to the cathode and anions to the anode, Figure 8. As

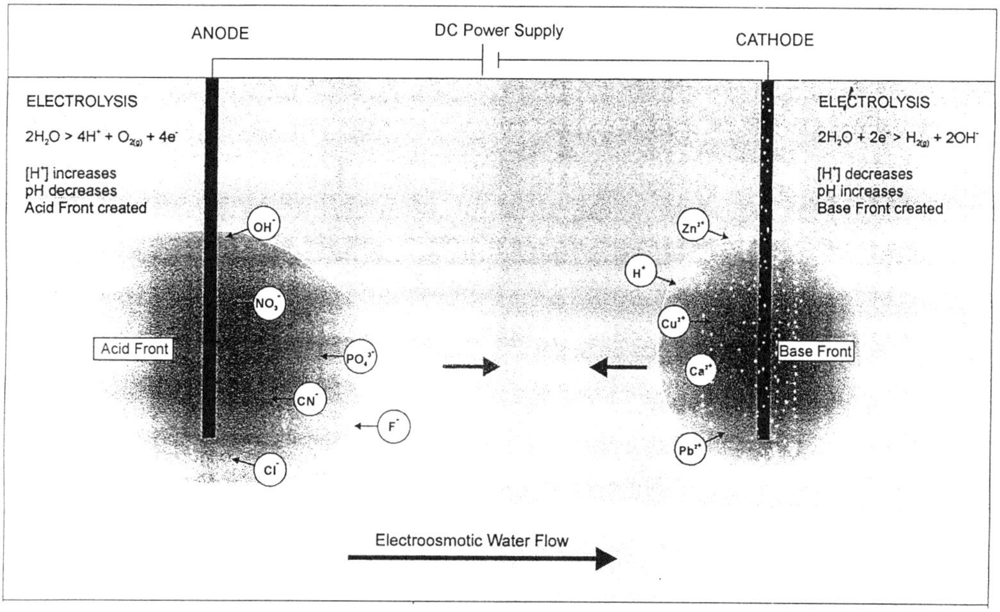

Figure 8: Electrode reactions and electrokinetic transport of pore water (electroosmosis) and ions (ion migration)

the ions migrate they carry their hydration water with them, and exert a frictional force on the water around them. Hence, there is a flow of water to both the anode and the cathode. In order to maintain a charge neutrality, however, there are more cations than anions in the pore fluid of a soil containing negatively charged clay particles. Therefore, there is a net flow of water to the cathode, Gray & Mitchell (1967). This electroosmotic flow depends upon the applied voltage gradient and the electroosmotic permeability of the soil.

The electroosmotic flow of water through a soil can be expressed in the form of Darcy's equation for water flow:

$$q_A = k_e i_e A$$

where q_A is the flow rate, i_e is the potential gradient $\Delta V/\Delta L$, A is the cross-sectional area and k_e is the coefficient of electroosmotic hydraulic conductivity.

The electroosmotic permeability (k_e) for most soils lies in the range $1x10^{-9}$ to $1x10^{-8}$ m²/Vs, Lageman *et al* (1989), and is independent of pore size, Casagrande (1949). As an example of the effectiveness of electroosmosis for the movement of water through fine grained soils consider a sample of white silt (Grade E Kaolinite) with:

- Cross-sectional area = 0.04 m²
- Hydraulic permeability = $5x10^{-9}$ m/s
- Applied voltage gradient = 50 V/m
- Resulting electroosmotic flow rate = 48 ml/hr = 0.0133 ml/s
- Electroosmotic permeability = 0.0133 / (50 * 0.04) * $(1x10^{6})$ = $6.65x10^{-9}$ m²/Vs
- The hydraulic gradient to yield the same flow = 0.0133 / $(5x10^{-9}$ * 0.04) * $(1x10^{6})$ = 66.5

Electromigration or Ion Migration The application of an electrical field across a soil mass causes migration of the free ions and ion complexes, which are present within the pore fluid, to the appropriate electrode, Lageman *et al* (1989), Hellawell (1994). The migration velocity for an electrical field of 1V/m is termed the effective ion mobility u_i* (m²/Vs).

Lageman *et al* (1989) report the average mobility of ions in soils as $5x10^{-8}$ m²/Vs, which is an order of magnitude greater than the electroosmotic permeability. Hence, anions can usually overcome the electroosmotic flow and migrate towards the anode. In the remediation of ionic species ion migration is the dominant transport mechanism, Hellawell (1994).

Electrophoresis When a DC electrical field is applied across a particulate suspension (colloids, clay particles, organics), charged particles in suspension are electrostatically attracted to one of the electrodes and are repelled from the other. Positively charged particles are attracted to the cathode and negatively charged particles are attracted to the anode. Most colloids are negatively charged and are therefore attracted to the cathode.

Electrophoretic mobilities reported by Lageman *et al* (1989) and by van Olphen (1977) are in the ranges $1x10^{-10}$ to $3x10^{-10}$ m²/Vs and $1x10^{-8}$ to $3x10^{-8}$ m²/Vs respectively. Hellawell (1994) attributes the discrepancy to differences in the moisture content of the medium. Electrophoresis has found applications in the densification of sludges and mine tailings, Lo *et al* (1991). Esrig (1968) reports that the process is inconsequential for most naturally occurring soils.

4.2 *Electrokinetic Geosynthetics*

The new EKGs can take the form of single materials, which are electrically conductive, or composite materials, in which at least one element is electrically conductive. They can be of the same basic form as present day filter, drainage, separator and reinforcement materials, but offer sufficient electrical conduction to allow the application of electrokinetic techniques for ground improvement.

There are a number of materials which can be used to produce electrically conductive geosynthetics. The principle conductive materials which are being considered and evaluated at Newcastle are:

Carbon Fibre Materials Carbon fibres are representative of graphite materials; they are electrically conductive (> 10^2 s/cm) and chemically stable. High modulus carbon fibres are expensive and their use is limited to specialist applications where high stresses, high temperatures and radiation exposure are expected. The lower modulus carbon fibres are cheaper, their production simpler and the raw materials more abundant, Ermolenko *et al* (1990).

Activated carbon fibres (ACFs) have an open microporous structure which provides sorption sites. They provide similar or higher sorption than activated carbon particles and their shape allows more rapid and greater sorption than is found with carbon particles, Ermolenko *et al* (1990). The low modulus materials have many of the features required for the production of electrically conductive geosynthetics.

Conductively Filled Polymers Organic polymers with all carbon backbones are insulators (eg polyethylene PE, polyvinylchloride PVC, ethylvinylacetate EVA) and can be used for electrical insulation, Campbell (1994). However, if carbon black, carbon fibres or finely divided metals are used as fillers in such organic polymers, then conductive (10^{-5} to 1 s/cm) composites may be created, Cowie (1991). In these composites it is the filler which conducts and not the polymer.

If carbon black is used as the conductive filler then a loading of between 20 and 30 wt % will be required to produce a suitable conductivity. The structure of the carbon black fillers plays an important role in determining their properties. There are two main types of carbon structure of interest, the Low structure which is a jumbled array of carbon spheres, and the High structure which is a chain of spheres giving a higher conductivity.

An activated carbon black or carbon fibre filler may also be introduced to increase the sorption characteristics of polymeric materials. Sorption is another potential function of EKGs.

Metallic Fibres Metal fibres, metallised fibres or metal coated fibres can be incorporated into the manufacturing processes for geotextiles (needle punching, weaving). Metal coated fibres have only low conduction (10^{-6} to 10^{-1} s/cm) and are not suitable for EKGs. However, the metal and metallised fibres are considered suitable. Metallised fibres have conductivities of 10^0 to 10^4 s/cm.

4.3 *Ground Improvement and Reinforcement using Electrokinetic Geosynthetics*

Electroosmotic treatment of soils generates negative pore pressures and thereby increases the effective stress, leading to consolidation of the soil. The end result is similar to that gained by the application of a surcharge; however, unlike vertically surcharging a soil mass, there are no stability problems. Electroosmosis may also be used to accelerate the dissipation of positive pore pressures resulting from vertical surcharging. Indeed, once the positive pore pressures have been dissipated electroosmosis can be continued for further consolidation, due to the generation of negative pore pressures. These processes increase the *strength* of the soil and the *bond* characteristics of the soil and the reinforcement.

In addition to the electroosmotic improvement of cohesive soils there is often an increase in the strength of the soils due to chemical cementing of the soil fabric and increases in the Atterberg limits of the soil, Mitchell (1991).

4.4 *Factors Affecting the Incorporation of EKGs in Reinforced Soil*

Incorporating the electroosmotic phenomenon with the traditional function of geosynthetics is a new concept. Some of the factors which need to be considered for its successful incorporation include:

Drainage During electroosmotic treatment, water flows from the soil into the drainage media and gases in the form of oxygen at the anode and hydrogen at the cathode are produced due to the electrolysis of water. Both the water and the gases must be removed from the soil by the drainage media. The drainage media must fulfil two basic functions:

(a) Filtration - to allow the free passage of water from the soil to the drainage system.
(b) Transmissivity - to transport water and gas out of the system without hindrance.

Electrode Configuration The flow of electric current through a saturated clay mass occurs primarily through the pore fluid. An efficient mode of current transmission from the electrodes to the soil is important as it affects the efficiency of the electroosmotic treatment. The higher the interface resistance, the lower will be the effective voltage gradient for electroosmotic consolidation; consequently the magnitude of negative pore pressure generated will be reduced. Factors affecting the soil/electrode interface resistance include:

(a) form of the electrodes;
(b) gas generation during the process;
(c) current densities.

Electrode configuration plays an important role in determining the magnitude of interface resistance. An important consideration is the presence of ground water in the pore of the geosynthetic which makes its whole surface an electrode. A major advantage of the use of a geosynthetic electrode is that the surface area can be much larger than traditional electrodes.

Figure 9a shows the distribution of the negative pore pressure during consolidation of the fill layer incorporating two EKG materials. In this case, drainage is at the cathode, hence the boundary conditions would be open cathode and closed anode. The maximum negative pore pressure due to electroosmotic consolidation will be generated at the anode, Esrig (1968). At the cathode, the fill material will consolidate under its self weight. The flow of water due to electroosmotic consolidation v_e and due to self weight v_h is downward towards the cathode.

Once consolidation is completed, reversing the polarity of the electrodes would result in consolidation of the soil in the vicinity of the cathode, Figure 9b.

The effect of placing successive layers of fill and conductive reinforcement can be seen in Figures 9c to e.

4.5 *Type of Soil Suitable for Treatment*

The magnitude of negative pore pressure generated at the anode at equilibrium is given by

$$u = \frac{k_e}{k_h}\gamma_w V \qquad (3)$$

where k_e is the electroosmotic permeability, k_h is the hydraulic permeability and V is the applied voltage.

Equation 3 indicates that for a given applied voltage, the lower the hydraulic permeability the higher would be the negative pore pressure generated. If the total stress does not change during electroosmotic consolidation then the magnitude of the effective stress is proportional to the negative pore pressure generated at the anode. The soils deemed suitable for electroosmotic consolidation are silts, clayey silt and silty clay, Mitchell (1993).

4.6 *Laboratory Tests - Consolidation*

A number of laboratory tests have been conducted to evaluate the use of conductive geotextiles as electrodes in electroosmotic consolidation. The tests using an electroosmotic cell simulated the reinforcement, electrodes and the drainage configuration expected in reinforced soil applications, Figure 10.

A number of geosynthetic electrodes were considered and compared with a perforated copper disk. The initial type 1 geosynthetic electrode was formed from a non-woven, needle punched geotextile. A copper wire of 300 mm length and 1 mm diameter was inserted at the centre of the geotextile to transmit current to the water in the pores of the material. Copper wire was chosen to inhibit the generation of oxygen at the anode. Any oxygen generated at the anode combined with the copper to form copper oxide which is a good

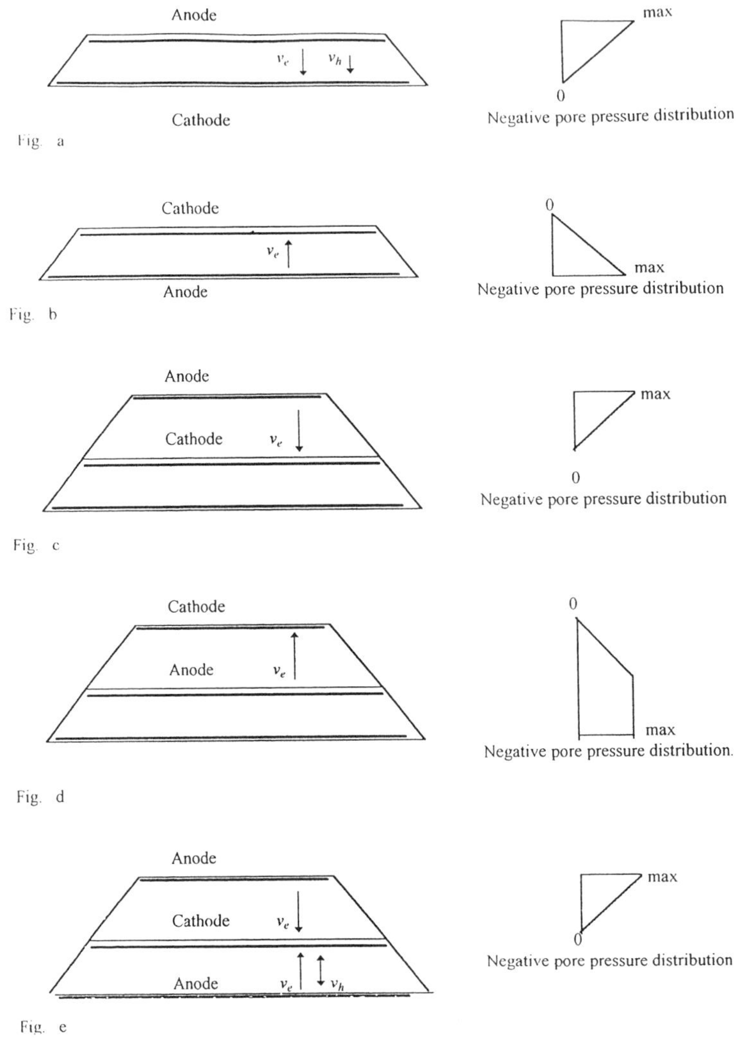

Figure 9

conductor of electricity and the generation of negative pore pressures at the anode were not affected, Lo *et al* (1991).

Commercially available Kaolin grade E was used in the tests. It was classified as silty clay with a plastic limit of 35% and a liquid limit of 55%. The electroosmotic permeability was $k_e = 3.4 \times 10^{-5}$ cm/sec per volt/cm and hydraulic permeability $k_h = 6 \times 10^{-5}$ cm/sec. By substituting the value of k_e and k_h in equation 3, the negative pore pressure generated due to electroosmotic consolidation, U=0.98V.

The soil sample for the tests was prepared from slurry at a moisture content of 1.5 times the liquid limit to ensure it would be fully saturated. Initially the soil was consolidated with an effective stress of 25 kPa.

The electroosmotic tests were conducted with vertical pressures of 50 and 100 kPa to simulate the self weight of the soil. During the tests, pore pressures, current and voltage variations across the soil sample were recorded. After treatment, tests were carried out to determine the moisture content and strength of the soil.

Type of Electrodes Figure 11 shows the applied voltage versus maximum pore pressure at the anode

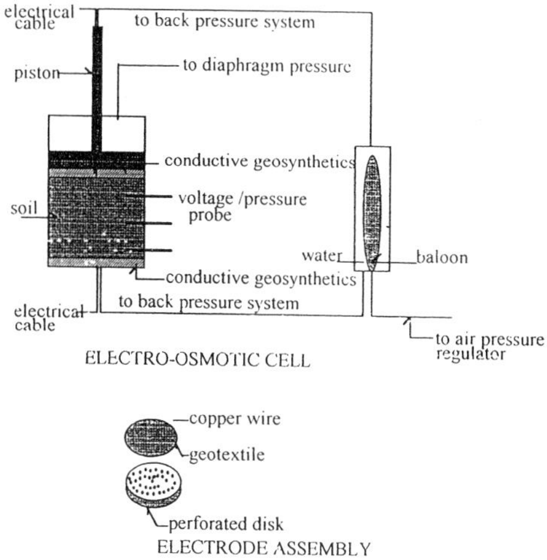

Figure 10: Schematic diagram of the electroosmotic cell and the electrode assembly

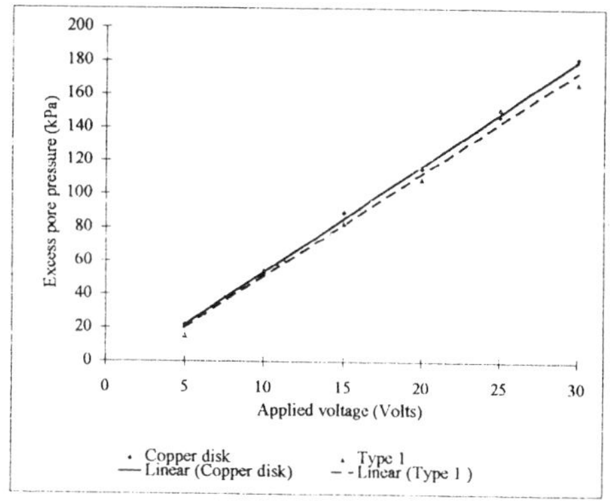

Figure 11

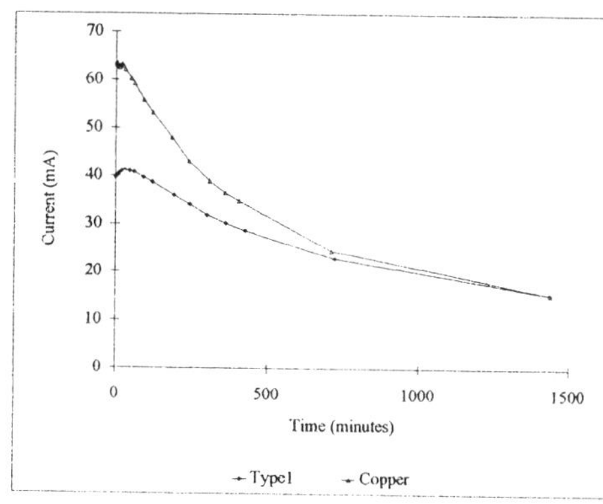

Figure 12

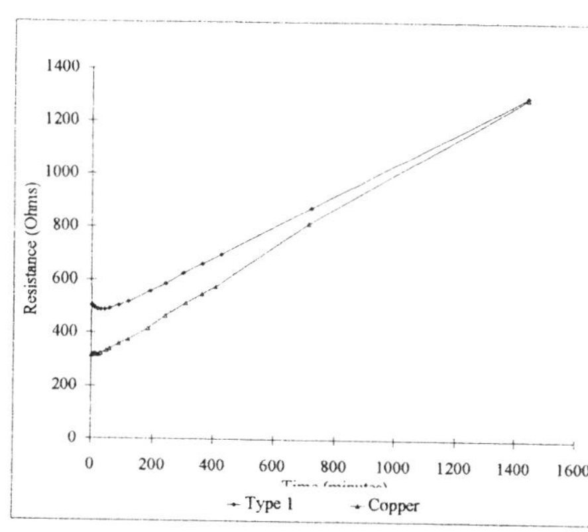

Figure 13

during electroosmotic consolidation with copper and the type 1 geosynthetic electrode. In this series of tests no vertical pressure was applied. It may be seen that the negative pore pressure varies linearly to the applied voltage for both types of electrodes. However, the two lines do not intersect at the origin but at some points along the x axis. The points of intersection at the x axes represent the voltage loss at the soil/electrode interface. The magnitude of the voltage loss is constant regardless of the applied voltage. The values of voltage loss were 1.51 volts and 1.78 volts for the copper and type 1 geosynthetic electrodes respectively.

Samples of variations of current and resistance during electroosmotic tests are shown in Figures 12 and 13 respectively. The applied voltage for the tests was 20 volts. The initial current and resistance for tests using copper disk electrodes were 63 mA and 310 Ohms while the value for type 1 geosynthetic electrodes were 40 mA and 500 Ohms. The soil resistance is the same for both tests.

Negative Pore Pressure Figure 14 shows the change in pore pressure at the anode during electroosmotic consolidation using type 1 geosynthetic electrodes with different applied voltages. Initially the back pressure was lowered by 50 kPa to simulate an applied vertical pressure. At the start of the test, the pore pressure at the anode decreased until it equalised the back pressure,

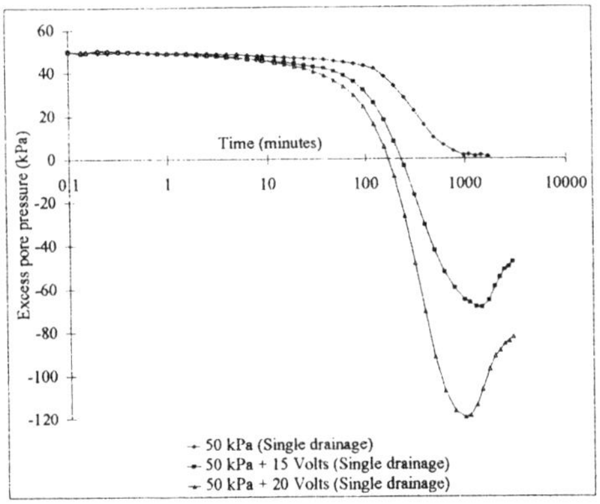

Figure 14

signifying full dissipation of the positive pore pressure. As the test continued, the pore pressure at the anode decreased to a value lower than the back pressure, signifying the generation of the negative pore pressure. Upon reaching a maximum value, the pore pressure increased before stabilising at a lower negative value. Figure 14 shows that the positive pore pressure generated by the vertical loading was dissipated in a shorter time by the incorporation of electroosmotic consolidation. The results concur with the theoretical solution by Wan *et al* (1976).

Vertical Strain The vertical strain-time curves are as shown in Figure 15. The electroosmotically induced vertical strain curve was similar to the consolidation due to vertical load. The vertical strain increased from 9.7% for consolidation with a normal load to 13% with the incorporation of electroosmotic consolidation.

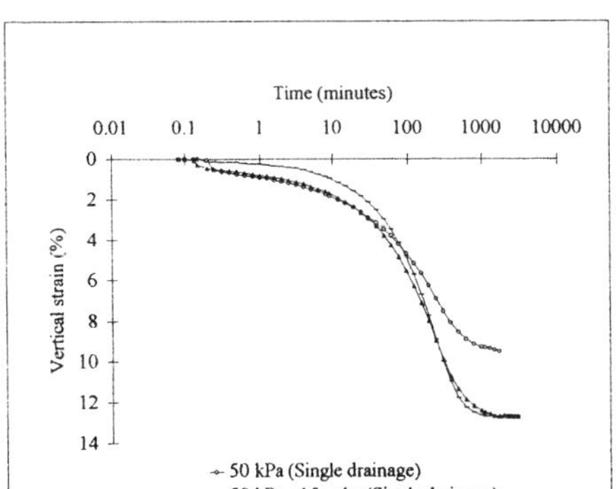

Figure 15

Moisture Content The moisture content profile is as shown in Figure 16. It indicates that moisture content for soil consolidated by a combination of vertical load and electroosmotic consolidation is lower than soil consolidated only with a vertical load. The shape of the moisture content distribution indicates that the lowest moisture content was at the centre of the specimen. However, the moisture content at the anode was still significantly reduced. The moisture content profile agrees with the finding of Lo *et al* (1991).

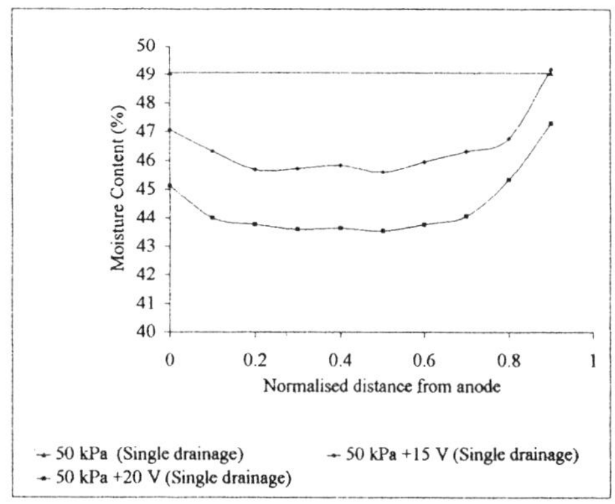

Figure 16

Shear Strength The undrained shear strength of the soil was determined at a depth of 30 mm from the surface. The test showed an increase in the shear strength up to 123% for an applied voltage of 20 volts, Table 5.

4.7 *Laboratory Tests - Soil-Reinforcement Bond*

The use of electrically conductive geosynthetic reinforcement can be shown to increase soil-reinforcement bond. A series of pullout tests have been undertaken at Newcastle University to illustrate this concept. The object of the tests was to study the effect of an electroconductive reinforcement on bond performance under undrained conditions. This represented the most severe case in the use of cohesive fill in a reinforced soil structure.

The tests were performed in a specially designed pullout cell and were conducted on soil strengthened by vertical overburden and also by a combination of overburden and electroosmotic consolidation.

Table 5 Increase in shear strength when using electrokinetic reinforcement

Type of Consolidation	*Shear Strength (kPa)*	*% Increase*
Soil consolidated with 50 kPa vertical pressure	11.17	-
Soil consolidated with 50 kPa vertical pressure and 15 volt applied voltage	19.16	71
Soil consolidated with 50 kPa vertical pressure and 20 volt applied voltage	25.52	123

Commercially available grade E kaolin was used for the tests. The soil was mixed at a moisture content of about 52% which was about 3% below the liquid limit.

A circular 300 mm diameter non-woven need punched geosynthetic was used as a drainage media for the cathode electrode. A 600 mm copper wire was threaded through it to make it electrically conductive. The anode was made of "Paraweb" reinforcement (33 mm width and 330 mm length) with copper wire woven into its surface to make it conductive. The embedment length for the pullout test of the Paraweb was approximately 285 mm. Paraweb was chosen as an example of an established impermeable geosynthetic reinforcing strip. The anode and cathode configuration is as shown in Figure 17.

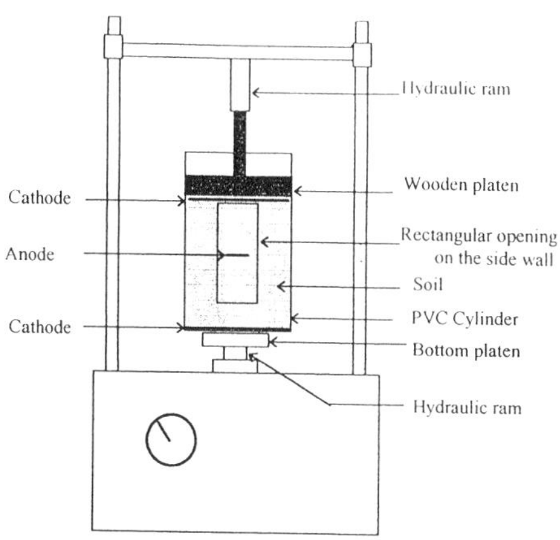

Figure 17: Consolidation set-up of the pullout test

A drainage layer which also served as the cathode was first placed at the bottom of the cell. The soil was then placed in the cell and compacted manually. The anode was inserted at mid height of the sample. Another layer of conductive geosynthetic was placed at the top of the soil sample forming a second cathode. Hence a two way drainage configuration was used to accelerate the consolidation process.

The soil was consolidated by the application of vertical pressure or by a combination of vertical pressure and electroosmotic consolidation. The vertical pressure was applied by a hydraulic jack capable of providing a constant vertical pressure. A potential difference of 30 volts was applied by means of a direct current power supply unit with a maximum output current of 1 ampere.

After the completion of the consolidation process, the cell was turned sideways and bolted to the platen of the frame of the triaxial testing machine as shown in Figure 18. The slit cover was then opened, exposing the geosynthetic reinforcement. The pullout tests were performed by lowering the platen at the rate of 1.5 mm per minute. The pullout resistance and the vertical displacement were automatically recorded by the load cell and the LVDT respectively. After the pullout tests, laboratory vane shear tests were carried out on the soil sample close to the anode. A small portion of the soil sample was also taken for the determination of the moisture content.

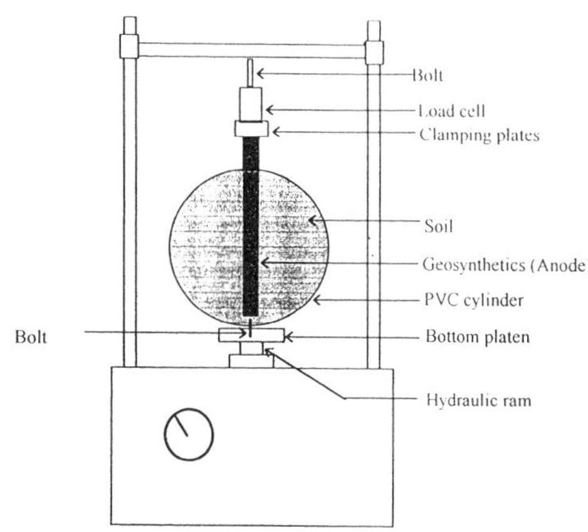

Figure 18: Set-up for the pullout test

Table 6 Percentage increase in shear and soil reinforcement bond strength when using electrokinetic geosynthetic materials

Consolidation pressure (kPa)	Increase in shear strength (%)	Increase in bond strength (%)
110	150	211
140	200	113
356	70	54

Bond Strength The increase in reinforcement-soil bond due to the use of electrically conductive reinforcement is shown in Table 6. The results show that the increase ranged from 54% to 211% at different overburden pressures.

5 CONCLUSIONS

Understanding the fundamental mechanisms involved in reinforced soil is the key to the development and correct use of soil reinforcements. Once the controlling parameters are established it is often possible to produce very effective improved reinforcements using established materials and manufacturing technologies.

The development of electrically conductive geosynthetic materials offers significant new application areas for geosynthetic products, ranging from accelerated consolidation of soft soils to soil decontamination and pollution control using bioremediation. The use of electrically conductive reinforcement in reinforced soil structures is beneficial as the concept is fully compatible with, and has a positive influence on, the parameters controlling the reinforcement mechanisms.

REFERENCES

Hassan, C A (1992) "The use of flexible transverse anchors in reinforced soil", PhD Thesis, University of Newcastle upon Tyne.

Hellawell, E (1994) "Modelling transport processes in soil due to hydraulic density and electrical gradients", PhD Thesis, Cambridge University, UK.

Heshmati, S (1993) "The action of geotextiles in providing combined drainage and reinforcement to cohesive soil", PhD Thesis, University of Newcastle upon Tyne.

Ingold, T S (1979a) "Some observations on failure mechanisms in reinforced clay", Proc. VI Panamerican Conference on Soil Mechanics and Foundation Engineering, Lima, Peru.

Ingold, T S (1979b) "Reinforced clay - a preliminary study using the triaxial apparatus", Proc. Int. Conf. on Soil Mechanics and Foundation Engineering, Brighton.

Jewell, R A (1992) "Strength and deformation in reinforced soil design", 4th Int. Conf. on Geotextiles, Geomembranes and Related Products, Dur Hoedt (ed), Rotterdam, pp.913-946, Balkema.

Jones, C J F P, Cripwell, J P and Bush, D I (1990) "Reinforced earth trial wall for Dewsbury ring road", Proc. Inst. Civil Engineering, Part 1 , Vol.88, pp.321-345.

Jones, C J F P and Hassan, C A (1992) "Reinforced soil formed using polymeric anchors", Earth Reinforcement Practice (eds. Ochiai, Hayashi and Otani), Vol. 1, pp.345-350, Balkema.

Lageman, R, Pool, W and Seffinga, G A (1989) "Electro-reclamation: state-of-the-art", In NATO/ CCMS 3rd Int. Conf. on Demonstration of Remedial Action Technologies for Contaminated Land and Groundwater, Nov. 6-9 1989, 115-136, Montreal, Canada.

Lee, K L (1976) "Reinforced earth - an old idea in a new setting", Int. Symp. New Horizons in Construction Materials, Envo. Pub. Co., Pennsylvania, Vol. 1 .

Lo, K Y, Ho, K S and Inculet, I I (1991) "Field tests of electroosmotic strengthening of soft sensitive clay", Canadian Geotechnical Journal, Vol.28, 74-83.

Lo, K Y, Inculet, I I and Ho, K S (1991) "Electro-osmotic strengthening of soft sensitive clays", Canadian Geotechnical Journal, 28: 62-73.

Milligan, V (1994) "First applications of electro-osmosis to improve friction pile capacity - three decades later", In Proc. 13th Int. Conf on Soil Mechanics and Foundation Engineering,

Jan. 5-10 1994, Vol.5, 34820, New Delhi, India. Balkema: Rotterdam, The Netherlands.

Mitchell, J K (1991) "Conduction phenomena: from theory to geotechnical practice", Geotechnique, Vol.41 (3), 299-340.

Mitchell, J K (1993) "Fundamentals of soil behavior", Wiley, New York.

Murray, R T (1977) "Research at TRRL to develop design criteria for reinforced earth", Symp. Reinforced Earth and Other Composite Soil Techniques, Heriot-Watt University, TRRL Sup. 457.

Murray, R T and Boden, B (1979) "Reinforced earth wall constructed with cohesive fill", Porc. Int. Conf on Soil Reinforcement, Paris.

Murray, R T, and Irwin, M J (1981) "A preliminary study of TRRL anchored earth", TRRL Supplementary Report 674, Transport and Research Road Laboratory, Crowthorne, UK.

Netlon Ltd (1996): Technical literature, Blackburn, UK.

Pasley, C W (1822) "Experiments in revetments", Vol.2, Murray, London.

Schlosser and Long (1974) "Recent results in French research on reinforced earth", Journal of the Construction Division, ASCE, Sept., pp.223-237.

Schlosser, F (1977) "Experience on reinforced earth in France", Symp. Reinforced Earth and other Composite Techniques, Heriot-Watt University, TRRL Sup. 457.

Schlosser, F and Elias, V (1978) "Friction in reinforced earth", Proc. ASCE Symp. Earth Reinforcement, Pittsburgh.

Skempton, A W (1954) "The pore pressure coefficients A and B", Geotechnique, Vol.4, pp.143-147.

Smith, M S, Bell, A L and Du Bois, D D (1979) "Embankment construction from marginal material", Proc. Int. Conf. on Soil Reinforcement, Paris.

Tan, S A Muhammad, N and Karunaratne, G P (1994) "Forming a thin sand seam on a clay slurry with the aid of a jute geotextile", Geotextiles and Geomembranes, 13(3): pp.147-163 .

Tatsuoka, F and Yamauchi, H (1986) "A reinforcing method for steep clay slopes using a non-woven geotextile", Geotextiles and Geomembranes, Vol.4, No.3-4, pp.241-268, Elsevier.

Tatsuoka, F (1992) "Roles of facing rigidity in soil reinforcing", In Earth Reinforcement Practice, Ochiai, Hayashi and Otani (eds), pp.831-870, Balkema.

Tatsuoka, F, Murata, O and Tateyama, M (1992) "Permanent geosynthetic-reinforced soil retaining walls used for railway embankments in Japan", Geosynthetic-Reinforced Soil Retaining Walls, Wu (ed), Balkema, pp.101-130 .

Tatsuoka, F, Tateyama, M and Koseki, J (1995) "Performance of geogrid-reinforced soil retaining walls during the Great Hanshin-Awaji Earthquake, January 17 1995", 1st Int. Conf. Earthquake Geotechnical Eng., IS-Tokyo 95, Nov.

Toh, C T, Chee, S K, Lee, C H and Wee, S H (1994) "Geotextile-bamboo fascine mattress for filling over very soft soil in Malaysia", Geotextiles and Geomembranes, 13 (6-7): 357-369.

van Olphen, H (1977) "An introduction to clay colloidal chemistry" (2nd Ed.), 318, Wiley Interscience: New York, USA.

Walkinshaw, J L (1975) "Reinforced earth construction", Dept. of Transportation FHWA Region 15, Demonstration Project No. 18.

Wan, T W and Mitchell, J K (1976) "Electro-osmotic consolidations of soils", Journal of Soil Mechanics and Foundation Division, ASCE, Vol. 102, (GT5), May.

Westergaard, H M (1938) "A problem of elasticity suggested by a problem in soil mechanics, soft material reinforced by numerous strong horizontal sheets", Harvard University,

Yamanouchi, T (1970) "Experimental study on the improvement of the bearing capacity of soft clay ground by laying a resinous net", Proc. Symp. Foundations Interbedded Sands, Perth, pp. 102-108.

Yano, K, Watari, Y and Yamanouchi, T (1985) "Earthwork on soft clay grounds using rope-netted fabrics", Proc. Recent Developments in Ground Improvement Techniques, Bangkok, Thailand, pp.225-237.

Zakaria, N-A B (1994) "Construction on supersoft soils using geogrids", PhD Thesis, University of Newcastle upon Tyne.

Keynote lectures

Earth Reinforcement, Ochiai, Yasufuku & Omine (eds) © 1997 Balkema, Rotterdam, ISBN 90 5410 833 9

Review of seismic design, analysis and performance of geosynthetic reinforced walls, slopes and embankments

R.J. Bathurst & M.C. Alfaro
Royal Military College of Canada, Kingston, Ont., Canada

ABSTRACT: The paper reviews published work related to properties and modelling of unsaturated cohesionless soils and geosynthetic reinforcement as they apply to geosynthetic reinforced soil walls, slopes and embankments subjected to seismic loading. Current analytical and numerical methods for analysis and design of structures constructed over competent foundations and subjected to seismic shaking are presented. Recent work by the writers and other researchers related to numerical simulation and experimental reduced-scale modelling is presented. The paper identifies the principal elements of North American and Japanese analysis and design practice. Observations of the seismic performance of actual geosynthetic reinforced walls and slopes after earthquakes is summarized. Special attention is paid to seismic analysis, design and performance of geosynthetic reinforced segmental (modular block) retaining wall structures that have seen a rapid growth in North America. A number of research needs are identified.

1 INTRODUCTION

The first analytical treatment of the influence of seismic-induced forces on the stability of earth retaining structures can be traced to the work of Sabro Okabe in his landmark paper of 1924. Since this seminal work there has been a large body of research on the development of analytical methods that consider the potentially large forces that exert additional destabilizing forces on earth retaining walls, slopes, dams and embankments during earthquakes. The vast majority of this work has been focussed on conventional earth structures. The analysis methods that have been proposed include: pseudo-static rigid body analyses that are variants of the original Mononobe-Okabe approach; displacement methods that originate with Newmark sliding block models; and, dynamic finite element (FE)/finite difference methods.

However, with the growing use of geosynthetics to reinforce walls, slopes and embankments, the need to extend current methods of analysis for conventional structures under seismic loading to geosynthetic reinforced systems in similar environments has developed. A concurrent need has been the requirement to select properties of the component materials that represent rapid and/or cyclic loading conditions.

2 SCOPE

The paper provides a review of selected published work related to the properties of cohesionless soil, geosynthetic reinforcement and facing components under cyclic loading and summarizes the important features of current analytical and numerical methods for the seismic analysis and design of geosynthetic reinforced soil walls, slopes and embankments. The scope of the paper is restricted to structures seated on firm foundations for which settlement and collapse of the foundation materials are not a concern. Non-surcharged structures with simple geometries are considered and the reinforced and retained soils are assumed to be homogeneous, unsaturated and cohesionless.

Many of the examples that highlight important issues related to seismic performance of geosynthetic reinforced soil retaining structures are taken from recent work by the writers and co-workers on seismic performance of reinforced segmental (modular block) retaining walls. These structures have gained wide popularity in North America due to their cost-effectiveness (Bathurst and Simac 1994). Nevertheless, these structures pose unique challenges to the designer for seismic loading conditions because of the modular dry-stacked construction of the facing column.

2 MATERIAL PROPERTIES

The properties of the components of geosynthetic reinforced soil structures may be influenced by rate of loading and cyclic loading response. This section reviews data and models that have been used by the writers, co-workers and others for the analysis, design and numerical simulation of structures under seismic loading. The complexity of the constitutive models discussed here ranges from relatively simple for limit equilibrium-based approaches, to relatively sophisticated for dynamic finite element and dynamic finite difference modelling.

2.1 *Cohesionless soils*

2.1.1 *Coulomb friction angles*

Pseudo-static, pseudo-dynamic and displacement (Newmark) methods introduced later in the paper describe cohesionless soil strength according to the Coulomb failure criterion. The selection of an appropriate value of soil friction angle, ϕ, becomes an issue in these methods particularly with respect to the choice of peak or residual strength, ϕ_{cv}, values. A review of the literature suggests that for dry cohesionless soils the rate of loading used in direct shear or triaxial tests has negligible effect on shear strength (Bachus et al. 1993). For example, Schimming and Saxe (1964) used a direct shear device to test Ottawa sand under both static and dynamic conditions. No significant difference in strength envelopes was recorded (Figure 1). Conventional practice using Newmark methods is to assume that the cohesionless soil friction angle does not change during an earthquake.

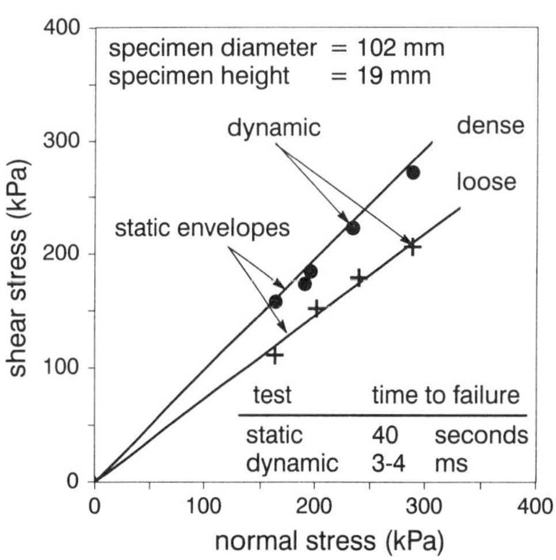

Figure 1. Results of direct shear tests on dry Ottawa sand (after Schimming and Saxe 1964).

Conventional practice in pseudo-static methods of analysis for retaining walls and slopes is to relate interface friction angles, δ, to the soil friction angle, ϕ. In static stability analyses, δ is often assumed to be equal to $2\phi/3$ for internal stability analyses (facing column/reinforced soil interface) and $\delta = \phi$ for external stability analyses (reinforced soil/retained soil interface, or between wedges in two-part wedge analyses). A value of $2\phi/3$ has been shown to be applicable for wall-soil interface friction based on small-scale shaking table tests of conventional gravity wall structures (Ishibashi and Fang 1987) and has been assumed to be also applicable for geosynthetic reinforced retaining wall structures (Bathurst and Cai 1995).

Peak friction angle values have been used in the proposed pseudo-static design methodology for segmental retaining walls proposed by Bathurst and Cai (1995). The choice of peak friction angle for seismic design is consistent with Federal Highway Administration (FHWA - Christopher et al. 1989) and National Concrete Masonry Association (NCMA - Simac et al. 1993) guidelines for static design of geosynthetic reinforced soil walls. Bonaparte et al. (1986) have recommended that residual friction angles, ϕ_{cv}, be used in seismic design of slopes based on reinforcement-strain compatibility requirements. Leshchinsky et al. (1995) have proposed using the residual soil strength for retaining walls but recognize that this is likely a conservative assumption for design. It appears that in practice the choice of peak or residual values is either prescribed or left to engineering judgement.

2.1.2 *Stress-strain models*

For more complex modelling using dynamic finite element codes, the shear behavior of soils under seismic loading can be simulated using available nonlinear cyclic constitutive relationships (Kramer 1996a). A model that has been used successfully by the senior writer and others adopts the Masing rule for hysteretic unloading and reloading of cohesionless soils and is illustrated in Figure 2 (Finn et al. 1986; Yogendrakumar et al. 1991, 1992; Cai and Bathurst 1995).

The relationship between soil shear stress, τ, and shear strain, γ_ε, for the initial loading phase (backbone curve) is assumed to be hyperbolic and given by:

$$\tau = f(\gamma_\varepsilon) = \frac{G_{max}\ \gamma_\varepsilon}{\left[1 + (G_{max}/\tau_{max})|\gamma_\varepsilon|\right]} \tag{1}$$

where: G_{max} = maximum shear modulus; and τ_{max} = maximum shear strength. The equation for the unloading curve from a point (γ_r, τ_r) at which the loading reverses direction is given by:

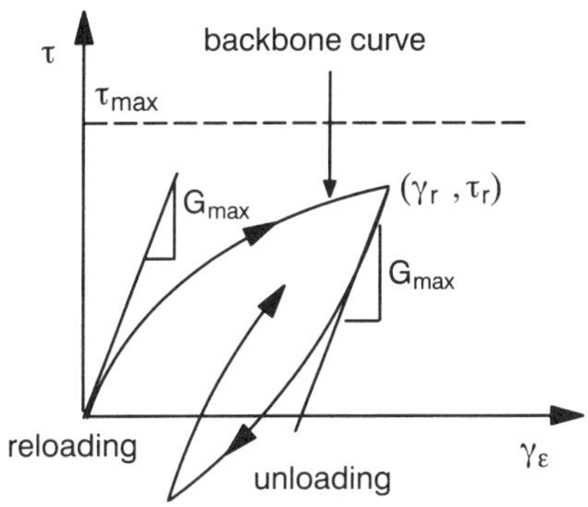

Figure 2. Nonlinear hysteretic loading paths.

$$\frac{\tau - \tau_r}{2} = f\left(\frac{\gamma_\varepsilon - \gamma_r}{2}\right) \tag{2}$$

or

$$\frac{\tau - \tau_r}{2} = \frac{G_{max}\ (\gamma_\varepsilon - \gamma_r)/2}{\left[1 + (G_{max}/2\tau_{max})|\gamma_\varepsilon - \gamma_r|\right]} \tag{3}$$

The shape of the unloading-reloading curve is shown in Figure 2. The tangent shear modulus, G_t, for a point on the backbone curve is given by:

$$G_t = \frac{G_{max}}{\left[1 + (G_{max}/\tau_{max})|\gamma_\varepsilon|\right]^2} \tag{4}$$

and at a stress point on an unloading or reloading curve:

$$G_t = \frac{G_{max}}{\left[1 + (G_{max}/2\tau_{max})|\gamma_\varepsilon - \gamma_r|\right]^2} \tag{5}$$

The response of the soil to uniform all-round pressure is assumed to be nonlinear elastic and dependent on the mean normal stress. Hysteretic behaviour, if any, is neglected in this mode. The tangent bulk modulus, B_t, is expressed in the form:

$$B_t = K_b\ P_a\ \left(\frac{\sigma_m}{P_a}\right)^n \tag{6}$$

where: K_b = bulk modulus constant; P_a = atmospheric pressure in units consistent with mean normal effective stress σ_m ; and n = bulk modulus exponent.

References to variations on the above model and other advanced constitutive models can be found in the textbook by Kramer (1996a).

2.2 *Geosynthetic reinforcement*

2.2.1 *In-isolation load-strain behaviour*

In-isolation monotonic and cyclic load testing of high density polyethylene (HDPE) and woven polyester (PET) geogrid reinforcement materials have been reported by Bathurst and Cai (1994). The results of constant rate of loading (monotonic loading tests) showed that HDPE geogrids were sensitive to the rate of loading while PET geogrids were less sensitive (Figure 3).

2.2.2 *In-isolation cyclic load testing*

In order to determine cyclic load parameters for reinforcement models used in dynamic finite element modelling, in-isolation cyclic load tests were carried out on typical polymeric geogrid reinforcement materials (Bathurst and Cai 1994). Example results are presented in Figure 4 for an HDPE geogrid. The cyclic load-strain behavior of typical HDPE and woven PET geogrid reinforcement materials exhibited two distinct features: 1) nonlinear hysteresis unload-reload loops; and 2) a load-strain cap that is tangent to all initial unload-reload hysteresis curves. The load-strain cap/hysteretic unload-reload model described earlier for soil materials has been modified for polymeric reinforcement materials by Yogendrakumar and Bathurst (1992) and is illustrated in Figure 5. The relationship between axial load and tensile strain for the load-strain cap (backbone curve) is expressed by:

$$T_a = \frac{J_i\ \varepsilon_a}{[\ 1\ +\ (\ J_i\ /\ T_{max}\ )|\varepsilon_a|\ ]} \tag{7}$$

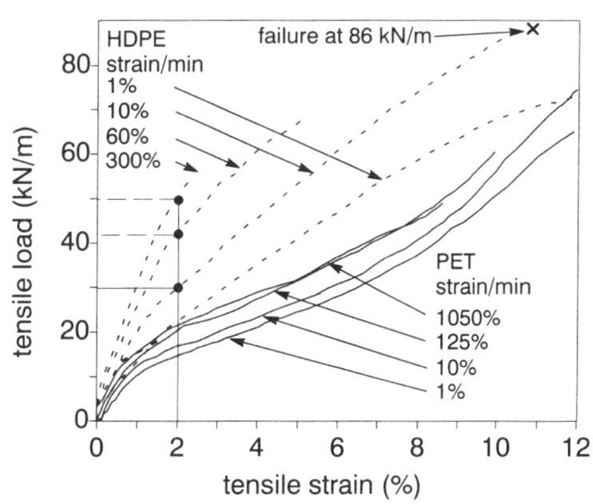

Figure 3. Influence of strain rate on monotonic load-extension behaviour of typical geogrid reinforcement products (after Bathurst and Cai 1994).

where: T_a = axial tensile load per unit width of specimen (e.g. kN/m); ε_a = axial strain; J_i = initial modulus; and T_{max} = extrapolated asymptotic ultimate strength of the reinforcement material. The data in Figure 6 show that the initial stiffness, J_i, and the shape of the load-strain cap are sensitive to frequency of loading for the HDPE geogrid while the PET geogrid with a similar index strength is not. During an unload-reload cycle, the reinforcement model is assumed to follow the Masing rule. The equation for the unloading curve from point $A(\varepsilon_r, T_r)$ or, for the reloading curve from point B at which the load reverses direction is given by:

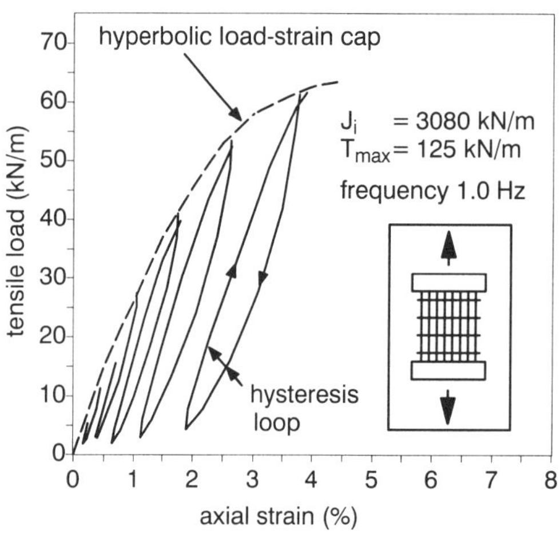

Figure 4. In-isolation cyclic load test on an HDPE geogrid (Cai and Bathurst 1995).

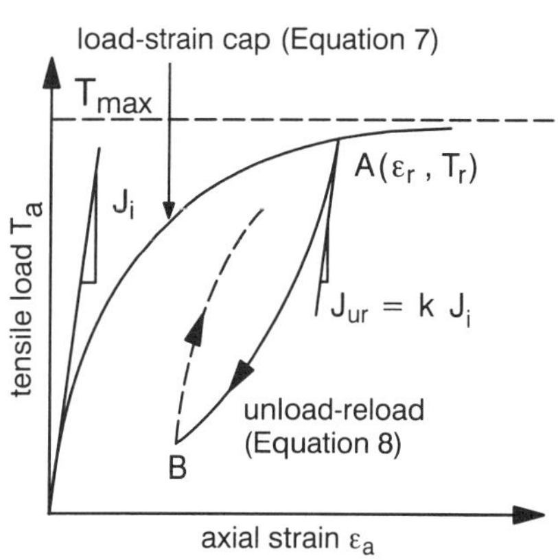

Figure 5. Cyclic unload-reload model for polymeric reinforcement (Yogendrakumar and Bathurst 1992).

$$\frac{T_a - T_r}{2} = \frac{J_{ur}\,(\varepsilon_a - \varepsilon_r)/2}{[1 + (\,J_{ur}\,/2T_{max}\,)|\varepsilon_a - \varepsilon_r|]} \quad (8)$$

where: J_{ur} = unload modulus defined in terms of the initial load modulus according to $J_{ur} = k\,J_i$, with k equal to a constant.

An implication to seismic design of the results of standard monotonic loading wide-width tensile tests and the cyclic load data reviewed here is that initial and secant stiffness values for HDPE geogrids used for static load environment design may greatly underestimate reinforcement stiffness and reinforcement tensile capacities for working stress design. Bonaparte et al. (1986) cautioned that the strain at rupture for HDPE geogrids will decrease with increasing rate of loading and hence influence the choice of rupture load in limit state design. Only one of the tests shown in Figure 3 was taken to rupture due to equipment limitations so possible rate effects cannot be quantified here. Reduction of rupture load capacity for HDPE geogrids under high rates of loading also has implications to Newmark sliding block methods of analysis where large cumulative displacements may be computed (see Section 3.3).

2.2.3 *In-soil reinforcement cyclic load testing*

Some guidance on the effect of soil confinement on geotextile load-strain deformation may be inferred from the results of in-soil tensile tests using monotonic constant rates of displacement. The work of McGown (1982), Ling et al. (1992), and Wilson-Fahmy et al. (1993), indicates that increasing confining pressure increases the modulus of needle-punched nonwoven geotextiles and may increase the ultimate strength as well. The in-air modulus and ultimate strength of woven geotextiles was shown to be unchanged due to soil confinement (Wilson-Fahmy et al. 1993).

Results of in-soil cyclic load testing of geosynthetic reinforcement materials is sparse. McGown et al. (1995) performed a series of low frequency in-soil cyclic load tests on a stiff uniaxial HDPE geogrid similar to that reported by Bathurst and Cai (1994). McGown et al. illustrated that stresses and permanent strains may be "locked-in" the reinforcement due to repeated tensile loading resulting in a stiffer reinforcement response than that for in-air tests.

Taken together, the implications to seismic design is that cyclic in-air tests of the type reported by Bathurst and Cai (1994) may represent a lower bound on reinforcement stiffness values (J_i and J_{ur} values) at working stress levels for stiff HDPE geogrids but confinement likely has a negligible influence on stiffness for woven geotextiles and geogrids.

2.3 *Interface properties*

Geosynthetic-soil interface sliding and pullout of reinforcement within anchorage zones are potential failure mechanisms in reinforced walls, slopes and embankments. A conventional approach is to quantify the shearing resistance at these interface locations by an interaction coefficient, C_i, that is defined as the ratio of the interface friction coefficient to soil friction coefficient ($C_i = \tan \phi_{ds} / \tan \phi$). The interaction coefficient is usually evaluated using a direct shear test and/or pullout test. These two tests differ significantly in loading path and boundary conditions and interaction coefficients for nominally identical specimen conditions may vary between tests (Juran et al. 1988).

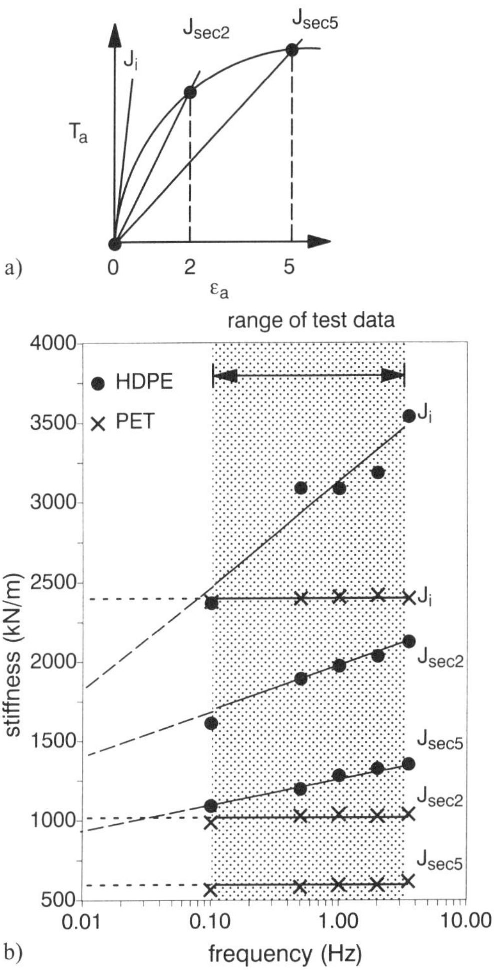

Figure 6. Load-strain cap secant stiffness versus frequency of loading for HDPE and PET geogrid specimens (after Bathurst and Cai 1994).

2.3.1 *Cohesionless soil/geosynthetic interface shear strength*

A large body of work has been reported on interface shear characteristics of soil/geosynthetic interfaces using a variety of direct shear methodologies and apparatuses (Takasumi et al. 1991). The work is restricted almost exclusively to monotonic loading. Myles (1982) reported values of sand/geotextile interface coefficients in the range of 0.97-0.81 for three different types of geotextiles. Miyamori et al. (1986) reported interaction coefficients in the range 0.72-0.87 for dry sand/nonwoven geotextile interfaces. Myles argues that loading rate effects are not a concern for cohesionless sands but recommends that residual interface shear strength should be used for design with geotextiles to be consistent with the notion that full mobilization of shear strength in reinforcement applications occurs at large geotextile strains. Cancelli et al. (1992) reported interaction coefficients in the range of 1.04-1.12 for a number of different stiff HDPE geogrids in combination with sand and gravel. Cancelli et al. argue that interface shear for geogrids is controlled by soil/soil interface shear strength.

There are no published reports of cyclic interface direct shear tests on geotextiles and geogrids. However, a limited number of repeated direct shear tests on a single specimen of HDPE sheet in combination with Ottawa sand at low confining pressure showed that there was no reduction in interface shear strength with number of shear applications (O'Rourke et al. 1990).

Based on the data presented above and the expectation that soil/soil interface shear capacity for dry cohesionless soils is independent of rate of loading, it is reasonable to use results of monotonic loading direct shear tests for limit equilibrium-based seismic design.

Fakharian and Evgin (1995) describe the results of cyclic shear tests between sand and fine steel mesh surfaces which showed that monotonic and cyclic direct shear tests gave the same values of peak and residual interface shear strength. However, under cyclic shear conditions there was evidence that peak and post-peak behaviour may occur at displacement amplitudes that are less than the displacement required to fail the interface under monotonic loading conditions. The applicability of this result to geotextile/soil interfaces has not been investigated.

2.3.2 *Cohesionless soil/geosynthetic pullout*

The simplest pullout model in limit equilibrium-based methods of analysis takes the form (e.g. Public Works Research Institute, PWRI 1992):

$$T_{pull} = 2\ L_a C_i\ \sigma_v\ \tan\ \phi \qquad (9)$$

where: T_{pull} = pullout capacity; L_a = anchorage length; σ_v = vertical stress acting over the anchorage length; ϕ = friction angle of the soil; and C_i = interaction coefficient that is interpreted from the results of pullout tests. In the USA a combination of terms are used to calculate default values of C_i based on the type of geosynthetic, aperture size, and d_{50} of the confining soil. The reader is referred to FHWA (1996) guidelines for details.

A large amount of data can be found on the pullout behaviour of geotextiles and geogrids in combination with cohesionless soils (e.g. Farrag 1990). Bachus et al. (1993) reported the results of constant rate of displacement (static) pullout test results on four different geogrids in sand. Most tests gave interaction coefficient values equal to or slightly in excess of 1.0. Increasing the rate of loading from 1 to 150 mm/min did not result in significant changes in interaction coefficient values.

Relatively few investigations have been performed that examined the effect of repeated tensile load application using conventional pullout box devices. Bathurst and McLay (1996) carried out large-scale repeated load pullout tests on 1.6 m long specimens of a stiff uniaxial HDPE geogrid in combination with a standard #40 laboratory silica sand. Tensile loads were applied to the specimens while subject to constant surcharge pressures ranging from 25 to 73 kPa. The cyclic load amplitude ranged from about 25 to 50% of the index strength of the material. Pullout of the specimens was not achieved with the cyclic load amplitudes and overburden pressures used in this test series - even after 90,000 load applications in some tests. The mobilized length of reinforcement was observed to increase linearly with log number of load applications. Full shear mobilization along the 1.6 m lengths of reinforcement was not observed in tests with overburden pressures greater than 25 kPa. The rate of displacement of the front (in-soil) end of the reinforcement was observed to diminish linearly with the number of load applications (log-log scale) indicating that the anchorage system was intrinsically stable under repeated loading. Qualitative features of the test program by Bathurst and McLay (1996) are in agreement with similar work by Hanna and Touahmia (1991). Min et al. (1995) and Yasuda et al. (1992) carried out repeated load pullout tests on stiff uniaxial and biaxial polyolefin geogrids subject to constant surcharge pressure. Raju (1995) carried out cyclic load tests on both uniaxial HDPE and woven PET geogrids at small strain amplitudes representative of working load levels in the field. Raju and Yasuda et al. report that the magnitude of peak cyclic load to cause pullout failure is greater than the load required for static pullout failure. Min et al. (1995) carried out repeated load pullout tests using a biaxial polypropylene (PP) geogrid and concluded that the interaction coefficient, C_i, was reduced by about 20% due to repeated loading compared to the interaction coefficient back-calculated from single load pullout tests.

The conflicting results with respect to C_i for limit equilibrium calculations may be attributed to the interpretation of anchorage length used to back-calculate interaction coefficient values. In addition, the interpretation of pullout results is sensitive to test size, setup and execution (Juran et al. 1988). Finally, it can be noted that AASHTO (1996) interims and FHWA (1996) guidelines recommend a reduction in pullout interaction coefficient to 80% of the values used for static design. This recommendation appears to be based on results of pullout tests on steel strip reinforcement. However, this reduction is more than compensated by AASHTO and FHWA recommendations that permit factors of safety against pullout failure in limit equilibrium-based design to be reduced to 75% of static design values.

2.3.3 *Facing connection and interface shear tests*

Geosynthetic reinforced segmental retaining walls comprise dry-stacked columns of modular concrete blocks which may be solid or infilled with granular soil. The connection between the facing column and reinforced soil mass is typically formed by extending the reinforcement layers between facing units to the front face of the wall. This connection must carry greater loads during seismic shaking and additional shear forces may be transmitted between modular block units. The performance of the connection between dry-stacked modular concrete units and interface shear between facing units with and without the inclusion of a geosynthetic reinforcement layer can be evaluated by adapting test protocols originally proposed by the senior writer and co-workers (Simac et al. 1993; Bathurst and Simac 1993) for static load environments. Example test results for a particular connection system prior to and after (load controlled) cyclic loading is illustrated in Figure 7. The reinforcement in this particular test was a woven PET geogrid and the block was a solid concrete unit with a continuous concrete shear key. A constant rate of displace-

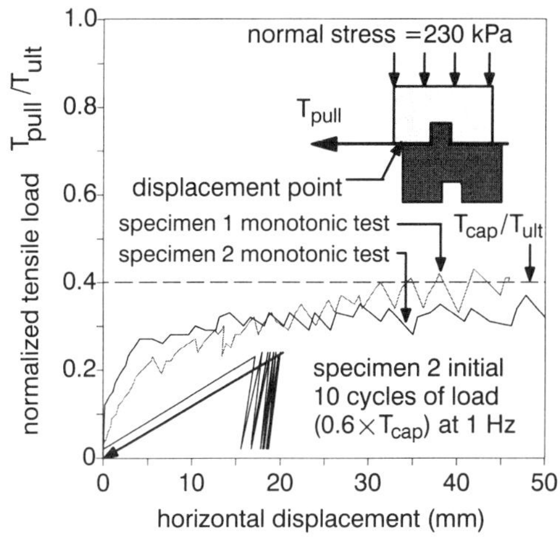

Figure 7. Cyclic load test on woven PET geogrid/ solid block connection (T_{ult} = index strength of geogrid using ASTM 4595).

ment load test was carried out on a virgin specimen of geogrid. A second identical test wascarried out after the connection had been subjected to 10 load cycles at 60% of the ultimate connection capacity (T_{cap}). For this particular system there was no degradation of the connection based on ultimate strength or capacity after 18 mm displacement measured at the back of the block. This result cannot be assumed for all block-geosynthetic systems on the market today, or at lower normal stresses. Further research on repeated load connection testing is required.

Repeated load interface shear tests can also be carried out using the NCMA (Simac et al. 1993) methodology for block/block shear. Static testing shows that interface shear behaviour can be influenced by the presence of a geosynthetic layer. Cai and Bathurst (1996a) assumed that static interface shear values were reasonable in sliding block analyses for systems that provide positive interlock in the form of shear keys, pins and other forms of connectors (Section 3.3.3).

2.3.4 *Other tests*

Shaking table tests used to measure the dynamic interface coefficient for geotextile/geomembrane interfaces have been reported by Zimmie et al. (1994). This technique offers possibilities for the characterization of interface shear properties for soil/geotextiles under light surcharges since the frequency of horizontal shear loading can be chosen to match the frequency and duration of typical seismic events (Figure 8). The

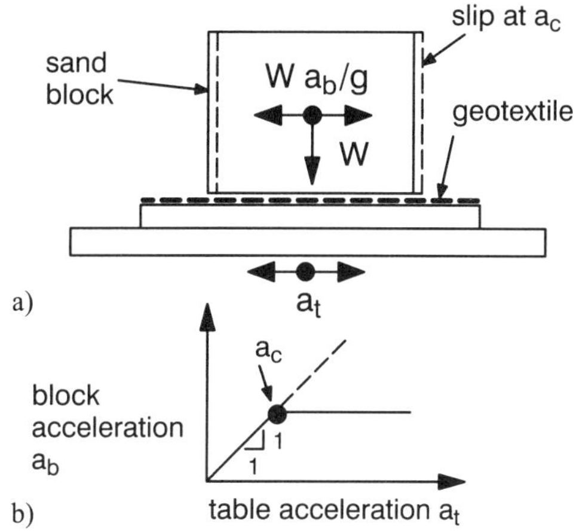

Figure 8. Dynamic interface shear test using shaking table.

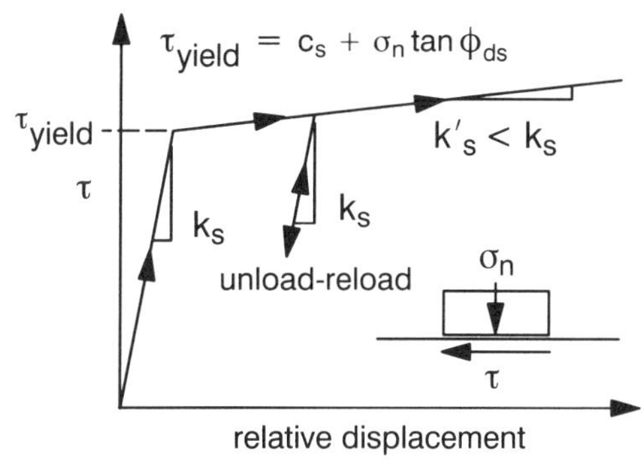

Figure 9. Interface slip model.

critical acceleration, a_b, required to initiate slip can be used to calculate interface friction coefficients according to $C_i = \tan\phi_{ds} = a_c/g$. To examine interface shear resistance at greater surcharges, Zimmie et al. placed a shaking table apparatus in a centrifuge.

2.3.5 *Interface shear-displacement modelling*

The shear transfer at reinforcement/soil, soil/facing unit interfaces (or in the case of segmental retaining walls, reinforcement/concrete block, and block/ block) can be modelled in dynamic finite element codes using a slip element model proposed by Goodman et al. (1968) (see Figure 9). The failure (yield) state of the slip element is assumed to obey the Mohr-Coulomb criterion where: τ_{yield} = shear strength at which slip occurs for the first time; c_s = apparent cohesion; σ_n = normal stress; and ϕ_{ds} = interface friction angle at the yield state. When the applied shear stress exceeds the yield strength, the shear stiffness of the slip element is reduced to a fraction of the original value and slip is initiated. In the normal direction, the stress, σ_n, is assumed to vary linearly with the average relative nodal displacement. Separation of the contact interface is assumed to occur when the normal stress, σ_n, is tensile (which may occur at facing column/soil interfaces). The interpretation of physical test results to obtain example property values can be found in the paper by Cai and Bathurst (1995).

3 SEISMIC ANALYSIS APPROACHES

Analytical and numerical approaches for the seismic analysis of reinforced walls, slopes and embankments can be divided into the following categories: 1) pseudo-static methods; 2) displacement methods; and 3) dynamic finite element/finite difference methods. In the current paper, global stability modes of failure for walls are not addressed.

3.1 *Pseudo-static methods*

Pseudo-static methods extend conventional limit-equilibrium methods of analysis for earth structures to include destabilizing body forces that are related to assumed horizontal and vertical components of ground acceleration.

3.1.1 *Mononobe-Okabe approach*

Pseudo-static rigid body approaches that use the Mononobe-Okabe (M-O) method to calculate dynamic earth forces (Okabe 1924; Mononobe 1929) acting on earth retaining structures (typically walls) are well established in geotechnical engineering practice (e.g. Seed and Whitman 1970; Richards and Elms 1979). The M-O method can be recognized as an extension of the classical Coulomb wedge analysis. The total active earth force, P_{AE}, imparted by the backfill soil is calculated as (Seed and Whitman 1970):

$$P_{AE} = \frac{1}{2}(1 \pm k_v)K_{AE}\gamma H^2 \quad (10)$$

where: γ = unit weight of the soil; and H = height of the wall. The application of force, P_{AE}, against the facing column of a segmental retaining wall structure is illustrated in Figure 10. The total earth pressure coefficient, K_{AE}, can be calculated as follows:

$$K_{AE} = \frac{\cos^2(\phi + \psi - \theta)/\cos\theta\cos^2\psi\cos(\delta - \psi + \theta)}{\left[1 + \sqrt{\dfrac{\sin(\phi+\delta)\sin(\phi-\beta-\theta)}{\cos(\delta - \psi+\theta)\cos(\psi+\beta)}}\right]^2} \quad (11)$$

where: ϕ = peak soil friction angle; ψ = wall/slope face inclination (positive in a clockwise direction from the vertical); δ = mobilized interface friction angle at the back of the wall (or back of the reinforced soil zone); β = backslope angle (from horizontal); and θ = seismic inertia angle given by:

$$\theta = \tan^{-1}\left(\frac{k_h}{1 \pm k_v}\right) \quad (12)$$

Quantities k_h and k_v are horizontal and vertical seismic coefficients, respectively, expressed as fractions of the gravitational constant, g. Seed and Whitman (1970) decomposed the total (active) earth force, P_{AE}, calculated according to Equations 10 and 11 into two components representing the static earth force component, P_A, and the incremental dynamic earth force due to seismic effects, ΔP_{dyn}. Hence:

$$P_{AE} = P_A + \Delta P_{dyn} \quad (13)$$

or

$$(1 \pm k_v)K_{AE} = K_A + \Delta K_{dyn} \quad (14)$$

where: K_A = static active earth pressure coefficient; and ΔK_{dyn} = incremental dynamic active earth pres-

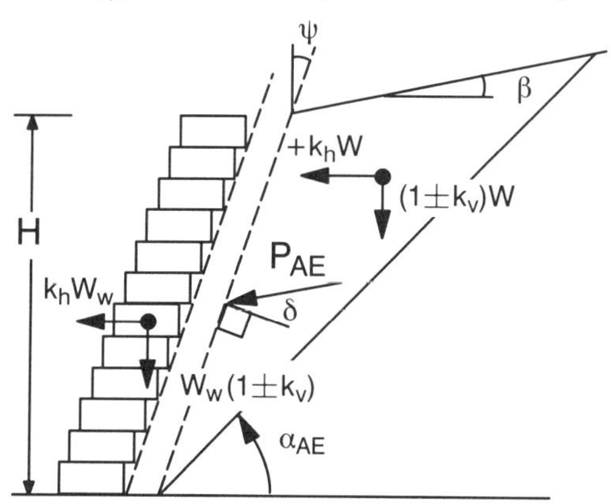

Figure 10. Forces and geometry used in pseudo-static seismic analysis of segmental retaining walls.

sure coefficient. It should be noted that the partitioning of forces according to Equation 13 is not strictly correct since the failure wedge corresponding to P_A will become shallower with increasing magnitude of k_h and hence influence the magnitude of P_A. Closed-form approximate solutions for the orientation of the critical planar surface from the horizontal, α_{AE}, have been reported by Okabe (1924) and Zarrabi (1979). These solutions can be expressed as follows:

$$\alpha_{AE} = \phi - \theta + \tan^{-1}\left[\frac{-A + D}{E}\right] \qquad (15)$$

where:

$$A = \tan(\phi - \theta - \beta)$$

$$D = \sqrt{A[A + B][BC + 1]}$$

$$E = 1 + \left[C\left(A + B\right)\right]$$

$$B = 1/\tan(\phi - \theta + \psi)$$

$$C = \tan(\delta + \theta - \psi) \qquad (16)$$

Equation 15 can be used to calculate the orientation of the assumed active failure plane within the reinforced soil mass and in the retained soil.

Bathurst and Cai (1995) have proposed the total active earth pressure distribution illustrated in Figure 11 for external, internal and facing stability analyses of reinforced segmental retaining walls. The normalized point of application of the resultant total earth force varies over the range $1/3<m<0.6$ depending on the magnitude of ΔK_{dyn}. The assumed pressure distribution is based on a review of the literature for conventional gravity retaining wall structures in North America where the dynamic increment is typically taken as acting at 0.6H above the base of the wall. The total pressure distribution is identical to that recommended for the design of flexible anchored sheet pile walls under seismic loads (Ebling and Morrison 1993). In the absence of ground acceleration, the distribution reduces to the triangular active earth pressure distribution due to soil self-weight.

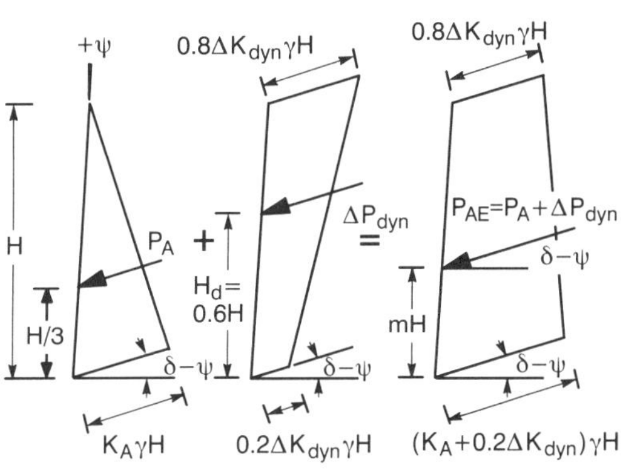

Figure 11. Calculation of total earth pressure distribution due to soil self-weight (Bathurst and Cai 1995).

3.1.2 *External stability calculations for walls*

External stability calculations for factors of safety against base sliding and overturning of geosynthetic reinforced retaining walls are similar to those carried out for conventional gravity structures. For reinforced structures, the gravity mass is taken as the composite mass formed by the reinforced soil zone. For segmental retaining walls the gravity mass includes the facing column since it may comprise a significant part of the gravity mass particularly for low height structures (and hence generate additional inertial forces during a seismic event). The earth pressure distribution shown in Figure 11 is used to calculate the destabilizing forces in otherwise conventional expressions for the factor of safety against sliding along the foundation surface and overturning about the toe of the structure. The simplified geometry and body forces assumed in these calculations for the case of segmental retaining walls are illustrated in Figure 12. The term W_R in the figure is the weight of the reinforced zone plus the weight of the facing column used to calculate resisting terms in factor of safety expressions for base sliding and overturning. The quantity P_{IR} denotes the horizontal inertial force due to the gravity mass used in external stability factor of safety calculations. Different strategies have been proposed

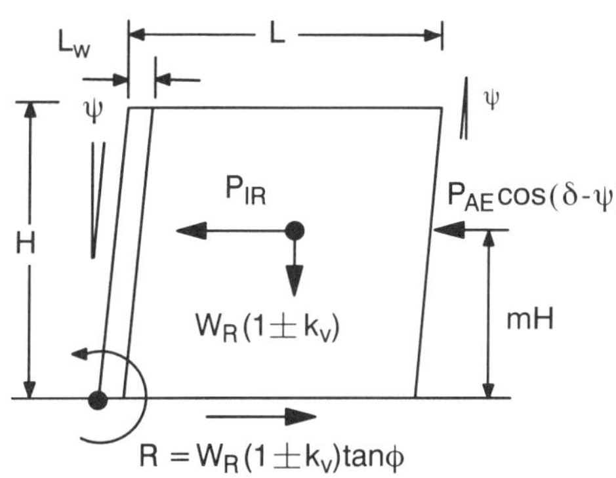

Figure 12. Forces and geometry for external stability calculations for base sliding and overturning.

in North America to compute $P_{IR} < k_h W_R$ to ensure reasonable designs. The justification is based on the expectation that horizontal inertial forces induced in the gravity mass and the retained soil zone will not reach peak values at the same time during a seismic event. Christopher et al. (1989) proposed the following expression for horizontal backfills:

$$P_{IR} = 0.5\eta k_h \gamma H^2 \tag{17}$$

where $\eta = 0.6$ based on recommendations for reinforced walls that use steel reinforcement strips (Segrestin and Bastick 1988). Cai and Bathurst (1995) proposed an expression that gives similar results for typical L/H ratios for segmental walls:

$$P_{IR} = \eta k_h W_R \tag{18}$$

where $\eta = 0.6$. AASHTO (1996) interim guidelines propose that P_{IR} be calculated using Equation 17 with $\eta = 1$ and that the external dynamic active earth force component, ΔP_{dyn}, be reduced by 50%. North American practice is to reduce dynamic static factors of safety against sliding and overturning to 75% of the static factor of safety values in recognition of the transient nature of seismic loading.

Factors of safety are also reduced in Japan (PWRI 1992; GRB 1990; Koga and Washida 1992). However, factor of safety calculations for wall base sliding in Japan do not consider any reduction in inertial force, P_{IR} (i.e. Equation 18 is used with $\eta = 1$). In order to further reduce conservativeness in the Japanese approach for base sliding, Fukuda et al. (1994) have proposed ignoring the dynamic force increment, ΔP_{dyn}, and to restrict seismic loading contributions to the gravity mass term, P_{IR}, only. Overturning criteria for walls are restricted to ensuring that the resultant force acting at the base of the reinforced mass, W_R, falls within L/3 of the base midpoint for walls subject to earthquake. FHWA (1996) guidelines for geosynthetic reinforced walls also omit overturning as a potential failure mode for geosynthetic reinforced soil walls.

3.1.3 *Two-part wedge failure mechanism*

The general solution for a trial two-part wedge failure mechanism with the slope subject to horizontal and vertical acceleration components is illustrated in Figure 13. The horizontal and vertical forces P_1 and V_1 acting on wedge 2 from wedge 1 are, respectively:

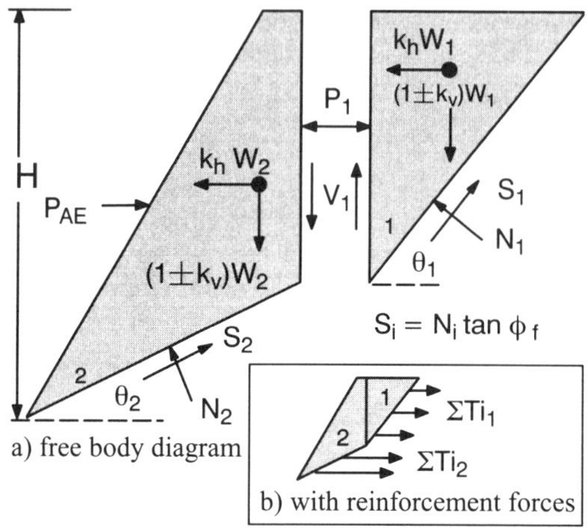

Figure 13. Two-part wedge analysis.

$$P_1 = \frac{(1 \pm k_v)W_1 + B_1 A_1 k_h W_1}{\lambda \tan\phi_f + B_1 A_1} \tag{19}$$

$$V_1 = \lambda\ P_1 \tan\phi_f \tag{20}$$

where:

$$A_1 = \frac{1}{\sin\theta_1 - \tan\phi_f \cos\theta_1} \tag{21}$$

$$B_1 = \tan\phi_f \sin\theta_1 + \cos\theta_1 \tag{22}$$

The quantity λ is the inter-wedge shear mobilization ratio and varies over the range $0 \leq \lambda \leq 1$. Parameter ϕ_f is the factored soil friction angle expressed as:

$$\phi_f = \tan^{-1}(\tan\phi / FS) \tag{23}$$

The horizontal out-of-balance force, P_{AE}, is calculated as:

$$P_{AE} = P_1 + k_h W_2 - B_2 A_2 \left[(1 \pm k_v)W_2 + V_1\right] \tag{24}$$

where:

$$A_2 = \frac{1}{\tan\phi_f \sin\theta_2 + \cos\theta_2} \tag{25}$$

$$B_2 = \tan\phi_f \cos\theta_2 - \sin\theta_2 \tag{26}$$

By setting FS = 1 (i.e. $\phi = \phi_f$) an equivalent total active earth pressure coefficient for the slope can be calculated as:

$$K_{AE} = 2P_{AE}/\gamma H^2 \tag{27}$$

for the most critical trial geometry (i.e. trial search that yields a maximum value for P_{AE}). This approach has been used by Bonaparte et al. (1986) to produce seismic design charts for geosynthetic reinforced soil slopes. The total required design strength of the horizontal layers of reinforcement is taken as $\Sigma T_i = P_{AE}$. The two-part wedge approach with $\lambda=0$ is used by the Geogrid Research Board (GRB 1990) to calculate K_{AE} according to Equations 24 and 27 for internal stability calculations.

The two-part wedge analysis degenerates to a single wedge analysis by restricting trial searches to $\theta_1 = \theta_2$ and setting $\lambda = 0$. All three solutions (M-O, single and two-part wedge) give the same solution for the horizontal component of total earth force when $\lambda = \psi = 0$.

An alternative strategy that extends the general approach used by Woods and Jewell (1990) for static loaded slopes to the seismic case (Bathurst 1994) is to rewrite Equation 24 as:

$$P_{AE} = P_1 - \frac{B_1 A_1 \sum T_{i1}}{\lambda \tan\phi_f + B_1 A_1} + k_h W_2 - \sum T_{i2} - B_2 A_2 \left[(1 \pm k_v) W_2 + V_1\right] \tag{28}$$

The factor of safety for a given two-part wedge geometry corresponds to the value of FS that yields $P_{AE} = 0$. The factor of safety for a slope corresponds to the minimum value of FS from a search of all potential failure geometries. It should also be noted that in this approach the same global FS is applied to the reference design tensile strength of the reinforcement and pullout capacity defined by Equation 9. Equation 28 illustrates that the value of FS against collapse is independent of the location of the reinforcement layers for $\lambda=0$.

3.1.4 *Log spiral failure mechanism*

Log spiral failure mechanisms (Figure 14) have been used to calculate the out-of-balance force to be carried by horizontal reinforcement layers in slopes and walls under seismic loading (Leshchinsky et al. 1995). An advantage of this method is that moment equilibrium is also satisfied (i.e. the problem is statically determinate). The trace of a log spiral surface is given by:

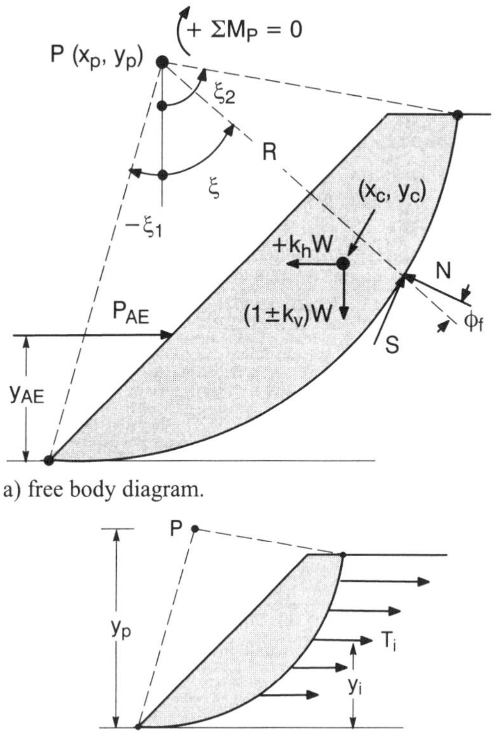

Figure 14. Log spiral analysis.

$$R = A \exp\left(-\tan\phi_f \times \xi\right) \tag{29}$$

For an assumed surface (i.e. for any three independent parameters defining a log spiral, x_p, y_p and A), the moment equilibrium equation about the pole, P, can be explicitly written as:

$$\sum M_P = (1 \pm k_v)\, W\, (x_c - x_p) + k_h W\, (y_p - y_c) - P_{AE}(y_P - y_{AE}) = 0 \tag{30}$$

Note that the moment about the log spiral pole is independent of the distribution of normal and shear stresses over the log spiral because its resultant must pass through the pole. The point of application of the components of seismic inertial forces is taken at the centre of the failure mass. The critical mechanism corresponds to the trace that yields the maximum value of P_{AE} required to satisfy Equation 30. Clearly the point of application, y_{AE}, of the equivalent out-of-balance horizontal force P_{AE} will influence the magnitude of P_{AE}. Here it is assumed *a priori* that $y_{AE}=H/3$. The calculation of an equivalent dynamic active earth pressure, K_{AE}, now follows Equation 27 with FS = 1 (i.e. $\phi=\phi_f$).

In practice, the factor of safety against collapse of a reinforced slope can be determined by replacing $P_{AE}(y_p - y_{AE})$ with $\Sigma T_i (y_p - y_i)$ in Equation 30 and finding the minimum value for FS from a search of all potential failure geometries that yields $\Sigma M_P = 0$. This value corresponds to the minimum factor of safety for the reinforced soil slope (Leshchinsky 1995). The formulation of Equation 30 illustrates that the FS against collapse is a function of the location of the reinforcement layers.

3.1.5 *Circular slip failure mechanism*

Conventional methods of slices can be modified to account for the additional restoring moment due to reinforcement layers. The general case can be referred to Figure 15. Moment equilibrium leads to the following equation to calculate the factor of safety FS against collapse:

$$FS = \frac{M_R + \Delta M_R}{M_D} \qquad (31)$$

where: M_R = moment resistance due to soil shear strength; ΔM_R = increase in moment resistance due to the reinforcement; and M_D = driving moment. Introducing k_v into the derivations for Bishop's Simplified Method (e.g. Fredlund and Krahn 1976), results in the driving moment calculated as:

$$M_D = \sum W[(1 \pm k_v)R\sin\alpha + k_h y] \qquad (32)$$

The moment resistance due to cohesionless soil shear strength is:

$$M_R = (1 \pm k_v)R\sum\left[\frac{W\tan\phi\ \sec\alpha}{1 + \tan\alpha\ \tan\phi_f}\right] \qquad (33)$$

The additional resisting moment due to the tensile capacity of the reinforcement is calculated as:

$$\Delta M_R = R\sum T_i \cos(\psi_i - \delta_i) \qquad (34)$$

The summation term in Equation 34 considers the available reinforcement tensile force in each layer (lesser of tensile reinforcement strength based on over-stressing or pullout) and the orientation, δ_i, of the force with respect to the horizontal. For flexible geosynthetic reinforcement products the restoring force, T_i, can be argued to act tangent to the slip surface at incipient collapse of the slope. This assumption leads to the summation term in Equation 34 becoming $\Sigma T_i R$. This approach is used in FHWA (1996) guidelines together with $k_v = 0$. It is important to note that in the above formulation the influence of reinforcement capacity, T_i, and horizontal acceleration term, k_h, on base sliding resistance is not considered.

An alternative strategy is to modify the "Ordinary Method" (e.g. Fredlund and Krahn 1976). In this approach, equations for vertical and horizontal equilibrium of slices include forces due to acceleration components and reinforcement forces. Hence, these parameters directly affect base sliding resistance. The resisting moment term in Equation 31 becomes:

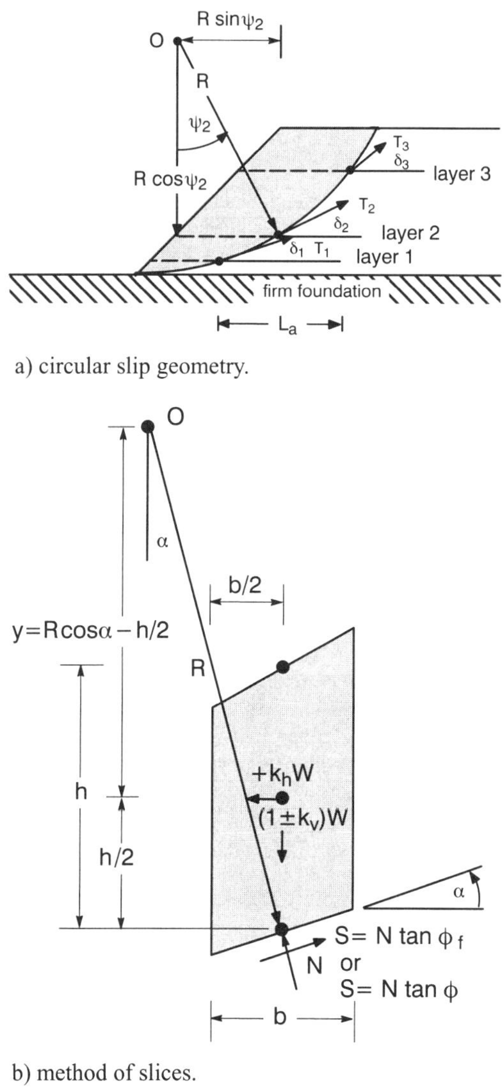

Figure 15. Circular slip analysis.

$$M_R = R \sum \left[(1 \pm k_v) W \cos\alpha - k_h W \sin\alpha \right] \tan\phi \quad (35)$$

and the incremental resisting moment due to reinforcement layers is:

$$\Delta M_R = R \sum T_i \left[\cos(\psi_i - \delta_i) + \sin(\psi_i - \delta_i) \tan\phi \right] \quad (36)$$

where the summation term in Equation 36 is with respect to reinforcement layers. An advantage of the modified "Ordinary Method" is that Equation 31 is a linear function of FS. This approach is used by PWRI (1992) in Japan with $\delta_i = 0$ for retaining walls and $\delta_i = \psi_i$ for slopes. In the Japanese approach, the distribution of total reinforcement load is assumed to be uniform with depth for slopes less than 45° from horizontal. For steeper slopes, including walls, the static portion of required reinforcement load is assumed to increase linearly with depth below the crest while the additional seismic portion is assumed to be distributed uniformly. FHWA (1996) guidelines allow the global factor of safety, FS, to be as low as 1.1 for seismic design of slopes using pseudo-static methods.

3.1.6 *Pseudo-dynamic earth pressure theory*

A pseudo-dynamic earth pressure theory has been proposed by Steedman and Zheng (1990) to account for the influence of phase difference over the height of a vertical retaining wall. The approach recognizes that a base acceleration input will propagate up through the retained soils at a speed that corresponds to the shear velocity of the soil. The general approach has been extended to the case of cohesionless slopes by Sabhahit et al. (1996). Introducing an interface friction angle, δ, and setting $k_v = 0$ leads to a further refinement (Figure 16). The horizontal acceleration is assumed to vary as:

$$a(z,t) = a_o \sin\left[\omega \left(t - \frac{H - z}{V_s} \right) \right] \quad (37)$$

where: ω = angular frequency; V_s = shear wave velocity of the cohesionless soil; a_o = peak base acceleration; and t = time. Horizontal slices of the assumed failure wedge with linear failure surface, α, have the incremental mass:

$$m(z) = \frac{\gamma}{g}(H - z)(\cot\alpha - \tan\psi)dz \quad (38)$$

The total active earth force is computed as:

$$P_{AE}(t) = \frac{Q_h(t)\cos(\alpha - \phi)}{\cos(\delta - \alpha + \phi)} + \frac{W\sin(\alpha - \phi)}{\cos(\delta - \alpha + \phi)} \quad (39)$$

where:

$$Q_h(t) = \int_0^H m(z)\, a(z,t)\, dz \quad (40)$$

The calculation of an equivalent dynamic coefficient of earth pressure, K_{AE}, follows from Equation 27. The pseudo-dynamic approach leads to values of $P_{AE}(t)$ that in the limit $V_s \rightarrow \infty$ gives the pseudo-static value according to M-O theory. The pseudo-dynamic approach allows the location H_d of the dynamic force increment ΔP_{dyn} (the first term in Equation 39) to be determined numerically for a range of base motion frequencies. The solution is independent of soil friction angle, ϕ, and slope angle, ψ, but is dependent on shear velocity (soil density) and period, T, of the assumed sinusoidal horizontal acceleration function.

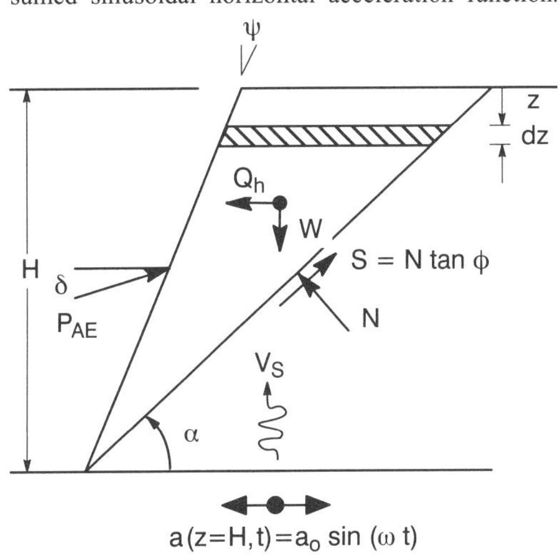

Figure 16. Pseudo-dynamic method.

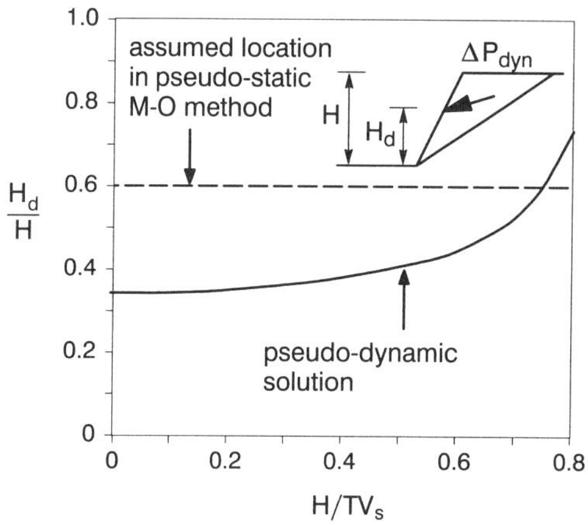

Figure 17. Point of application of dynamic force increment.

The results of calculations are illustrated in Figure 17 and show that for low frequency excitation the point of application is at $H_d = H/3$ above the toe of the soil mass but will increase at higher frequencies. It appears that the pseudo-static M-O method is conservatively safe for overturning/base eccentricity design calculations for a wide range of base motion frequencies.

3.1.7 *Comparisons between selected pseudo-static methods*

A comparison of total active earth forces calculated using wedge and log-spiral pseudo-static methods is illustrated in Figure 18a for frictionless soil/facing interfaces ($\delta = 0$). In these calculations, fully mobilized inter-wedge friction was assumed ($\lambda = 1$) and the point of equivalent total earth force application was taken as H/3. Pseudo-dynamic pressure theory results are not included since there are additional parameters required in this method. Some results of parametric studies using pseudo-dynamic theory can be found in the paper by Sabhahit et al. (1996). Figure 18a shows that for vertical faced slopes and walls (ψ=0) the magnitude of P_{AE} is the same. However, for shallow slopes, there can be a significant difference between methods. In particular, the M-O method may be non-conservative at high horizontal ground accelerations. For walls, the choice of earth pressure theory is not a concern, but for slopes the choice of theory must be carefully considered. In conventional tie-back methods of design it is necessary that reinforcement lengths extend beyond the assumed active failure volume in order that pullout resistance is available for each layer. This is of particular concern towards the top of reinforced wall and slope structures. All pseudo-static

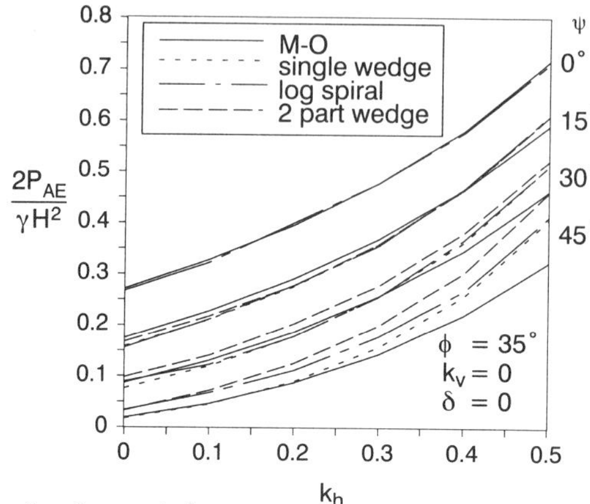

a) active earth force.

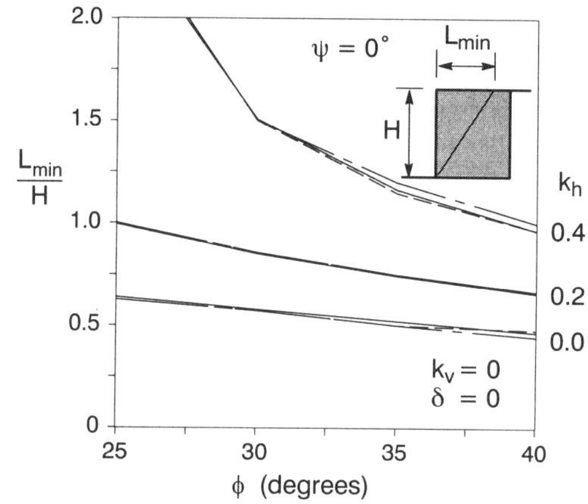

b) maximum width of failure volume(vertical face).

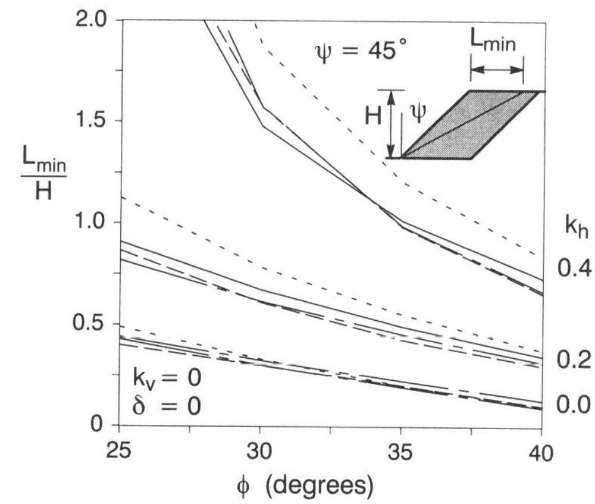

c) maximum width of failure volume (sloped face).

Figure 18. Comparison of wedge and log spiral pseudo-static methods (L_{min} = minimum length of reinforcement to contain failure volume. Note: L_{min} may not be at the top of the reinforced mass).

methods consistently predict that the minimum required reinforcement length will increase with increasing horizontal ground acceleration (Figures 18b,c) and hence reinforcement lengths may have to be increased for reinforced soil structures, particularly towards the crest. The observed cracking at the back of the reinforced soil mass in some wall structures has been attributed to this deficiency in post-earthquake surveys reported in the literature (Section 6).

3.1.8 *Internal stability calculations for walls and slopes*

Pseudo-static/dynamic methods for walls which involve an assumed distribution of internal earth pressure (e.g. Figure 11) require that each reinforcement layer carry a portion of the integrated earth pressure over a contributory area, S_v, as illustrated in Figure 19. The magnitude of tensile force must not exceed the allowable design load in the reinforcement based on tensile over-stressing, facing connection strength and pullout capacity of the layer. In North American practice, factors of safety against these modes of failure are reduced to values that are typically 75% of static values. Figure 19 also demonstrates that the inertial force due to the contributory portion of the facing column should be added to the reinforcement forces under seismic loading in the case of segmental walls. An important implication of the assumed earth pressure distribution using the pseudo-static M-O method described earlier is that the relative proportion of load to be carried by the reinforcement layers closest to the crest of a wall with uniform spacings increases with

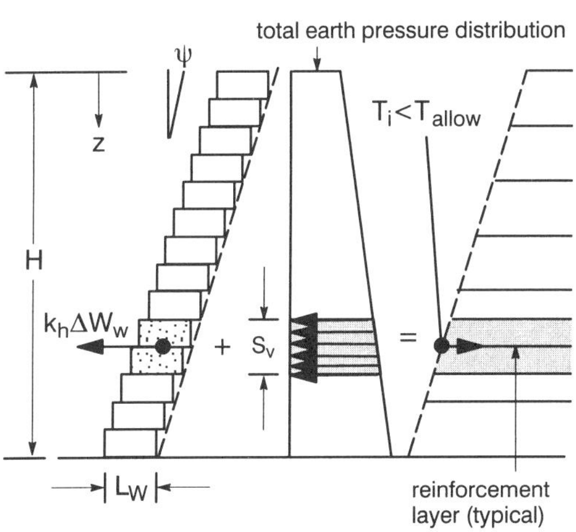

Figure 19. Calculation of tensile load, T_i, in a reinforcement layer due to dynamic earth pressure and wall inertia for segmental retaining walls (Bathurst and Cai 1995).

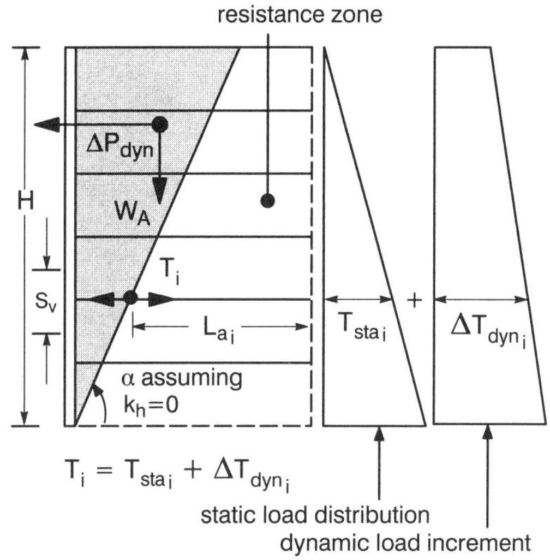

Figure 20. Calculation of tensile load, T_i, in a reinforcement layer for reinforced soil walls with extensible reinforcement using FHWA (1996) method.

increasing horizontal acceleration. This may require a greater number of layers towards the top of the wall than is required for static load environments. A similar conclusion was reached by Vrymoed (1989) using a contributory area approach that assumes that the inertial force carried by each reinforcement layer increases linearly with height above the toe of the wall for equally spaced reinforcement layers. Bonaparte et al. (1986) applied the contributory area method to walls and slopes but recommended a uniform distribution for the dynamic earth pressure increment (i.e. $H_d = 0.5H$ in Figure 11). Nevertheless, Bonaparte et al. concluded that the combination of higher available reinforcement strength and reduced factors of safety used for seismic loading cases will often result in no requirement to increase the number of reinforcement layers required for static loading cases. FHWA (1996) guidelines use the procedure shown in Figure 20 to assign reinforcement forces for over-stressing and pullout calculations. In this method, the static earth force, P_A, is calculated using Rankine earth pressure theory with a Rankine failure plane ($\alpha = \pi/4 + \phi/2$) for vertical walls, and Coulomb theory with a Coulomb angle according to Equation 15 (using $k_h = k_v = 0$) for walls with a facing batter greater than 10°. The dynamic earth pressure force is calculated as $\Delta P_{dyn} = k_h W_A$, where W_A is the weight of the static internal failure wedge. The distribution of the dynamic tensile reinforcement load increment, ΔT_{dyn}, is weighted based on total anchorage length in the resistance zone according to:

$$\Delta T_{dyn_i} = \Delta P_{dyn} L_{a_i} / \sum_{j=1}^{N} L_{a_j} \quad (41)$$

where: N = number of reinforcement layers; and L_a = anchorage length. This approach leads to redistribution of dynamic force to the lower reinforcement layers for internal stability calculations in structures with uniform reinforcement length. This strategy is based on the results of finite element modelling of reinforced walls that used (inextensible) steel strips (Segrestin and Bastick 1988). However, the dynamic increment force distribution shows the opposite trend to that used for external stability calculations in the same FHWA (1996) guidelines (see Figure 11). Although not demonstrated in the current paper, it is clear that the FHWA method is the least conservative for design of reinforcement forces, T_i, of all the methods reviewed. Furthermore, the FHWA approach is less likely to result in an increased number of reinforcement layers at the top of reinforced wall structures and increased reinforcement lengths to accommodate shallower internal failure surfaces with increasing horizontal acceleration which is often the case using a rigorous interpretation of M-O theory.

3.2 *Selection of seismic coefficients*

In conventional pseudo-static methods of analysis the choice of horizontal seismic coefficient, k_h, for design is related to a specified horizontal peak ground acceleration for the site, a_h. The relationship between a_h and a representative value of k_h is nevertheless complex and there does not appear to be a general consensus in the literature on how to relate these parameters. For example, Whitman (1990) reports that values of k_h from 0.05 to 0.15 are typical values for the design of conventional gravity wall structures and these values correspond to 1/3 to 1/2 of the peak acceleration of the design earthquake. Bonaparte et al. (1986) used $k_h = 0.85 a_h/g$ to generate design charts for geosynthetic reinforced slopes under seismic loading using the two-part wedge method of analysis. However, the results of FE modelling of reinforced soil walls by Segrestin and Bastick (1988), Cai and Bathurst (1995) and limited half-scale experimental work (Chida et al. 1982) has shown that the average acceleration of the composite soil mass may be equal to or greater than a_h depending on a number of factors. Factors include: magnitude of peak ground acceleration; predominant modal frequency of ground motion; duration of motion; height of wall; and stiffness of the composite mass. Current FHWA guidelines use an equation proposed by Segrestin and Bastick (1988) that relates k_h to a_h according to:

$$k_h = (1.45 - a_h/g)(a_h/g) \quad (42)$$

This formula results in $k_h > a_h/g$ for $a_h < 0.45g$. However, as clearly stated by Segrestin and Bastick, Equation 42 should be used with caution because it is based on the results of FE modeling of steel reinforced soil walls up to 10.5 m high that were subjected to ground motions with a very high predominant frequency of 8 Hz. The results of FE modeling reported by Cai and Bathurst (1995) for a 3.2 m high geosynthetic reinforced segmental retaining wall with $a_h = 0.25g$ and a predominant frequency range of 0.5 to 2 Hz gave a distribution of peak horizontal acceleration through the height of the composite mass and retained soil that was for practical purposes uniform and equal to the base peak input acceleration. These observations are consistent with the results of Chida et al. (1982) who constructed 4.4 m high steel reinforced soil wall models and showed that the average peak horizontal acceleration in the soil behind the walls was equal to the peak ground acceleration for ground motion frequencies less than 3 Hz.

The general solutions to pseudo-static methods of analysis admit both vertical and horizontal components of seismic-induced inertial forces. The choice of positive or negative k_v values will influence the magnitude of dynamic earth forces calculated using Equations 10 and 11. In addition, the resistance terms in factor of safety expressions for internal and external stability of walls and slopes that include the vertical component of seismic force may be influenced by the choice of sign for k_v. An implicit assumption in many of the papers on pseudo-static design of conventional gravity wall structures reviewed by the writers is that the vertical component of seismic body forces acts upward. However, the designer must evaluate both positive and negative values of k_v to ensure that the most critical condition is considered in dynamic stability analyses if non-zero values of k_v are assumed to apply. For example, Fang and Chen (1995) have demonstrated in a series of example calculations that the magnitude of P_{AE} may be 12% higher for the case when the vertical seismic force acts downward ($+k_v$) compared to the case when it acts upward ($-k_v$). Nevertheless, selection of a non-zero value of k_v implies that peak horizontal and vertical accelerations are time coincident which is an unlikely occurrence in practice. The assumption that peak vertical accelerations do not occur simultaneously with peak horizontal accelerations is made in the current FHWA and AASHTO guidelines for the seismic design of mechanically stabilized soil retaining walls and in Japan (PWRI 1992). Seed and Whitman (1970) have suggested that $k_v = 0$ is a reasonable assumption for the practical design of conventional gravity structures using pseudo-static methods. Wolfe et al. (1978) studied the effect of combined horizontal and vertical ground acceleration on the seismic stability of reduced-scale model Reinforced Earth walls using shaking table tests. They concluded that the vertical component of the seismic motion may be disregarded in terms of practical seismic stability design. Their conclusion can also be argued to apply to geosynthetic reinforced walls. Nevertheless, significant vertical accelerations may occur at sites located at short epicentral distances and engineering judgement must be exercised in the selection of vertical and horizontal seismic coefficients to be used in pseudo-static seismic analyses.

In order to address specific concerns raised by Allen (1993) related to facing stability of geosynthetic reinforced segmental retaining walls during a seismic event that includes vertical ground accelerations, parametric analyses were carried out by Bathurst and Cai (1995) to investigate the combined effect of horizontal and vertical acceleration using the range $k_v = -2k_h/3$ to $+2k_h/3$. The upper limit on the ratio k_v to k_h is equal to the calculated ratio of peak vertical ground acceleration to peak horizontal ground acceleration from seismic data recorded in the Los Angeles area (Stewart et al. 1994). The results are shown in Figure 21 and illustrate that for $k_h < 0.35$ the effect on total dynamic earth pressure is not significant.

Madabhushi (1996) has identified polarization of ground motion in preferred directions and different arrival times of vertical and horizontal ground motions as having a possible influence on the selection of design ground acceleration values for soil reinforced structures.

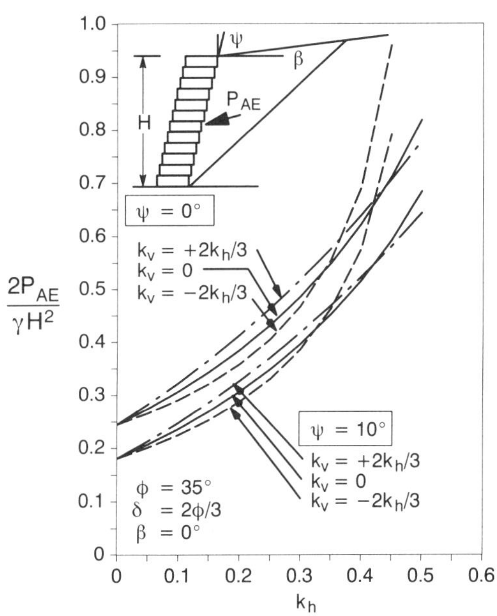

Figure 21. Influence of seismic coefficients, k_h and k_v, and wall inclination angle, ψ, on dynamic earth force, P_{AE} (Bathurst and Cai 1995).

In practice, the final choice of k_h may be based on local experience, and/or prescribed by local building codes or other regulations. The magnitude of a_h for a particular location in the USA can be found in AASHTO (1992) and NEHRP (1994) guidelines. Similar data can be found in the CFEM (1993) for Canada. Readers may refer to the book by Paz (1994) for information on seismic codes for most other countries. The textbooks by Kramer (1996a) and Okamoto (1984) and agency documents by AASHTO and NEHRP provide valuable information on the effect of foundation conditions on attenuation or amplification of bedrock source ground motion.

Finally, FHWA (1996) guidelines for reinforced soil wall structures caution that pseudo-static design methods should be restricted to sites with peak horizontal ground accelerations not exceeding 0.29g. For more intense earthquakes, large structure displacements may occur and the services of a specialist are recommended. For reinforced soil slopes which are flexible structures, FHWA (1996) guidelines allow peak horizontal ground acceleration values published by AASHTO (1992) to be reduced by 50%.

3.3 *Displacement methods*

As with all limit equilibrium methods of analysis, the pseudo-static approach cannot explicitly include wall or slope deformations. This is an important shortcoming since failure of geosynthetic reinforced soil walls, in particular, may be manifested as unacceptable movement without structural collapse. The permanent displacement of a geosynthetic reinforced soil structure due to horizontal sliding/shear mechanisms can be estimated using one of two general approaches as described below.

3.3.1 *Newmark's method*

For a given input acceleration time history, Newmark's double integration method for a sliding mass can be used to calculate permanent displacement (Newmark 1965). According to Newmark theory, a potential sliding body is treated as a rigid-plastic monolithic mass under the action of seismic forces. Permanent displacement of the mass takes place whenever the seismic force induced on the body (plus the existing static force) overcomes the available resistance along a potential sliding/shear surface. Newmark's method requires that the critical acceleration,

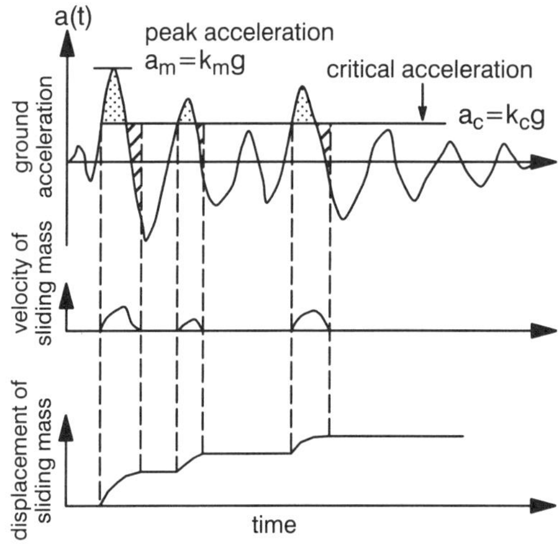

Figure 22. Calculation of permanent displacements (unidirectional displacement) using Newmark's method.

k_c, to initiate sliding or shear failure be determined for each translation failure mechanism. The value of k_c can be determined by searching for values of k_h that give a factor of safety of unity in pseudo-static factor of safety expressions. The critical acceleration is then applied to the horizontal ground acceleration record at the site and double integration is performed to calculate cumulative displacements as illustrated in Figure 22 where: g = gravitational constant; a(t) = horizontal ground acceleration function with time t; $a_m = k_mg$ is the peak value of a(t); and $a_c = k_cg$ is the critical horizontal acceleration of the sliding block. For a given ground acceleration time history and a known critical acceleration of the sliding mass, the earthquake induced displacement is calculated by integrating those portions of the acceleration history that are above the critical acceleration and those portions that are below until the relative velocity between the sliding mass and the sliding base reduces to zero.

A number of researchers have postulated that the critical acceleration value to initiate slip should be based on the peak shearing resistance of the soil (e.g. peak ϕ) but thereafter residual strength values should be used (e.g. Elms and Richards 1990; Chugh 1995). Alternatively, conservative (for design) estimates of seismic induced displacements should be based on residual strength values if a single value of ϕ is adopted to simplify analyses.

3.3.2 *Empirical approaches*

If the input acceleration data at a site is specified by characteristic parameters such as the peak ground acceleration and the peak ground velocity, then empirical methods that correlate the expected permanent displacement to the characteristic parameters of the earthquake and a critical acceleration ratio for the structure are required. Alternatively, if the tolerable permanent displacement of the structure is specified, based on serviceability criteria, the wall can then be designed using an empirical method so that expected permanent displacements do not exceed specified values. Newmark's sliding block theory has been widely used to establish empirical relationships between the expected permanent displacement and characteristic seismic parameters of the input earthquake by integrating existing acceleration records. The critical acceleration ratio, which is the ratio of the critical acceleration, k_cg, of the sliding block to the peak horizontal acceleration, k_mg, of the earthquake, has been shown to be an important parameter that affects the magnitude of the permanent displacement. Thus, the seismic displacement of a potential sliding soil mass computed using Newmark's theory has been traditionally correlated with the critical acceleration ratio, k_c/k_m, and other representative characteristic seismic parameters such as the peak ground acceleration, k_mg, the peak ground velocity, v_m, and the predominant period, T, of the acceleration spectrum (e.g. Newmark 1965; Sarma 1975; Franklin and Chang 1977).

Cai and Bathurst (1996b) have reformulated a number of existing displacement methods based on non-dimensionalized displacement terms that are common to the methods, and divided them into two separate categories based on the characteristic seismic parameters referenced in each method. Example relationships between the dimensionless displacement term, $d/(v_m^2/k_mg)$, where d is the actual expected permanent displacement, and the critical acceleration ratio are shown in Figure 23. Other curves are available in the literature but it should be noted that any empirical curve will be influenced by the earthquake data that is used to establish the curve and the interpretation of the original data.

3.3.3 *Example applications*

Newmark methods have been applied to unreinforced slopes (Chang et al. 1984). Vrymoed (1989) used the Newmark method to estimate the cumulative base sliding displacement of a rectangular reinforced soil mass for a single cycle of base acceleration record. Ling et al. (1996a, b) have proposed a method to calculate reinforcements loads and anchorage lengths under horizontal seismic loads using a two-part wedge sliding block model. Cai and Bathurst (1996a) demonstrated the application of Newmark's method and

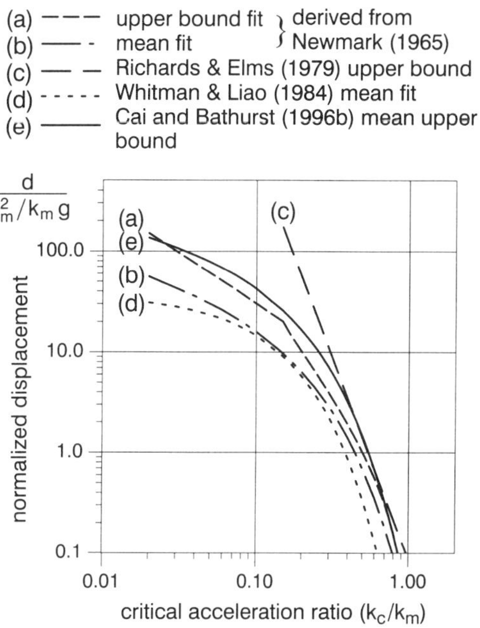

Figure 23. Summary of proposed relationships between non-dimensionalized displacement term and critical acceleration ratio (after Cai and Bathurst 1996b).

empirical approaches to geosynthetic reinforced soil segmental retaining walls. Analyses are restricted to horizontal sliding or shear mechanisms: i.e. 1) external sliding along the base of the total structure which includes the reinforced soil mass and the facing column (Figure 12); 2) internal sliding along a reinforcement layer and through the facing (Figure 24a) and; 3) interface shear between facing units with or without the presence of a geosynthetic inclusion (Figure 24b). A summary of calculation results for the geosynthetic reinforced soil wall structure shown in Figure 25 is given in Table 1 assuming $\phi=35°$. The material properties for the facing units have been taken from large-scale laboratory tests carried out at the Royal Military College of Canada (RMCC). The block-geosynthetic interface shear properties (a_u, λ_u) were selected to represent a system with relatively low interface shear capacity in order to generate a worst case set of displacement predictions. The E-W (90°) horizontal ground acceleration component recorded at Newhall Station (California Strong Motion Instrumentation Program) during the 17 January 1994 Northridge earthquake (M=6.7) was used as the input earthquake data. The record shows a peak horizontal ground acceleration of $k_m=0.60$. The total permanent displacement at the wall face at each elevation from the initial static position was estimated by adding the layer displacement to the cumulative displacement below that layer. The

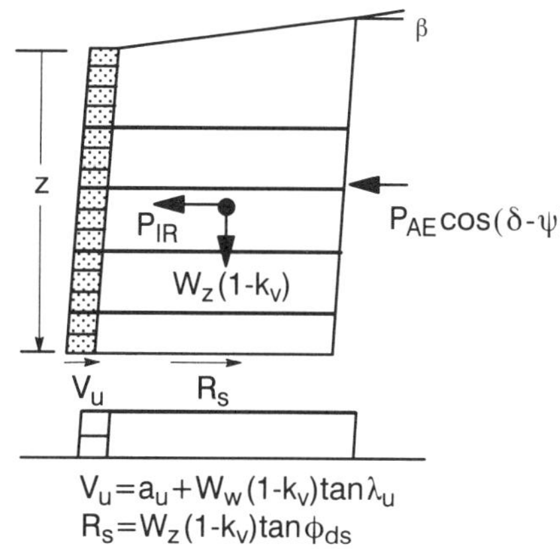

a) internal sliding.

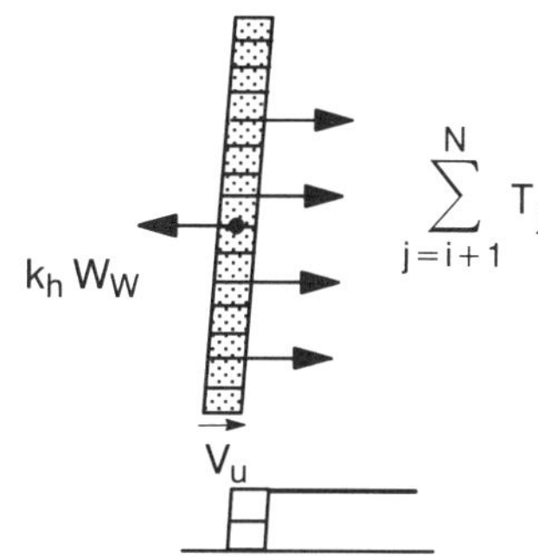

b) facing column shear.

Figure 24. Newmark's sliding block method applied to geosynthetic reinforced segmental retaining wall structures.

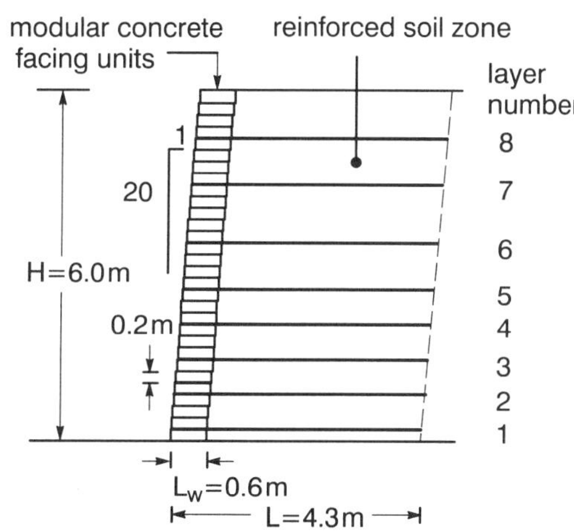

Figure 25. Geogrid reinforced segmental soil retaining wall used in displacement method example (after Cai and Bathurst 1996a).

layer displacement was taken as the larger of the column shear displacement or internal sliding at that layer. The data in Table 1 shows that large displacements are possible at the top of the wall using $k_h = k_m = 0.6$ in the pseudo-static seismic stability analysis. This is an extreme loading condition that was used to illustrate the general approach. Similar calculations with a higher quality fill (i.e. $\phi=40°$) resulted in displacements that were restricted to the top two facing courses. Furthermore, analyses with better block-geosynthetic properties resulted in insignificant or no displacements at all elevations. This last result is consistent with observations made at the site of two segmental retaining walls after the Northridge earthquake that showed no detectable shear movement of the facing column units despite significant horizontal ground accelerations estimated to be as high as 0.5g (Bathurst and Cai 1995). The table illustrates that the order of magnitude accuracy of the empirical method (compared to Newmark's method) is satisfied for all large displacement results. Predicted displacements must be viewed as order-of-magnitude estimates rather than accurate predictions. Engineering judgement plays an important role in the interpretation of results using any empirical approach.

Table 1. Total permanent displacement considering all displacement mechanisms.

Layer	displacement (mm) Newmark	Empirical
8	154*	206*
7	47*	70*
6	29*	49*
5	25	41
4	25	41
3	25	41
2	24	36
1	21	29
Base sliding	11	15

Note: * controlling mechanism is facing shear, otherwise internal sliding controls.

3.4 *Finite element modelling*

The attraction of properly formulated finite element methods is that they can implement complex models for the component materials such as nonlinear cyclic behaviour of the soil and reinforcement materials using models such as those described in Section 2.

3.4.1 *Slopes and embankments*

The dynamic response of a reinforced and unreinforced soil slope with c–ϕ properties resting on a firm foundation was determined by the senior writer and co-workers using a modified version of the TARA-3 program (Finn et al. 1986). The slopes were 12 m high with a side slope of 1:1. One slope was lightly reinforced with polymeric reinforcements 12 m in length. The reinforcement layers were placed with a vertical spacing of 2 m.

The finite element representation of the reinforced soil slope is shown in Figure 26a. The slope was subjected to the first 9.60 seconds of the N-S component of the 1940 El Centro earthquake scaled to 0.2g. The base was assumed to be rigid and the nodes on the left and right vertical boundary were supported on horizontal rollers for the dynamic analysis. A static analysis was first conducted to establish the stress-strain field prior to the earthquake excitation. The program simulated the incremental construction process of the slope. The results of the analyses showed that the dynamic displacement-time response of the slope was not significantly influenced by the presence of the re-

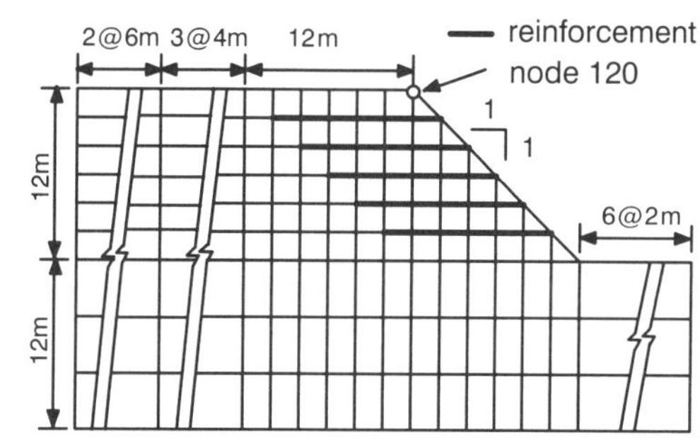

a) finite element representation of reinforced soil slope.

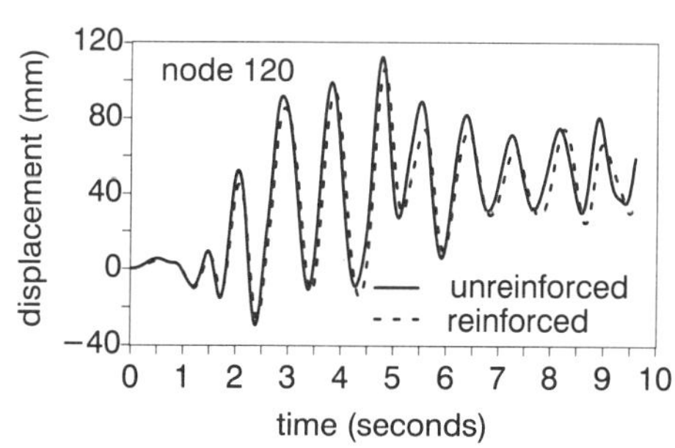

b) horizontal displacement-time history at slope crest.

Figure 26. Dynamic finite element analysis of reinforced soil slope (after Yogendrakumar et al. 1991).

inforcement (Figure 26b) for the duration of shaking applied. However, it must be noted that a perfect bond was assumed between the reinforcement and the soil (which is consistent with $C_i = 1$ for many geogrid reinforcement products in cohesionless soils) and hysteresis of the reinforcement was not considered. These results suggest that reinforcement may not reduce seismic-induced displacements of slopes that do not require reinforcement to prevent collapse. This particular result deserves further investigation.

3.4.2 *Walls*

Finite element modelling has been used to to gain insight into the behaviour of geosynthetic reinforced soil walls (e.g. Rowe and Ho 1992; Karpurapu and Bathurst 1995; Wu 1992).

The use of dynamic finite element (FE) modelling for reinforced earth structures is much more limited. Segrestin and Bastick 1988 and Yogendrakumar et al. 1991 used the programs SUPERFLUSH and TARA-3 respectively, to study the seismic response of reinforced soil walls that used inextensible reinforcement (steel strips). Bachus et al. 1993 used the program DYN3D to simulate the blast response of geosynthetic reinforced soil walls constructed with incremental concrete panel facings.

The results of the FE parametric analyses of RECO systems (e.g. Segrestin and Bastick 1988) has been the principal source of analysis and design guidelines for reinforced soil walls with inextensible (steel strip) reinforcement. Because of the lack of similar parametric data for extensible reinforced structures the data for simulated RECO walls has been adopted by FHWA and AASHTO agencies in the USA for design of geosynthetic reinforced soil walls.

Cai and Bathurst (1995) carried out dynamic finite element modelling of geosynthetic reinforced segmental retaining walls in order to investigate the entire load-deformation response of an example system under simulated earthquake loads. The modified TARA-3 code mentioned in the previous section was used together with the hysteretic soil and reinforcement models described in Sections 2.1 and 2.2. The results of large-scale interface shear and connection tests were used to provide parameters for the modelling of the facing column. The interface shear capacities that were used are considered to be relatively poor for segmental retaining wall systems, based on a large amount of test data gathered at RMCC. The base reference acceleration-time history used is a scaled El Centro 1940 earthquake record. Spectrum analysis of the input acceleration record gives a dominant frequency range of 0.5 to 2 Hz. Predicted cumulative lateral deformations through the height of the facing column at the end of two scaled base input records are illustrated in Figure 27. The relative displacements are largest at reinforcement elevations where locally greatest interface shear loadings occurred. While the potential for interface shear leading to collapse of these structures is clear, it is worth noting that the vertical out-of-alignment is less than 1% of the height of the wall. In practice this amount of relative displacement is within the limits usually achieved during construction (Ba-

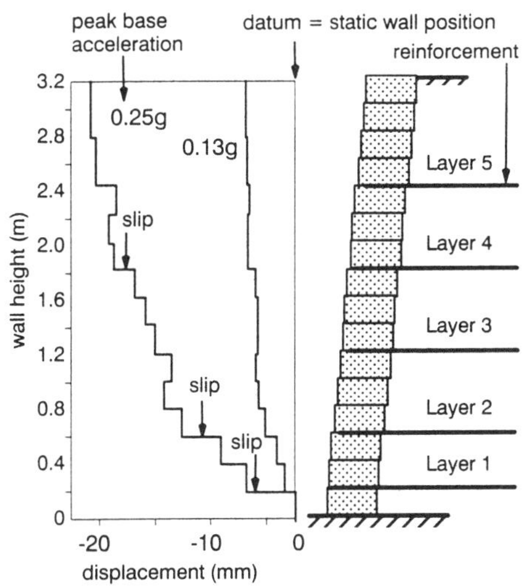

Figure 27. Facing column lateral displacement at end of excitation history (Cai and Bathurst 1995).

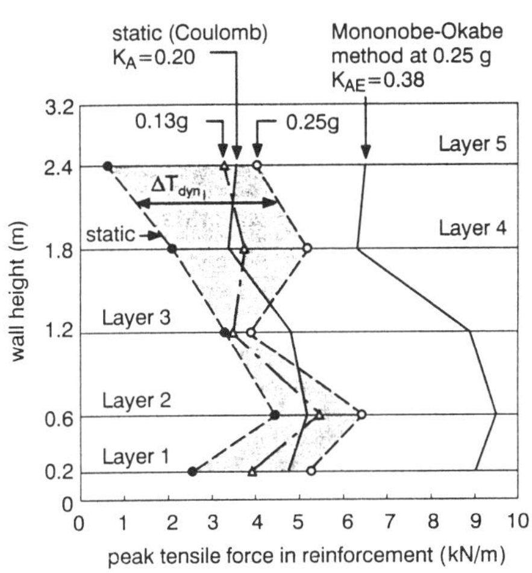

Figure 28. Distribution of peak reinforcement forces (after Cai and Bathurst 1995).

thurst et al. 1995) and, hence, from practical considerations may be judged to be insignificant. The results suggest that for the range of peak accelerations and duration of excitation applied to this low height wall, the structure performed well despite relatively poor interface shear characteristics. An important observation made by the writers was that reinforcement forces predicted by the FE model were consistently lower than those computed using the pseudo-static M-O approach as illustrated in Figure 28. This result is consistent with the opinion of many practitioners that M-O theory is conservative for routine soil retaining wall structures. In addition, for this low height wall the maximum incremental dynamic reinforcement forces were observed towards the top of the wall which is consistent with the pseudo-static model proposed by Bathurst and Cai (1995) for segmental retaining walls with extensible reinforcements. Finally, the data in Figure 28 shows that reinforcement loads were low even under seismic shaking and likely well within limits expected for reinforcement over-stressing.

3.5 *Dynamic finite difference modelling*

Seismic response of reinforced walls, slopes and embankments can be analyzed using explicit dynamic finite difference methods such as the FLAC computer program developed by Itasca Consulting Group (1996). FLAC (Fast Lagrangian Analysis of Continua) is based on the Lagrangian calculation scheme that is well suited for modelling large distortions and material collapse. Complete descriptions of the numerical formulation are reported by Cundall and Board (1988). Several built-in constitutive models are available in the FLAC package and can be easily modified by the user. For example, geosynthetic reinforcement layers can be represented as either cable, beam or pile structures. One advantage of using FLAC in seismic analysis is the simplicity of applying seismic loading anywhere within the problem domain.

Example preliminary FLAC analyses for reinforced soil slopes in which the reinforcement layers have been modelled using cable elements are shown in Figures 29 and 30. The duration of base shaking for the slope in Figure 30 was 2.5 seconds with a horizontal sinusoidal base acceleration having a peak amplitude of 0.6g and a frequency of 2 Hz. Results of preliminary analyses conducted by the writers using the slope in Figure 29 show that the behavior of the slope is very dependent on the stiffness of the foundation materials (soil or rock). For example, the effectiveness of reinforced soil masses to minimize cumulative

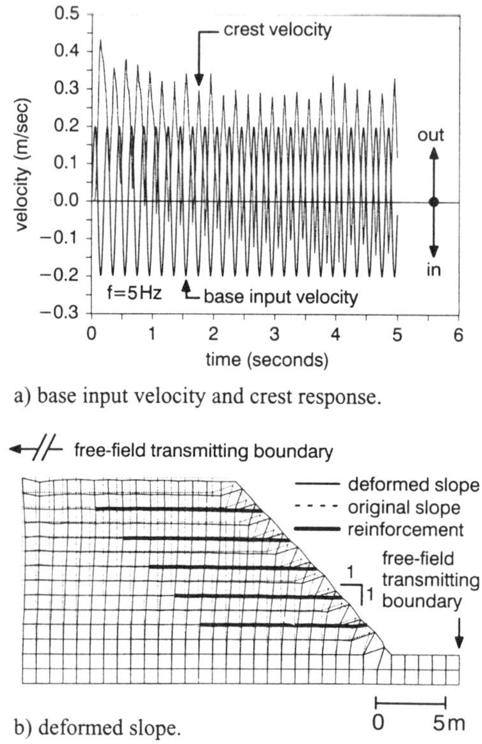

Figure 29. Example FLAC analysis of reinforced soil slope.

lateral displacements during horizontal ground shaking increases with decreasing depth to bedrock.

4 MODEL TESTING

Model tests for seismic studies fall into two categories: 1) reduced-scale shaking table tests; and 2) centrifuge tests subjected to base shaking. Both shaking table and centrifuge model tests share certain drawbacks, among the most recognized of which are similitude and boundary effects.

4.1 *Shaking table tests*

The advantage of shaking table tests is that they are relatively easy to perform. The principal disadvantages are related to problems of similitude between reduced-scale models and equivalent prototype scale systems (Fairless 1989). Similitude rules have been proposed by Sugimoto et al. (1994) and Telekes et al. (1994). Of particular concern is the difficulty of 1g

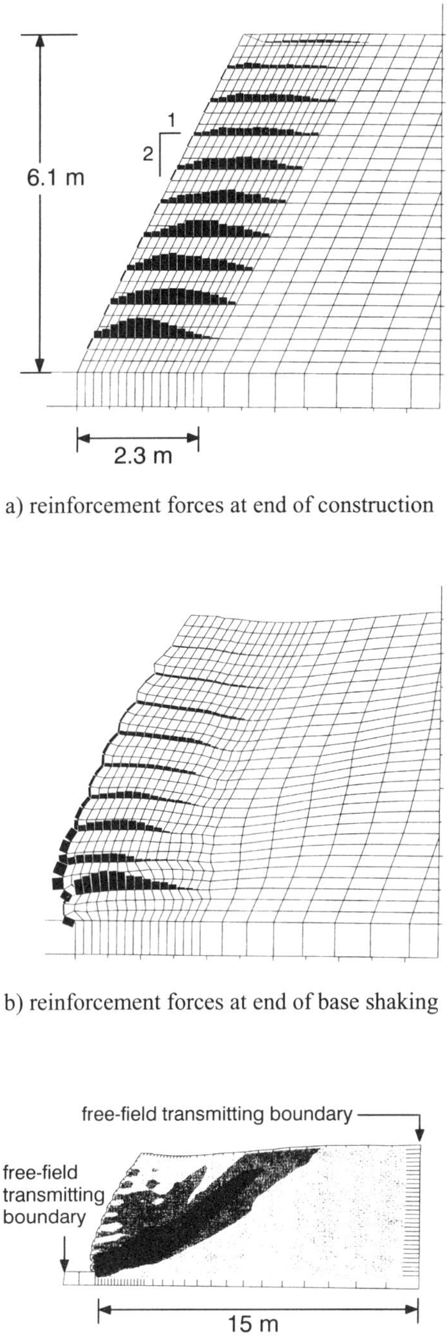

a) reflection forces at end of construction

b) reinforcement forces at end of base shaking

c) shear strains at end of base shaking

Figure 30. Example FLAC analysis of reinforced soil slope with wrap-around facing (Kramer 1996b).

models to scale nonlinear soil strength and stress-strain properties that vary with confining pressure. An important consequence of these difficulties is that failure mechanisms observed in reduced-scale models may be different from those observed at the prototype scale.

Nevertheless, the summary of investigations given in Table 2 identify important performance features of reinforced soil structures under dynamic loading. Most investigators have noted amplification of base input acceleration over the height of structures particularly at the top of the structures.

These observations give support to design methodologies that either incorporate empirical acceleration profiles directly (Steedman and Zeng 1990) or indirectly (Bathurst and Cai 1995) and lead to the requirement in some cases to increase the number and length of reinforcement layers close to the top of reinforced wall structures based on limit equilibrium design.

Bathurst et al. (1996) and Pelletier (1996) have reported the results of a series of shaking table tests that examined seismic resistance of model reinforced segmental retaining walls. The tests were focused on the influence of interface shear properties on facing column stability which was identified as an important design consideration based on pseudo-static methods of analysis (Bathurst and Cai 1995). A set of 1/6 scale model walls were constructed inside a plexiglas box and were 2400 mm long by 1400 mm wide by 1020 mm high. Similitude rules proposed by Iai (1989) were used to scale the model components and geometry. A typical test configuration is illustrated in Figure 31. The models were constructed with concrete blocks 100 mm wide (toe to heel) x 160 mm wide x 34 mm high. Five layers of a weak geogrid (HDPE bird fencing) were used to model the reinforcement. The backfill was a standard laboratory silica #40 sand prepared at a relative density of 67%. The four test configurations used are summarized in Table 3. The differences between tests are related to interface shear capacity and wall batter. Interfaces identified as frictional in Table 3 derive shear capacity solely from sliding resistance at the interface. These interfaces represent a very poor facing column detail with respect to shear capacity. In two of the tests the interfaces were fixed at some locations in order to simulate systems with high shear capacity at all or selected facing column interfaces (i.e. positive interlock due to effective shear keys, pins or other types of connectors).

Each test was subjected to a staged increase in base input motion resulting in the acceleration-time record shown in Figure 32. The base input frequency was kept constant at 5 Hz. At the prototype scale this frequency corresponds to 2 Hz.

Table 2. Summary of shaking table studies.

Reference	Model details	Observed behavior and implications to design and analysis
Koga et al. 1988; Koga and Washida 1992	1.0-1.8 m high models with vertical and inclined slopes at 1/7 scale. Sandbags with wrap-around facing. Nonwoven geotextile, plastic nets and steel bars with sandy silt backfill.	Deformations decreased with increasing reinforcement stiffness and density and decreasing face slope angle. Failure volumes were shallower for reinforced structures. Relative reduction in deformation of reinforced structures compared to unreinforced structures increased with steepness of the face. Circular slip method agrees well with experimental results except for steep face models.
Murata et al. 1994	2.5 m high 1/2 scale model walls with gabion/rigid concrete panel walls. Geogrid with dry sand backfill. Horizontal shaking using sinusoidal and scaled earthquake record. Base accelerations up to 0.5g at 3.4 Hz.	Increase in reinforcement forces due to shaking was very small. Reinforcement loads increased towards the front of the wall. Acceleration amplification was negligible up to mid-height of wall but increased to about 1.5 at the top. Amplification behavior was similar for reinforced and unreinforced zones. The reinforced zone behaved as a monolithic body. Sinusoidal base input resulted in greater deformations than scaled earthquake record. Rigid facing adds to wall seismic resistance.
Sugimoto et al. 1994; Telekes et al. 1994	1.5 m high model embankment with sand bags and wrap-around slope surface. Geogrid reinforcement with sand backfill. Model scales 1/6 and 1/9. Sinusoidal and scaled earthquake record. Base acceleration up to 0.5g at 40 Hz.	Reinforced models more stable than unreinforced. Proposed similitude rules for small and large strain deformation modelling. Largest amplification recorded at crest of models. Failure of structures was progressive from top of structure downward. Reinforcement forces increased linearly with acceleration up to start of failure. Failure mechanism difficult to predict using proposed scaling rules. Under seismic loading conditions, there was a tendency for shallow slopes to fail compared to steeper ones. Scale effects due to vertical stress and apparent cohesion of backfill soil influenced the relative performance of steep faced and shallow faced models.
Budhu and Halloum 1994	0.72 m high model wall with wrap-around facing. Geotextile with dry sand backfill. Base acceleration in increments of 0.05g at 3 Hz.	Sliding progressed with increasing acceleration from the top geotextile/sand interface to the bottom layer. No consistent decreasing trend of critical acceleration was observed with increasing spacing to length ratio. Critical acceleration proportional to the soil/geotextile interface friction value.
Sakaguchi et al. 1992; Sakaguchi 1996	1.5 m high model walls. One wrap-around and four unreinforced rigid concrete panel walls. Geogrid with dry sand backfill. Sinusoidal loading with base acceleration up to 0.72g at 4 Hz.	Wrap-around wall behaved as a rigid body and failed at a higher acceleration than unreinforced structures. However, at smaller accelerations (due to stiff facing panels) the displacements of the unreinforced structures were less. A base input acceleration of 0.32g delineated stable wall performance from yielding wall performance for the reinforced structure. Residual strains were greatest closest to the face. Concluded that more rigid light-weight modular block facings may be effective in reducing reinforcement loads.

The influence of interface shear capacity and facing batter can be seen in Figure 33. The vertical wall with fixed interface construction (high shear capacity at each interface) required the greatest input acceleration to generate large wall displacements during staged shaking (Test 4). The vertical wall with poor interface shear at all facing unit elevations performed worst (Test 1). However, the resistance to wall displacement was improved greatly for the uniformly weakest interface condition by simply increasing the wall batter (Test 3). The vertical wall with poor interface properties only at the geosynthetic layer elevations (Test 2) gave a displacement response that fell between the results of walls constructed with uniformly poor interface shear properties (Test 1) and the nominally identical structure with uniformly good interface shear properties (Test 4). The resistance of the facing column to horizontal base shaking improved with increasing shear capacity between dry-stacked modular blocks or by increasing the wall batter.

The results of this study confirmed that measured accelerations were not uniform throughout the soil-wall system. Large acceleration amplifications as high as 2.2 were recorded particularly at the top of the unreinforced portion of the facing column. Observed critical accelerations to cause failure of the wall models were compared to predictions based on the analysis method proposed by Bathurst and Cai (1995). The measured peak acceleration at the middle wall height or at the top of the backfill surface was shown to give more accurate estimates of critical acceleration to be used in pseudo-static analysis. The estimated total load in the reinforcement layers was estimated to be only a very small percentage of the tensile capacity of the reinforcement layers. The test results showed that while critical accelerations to cause incipient collapse of the wall models could be predicted reasonably well the actual failure mechanism was difficult to predict. For example, pullout of the top reinforcement layer was identified as a critical mechanism when in fact the observed failure mechanism was toppling of the top unreinforced facing column.

4.2 *Combined centrifuge/shaking table tests*

The scaling difficulties identified for 1g shaking table tests can be overcome theoretically using centrifuge

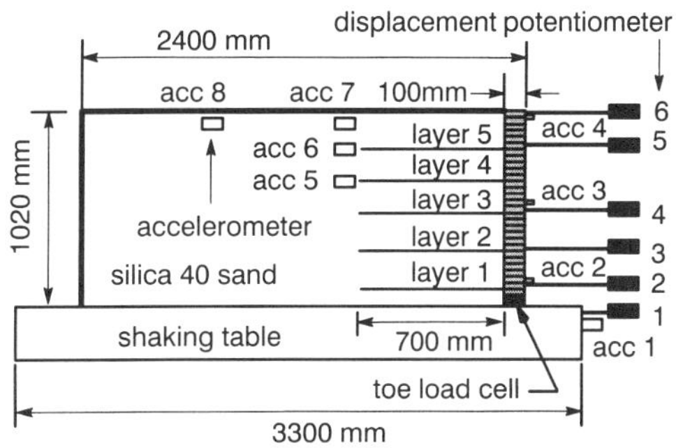

Figure 31. Example shaking table model of reinforced segmental retaining wall.

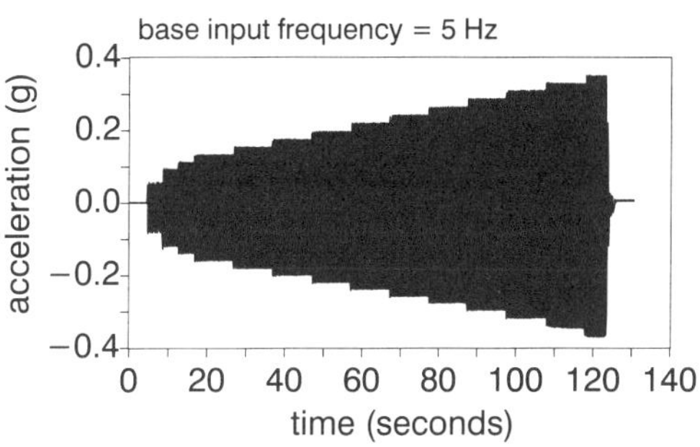

Figure 32. Base input acceleration record for shaking table tests.

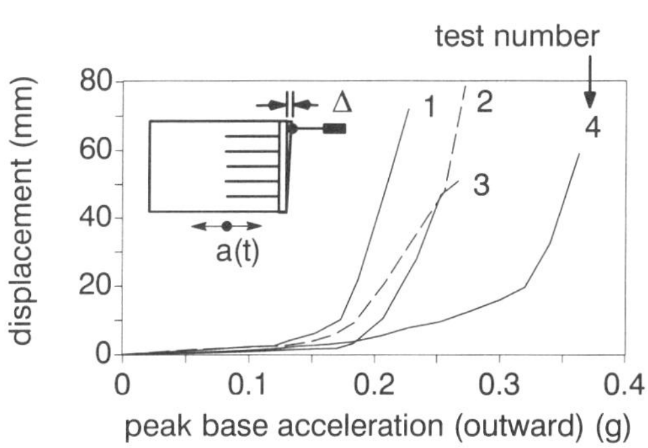

Figure 33. Displacement close to top of wall versus peak base input acceleration.

testing. Sakaguchi et al. (1992, 1994) and Sakaguchi (1996) mounted a shaking table on a centrifuge apparatus. Sakaguchi and co-workers examined the response of a segmental retaining wall constructed with light-weight facing units. Three 150 mm high models were accelerated in the centrifuge to simulate walls 4.5 m high. Three different geotextiles were used having a range of tensile strengths, and walls were built with three different reinforcement lengths. The qualitative performance of the centrifuge tests was similar to that recorded for the 1g shaking table tests (see Table 2). The results showed that up to a limiting value of reinforcement length ($L/H \approx 2/3$) there was a corresponding reduction in wall displacements with increasing base acceleration. Geotextile strength for the range of materials used did not influence wall deformation.

Table 3. Model test configurations (Bathurst et al. 1996).

Test No.	Facing batter	Block-block interface	Block-geosynthetic interface
1	vertical	frictional	frictional
2	vertical	fixed	frictional
3	8 degrees*	frictional	frictional
4	vertical	fixed	fixed

* from vertical

5 GEOFOAM SEISMIC LOAD BUFFERS

The generic term geofoam has recently entered geosynthetics terminology to describe expanded foams used in geotechnical applications (Horvath 1995). Horvath proposed that geofoam panels could be used against rigid wall structures (e.g. basement walls) to reduce seismic-induced stresses that would otherwise over-stress rigid wall structures.

To the best of the writers' knowledge, the first application of this technology in North America was recently reported by Inglis et al. (1996). Panels of low density expanded polystyrene (EPS) from 450 to 610 mm thick were placed against rigid basement walls up to 9 m in height at a site in Vancouver, British Columbia. Analyses using program FLAC showed that a 50% reduction in lateral loads could be expected (Figure 34) during a seismic event compared to a rigid wall solution. The design challenge using this technique is to optimize the thickness of the buffer panels for a candidate geofoam material so that the horizontal compliance under peak loading is just sufficient to minimize lateral earth pressures without excessive lateral deformations. In addition, the ideal properties of the geofoam are adequate compressive stiffness under static loading conditions but with a compressive yield plateau that will be just exceeded under the design seismic lateral stresses. Horvath has recognized that the technique described here may be an economical solution to the problem of retrofitting existing rigid wall structures that do not satisfy modern seismic design codes.

6 OBSERVED PERFORMANCE

6.1 *Northridge Earthquake 1994, Loma Prieta Earthquake 1989 (California, USA)*

Sandri (1994) conducted a survey of reinforced segmental retaining walls greater than 4.5 m in height in the Los Angeles area immediately after the Northridge Earthquake of 17 January 1994 (moment magnitude = 6.7). The results of the survey showed no evidence of visual damage to 9 of 11 structures located within 23 to 113 km of the earthquake epicenter. Two structures (Valencia and Gould Walls) showed tension cracks within and behind the reinforced soil mass that were clearly attributable to the results of seismic loading. Bathurst and Cai (1995) analyzed both structures and noted that minor cracking at the back of the reinforced soil zone could be attributed to the flattening of the internal failure plane predicted using M-O theory.

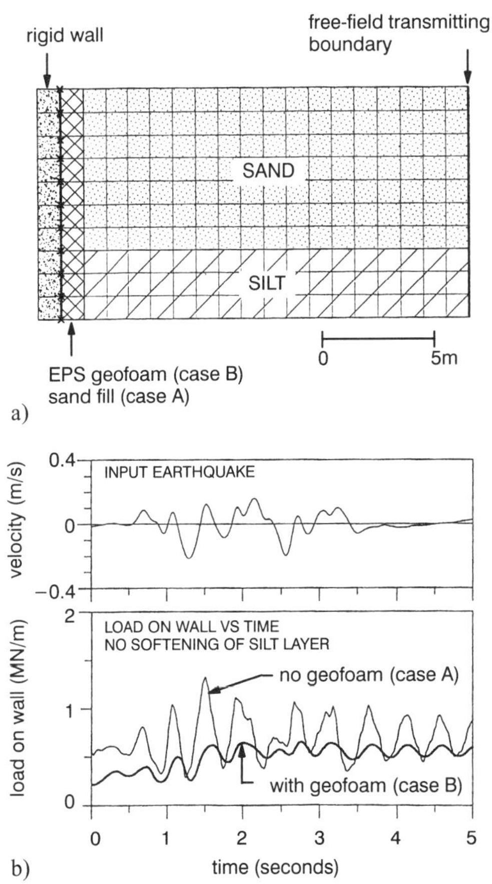

Figure 34. Results of FLAC analyses on seismic load reduction using geofoam buffer (after Inglis et al. 1996).

The facing columns for all walls were intact even though peak horizontal ground accelerations as great as 0.5g were estimated at one site.

A similar survey of three geosynthetic reinforced walls and four geosynthetic reinforced slopes by White and Holtz (1996) after the same earthquake revealed no visual indications of distress. Stewart et al. (1994) report that slope indicator measurements at the toe of a 24 m high geogrid reinforced slope which was estimated to have sustained peak horizontal ground accelerations of 0.2g showed no movement. Some unreinforced crib walls and unreinforced segmental walls were observed to have developed cracks in the backfill during the same survey by Stewart et al.. They concluded that concrete crib walls may not perform as well as more flexible retaining wall systems under seismic loading. Similar good performance of several geosynthetic reinforced soil walls and slopes during the 1989 Loma Prieta earthquake (Richter magnitude = 7.1) was reported by Eliahu and Watt (1991) and Collin et al. (1992).

6.2 *Great Hanshin Earthquake 1995 (Kobe, Japan)*

Tateyama et al. (1995) reported on the seismic performance of traditional unreinforced wall structures after the Great Hanshin earthquake of 17 January 1995 (moment magnitude = 6.8). Concrete and masonry walls suffered serious failures, including collapse. Conventional reinforced concrete cantilever structures suffered some cracking and limited displacement.

Tatsuoka et al. (1995, 1996) reported on the performance of a 6.2 m high geosynthetic reinforced soil retaining wall with a full height rigid facing construction. The peak ground acceleration at the site was estimated to have been as great as 0.7g. The structure was observed to have moved 260 mm at the top and 100 mm at ground level but was otherwise undamaged. Tatsuoka et al. conclude that shortening of the reinforcement lengths due to site constraints was a likely cause of the observed tilting of the wall.

Nishimura et al. (1996) surveyed 10 geogrid reinforced soil walls and steepened slopes after the same event. All structures survived the earthquake even though peak ground accelerations were estimated in the range of 0.3 to 0.7g. Nishimura et al. determined critical accelerations for these structures using GRB (1990) and PWRI (1992) methods of analysis and found that predicted critical acceleration coefficient (k_h) values were as low as 0.1. They concluded that both methods are very conservative. Where minor damage was observed it was related in one instance to minor separation between an unattached concrete facing column and in the other case there was cracking at the back of the reinforced soil mass although this last observation may be the result of poor base foundation conditions. Results of stability calculations using GRB and PWRI methods led Nishimura et al. to conclude that the length of reinforcement layers at the top of the reinforced soil structures should be increased in order to capture critical failure volumes generated under even modest horizontal seismic accelerations.

7 CONCLUSIONS

Largely qualitative observations of the performance of geosynthetic reinforced slopes and walls in both the USA and Japan suggest that these structures perform well during seismic events when located on competent foundation soils and above the water table. The relatively flexible nature of reinforced soil walls constructed with extensible and inextensible reinforcement is routinely cited as the reason for the good performance of these structures during a seismic event. Nevertheless, the geotechnical engineer requires seismic design tools and representative component properties for geosynthetic reinforced soil walls and slopes in order to optimize design of these structures in seismic environments. The review of the literature and the work by the writers and co-workers leads to the following conclusions and research needs:

1. The depth, strength and stiffness of the foundation soil may have a greater influence on the internal and external stability of reinforced soil slopes and walls than the design of the reinforced mass in isolation. Parametric analyses are required to investigate the influence of foundation condition on seismic performance.
2. The design methodologies that are currently used in the USA for geosynthetic reinforced soil walls have been based largely on the results of numerical modelling of reinforced structures constructed with inextensible reinforcement (steel strips). Similar studies are required to confirm that the general approach is valid for relatively less stiff geosynthetic reinforced soil wall structures. Further numerical and experimental work is required to investigate the validity of pseudo-static analysis methods that predict increased reinforcement lengths at the top of reinforced walls and slopes.
3. Ground motion amplification (or attenuation) through retained soils plays a major role in generating additional dynamic loads on geosynthetic reinforcement and wall facing components. More work is required to offer guidance on the appropriate distribution of incremental seismic forces to be applied to extensible reinforcing elements and to establish the influence of system stiffness (i.e. the combined effect of reinforcement stiffness, number of reinforcement layers, facing stiffness and height of structure) on this distribution. Numerical models calibrated against the results of carefully conducted large shaking table tests or small-scale centrifuge tests are possible research strategies to meet this goal.
4. A number of design methodologies have been proposed in the USA and Japan for the seismic design of walls and slopes that can lead to important differences in the required number/strength, location and length of reinforcement layers. Comparative analyses should be carried out to examine the relative conservativeness (or non-conservativeness) of the proposed methodologies.
5. Geosynthetic reinforced segmental retaining walls in seismic areas offer unique challenges to the designer because of their modular facing column construction. These structures involve analyses not required for other retaining wall systems. The experience of the senior writer is that the economic potential of these systems in seismic areas will not be fully realized until confidence is developed through proven design methodologies for these structures.
6. The design engineer will continue to be attracted to relatively simple seismic design tools based on pseudo-static and displacement methods for the design and analysis of routine walls and slopes under modest seismic loads. Nevertheless, the results of sophisticated numerical models carried out by experienced modelers offers the possibility of refining simple models to minimize unwarranted conservativeness.

8 ACKNOWLEDGEMENTS

The funding for the work reported in the paper was provided by the Department of National Defence (Canada) through an Academic Research Program (ARP) grant and a research contract from Directorate Infrastructure Support (DIS/DND) awarded to the senior writer. The writers thank Professors H. Ochiai, R.D. Holtz, T. Akagi, and F. Tatsuoka and Messrs. J. DiMaggio and J. Nishimura for provision of many useful references, and Professor S.L. Kramer for permission to publish results of FLAC analyses carried out at the University of Washington, WA, USA. The contribution of former post-doctoral research associates Drs. Z. Cai and M. Yogendrakumar to the research program at RMCC is also gratefully acknowledged as are the efforts of former graduate students Capt. M. McLay and Capt. M. Pelletier. The writers would also like to thank Mr. M. Simac and Mr. T. Allen for many fruitful discussions on the general topic of segmental walls, and the efforts of Mr. P. Clarabut who assisted with much of the experimental work carried out at RMCC.

REFERENCES

Allen, T.M. 1993. Issues regarding design and specification of segmental block-faced geosynthetic walls. *Transportation Research Record*, 1414, pp. 6-11.

AASHTO 1992; and 1996 Interims. *Standard specifications for highway bridges*. American Association of State Highway and Transportation Officials, Washington, D.C., USA.

Bachus, R.C., Fragaszy, R.J., Jaber, M., Olen, K.L., Yuan, Z. and Jewell, R. 1993. *Dynamic response of reinforced soil systems*. Report ESL-TR-92-47, Engineering Research Division, US Department of the Air Force Civil Engineering Support Agency, March 1993, Vol. 1, 230 p., Vol. 2, 227 p.

Bathurst, R.J. 1994. *Reinforced soil slopes and embankments*. Technical Notes for Computer Programs GEOSLOPE and GEOPLOT, 17 p.

Bathurst, R.J. and Cai, Z. 1995. Pseudo-static seismic analysis of geosynthetic-reinforced segmental retaining walls. *Geosynthetics International*, Vol. 2, No. 5, pp. 787-830.

Bathurst, R.J. and Cai, Z. 1994. In-isolation cyclic load-extension behavior of two geogrids. *Geosynthetics International*, Vol. 1, No. 1, pp. 3-17.

Bathurst, R.J., Cai, Z. and Pelletier, M.J. 1996. Seismic design and performance of geosynthetic reinforced segmental retaining walls. *Proc. 10th Annual Symp. of the Vancouver Geotechnical Society*, Vancouver, BC, 26 p.

Bathurst, R.J. and McLay, M.J. 1996. Repeated load pullout testing of a HDPE geogrid. *Submitted for publication*.

Bathurst, R.J. and Simac, M.R. 1994. Geosynthetic reinforced segmental retaining wall structures in North America. Keynote Paper, *Proc. 5th Int. Conf. on Geotextiles, Geomembranes and Related Products*, Singapore, pp. 31-54.

Bathurst, R.J. and Simac, M.R. 1993. Laboratory testing of modular unit-geogrid facing connections. STP 1190 Geosynthetic Soil Reinforcement Testing Procedures (S.C.J. Cheng, Ed.), *American Society for Testing and Materials (Special Technical Publication)*, pp. 32-48.

Bathurst, R.J., Simac, M.R. and Sandri, D. 1995. Lessons learned from the construction performance of a 14m high segmental retaining wall. *Proc. of Geosynthetics: Lessons Learned from Failures*, (J.P. Giroud, Ed.), Nashville, Tennessee, February 1995, pp. 21-34.

Bonaparte, R., Schmertmann, G.R. and Williams, N.D. 1986. Seismic design of slopes reinforced with geogrids and geotextiles. *Proc. 3rd Int. Conf. on Geotextiles*, Vienna, Austria, Vol. 1, pp. 273-278.

Budhu, M. and Halloum, M. 1994. Seismic external stability of geotextile reinforced walls. *Proc. 5th Int. Conf. on Geotextiles, Geomembranes and Related Products*, Singapore, Vol. 1, pp. 529-532.

Cai, Z. and Bathurst, R.J. 1996a. Seismic-induced permanent displacement of geosynthetic reinforced segmental retaining walls. *Canadian Geotechnical J.*, Vol. 31, pp. 937–955.

Cai, Z. and Bathurst, R.J. 1996b. Deterministic Sliding block methods for estimating seismic displacements of earth structures. *Soil Dynamics and Earthquake Engineering*, Vol. 15, pp. 255-268.

Cai, Z. and Bathurst, R.J. 1995. Seismic response analysis of geosynthetic reinforced soil segmental retaining walls by finite element method. *Computers and Geotechnics*, Vol. 17, No. 4, pp. 523-546.

Canadian Foundation Engineering Manual (CFEM) 1992. 3rd Edition, *Canadian Geotechnical Society*, 512 p.

Cancelli, A., Rimoldi, P. and Togni, S. 1992. Frictional characteristics of geogrids by means of direct shear and pullout tests. *Earth Reinforcement Practice* (Ochiai, Hayashi and Otani, Eds.), Balkema, Proc. Int. Symp. on Earth Reinforcement Practice, Fukuoka, Japan, Vol. 1, November 1992, pp. 51-56.

Chang, C.J., Chen, W.F. and Yao, J.T. 1984. Seismic displacement in slopes by limit analysis. *J. of Geotechnical Engineering, ASCE*, Vol. 110, No. 7, pp. 860-875.

Chida, S., Minami, K. and Adachi, K. 1982. *Test de stabilité de remblais en Terre Armée* (translated from Japanese).

Christopher, B.R., Gill, S.A., Giroud, J.P., Juran, I., Schlosser, F., Mitchell, J.K. and Dunnicliff, J. 1989. *Reinforced soil structures. Volume I. Design and construction guidelines*. Report No. FHWA-RD-89-043, Washington, DC., USA, Nov. 1989, 287 p.

Chugh, A.K. 1995. Dynamic displacement analysis of embankment dams. *Geotechnique*, Vol. 45, No. 2, pp. 295-299.

Collin, J.G., Chouery-Curtis, V.E. and Berg, R.R. 1992. Field observations of reinforced soil structures under seismic loading. *Earth Reinforcement Practice* (Ochiai, Hayashi and Otani, Eds.), Balkema, Proc. Int. Symp. on Earth Reinforcement Practice, Fukuoka, Japan, Vol. 1, November 1992, pp. 223-228.

Cundall, P. and Board, M. 1988. A microcomputer program for modelling large-strain plasticity problems. *Proc. 6th Int. Conf. on Numerical Methods in Geomechanics,* Balkema, Vol. 3, pp. 2101-2108.

Ebling, R.M. and Morrison, E.E. 1993. *The seismic design of waterfront retaining structures.* Naval Civil Engineering Laboratory Technical Report ITL-92-11 NCEL TR-939, Port Huenene, CA, USA, 329 p.

Eliahu, U. and Watt, S. 1991. Geogrid-reinforced wall withstands earthquake. *Geotechnical Fabrics Report,* IFAI, St. Paul, MN, USA, Vol. 9, No. 2, pp. 8-13.

Fang, Y.-S. and Chen, T.-J. 1995. Modification of Mononobe-Okabe theory. *Géotechnique,* Vol. 45, No. 1, pp. 165-167.

Fairless, G.J. 1989. *Seismic performance of reinforced earth walls.* Research Report, Department of Civil Engineering, University of Canterbury, New Zealand, September 1989, 312p.

Fakharian, K. and Evgin, E. 1995. Simple shear versus direct shear tests on interfaces during cyclic loading. *Proc. 3rd Int. Conf. on Recent Advances in Geotechnical Engineering and Soil Dynamics,* Vol. 1, 2–7 April 1995, St. Louis, MO, pp. 13–16.

Farrag, K. 1990. Interaction properties of geogrids in reinforced soil walls - testing and analysis. *Ph.D. Thesis,* Louisiana State University, Baton Rounge, LA, 266 p.

Federal Highway Administration (FHWA) 1996. *Mechanically stabilized earth walls and reinforced soil slopes design and construction guidelines.* FHWA Demonstration Project 82, (Elias, V. and Christopher, B.R.), Washington, DC., USA, 364 p.

Finn, W.D.L., Yogendrakumar, M. and Yoshida, N. 1986. *TARA-3: a program to compute the response of 2-D embankment and soil-structure interaction systems to seismic loading.* Department of Civil Engineering, University of British Columbia, Vancouver, Canada.

Franklin, A.G. and Chang, F.K. 1977. *Permanent displacement of earth embankments by Newmark sliding block analysis.* Misc. Paper S-71-17, Soil and Pavements Laboratory., US Army Eng. Waterways Expt. Station., Vicksburg, Miss., Nov. 1977.

Fredlund, D.G. and Krahn, J. 1976. Comparison of slope stability methods of analysis. *Canadian Geotechnical J.,* Vol. 14, pp. 429-439.

Fukuda, N., Tajiri, N., Yamanouchi, T., Sakai, N. and Shintani, H. 1994. Applicability of seismic design methods to geogrid reinforced embankment. *Proc. 5th Int. Conf. on Geotextiles, Geomembranes and Related Products,* Singapore, Vol. 1, pp. 533-536.

Geogrid Research Board (GRB) 1990. *Geogrid construction method guidelines.* Fukuoka, Japan, Vol. 1, 244 p., Vol. 2, 150 p. (in Japanese).

Goodman, R.E., Taylor, R.L. and Brekke, T.L. 1968. A model for the mechanics of jointed rock, *J. SMFE Div.,* ASCE, Vol. 94, pp. 637-659.

Hanna, T.H. and Touahmia, M. 1991. Comparative behavior of metal and Tensar geogrid strips under static and repeated loading. *Proc. Geosynthetics '91,* Atlanta, Georgia, Vol. 2, pp. 575-585.

Horvath, J.S. 1995. *Geofoam geosynthetic.* Horvath Engineering, Scarsdale, NY, USA, 231 p.

Iai, S. 1989. Similitude for shaking tests on soil-structure-fluid models in 1g gravitational fields. *Soils and Foundations,* Vol. 29, No. 1, pp. 105-118.

Inglis, D., Macleod, G., Naesgaard, E. and Zergoun, M. 1996. Basement wall with seismic earth pressures and novel expanded polystyrene foam buffer layer. *Proc. 10th Annual Symp. of the Vancouver Geotechnical Society,* Vancouver, B.C., 18 p.

Ishibashi, I. and Fang, Y.-S. 1987. Dynamic earth pressures with different wall movement modes. *Soils and Foundations,* JSSMFE, Vol. 27, No. 4, pp. 11-22.

Itasca Consulting Group 1996. *FLAC: Fast Lagrangian Analysis of Continua, version 3.3.* Itasca Consulting Group, Inc. MN, USA.

Juran, I., Knochenmus, G., Acar, Y.B. and Arman, A. 1988. Pullout response of geotextiles and geogrids (synthesis of available experimental data). *Proc. Symposium on Geosynthetics for Soil Improvement,* ASCE Geotechnical Publication 18, pp. 92-111.

Karpurapu, R. and Bathurst, R.J. 1995. Behavior of geosynthetic reinforced soil retaining walls using the finite element method. *Computers and Geotechnics,* Vol. 17, No. 3, pp. 279-299.

Koga, Y. and Washida, S. 1992. Earthquake resistant design method of geotextile reinforced embankments. *Earth Reinforcement Practice,* (Ochiai, Hayashi & Otani, Eds.), Balkema, Proc. of the Int. Symp. on Earth Reinforcement Practice, IS-Kyushu'92, Fukuoka, Japan, November 1992, pp. 255-259.

Koga, Y., Itoh, Y., Washida, S. and Shimazu, T. 1988. Seismic resistance of reinforced embankment by model shaking tests. *Theory and Practice of Earth Reinforcement,* (Yamanouchi, Miura and Ochiai, Eds.), Balkema, Proc. Int. Geotechnical Symp. on Theory and Practice of Earth Reinforcement, IS-Kyushu'88, Fukuoka, Japan, October 1988, pp. 413-418.

Kramer, S.L. 1996a. *Geotechnical earthquake engineering.* Prentice-Hall, New Jersey, USA, 653 p.

Kramer, S.L. 1996b. (personal communication)

Leshchinsky, D. 1995. *Design procedure for geosynthetic reinforced steep slopes*. Technical Report REMR-GT-120 (Temporary Number), Waterways Experiment Station, US Army Corps of Engineers, Vicksburg, Miss., USA, 67 p.

Leshchinsky, D., Ling, H.I. and Hanks, G.A. 1995. Unified design approach to geosynthetic reinforced slopes and segmental walls. *Geosynthetics International,* Vol. 2, No. 5, pp. 845-881.

Ling. H.I., Leshchinsky, D. and Perry, E.B. 1996a. A new concept on seismic design of geosynthetic-reinforced soil structures: permanent-displacement limit. *Earth Reinforcement Practice,* (Ochiai, Hayashi and Otani, Eds.), Balkema, Proc. of the Int. Symp. on Earth Reinforcement Practice, IS-Kyushu'96, Fukuoka, Japan, November 1996, 6 p.

Ling. H.I., Leshchinsky, D. and Perry, E.B. 1996b. Seismic design and performance of geosynthetic–reinforced soil structures, *Géotechnique* (in press)

Ling, H.I., Wu, J.T.H. and Tatsuoka, F. 1992. Short-term strength and deformation characteristics of geotextiles under typical operational conditions. *Geotextiles and Geomembranes,* Vol. 11, No. 2, pp. 185-219.

Madhabushi, S.P.G. 1996. Importance of strong motion in the design of earth reinforcement. *Earth Reinforcement Practice,* (Ochiai, Hayashi and Otani, Eds.), Balkema, Proc. of the Int. Symp. on Earth Reinforcement Practice, IS-Kyushu'96, Fukuoka, Japan, November 1996, 6 p.

McGown, A., Andrawes, K.Z. and Kabir, M.H. 1982. Load-extension testing of geotextiles confined in soil. *Proc. 2nd Int. Conference on Geotextiles,* Las Vegas, Vol. 3, pp. 793-798.

McGown, A., Yogarajah, I., Andrawes, K.Z. and Saad, M.A. 1995. Strain behavior of polymeric geogrids subjected to sustained and repeated loading in air and in soil. *Geosynthetics International,* Vol. 2, No. 1, pp. 341-355.

Min, Y., Leshchinsky, D., Ling, H.I. and Kaliakin, V.N. 1995. Effects of sustained and repeated tensile loads on geogrid embedded in sand. *Geotechnical Testing J.,* ASTM, Vol. 18, No. 2, pp. 204-235.

Miyamori, T., Iwai, S. and Makiuchi, K. 1986. Frictional characteristics of non-woven fabrics. *Proc. 3rd Int. Conference on Geotextiles,* Vienna, Austria, Vol. 3 pp. 701-705.

Mononobe, N. 1929. Earthquake-proof construction of masonry dams. *Proc. World Engineering Conf.,* Vol. 9, 275 p.

Murata, O., Tateyama, M. and Tatsuoka, F. 1994. Shaking table tests on a large geosynthetic-reinforced soil retaining wall model. *Recent Case Histories of Permanent Geosynthetic-Reinforced Soil Walls* (Tatsuoka and Leshchinsky, Eds.), Proc. Seiken Symp., Tokyo, Japan, pp. 289-264.

Myles, B. 1982. Assessment of soil fabric friction by means of shear evaluation. *Proc. 2nd Int. Conference on Geotextiles,* Las Vegas, Nevada, Vol. 3, pp. 787-791.

National Earthquake Hazards Reduction Program (NEHRP) 1994. *Recommended provisions for seismic regulations for new buildings.* Building Seismic Safety Council, Washington, D.C. USA, Part 1, 290 p., Part 2, 335 p.

Newmark, N.M. 1965. Effect of earthquakes on dams and embankments. *Geotechnique*, Vol. 15, No. 2, pp. 139-159.

Nishimura, J. Hirai, T., Iwasaki, K., Saito, Y. and Morishima, M. 1996. Earthquake resistance of geogrid-reinforced soil walls based on a study conducted following the southern Hyogo earthquake. *Earth Reinforcement Practice,* (Ochiai, Hayashi and Otani, Eds.), Balkema, Proc. Int. Symp. on Earth Reinforcement Practice, IS-Kyushu'96, Fukuoka, Japan, November 1996, 6 p.

Okabe, S. 1924. General theory on earth pressure and seismic stability of retaining wall and dam. *Doboku Gakkaishi - J. of the Japan Society of Civil Engineers*, Vol. 10, No. 6, pp. 1277-1323.

Okamoto, S. 1984. *Introduction to earthquake engineering.* University of Tokyo Press, Tokyo, Japan, 629 p.

O'Rourke, T.D., Druschel, S.J. and Netravali, A.N. 1990. Shear strength characteristics of sand-polymer interfaces. *J. of Geotechnical Engineering,* ASCE, Vol. 116, No. 3, pp. 451-469.

Paz, M. 1994. *International handbook of earthquake engineering.* Chapman and Hall, NY, USA, 545 p.

Pelletier, M.J. 1996. Investigation of the seismic resistance of reinforced segmental walls using small-scale shaking table testing. *M.Eng. Thesis,* Department of Civil Engineering, Royal Military College of Canada, Kingston, ON, 224 p.

Public Works Research Institute (PWRI) 1992. *Design and construction manual for reinforced soil structures using geotextiles.* Internal Report No. 3117, Public Works Research Institute, Ministry of Construction, Tsukuba, Japan, 404 p. (in Japanese).

Raju, M. 1995. Monotonic and cyclic pullout resistance of geosynthetics. *Ph.D. Thesis* The University of British Columbia, Vancouver, BC, 260 p.

Richards, R. and Elms, D.G. 1979. Seismic behavior of gravity retaining walls. *J. of the Geotechnical Engineering Division*, ASCE, Vol. 105, No. GT4, pp. 449-464.

Rowe, R.K. and Ho, S.K. 1992. A review of the behavior of reinforced soil walls. *Earth Reinforcement Practice,* (Ochiai, Hayashi & Otani, Eds.), Balkema, Proc. of the Int. Symp. on Earth Reinforcement Practice, IS-Kyushu'92, Fukuoka, Japan, November 1992, pp. 801-830.

Sabhahit, N., Madhav, M.R. and Basudhar, P.K. 1996. Seismic analysis of nailed soil slopes - a pseudo-dynamic approach. *Earth Reinforcement Practice,* (Ochiai, Hayashi and Otani, Eds.), Balkema, Proc. of the Int. Symp. on Earth Reinforcement Practice, IS-Kyushu'96, Fukuoka, Japan, November 1996, 4 p.

Sakaguchi, M. 1996. A study of the seismic behavior of geosynthetic reinforced walls in Japan. *Geosynthetics International,* Vol. 3, No. 1, pp. 13-30.

Sakaguchi, M., Muramatsu, M. and Nagura, K. 1992. A discussion on reinforced embankment structures having high earthquake resistance. *Earth Reinforcement Practice,* (Ochiai, Hayashi & Otani, Eds.), Balkema, Proc. of the Int. Symp. on Earth Reinforcement Practice, IS-Kyushu'92, Fukuoka, Japan, November 1992, pp. 287-292.

Sakaguchi, M., Yamada, K. and Tanaka, M. 1994. Prediction of deformation of geotextile reinforced walls subjected to earthquakes. *Proc. 5th Int. Conf. on Geotextiles, Geomembranes and Related Products*, Singapore, Vol. 1, pp. 521-524.

Sandri, D. 1994. Retaining walls stand up to the Northridge earthquake. *Geotechnical Fabrics Report*, IFAI, St. Paul, MN, USA, Vol. 12, No.4, pp. 30-31 (and personal communication).

Sarma, S.K. 1975. Seismic stability of earth dams and embankments. *Geotechnique*, Vol. 25, No. 4, pp. 743-761.

Schimming, B.B. and Saxe, H.C. 1964. Inertial effects of soil strength criteria. *Proc. Symposium on Soil-Structure Interaction,* University of Arizona, Tucson, Arizona, pp. 118-128.

Seed, H.B. and Whitman, R.V. 1970. Design of earth retaining structures for dynamic loads. *ASCE Specialty Conference: Lateral Stresses in the Ground and Design of Earth Retaining Structures*, Ithaca, NY, pp. 103-147.

Segrestin, P. and Bastick, M.J. 1988. Seismic design of reinforced earth retaining walls - the contribution of finite element analysis. *Theory and Practice of Earth Reinforcement*, (Yamanouchi, Miura, and Ochiai, Eds.), Balkema, Proc. Int. Geotechnical Symp. on Theory and Practice of Earth Reinforcement, IS-Kyushu'88, Fukuoka, Japan, October 1988, pp. 577-582.

Simac, M.R., Bathurst, R.J., Berg, R.R., and Lothspeich, S.E. 1993. *National Concrete Masonry Association segmental retaining wall design manual.* Earth Improvement Technologies, March 1993, 250 p.

Steedman, R.S. and Zeng, X. 1990. The influence of phase on the calculation of pseudo-static earth pressure on a retaining wall. *Geotechnique*, Vol. 40., No. 1, pp. 101-112.

Stewart, J.P., Bray, J.D., Seed, R.B. and Sitar, N. 1994. Preliminary Report on the Principal Geotechnical Aspects of the January 17, 1994 Northridge Earthquake, *Report No. UCB/EERC-94/08*, University of California at Berkeley, Earthquake Engineering Research Center, June 1994, 245p.

Sugimoto, M., Ogawa, S. and Moriyama, M. 1994. Dynamic characteristics of reinforced embankments with steep slope by shaking model tests. *Recent Case Histories of Permanent Geosynthetic-Reinforced Soil Walls* (Tatsuoka and Leshchinsky, Eds.), Proc. Seiken Symp., Tokyo, Japan, pp. 271-275.

Takasumi, D.L., Green, K.R. and Holtz, R.D. 1991. Soil-geosynthetics interface strength characteristics: a review of state-of-the-art testing procedures. *Proc. Geosynthetics '91,* Atlanta, Georgia, Vol. 1, pp. 87-100.

Tateyama, M., Tatsuoka, F., Koseki, J. and Horii, K. 1995. Damage to soil retaining walls for railway embankments during the Great Hanshin-Awaji Earthquake, January 17, 1995. *Earthquake Geotechnical Engineering* (Ishihara, K. Ed.), Balkema, Proc. 1st Int. Conf. on Earthquake Geotechnical Engineering, IS-Tokyo '95, Tokyo, Japan, November 1995, pp. 49-54.

Tatsuoka, F., Koseki, J. and Tateyama, M. 1995. Performance of geogrid-reinforced soil retaining walls during the Great Hanshin-Awaji Earthquake, January 17, 1995. *Earthquake Geotechnical Engineering* (Ishihara, K. Ed.), Balkema, Proc. 1st Int. Conf. on Earthquake Geotechnical Engineering, IS-Tokyo '95, Tokyo, Japan, November 1995, pp. 55-62.

Tatsuoka, F., Tateyama, M., Uchimura, T. and Koseki, J. 1996. Geosynthetic-reinforced soil retaining walls as important permanent structures. *Mercer Lecture 1996-97*, 28 p.

Telekes, G., Sugimoto, M. and Agawa, S. 1994. Shaking table tests on reinforced embankment models. *Proc. 13th Int. Conf. on Soil Mechanics and Foundation Engineering,* New Delhi, India, pp. 649-654.

Vrymoed, J. 1989. Dynamic stability of soil-reinforced walls. *Transportation Research Record* 1242, Washington, DC, USA, pp. 29-38.

White, D.M. and Holtz, R.D. 1996. (draft) Performance of geosynthetic-reinforced slopes and walls during the Northridge, California earthquake of January 17, 1994. *Earth Reinforcement Practice,* (Ochiai, Hayashi and Otani, Eds.), Balkema, Proc. Int. Symp. on Earth Reinforcement Practice, IS-Kyushu'96, Fukuoka, Japan, November 1996.

Whitman, R.V. 1990. Seismic design and behavior of gravity retaining walls. *ASCE Specialty Conference: Design and Performance of Earth Retaining Structures, ASCE Geotechnical Special Publication No. 25*, Cornell University, Ithaca, NY, pp. 817-842.

Whitman, R.V. and Liao, S. 1984. Seismic design of gravity retaining walls, *Proc. 8th World Conference on Earthquake Engineering*, San Francisco, Vol. 3, pp. 533–540.

Wilson-Fahmy, R.F., Koerner, R.M. and Fleck, J.A. 1993. Unconfined and confined wide width tension testing of geosynthetics. *Geosynthetic Soil Reinforcement Testing Procedures,* ASTM STP 1190 (S.C.J. Cheng, Ed.), ASTM, Philadelphia, pp. 49-63.

Wolfe, W.E., Lee, K.L., Rea, D. and Yourman, A.M. 1978. The effect of vertical motion on the seismic stability of reinforced earth walls. *Proc. ASCE Symp. on Earth Reinforcement*, Pittsburgh, PA, USA, April 1978, pp. 856-879.

Woods, R.I. and Jewell, R.A. 1990. A computer design method for reinforced soil structures. *Geotextiles and Geomembranes,* Vol. 9, No. 3, pp. 233-259.

Wu, J.T.H. (Editor) 1992. *Geosynthetic-Reinforced Soil Retaining Walls*, Proc. of the Int. Symp. on Geosynthetic-Reinforced Soil Retaining Walls, Denver, CO, USA, 8-9 August 1991, Balkema, 375p.

Yasuda, S., Nagase, H. and Marui, H. 1992. Cyclic pullout tests of geogrids in soils. *Earth Reinforcement Practice* (Ochiai, Hayashi & Otani, Eds.), Balkema, Proc. Int. Symp. on Earth Reinforcement Practice, IS-Kyushu'92, Fukuoka, Japan, November 1992, pp. 185-190.

Yogendrakumar, M. and Bathurst, R.J. 1992. Numerical simulation of reinforced soil structures during blast loads. *Transportation Research Record 1336,* TRB, pp. 1-8.

Yogendrakumar, M., Bathurst, R.J. and Finn, W.D.L. 1992. Dynamic response analysis of a reinforced soil retaining wall. *J. of Geotechnical Engineering*, ASCE, Vol. 118, No. 8, pp. 1158-1167.

Yogendrakumar, M., Bathurst, R.J. and Finn, W.D.L. 1991. Response of reinforced soil slopes to earthquake loadings. *Proc. 6th Canadian Conf. on Earthquake Engineering,* Toronto, pp. 445-452.

Zarrabi, K. 1979. Sliding of gravity retaining wall during earthquakes considering vertical acceleration and changing inclination of failure surface. *M.Sc. Thesis,* Department of Civil Engineering, Massachusetts Institute of Technology, Cambridge, MA, USA, 140 p.

Zimmie, T.F., De, A. and Mahmud, M.B. 1994. Centrifuge modelling to study dynamic friction at geosynthetic interfaces. *Proc. 5th Int. Conf. on Geotextiles, Geomembranes and Related Products,* Singapore, Vol. 1, pp. 415-418.

Earth Reinforcement, Ochiai, Yasufuku & Omine (eds) © 1997 Balkema, Rotterdam, ISBN 90 5410 833 9

Application of geosynthetics to embankment on soft ground and reclamation using soft soils

H. Miki
Public Works Research Institute, Ministry of Construction, Japan

ABSTRACT: This paper first outlines how to apply geosynthetics to embankments on soft ground and then positions this reinforced earth method in the array of measures used to improve soft ground for embankments based on a feasibility study. Concerning the two most suitable methods of application, namely, the soft ground surface layer reinforcing method using geogrid in combination with vertical drains and the method of preventing settlement using geosynthetics in combination with the deep mixing method, the design concept and method of estimating deformation are outlined. In addition, as a method of reclaiming and quickly using land made of low-quality soils from construction sites, the concept, design and execution of a new method of reclamation using horizontal drains in combination with sand mounds to lower the water level are outlined.

1 OUTLINE OF METHOD FOR APPLICATION OF GEOSYNTHETICS TO EMBANKMENTS ON SOFT GROUND

It has been said that the Japanese geotextile method originated with the method of reinforcing soft ground surface layers by means of an earth cover; the sheet method began to be used as a treatment method for soft ground surface layers for land reclaimed by means of pump sledging in the 1960s.

In the West, on the other hand, non-woven geotextiles were first widely used for preparation of roads on soft ground located in wetlands. Non-woven geotextiles were used more often in wetlands than in reclaimed land, and the method of using non-woven geotextile—which functioned as an excellent boundary separator—was developed.

In Japan, the method of using sheets in combination with rope nets and the method for net-matting embankments using geonets were developed subsequent to the sheet method; the method using sheets in combination with the traditional bamboo frame has proven valuable as a method for reinforcing embankments on extremely soft ground. These methods have also been used to reinforce soft ground under the sea, for example, when harbor facilities such as breakwaters and seawalls are built

Applications of geosynthetics for embankments on soft ground, roughly classified into four types, are listed in Table 1.

1.1 *Soft ground surface layer reinforcing method for earth cover*

(1) Outline of the reinforcing method

In the soft ground surface layer reinforcing method for earth cover, in order to ensure trafficability on soft ground where it is almost impossible to put people or construction machines, geosynthetic reinforcing materials are laid on the ground surface, and the ground is then reinforced using the horizontal continuity and tensile strength of the materials; earth is then spread over the ground. The tensile force generated by the load of covered earth will either serve as the tension member of the beam or alternatively, will produce a hammock effect as the part where the earth is placed sinks and the ground level is raised on the periphery because the edges are supported, resulting in reinforcement of the ground.

In the sheet, net matting, and rope/sheet methods, the tensile force produced at the edges is handled through friction between the ground and the reinforcing materials or by fixing it to an anchor pile embedded in the embankment. The bamboo frame/sheet method is one in which the covered earth is supported by allowing the rigidity of the bamboo frame to disperse the load of covered earth so that the load is evenly distributed.

The reinforcement work consists of installing reinforcing materials and spreading the earth used to cover the ground. Even on ground where people are not able to stand, reinforcing materials may be laid by a variety of methods, including one that requires reinforcing materials to be laid manually after they are spread and scaffolding plates are placed on them, a

Table 1. Application of geosynthetics to soft ground

Applications	Method and materials used	
Reinforcement of soft ground surface layers for earth covers • To ensure trafficability on soft ground where it is difficult for people and motor vehicles to pass • To ensure sufficient bearing capacity to support light-weight objects	Sheet method Net matting method Sheet/rope net method Sheet/bamboo frame method	(Woven geotextile, etc.) (Geonet) (Geo grid) (Woven geotextile, etc., ropes) (Woven geotextile, etc., bamboo)
Reinforcement of soft ground surface layers for embankments • To stabilize embankments on soft ground requiring more strength than that afforded by soft ground with a reinforcing earth cover; and also for control of differential settlement.	Plane reinforcement Single reinforcement material Multiple reinforcing materials Solid reinforcing	(Geotextile or geogrid) (Reinforcing bars with anchor plates) (Reinforcing bars with anchor plates) (Mattress), (Geocell)
Reinforcement of soft ground surface layers by means of geo-synthetics used in combination with piles • For high, permanent embankments on soft ground having the same strength as soft ground that needs earth cover for reinforcement	Pile net method Deep mixing pile/net method	(Wire mesh, geotextile, etc.) (Geotextile, etc.)
Reinforcement of soft ground for base courses and subgrade	(Roads) Reinforcement of bearing capacity of base courses Reinforcement of bearing capacity of subgrade (Railroads) Reinforcement of bearing capacity of base courses Prevention of mud pumping on base courses	 (Geogrid and woven geotextile) (Geo-cells, non-woven geotextile, woven geotextile, geomembrane)

method in which a winch is installed on the opposite bank to pull and lay the reinforcing materials, and a method in which the ground is covered with water, and reinforcing materials are loaded onto a barge and are subsequently placed at the bottom of the water as the barge moves about.

(2) Design concept

If reinforcing materials and a load of earth cover are applied to soft ground with a high water content (such as reclaimed land where cohesive soil dredged from the bottom of the sea is used as reclamation material) the part directly below the load tends to sink and the ground around the part rises. Models have been developed to analyze such ground deformation phenomena, and different design methods have been proposed as a result. However, since deformation phenomena differ greatly depending on soil and working conditions at the site, it is difficult to fully incorporate all the details underlying the supporting mechanism, and therefore various constants used in the formulae to calculate bearing capacity are determined based on measurements of behavior at the site and on the results of indoor experiments.

The following proposals have been made with regard to formulae used to calculate bearing capacity and settlement.

- Method based on the theory of ground bearing capacity: The tensile force and the effect of pressing down on the ground and the effects resulting from settlement and rising on embedment are taken into consideration, based on the ground bearing capacity formula.
- Method using a combination of soil reaction coefficients and the cable theory: After a load is applied, a basic formula is obtained from the deformed shape of the stabilized reinforcing material using cable theory, and the tensile force of the reinforcing material is calculated from its relation to the amount of settlement.
- Method based on slab theory: In this method, the slab theory is applied on the assumption that sandy soil—used as earth cover— integrated with reinforcing material with relatively large meshes assumes a rigid slab-like shape.
- Method based on pneumatic membrane theory: In this method, an analytical theory is applied when extremely soft ground is considered as a liquid, and air pressure acts on the membrane structure whose periphery is confined.

Of all these methods, the method based on the bearing capacity theory is used comparatively often. In this method, a model is developed for the ground, and the bearing capacity is expressed as the sum of four components.

Bearing capacity of conventional ground:

$$q_1 = c \cdot Nc \tag{1.1}$$

Bearing capacity resulting from tensile force generated at both ends of the reinforcing material:

$$q_2 = 2 \cdot T \cdot \sin\theta / B \tag{1.2}$$

Effect of reinforcing material pressing down the ground:

$$q_3 = T \cdot Nq / r \tag{1.3}$$

Embedment effect resulting from settlement and rising:

$$q_4 = r_t \cdot Df \tag{1.4}$$

Where c: denotes the cohesion of soft ground (kPa) {tf/m^2}, Nc, Nq: the bearing capacity factor, T: the tensile force of the reinforced material, θ the angle formed by the reinforcing material and the horizontal surface at the end of the load (°), B: the load width (m), r: the radius of the deformed shape of the ground near the load when the shape is considered as circular (m), r_t: the weight per unit volume (kPa){tf/m^2}, and Df: the amount of settlement of the soft ground (m).

Thus, the ultimate bearing capacity is calculated as follows.

$$q_d = q_1 + q_2 + q_3 + q_4$$
$$= c \cdot Nc + 2 \cdot T \cdot \sin\theta / B + T \cdot N_q / r + r_t \cdot Df \quad (1.5)$$

T in this formula is determined based on the measurements taken at the site and the indoor experimental values as there are three unknown constants generated by the state of deformation of the ground, namely, r, θ, and Df if the strength of the material to be used is employed.

As to the design procedure, ground strength is determined first of all, and then the partition embankment and the earth cover—in the block surrounded by the partition embankment—are designed. When designing the earth cover, a combination of both the earth covering method and reinforcing materials suitable to the site conditions are selected; the earth cover, the layer thickness in each state, the reinforcing materials and work method are all determined.

(3) Tensile strength of reinforcing materials reviewed in examples of reinforcing work

Figure 1 shows the relationship between ground cohesion and the tensile strength of the reinforcing material used at the site with all data cited from examples of reinforcing work. For extremely soft ground with a the cohesion $c \leqq 0.1$ tf/m^2, a reinforcing material with the tensile strength $T = 7 \sim 11$ tf/m is used. The type of reinforcing material is rope, bamboo frame and sheet, or geogrid; there is no case in which a sheet alone is used. This is because it is assumed that since the sheet lacks rigidity, part of the ground sinks when earth is spread, and this settlement cannot be controlled.

On the other hand, if cohesion $c > 0.1$tf/m,the tensile strength of the reinforcing material ranges very widely from 1tf/m to 11tf/m.This indicates that a particular reinforcing material is apparently determined considering not only the ground cohesion but also the shape of each block of earth cover surrounded by the partition embankment as well as the method used for spreading the earth. As for the use of the sheet, a sheet with a smaller tensile strength tends to be used as the ground cohesion increases, but a geonet with a relatively narrow range of cohesion from 0.1tf/m^2 to 0.5tf/m^2 is used regardless of the maximum tensile strength of 2.0tf/m or so. It is assumed that this is closely related also to the large tensile elongation.

1.2 *Soft ground surface layer reinforcing method for embankments*

(1) Outline of the reinforcing work

The main objective of the above-mentioned "soft ground reinforcing for earth cover" is chiefly to ensure trafficability for motor vehicles used for the reinforcing work or for placing light-weight structures on the ground by utilizing a stabilizing earth cover of good

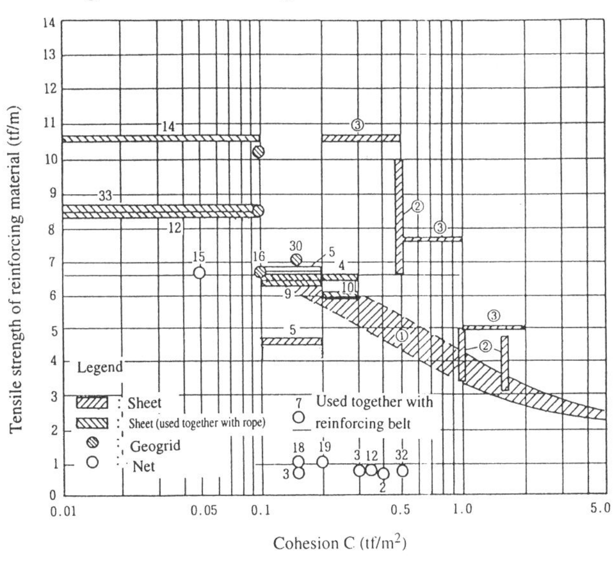

Fig. 1. Relation between ground strength and tensile strength of reinforcing material [1)]

quality over the geotextile laid on the soft ground surface. Therefore, the uneven thickness of the embankment or any differential settlement is tolerated because of its temporary nature.

On the other hand, the main objective of "soft ground reinforcing for embankment" is to stabilize the embankment and minimize the possibility of differential settlement by reinforcing the surface layer of the soft ground as well as the part closer to the bottom of the embankment. Therefore, the soft ground is required to have a certain degree of strength.

Application methods for geotextile can be roughly classified into 1) the method in which geogrid or some other material is laid on the surface layer of the foundation ground or on the lower layer of the embankment to ensure stability of the embankment with respect to sliding failure that occurs in the foundation ground (reinforcing material laying method); and 2) the method in which the foundation ground is reinforced by using a geogrid to construct grid frames in the surface layer of the foundation ground and putting filling material into the frames (mattress method).

(2) Design concepts

In designing 1) the reinforcing material laying method, stability with respect to the following three general failure modes is examined.[2)]

A) Excessive settlement or deformation due to insufficient bearing capacity of the foundation ground (see Fig. 2(a))
B) Sliding failure that cuts through the geotextile and passes into the foundation ground (see Fig. 2(b))
C) Embankment Sliding on geo-textile (see Fig.2 (c))

The basic design procedure is shown in Fig. 3.

First of all, whether excessive settlement or deformation is likely to occur is determined based on the value of safety factor *Fs* used in the circular arc sliding calculation at the time of reinforcing. It is generally accepted that if the value of *Fs* is below 1.0, excessive settlement or deformation as shown in Fig. 2(a) occurs and the effects of geotextiles are not significant in many cases. Therefore, improvements in the foundation ground itself must be examined initially.

In cases where a type A) failure is not likely to occur, but a type B) failure is more likely to occur ($1.0 \leqq Fs < 1.2 \sim 1.3$), geotextiles can be laid on the surface of the foundation ground (or in the lower section of the embankment) to compensate for the insufficient safety factor with the tensile strength of geotextiles.

The safety factor with respect to sliding failure in considering geo-textiles is obtained from the following formula. (See Fig. 4)

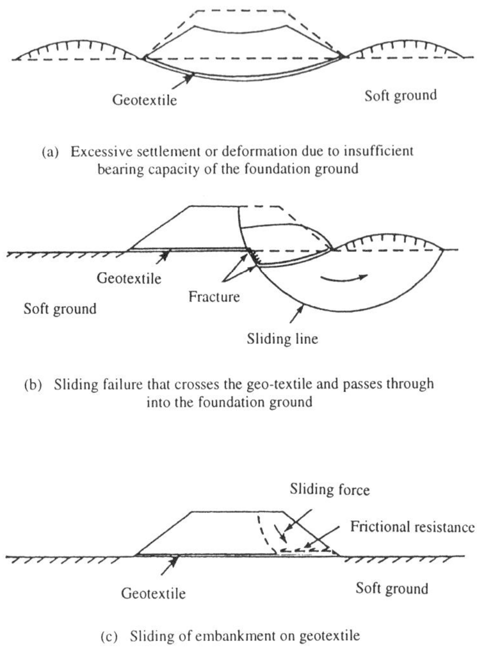

Fig. 2. Modes of failure in embankment on soft ground reinforced with geotextile

$$Fs = \frac{Mr + \Delta Mr}{M_d} \tag{1.6}$$

$$M_d = R \bullet \sum (W \bullet \sin \theta) \tag{1.7}$$

$$Mr = R \bullet \sum (c \bullet l + W \bullet \cos \theta \bullet \tan \phi) \tag{1.8}$$

$$\Delta Mr = R \bullet T_A \bullet \cos \alpha \text{ or } R \bullet T_A \tag{1.9}$$

Where, M_d : Sliding moment
Mr : Resisting moment of soil
ΔMr : Resisting moment of geotextile
T_A : Design tensile strength of geotextile, which must be determined from the tensile test but should be such that the tensile elongation in consideration of creep is kept below 10% or so.

A type C) failure is likely to occur when the frictional resistance between the embankment and the geotextile is small. Therefore, it is necessary to determine if the frictional resistance between the geotextile and the embankment is sufficient with respect to the sliding force of the embankment.

The safety factor with respect to embankment sliding on geotextile is determined from the following formula. (See Fig. 5)

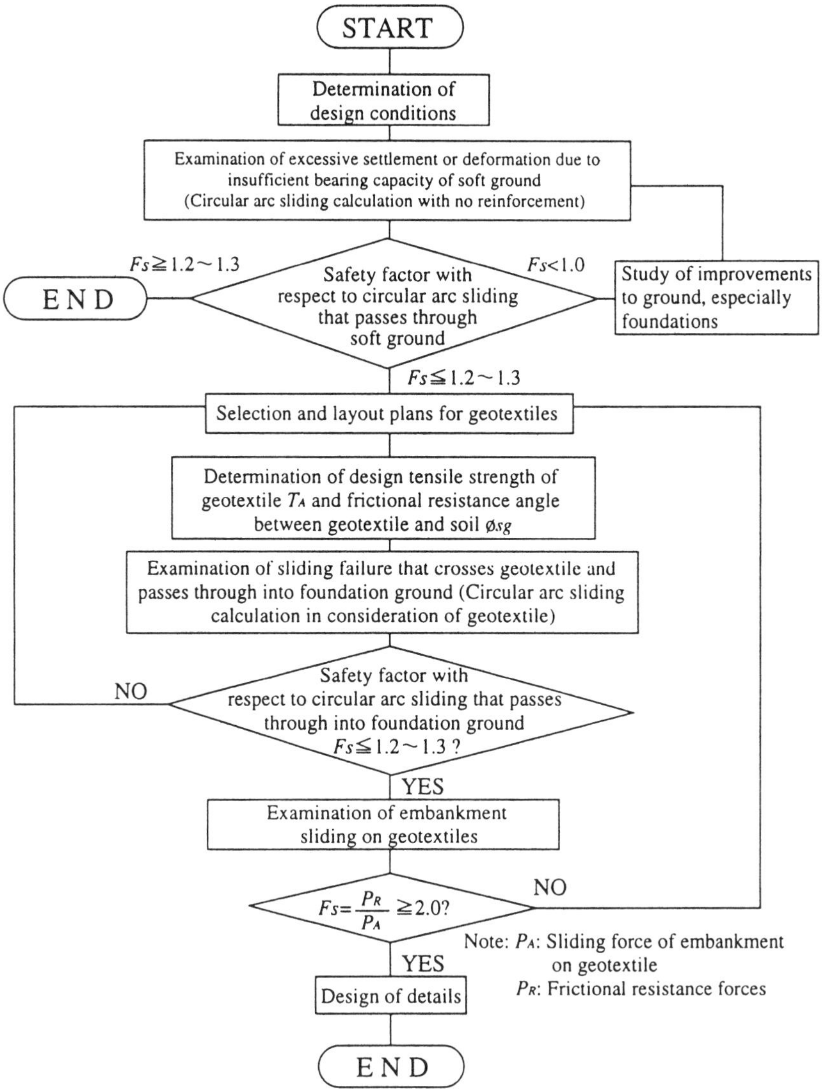

Fig. 3. Examination procedure undertaken when geotextiles are used to reinforce embankments on soft ground[2)]

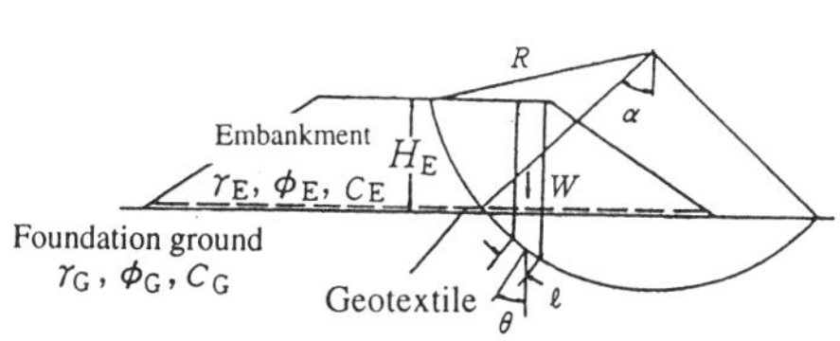

Fig. 4. Stability calculation with respect to sliding failures which pass through into the foundation ground[2)]

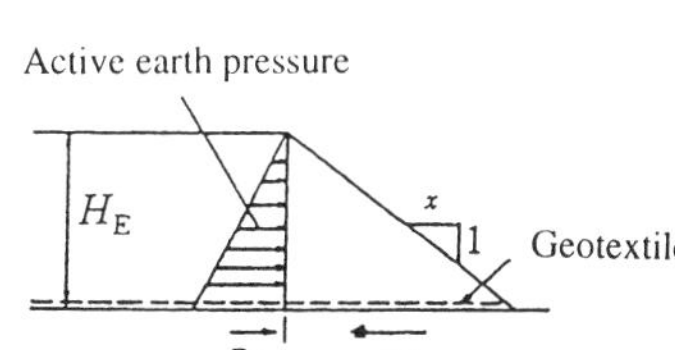

Fig. 5. Stability calculation with respect to sliding of embankment on geotextiles[2)]

$$Fs = \frac{P_R}{P_A} \qquad (1.10)$$

$$P_A = \frac{1}{2} r_E \cdot H_E^2 \cdot K_A \qquad (1.11)$$

$$P_R = \frac{1}{2} r_E \cdot x \cdot H_E^2 \cdot \tan \phi_{SE} \qquad (1.12)$$

Where, P_A : Sliding force of embankment on geotextile (resultant force of active earth pressure)
P_R : Frictional resistance force
K_A : Coefficient of active earth pressure
ϕ_{sg} : Frictional resistance angle between the geotextile and the soil
x : Slope gradient

Formulas (1.10), (1.11) and (1.12) are used to determine whether the safety factor with respect to sliding of embankment on geotextile is greater or less than 2.0.

1.3 *Soft ground surface layer reinforcing method using geo-textiles with piles*

The above-mentioned soft ground surface layer reinforcing method for embankments is applied to soft ground with a relatively high strength, and reinforcing methods used to build full-scale, permanent embankments on soft ground with extremely low strength include the pilenet method in which geotextiles are used in combination with piles, or piles used in the deep mixing method of soil stabilization and a method in which geogrid and the deep mixing method of soil stabilization are used together.

(1) Pile net method

The pilenet method was developed to be applied to the peat layer and soft cohesive soil layer that have a high water content, high compressibility and low strength, as found in Hokkaido and along the Ariake coast in Kyushu. As shown in Fig. 6, wooden piles are driven into soft ground in groups; the top of each pile is linked by reinforcing bars, and a sand mat is spread over them. One or more geotextiles are then laid inside the sand mat in the form of a sandwich to increase the confining effect, and a full-scale embankment is constructed.[3)~5)]

Since this method uses long piles to enhance rigidity quite deep into the soft layer, it prevents or greatly reduces the lateral flow.

In the pile net method, the distance between piles D appropriate to produce the group pile effect is determined from the following formula.

$$D = 1.5\sqrt{r \times l} \qquad (1.13)$$

Where r is the radius of the pile, and l the length of the pile.

From the above formula, if the radius of the pile is 10 cm, the distance between piles will be about 1.1 meters when the length of the pile is 5 meters, and 1.6 meters when the length of the pile is 10 meters. The length of the pile l must be determined such that the ultimate bearing capacity per pile Rd is obtained for a group of piles, and the value of the safety factor, which is the ratio of the ultimate bearing capacity for a group of piles to the load applied to each pile Rw, is more than 1.3. The connecting reinforcing bars must have a deflection angle of 30 degrees (θ=30°) so that T equals P as shown in Fig. 7. The tensile force T applied to the reinforcing bar is obtained from the following formula, and the type and the diameter of the reinforcing bar are determined so that S is equal to or greater than T.

$$T = \frac{0.5 \cdot Rw}{2 \cdot m} \qquad (1.14)$$

Where, T : the tensile force per reinforcing bar,
Rw : the load applied to each pile, and
m : the number of reinforcing bars used (single m = 1, double m = 2).

$$S = \tau_t \cdot \frac{\pi d^2}{4} \qquad (1.15)$$

Where, S : the allowable stress of the reinforcing bar,
τ_t : the allowable stress
SR24: 1400kgf/cm^2
SD30: 1600kgf/cm^2), and
d : the diameter of the reinforcing bar.

$$S \geqq T \qquad (1.16)$$

Geotextile laid on the upper part of the piles prevents the embankment material from slipping through and distributes the load of the embankment evenly to the lower ground and the reinforcing bars; geotextiles with excellent contact with the ground, air permeability and weather-resistance are used. In some cases, the tensile force T_g applied to the geotextile is determined from the following formula for conveniences' sake since the actual behavior is not known.

$$T_g = \frac{\text{Overburden load applied to the space between the reinforcing bars}}{\text{The total length of the reinforcing bars supporting the geotextile}}$$

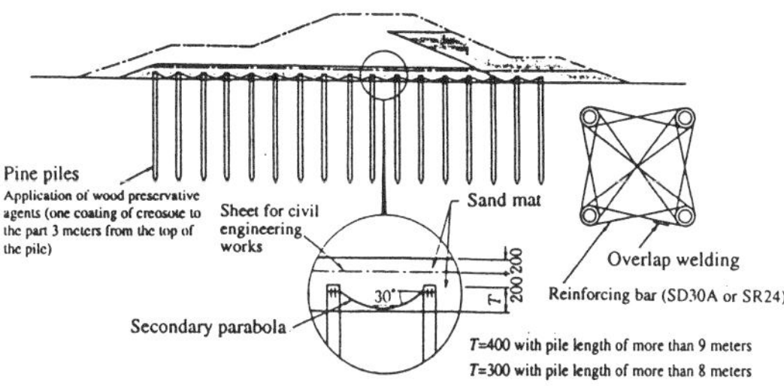

Fig. 6. Concept of the pile net method

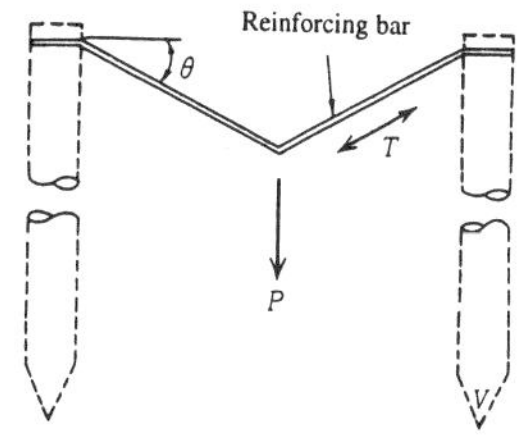

Fig. 7 Deflection angle θ of connecting reinforcing bars

When selecting a geotextile, a geotextile with a tensile strength of more than the value obtained in consideration of the safety factor for T_g (normally 1.2) is used.

As for the amount of settlement of the ground on which the pilenet is constructed, the portion surrounded by the piles is considered as a form of integrated caisson, and the ground settlement caused by the embankment load must be determined only for the section deeper than the tip of the pile inside the soft layer.

As for the drainage side required in the calculation of consolidation, the tip of the wooden pile is regarded as the drainage side in some cases on the assumption that, although the wooden pile itself does not have a drainage function, water will drain through the peripheral sides of the wooden pile. A simple calculation method such as this tends to produce a larger value than the actual measured amount of settlement, but this does not pose any problem for practical purposes, and is considered as generally acceptable.

(2) Method using geogrid in combination with deep mixing method of soil stabilization

Based on a concept similar to the pilenet method, a method is proposed in which a pile-type improvement is made to the entire section below the embankment by means of the deep mixing method of soil stabilization that offers a low improvement rate (about 10~30%). The lack of stability with respect to circular arc sliding is compensated for by a high-strength geogrid; geotextile is used to solve the problem of differential settlement that develops between the improved columns and the non-improved sections.

The design method is not yet fully established, but it is anticipated that this method will be made practicable as an economical and new solution to soft ground reinforcing. This method will be discussed in detail in Section 4.

1.4 *Soft ground reinforcing method for base courses and subgrade*

According to the guidelines governing asphalt pavement, if the design CBR for the subgrade section is less than 3, the soil is to be replaced with soil of good quality and a stabilization method is applied in many cases. When geotextile is used, geotextile is laid between the base course and subgrade (or in the base course) to reinforce the ground by preventing separation and confinement effects and thus compensating for insufficient subgrade bearing capacity. As to the design method, when geotextile is laid between the subgrade and the base course, the thickness of the lower base course must be determined so that the total amount of deformation of the pavement structure and the pavement surface will be the same as that of the normal subgrade (CBR $\geqq$ 3). When geotextile is laid inside the base course, the fundamental assumption is that the deformation of the paved road surface decreases because the elastic modulus of the base course around the geotextile increases, and thus the design must be predicated based on the multi-layer elasticity theory. Both cases use the results of indoor experiments, and verification through experiments at the site are desirable before the methods are used. Geotextile is used for unpaved roads such as temporary roads and access roads with low traffic volumes, and various design methods have been proposed. Basically, the thickness of the base course is determined so that the ultimate bearing capacity of the subgrade and the stress applied to the base course are well-balanced.

The objective in using geotextile in the reinforcement of railroad track beds is to keep the ballast well maintained. It is used to prevent intrusion and mud-pumping on the ballast track bed as well as to prevent deformation of the ballast track bed due to running trains. The surface of the track bed is covered by geotextile, such as a sheet, to prevent rain water from intruding on the track bed, and thus preventing the mud-pumping phenomenon. Replacement methods and stabilization methods are also often used to improve the bearing capacity of the track bed as in the case of roads when the value of K_{30} (obtained in the plate loading test) is less than 7 kgf/cm^3. There is also a method using geotextile in which the track bed is reinforced with geocell. Tests have proven that a track bed reinforced with geocell is effective with respect to repeated loading from running trains.

2 POSITIONING OF GEOSYNTHETIC REINFORCING METHODS IN THE ENTIRE SOFT GROUND IMPROVEMENT MEASURES

2.1 *Examination of low embankments and high embankments*

(1) Introduction

In recent years, the use of new soft ground improvement methods, such as the lightweight embankment method, and the embankment reinforcing method using geosynthetics, is increasing. However, it is not yet known how such new methods and conventional methods compare in terms of total cost including maintenance and repair expenses. Therefore, the authors designed various soft ground improvement methods for embankments on a trial basis, and examined applications of these different methods by comparing the construction cost including maintenance expenses, the work period, and the standing period by the end of which the residual settlement falls within the allowance.

(2) Outline of trial calculations

In the trial design, the improvement methods listed in Table 4 are designed in accordance with the design conditions shown in Table 3 for the specific cases shown in Fig. 8 and Table 2, and a comparative economic study of them was made. In the case of a low embankment (He=2m), the magnitude of the traffic equivalent load is set to 2.2tf/m^3, considering the settlement due to traffic loads.

(3) Results of the trial calculations

a) For low embankments with a relatively thin soft layer (see Fig. 9)

When the thickness of the soft layer is small, the surcharge method offers advantages since the method is inexpensive and has a short period during which the residual settlement falls within the allowance. However, the period during which the residual settlement falls within the allowance increases as the thickness of the soft layer increases, and thus it is better to promote the settlement by using a sand drain.

In addition, the surface layer stabilization method is effective if applied to ground with a thin soft layer because the depth of ground that can be improved is limited and the consolidation settlement below the improved layer cannot be controlled.

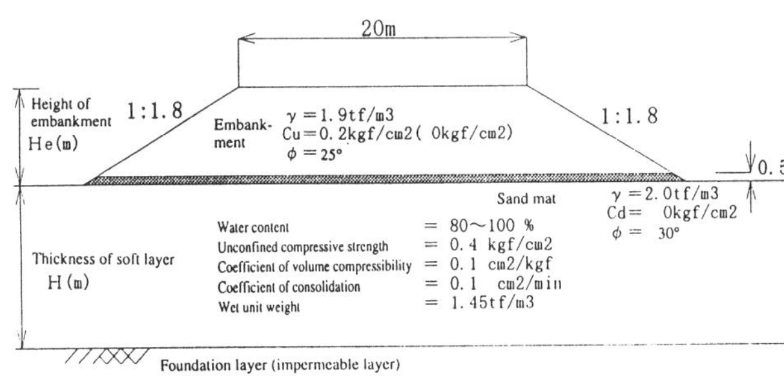

Fig. 8. Outline of soft ground and embankment; physical properties of the soil

Table 2. Design cases

Height of embankment He \ Thickness of soft layer H	5 m	10 m	20 m	30 m
2 m	Case 1 (A)	Case 2 (A)	Case 3 (B)	Case 4 (B)
5 m	-	-	Case 5 (C)	Case 7 (C)
8 m	-	-	Case 6 (C)	Case 8 (C)

b) For low embankments with a thick soft layer (see Fig. 10)

The surcharge method was found to be the most inexpensive, but the time required for the residual settlement to fall within the allowance is extremely long especially in a case involving thick soft layers, such as this case. Therefore, it is better to use the surcharge together with a sand drain because the time required for the settlement to fall within the allowance can be significantly reduced.

In the case where a non-treated embankment requires maintenance in many locations, the expense for overlay increases and thus the surcharge used in combination with the sand drain is less expensive. In addition, the time required for the residual settlement to fall within the allowance is much shorter if the surcharge is used together with the sand drain. Therefore, it is advisable that the residual settlement be controlled from the beginning by means of a sand drain, etc.

Although the work cost is on the high side due to material costs, the method is considered effective for roads requiring severe allowances for residual settlement and also for roads where it is difficult to block traffic to carry out repairs.

c) For high embankment with thick soft layer (see Fig. 11)

If the embankment is as high as about 5 meters, the sand drain method and the sand drain method used in combination with the embankment reinforcing method or the counterweight fill method is comparatively inexpensive and offers a number of advantages.

Table 3. Design conditions

Item	Design conditions
(1) Examination of sliding failure	Safety factor $Fs \geqq 1.1$ (at the completion of embankment) $Fs \geqq 1.2$ (after the completion of embankment) Earthquakes have not been factored in.
(2) Allowable residual settlement	Less than 30cm
(3) Consolidation settlement	The consolidation settlement is calculated at the place below the embankment center, and settlement is considered as a one-dimensional consolidation.
(4) Embankment work speed	5cm/day However, the speed is 30 cm/day for the deep mixing method of soil stabilization and sand compaction pile method, and 150 m^3/day for the lightweight embankment (EPS) method.
(5) Strength increase rate for soft layer	m=0.3. The maximum value for the degree of consolidation U that increases the strength due to consolidation is 90%.
(6) Overlay standard for non-treated embankment	The number of areas that need maintenance per 1 km is set to a maximum of 20 and a minimum of 6. An overlay is required once for every 3 cm of settlement.

Table 4. Improvement methods examined

Group	Improvement methods
A	• Surcharge method • Surcharge + sand drain method (hereinafter referred to as SD) • Surcharge + surface layer stabilization method
B	• Surcharge method • Surcharge + SD • Lightweight embankment method (using EPS blocks) • Non-treatment (overlay after service is considered, maximum maintenance) • Non-treatment (overlay after service is considered, minimum maintenance)
C	• Lightweight embankment method (using EPS blocks) • Deep mixing method of soil stabilization } SD used together • Counterweight fill method } • Sand compaction pile method (SCP)

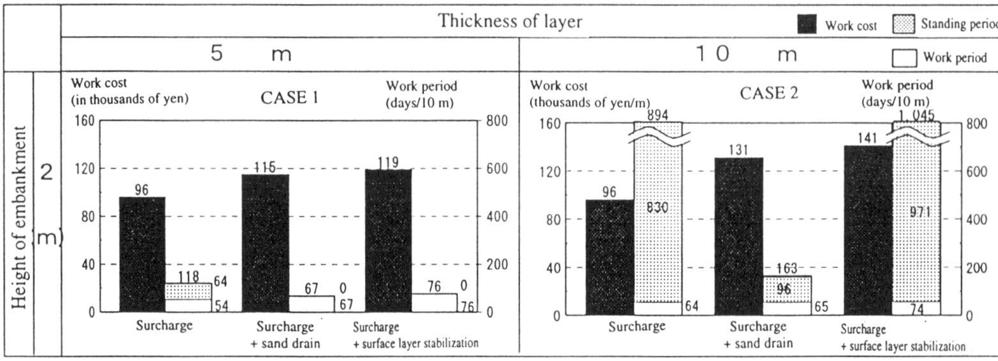

Fig. 9. Comparison group A (for low embankments with a relatively thin soft layer)

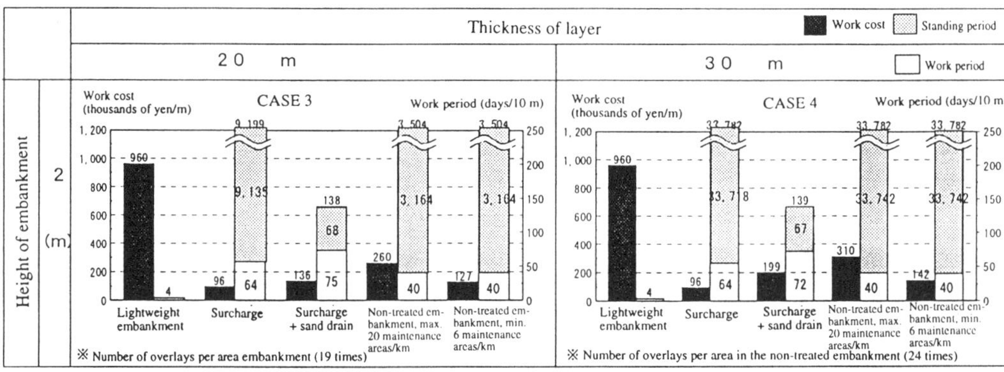

Fig. 10. Comparison group B (for low embankment with thick soft layer)

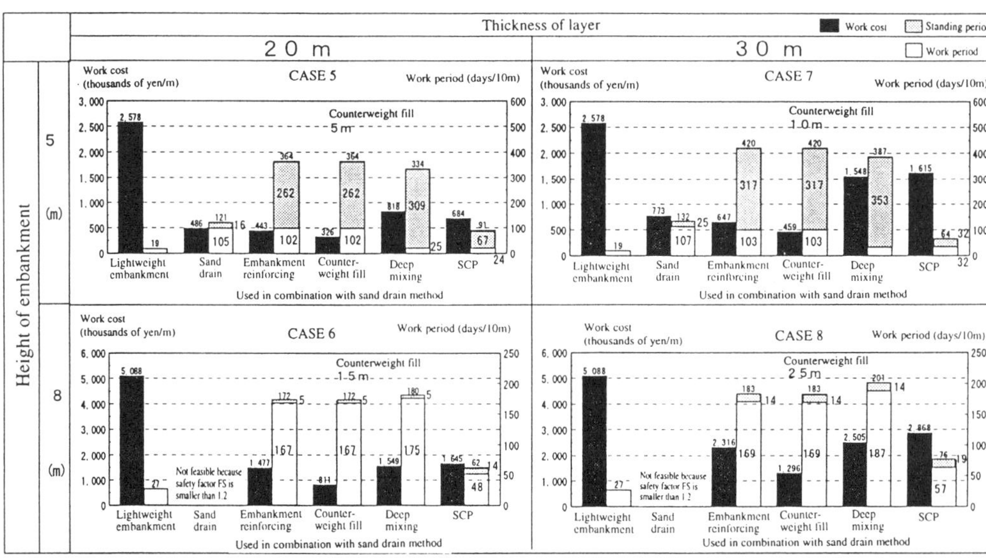

Fig. 11. Comparison group C (for high embankments with a thick soft layer)

However, if the sand drain method alone is used, the time required for the residual settlement to fall within the allowance can be reduced because drains are driven more densely compared with other combined methods (sand drain + embankment reinforcing, sand drain + counterweight fill), but the construction cost is somewhat higher. As the height of the embankment increases further, there are cases where the embankment cannot be stabilized by the sand drain method which enhances the strength of the embankment. Therefore, the sand drain method must be used together with another method (embankment reinforcing, counterweight fill, deep mixing method). In such cases, if sufficient land is secured, the sand drain + counterweight fill method is the most inexpensive and thus is preferred.

The embankment reinforcing method is the next most inexpensive method (as compared only with the counterweight fill), and is effective when used as a supporting method in cases where the safety factor offered by the sand drain method alone is insufficient.

The sand compaction pile (SCP) method offers a number of advantages, in that its work period is relatively short while the deep mixing method is also a superb method that has a limited effect on the surrounding environment. The lightweight embankment method is quite expensive compared with other methods, but the total amount of settlement is small and the work period is also very short. Therefore, this method can be used when the height of the embankment and the thickness of the soft layer are quite large, and a maintenance-free installation is required.

(4) Summary

For a low embankment with a relatively thin soft layer, the surcharge method is advantageous. However, if the thickness of the soft layer increases, the surcharge method should be used together with a sand drain to reduce the settlement period. For an embankment as high as about 5 meters, the sand drain method and the sand drain method combined with embankment reinforcing or counterweight fill are relatively inexpensive and offer advantages. As the height of the embankment increases, the sand drain method combined with other methods (embankment reinforcing, counterweight fill, deep mixing method) or the SCP method is required.

The work cost of the lightweight embankment method is rather high. However, this method is preferable when the height of the embankment and the thickness of the soft layer are very large, and a maintenance-free installation is required. The embankment reinforcing method is very effective when used in combination with the sand drain method for high embankments in cases where the safety factor provided by the drain method alone is not sufficient.

3 SOFT GROUND SURFACE LAYER REINFORCING METHOD USING GEOGRID IN COMBINATION WITH VERTICAL DRAIN

As the results of the feasibility study presented in Chapter 2 show, the soft ground surface layer reinforcing method is a good supporting method when the vertical drain method alone fails to provide a sufficient safety factor. This chapter gives an outline of embankment work for the Niigata-Nishi Bypass, in which the soft ground surface layer reinforcing method using geogrid was used to support the vertical drain. We also describe how the deformation behavior of the embankment and the tension of the geogrid were predicted through FEM elastoplastic consolidation analysis, and examine the applicability of the analysis method.

3.1 *Introduction*

The method features a simple calculation based on the rotational slip calculation method, though some aspects of the tension of the geogrid and the deformation behavior of the reinforced embankment have not been clarified yet. Field observations made at various sites have shown that the tension applied to the geogrid is quite small and the current design method allows us to design with a large safety margin. Therefore, if the tension applied to the geogrid can be estimated with practical accuracy in advance, a more reasonable and economical design method can be established.

This report describes the applicability of the analysis method, which we assessed by simulating the deformation behavior of a real embankment and the tension of the geogrid, using the FEM elastoplastic-consolidation analysis method for the Niigata-Nishi Bypass, where the vertical drain method is combined with a reinforcing method using geogrid.

3.2 *Outline of embankment work*

(1) Conditions of ground and embankment soil

The Niigata-Nishi Bypass is planned as part of Route 116, an ordinary national highway, and is a major expressway connecting the City of Niigata and the City of Kashiwazaki. The planned construction site is located on the flood plain of the Shinano River and is on soft ground. In addition, since most of the planned route is to be built on a high embankment, sliding failure and residual settlement after the highway is brought into service will be major problems.

Figure 12 shows the ground condition and the embankment shape of the cross-section of the Niigata-Nishi Bypass No. 113 which was subjected to the FEM elastoplastic-consolidation analysis.

The ground is soft ground with an N value ranging from 0 to 2 up to a depth of 6.5 meters, and a cone

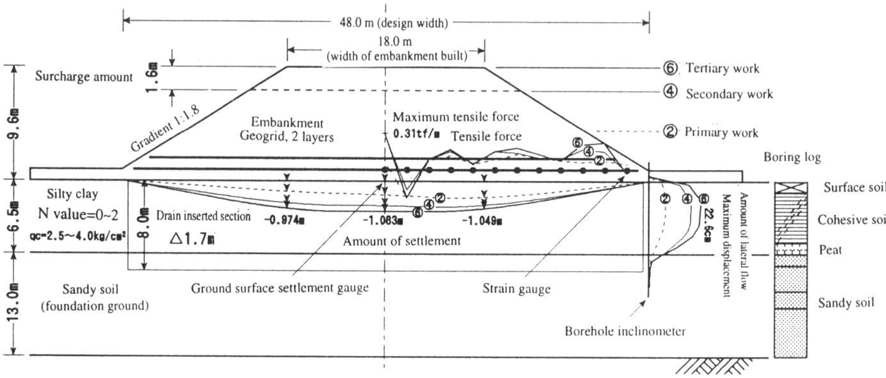

Fig. 12 Outline of cross-section of the Niigata-Nishi Bypass measurement point No.113, location of instruments and results of observation

penetration resistance ranging from 2.5 to 4.0 Kgf/cm². The embankment is high with a planned height of 8.0 meters, and if the embankment is built without reinforcement, the safety factor *Fs* for rotational slip that passes through the foundation ground will be 0.89, far below the required safety factor of 1.25.

We therefore studied the vertical drain method (1.7 m spacing), stage embankment method (5 cm/day) and the surcharge method (extra-fill 1.6 m) to reinforce the embankment. Even if these methods are used, *Fs* will be 1.15, slightly below the required factor, and thus the embankment cannot be fully stabilized. Therefore, we also used the geogrid reinforcing method, which gave the required safety factor *Fs* of 1.25. The safety factor of an embankment reinforced by geogrid is given by the following formula.

$$Fs = \frac{Mr + \Delta Mr}{M_d} \tag{3.1}$$

Where, M_d : Sliding moment
Mr : Resisting moment of soil
ΔMr : Resisting moment of geogrid
$\Delta Mr = R \bullet T_A \bullet \cos\theta \bullet n$
(see Fig. 13)

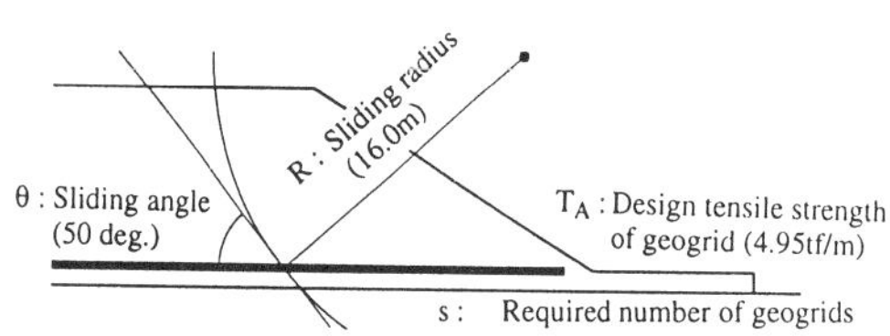

Fig. 13. Design of reinforcement method using geogrid

The required number of geo-grids is given by the following calculation:

$$1.25 = \frac{1033.467 + (16.0 \times 4.95 \times cos50 \times n)}{901.416} \tag{3.2}$$

n = 1.84, and thus the required number of geogrids is 2.

(2) Instrumentation for field observation

Various instruments for field observation over time are used. These instruments include ground surface settlement gauges installed at the natural ground surface (at three points; the center of the embankment, right and left top of the slope), a borehole inclinometer directly below the toe of the embankment slope (up to a depth of 12 meters at the toe of the embankment slope), and strain gauges placed on the surface of geogrid (at 12 locations at intervals of 2 meters).

(3) Results of field observation

The results of field observation are shown in Fig. 12. The figure indicates that settlement, lateral displacement and tension of the geogrid corresponding to approximately 80% of the final consolidation have developed by the secondary work. As for the tension, a maximum tension of only about 0.31 tf/m was generated, and this reflects the fact that the embankment was designed with a sufficient safety margin by this design method.

3.3 *Analysis of field observation results*

(1) FEM elastoplastic-consolidation analysis

The computer program (LADIS) was used to perform the FEM elastoplastic-consolidation analysis.

(2) Modeling of ground

The ground models used for the analysis was the Cam-

Table 5. Soft ground input constants

Name of stratum	Constitutive equation	Elastic modulus E (tf/m²)	Poisson's ratio γ	Cohesion C (tf/m²)	Internal friction angle ϕ	Compression index λ	Expansion index κ	Wet density (tf/m³)	Converted coefficient of permeability k (m/day)	Converted coefficient of permeability k (m/day)
Embankment	D.P.	1000	0.263	4.0	40.0	-	-	1.70	8.64×10^{-1}	→
Clay stratum 1	C.C.	-	0.400	-	-	0.600	0.060	1.59	2.16×10^{-4}	1.00×10^{-2}
Clay stratum 2	C.C.	-	0.400	-	-	0.680	0.068	1.56	1.40×10^{-4}	1.00×10^{-2}
Clay stratum 3	C.C.	-	0.336	-	-	4.800	0.480	1.35	2.16×10^{-4}	1.00×10^{-2}
Sand stratum 1	D.P.	340	0.300	0.0	35.0	-	-	2.00	8.64×10^{-2}	→
Sand stratum 2	D.P.	820	0.300	0.0	35.0	-	-	2.00	8.64×10^{-2}	→
Sand stratum 3	D.P.	1000	0.300	0.0	35.0	-	-	2.00	8.64×10^{-2}	→
Sand stratum 4	D.P.	1600	0.300	0.0	35.0	-	-	2.00	8.64×10^{-2}	→

Constitutive equation: D.P. = Drucker-Prager, C.C. = Cam-Clay
Coefficient of permeability: k = Coefficient of permeability after insertion of drain materials

Clay model for cohesive soil and the Drucker-Prager model for sandy soil, and the strength constant and other constants were determined by various soil tests (Table 5).

As for the section containing the vertical drains, assuming that the coefficient of permeability for the entire drain section will be improved by the insertion of drains, the average value of the coefficients of permeability calculated for the drain material and the natural ground based on the insertion interval is used as the coefficient of permeability (Fig. 14).

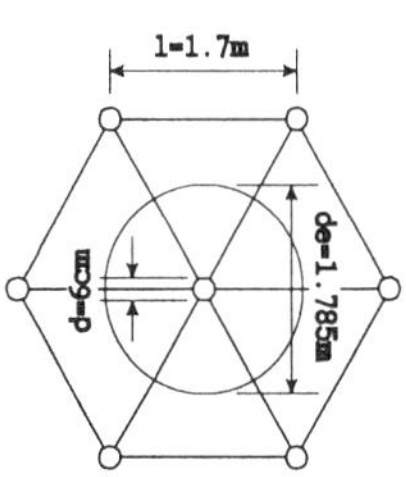

Interval at which drains are positioned: l = 1.7 (m)
Sand column equivalent diameter of drain: d = 0.05 (m)
Effective diameter of drain: de = 1.05d = 1.785 (m)
Coefficient of permeability for sand column: k' 8.64 (m/day)
Coefficient of permeability for soil: ka (m/day)

$$k' = \frac{(d^2\pi/4)\times ks + [(de^2\pi/4) - (d^2\pi/4)\}\times kc}{(de^2\pi/4)}$$

Fig. 14. Calculation of the average coefficient of permeability for ground with vertical drains

(3) Modeling of characteristics of friction between soil and geogrid

a) Modeling of geogrid

The geogrid, which is defined as a one-dimensional truss element that provides tension only, is regarded as having friction with the soil material around the geogrid (Fig. 15), and the elastic modulus, etc. were obtained from tensile tests (Table 6). The tension in the geogrid generated by slight sliding on the surface could thus be assessed.

Table 6. Geogrid input constants

Name of material	Elastic modulus E(tf/m²)	Yield stress (tf/m)	Sectional area m²
Adem F-10	80000	9.00	0.001

b) Joint element

To demonstrate the above-mentioned frictional relationship, a Goodman-type joint element was used (Fig. 15). This element is characterized by the fact that the shear modulus (frictional force) and the shear yield stress in the direction of axis increase in proportion to the overburden pressure (earth pressure) that acts vertically in the axial direction (Fig. 16).

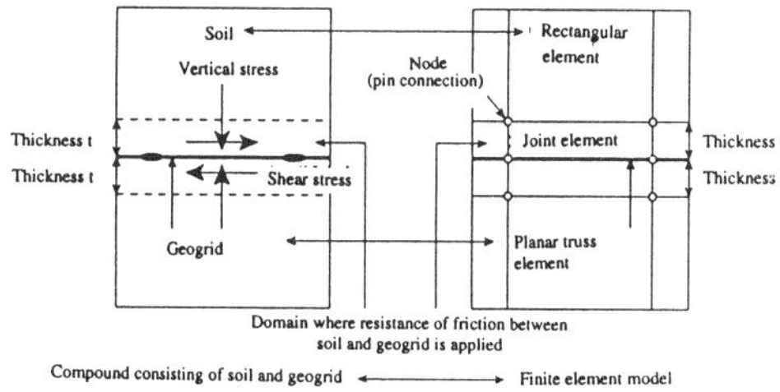

Fig. 15. Goodman-type joint element

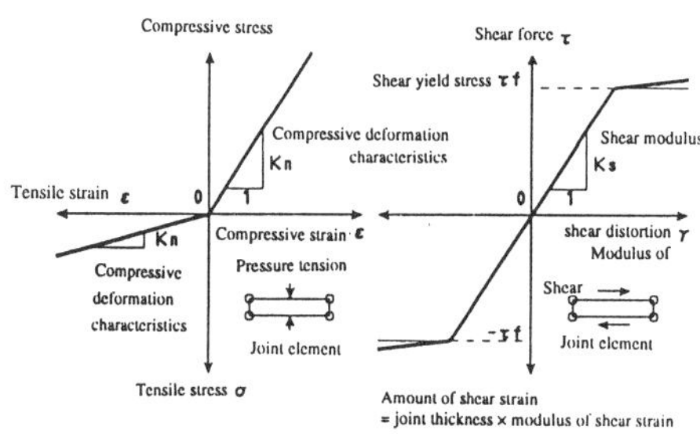

Fig. 16. Deformation characteristics of joint element

c) In-the-ground geo-grid pull-out test
To examine the frictional characteristics of the joint element, we conducted a geogrid pull-out test in the ground. An outline of the testing apparatus is shown below.

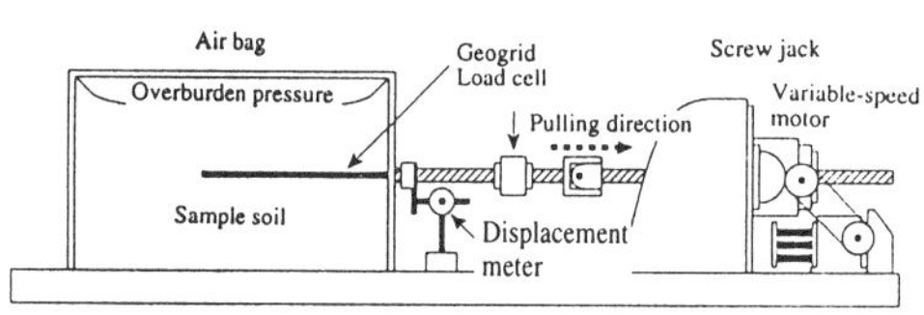

Fig. 17 In-the-ground geogrid pull-out testing apparatus

Assuming that the maximum height of the embankment at the site is just under 10 meters, a test was conducted for six cases, each with different overburden pressure, namely, 0.2, 0.6, 1.0, 1.4, 1.6 and 1.8 kg/cm^2, and geogrid with an embedded length of 45 cm was pulled out at a speed of 1 mm/min in each case. During each test, the strain of the geogrid was measured at 10 locations while real-time measurements of the amount and force of pull-out were taken. The relationship between the pull-out force on one side of the geogrid and the shear displacement for each overburden pressure is shown in Fig. 18. To plot this graph, the effective area method was used, in which only the part where friction is actually generated is evaluated.

The effective area method allows us to evaluate frictional force more accurately compared with the total area method in which the entire contact area is evaluated. From the displacement graph shown in Fig. 18, the values for the initial tangential gradient and the shear yield point were taken, and correlated with the overburden pressure as shown in Fig. 19. The figure shows the relationship between the overburden pressure *sn* and the shear modulus *Ks* as well as the shear yield stress *tf*.

(4) Results of analysis

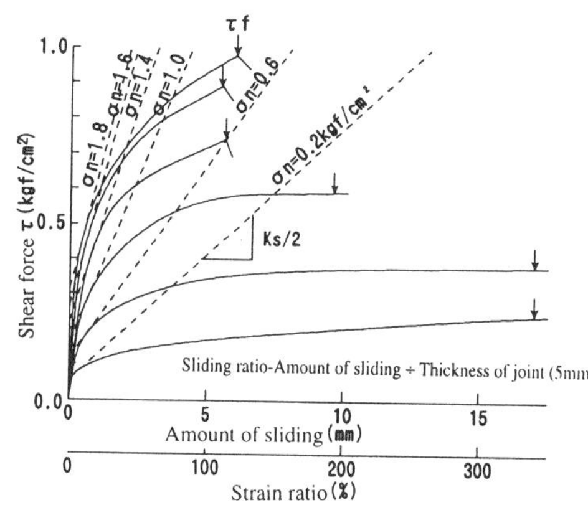

Fig. 18. Relationship between shear force and strain

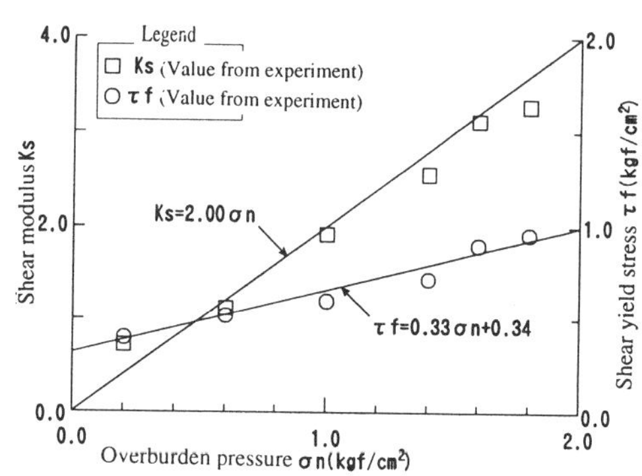

Fig. 19. Deformation characteristics of joint element

Figure 20 shows the amount of ground settlement for each stage of the embankment measured over time, Fig. 21 the state of lateral flow directly below the toe of the embankment slope, and Fig. 22 the distribution of tension generated in the geogrid. The circled numbers in the figures are for each work stage of the embankment in Fig. 20.

The settlement behavior evident in Fig. 20 is caused by the low coefficient of permeability for the ground, but the amount of the settlement did not sharply increase immediately after the embankment work in each state. Therefore, the amount of settlement remained constant through the standing period. The amount of settlement measured in the middle of the embankment showed a tendency very similar to the value obtained from the analysis, indicating that the actual settlement behavior can be predicted accurately in advance by using this analysis method.

As for the lateral flow shown in Fig. 21, lateral flow is likely to occur immediately after embankment work in each stage. A comparison of the predicted values with the values actually measured at stage (4) showed excellent correspondence, except that the predicted value was somewhat larger than the measured value. The analysis was thus fairly accurate.

As for the distribution of tension in Fig. 22, the measured values varied greatly from one measurement point to another, but the peak value of close to 0.3 tf/m of tension generated at the center of the embankment, which is important in the design, almost corresponded to the predicted value. We therefore consider that the maximum tension can be accurately predicted by the analysis. The reason for the variation of the measured values is thought to be that local stress concentration had a large effect because the field observation was conducted with strain gauges placed at intervals of 2 meters.

The above results showed that the behavior of the actual embankment can be modeled accurately through FEM analysis by appropriately representing the drained zone, etc., by evaluating the frictional characteristics of the geogrid by means of in-ground pull-out tests, and by using a joint element.

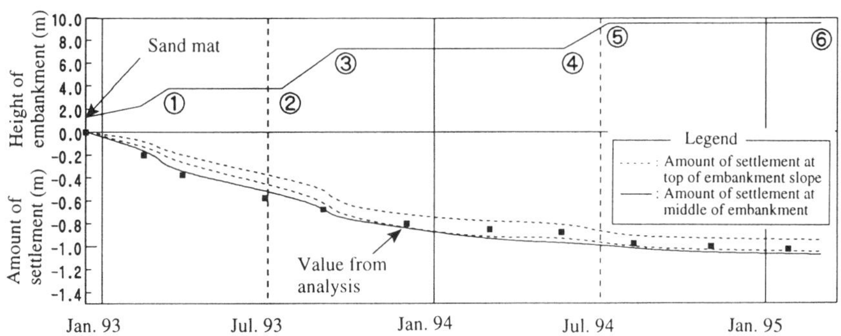

Fig. 20. Height of embankment and settlement curve

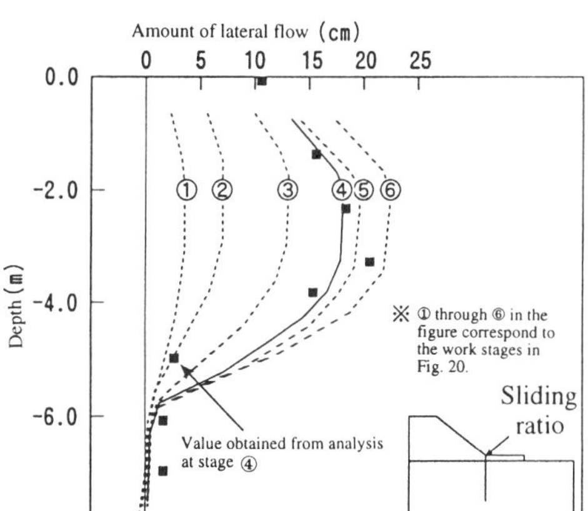

Fig. 21. Lateral flow directly below toe of embankment slope

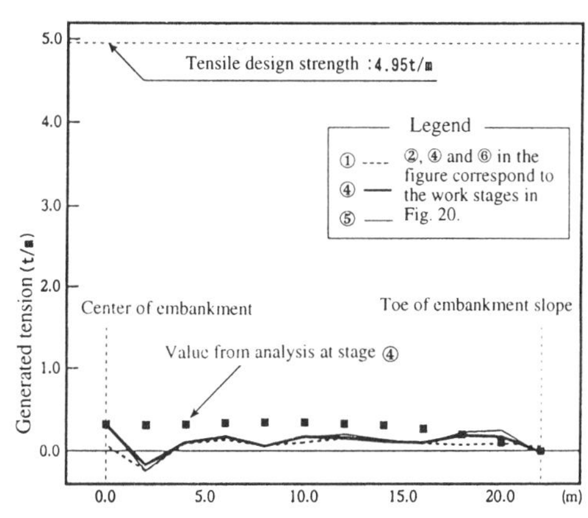

Fig. 22. Distribution of tension generated on geogrid

3.4 *Summary*

We attempted to reproduce the behavior of the site through FEM elastoplastic-consolidation analysis for the Niigata-Nishi Bypass in which a reinforcing method using geogrid is applied. For the analysis, we represented the coefficient of permeability for the natural ground by an average calculation for the intervals at which drains are embedded, conducted an in-the-ground pull-out test, and represented the geogrid using a joint element. The results showed that the actual measured values were similar to the analysis values for the settlement, lateral flow and generated tension, indicating that the behavior of the actual embankment is accurately predicted by the analysis method, and that the tension actually applied to the geogrid can be predicted.

In the future, we will apply this method to other sites, accumulate case study data, and study the possibility of establishing a more reasonable and economical design method by reflecting these results in the current design method, and thus design with a small safety margin.

4 SETTLEMENT CONTROL METHOD USING GEOGRID IN COMBINATION WITH DEEP-MIXING METHOD OF SOIL STABILIZATION

4.1 Outline of the method

Foundation ground is sometimes improved at a low improvement rate by the deep-mixing method of soil stabilization as shown in Fig.23 to reduce residual settlement in road embankments on soft ground. However, differential settlement sometimes develops between the improvement columns used for the deep-mixing method of soil stabilization and the unimproved part between the columns, causing cracks or unevenness at the top of the embankment. In this chapter, we propose a tentative design concept for laying geotextiles to reduce such differential settlement.[2)]

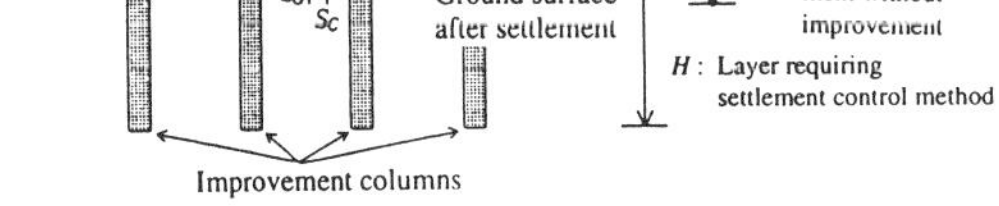

S_G : Differential settlement between improvement columns and unimproved section between columns (with geotextile)
S_c : Differential settlement between improvement columns and unimproved section between columns (without geotextile)

Fig. 23. Example of method of controlling differential settlement between improvement columns and unimproved part between columns by using geotextile

4.2 *Design method*

(1) Selection of materials

For the geotextile material, geogrid or a material with a smaller amount of tensile strain is desirable. Since this design method allows the tensile strain of the geotextile to be less than 1%, creep may not need to be considered for the safety factor of the material. As for the soil material around the place where the geotextile is laid, a material with large frictional resistance is desirable so that the tensile force of the geotextile can be fully utilized. Improper materials include those containing a large portion of fine particles or those that are so sharp and large grained that the geotextile may be damaged.

(2) Design conditions

1) Allowable settlement: The allowable values for total settlement S_{CO} and differential settlement S_{GO} are assumed to be 10~20 cm or smaller and 5 cm or smaller, respectively, based on the various guidelines.

2) Design load: Design load comprises embankment load and traffic load. The traffic load on a low-embankment road on soft ground will be evaluated in terms of embankment load equivalent to the portion affected by repeated traffic load based on long-term settlement (see the Road Earthwork Manual on soft ground improvement works).

3) Improvement columns: Improvement columns used for the deep-mixing method of soil stabilization will be positioned in a staggered manner or in a grid form, and provide column-based improvement. The distance between column centers should be about 1.5 ~ 3 m, and the columns must be embedded down to the bearing ground.

(3) Design procedure

The design procedure for the differential settlement control method using the deep-mixing method of soil stabilization in combination with geotextile is as follows.

1) Design conditions and load conditions such as design constants of the soft ground are determined.

2) The amount of settlement without the deep-mixing method of soil stabilization S_0 is examined, and if the required allowable value is not satisfied, improvement based on the deep-mixing method of soil stabilization is studied.

3) The load bearing capacity of the improvement column used for the deep-mixing method of soil

stabilization is examined, and the specification of the improvement column (strength of pile and improvement rate) is determined.

4) Based on the specification for ground improvement determined in step 3), the average amount of settlement of the improved ground S_T is reviewed. If the average settlement does not meet the target allowance, the improvement specification of the deep-mixing method of soil stabilization is re-examined.

5) The amount of differential settlement for the application of only the deep-mixing method of soil stabilization Sc is examined. If the allowance is not satisfied, the amount of differential settlement for combined application of the deep-mixing method and geotextile is examined, and the laying specification of the geotextile is determined.

6) The case where the improvement rate of the deep-mixing method is increased and the amount of differential settlement is reduced is compared with the case where geotextile and the deep-mixing method are used together for examination, and an economical method is then finalized.

(4) Design calculation method

This section explains the principles of the design calculation.

1) Examination of load-bearing capacity of improvement column and determination of specification of improvement column

The specification of the improvement column is examined using the following formula, assuming that all the overburden load above the top of the improvement column (embankment load and traffic load) is supported by the improvement column.

$$Fs \leqq \frac{q_{uck}}{\dfrac{\Delta p}{a_p}} \tag{4.1}$$

Where, Fs : Target safety factor (1.0~1.5),
q_{uck} : Design standard strength of pile (kgf/cm^2)
Δp : Overburden load (kgf/cm^2)
a_p : Improvement rate (0.1~0.3)

2) Method of calculating amount of settlement

a) Average amount of settlement of ground where deep-mixing method is used

The average amount of settlement of the composite ground S_T where the deep-mixing method is applied is obtained from the following formula, in consideration of the stress concentration on the improvement column, and assuming that the stress share ratio of the improvement column n is equal to the ratio of the coefficient of volume compressibility of the improvement column to the unimproved soil n.

$$S_T = \beta c \cdot So = \frac{1}{1 + a_p (n - 1)} \cdot So \tag{4.2}$$

Where, S_T : Average amount of settlement of improved ground (m)
βc : Settlement reduction ratio (ratio of reduction of increased stress applied to unimproved section)
a_p : Improvement rate
n : Stress share ratio($=\sigma_p/\sigma_c=n'=mv/m_p$=10~20)
mv : Coefficient of volume compressibility of unimproved section
m_p : Coefficient of volume compressibility of improvement column

b) Amount of differential settlement of composite ground where the deep-mixing method is used

The amount of differential settlement of the improvement column for the deep-mixing method and the unimproved section Sc is obtained from the following formula, using the amount of settlement of the pile top of the improvement column Sp and the amount of settlement of the unimproved section Sd. In this formula, n is 1/2 of the value used in 1) for simplicity, obtained from the comparison between FEM and actual measurements.

$$Sc = Sd - Sp \tag{4.3}$$

$$Sd = \beta c \cdot mv \cdot H \cdot \sigma \tag{4.4}$$

$$Sp = \beta c \cdot m_p \cdot H \cdot n \cdot \sigma \tag{4.5}$$

3) Amount of differential settlement for combined application of geotextile and deep-mixing method

When the deep-mixing method of soil stabilization and geotextile are used together, the amount of settlement that occurs between the improvement column and the unimproved section between columns S_G is calculated as follows.

$$S_G = \frac{Sc}{1 + 2 \cdot \alpha \cdot (Sc/\sigma)} \tag{4.6}$$

The proportional factor α in the above formula is determined from Fig. 24. Namely, it is determined from the tension stiffness of the geotextile EA, which will be assumed first, and from the pile pitch equivalent to the improvement rate a_p. The value of α is then substituted in Equation (4.6), and a trial calculation is continuously made until the standard value S_G is obtained.

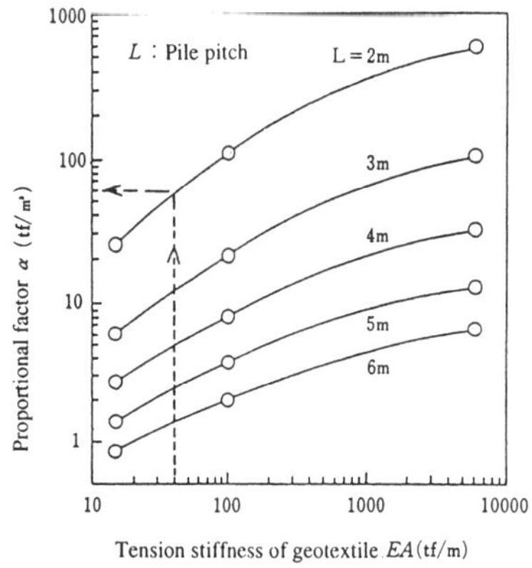

Fig. 24 Relationship between tension stiffness of geotextile and proportional factor α[2)]

4) Selection of material for geotextile
In this design method, the material is selected not based on the design tensile strength but on the stiffness obtained when the strain is about 1% because the strain level of the geotextile is low.

The amount of strain generated in the geotextile ε is estimated from the amount of differential settlement based on field observations, using the following formula.

$$\varepsilon = 0.15 \bullet S_G \qquad (4.7)$$

Then, from the amount of strain, the required tensile strength of the geotextile T is calculated, using the following formula, and geotextile with the required tensile strength or the number of layers requiring geotextile is determined.

$$T = \frac{E \bullet A \bullet \varepsilon}{100} \qquad (4.8)$$

4.3 *Future tasks*

(1) The formula (4.2) used to calculate the average settlement of the ground where the deep-mixing method is applied leads to excessive settlement compared with the measured values, and therefore needs to be reexamined. The values of S_T where the columns are embedded down to the bearing ground are so small that they can be ignored. Instead, a formula for calculating the settlement of the frictional pile type is required.

(2) An appropriate method of evaluating the stress share ratio, which is important when calculating the differential settlement between the improvement column used for the deep-mixing method and unimproved section, is required. Measurements taken so far indicate that the differential settlement Sc is so small that it can be ignored if the improvement rate is larger than15%, even when geotextile is not used in combination with the deep-mixing method.

(3) From the above, if the improvement rate of the deep-mixing method is larger than about 15%, the differential settlement between the improvement column and unimproved section is fairly small in many cases even when geotextile is not used. The possible combined use of geotextile should therefore be considered only when the improvement rate is less than 15%.

5 NEW RECLAMATION METHOD USING HORIZONTAL DRAINS IN COMBINATION WITH SAND MOUNDS TO LOWER WATER LEVEL

5.1 *Outline and characteristics of the method*

(1) Outline of the method

This method makes it possible to quickly start using land reclaimed with cohesive soil (class 3 or 4) from construction sites by accelerating the consolidation of the reclaimed ground with horizontal drains that are installed in it during the reclamation.

In order to load the ground, a multilayered sand mound is constructed inside the land-facing side of the revetment of the reclaimed land and a wellpoint or the like is used to lower the water level within the mound. This is the key characteristic of the method (see Figure 25).

The reclaimed ground should be an artificial island connected to the land with a road. Of the revetment surrounding it, the side facing the land should be of a cutoff type and the other three sides should be of optional sections. The revetment and the connecting road should all be completed before the reclamation starts.

In the reclamation, which is executed layer by layer, a multilayered sand mound is constructed inside the land-facing side of the revetment and, on each of the layers of fixed depth, horizontal drains are installed. On top, earth brought from another place is placed in a thinner layer. Finally, a wellpoint is driven into the multilayered sand mound from above the revetment facing the land.

In the loading procedure, first the inside water level is lowered as much as possible by opening the conduits passing through the land-facing side of the revetment. Then, the wellpoint is used to lower the water level within the mound so that a hydraulic gradient occurs in the horizontal drains, thus enabling drainage from the layers of reclamation soil.

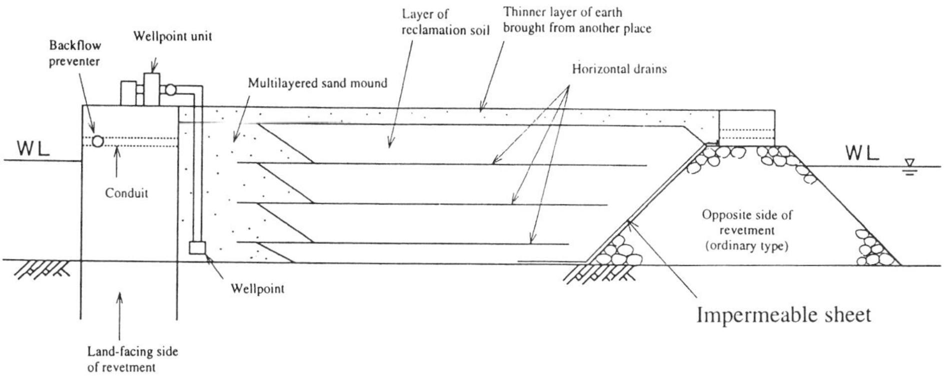

Fig. 25. Reclamation Method Using Horizontal Drains in Combination with Sand Mounds to Lower Water Level: Conceptual Illustration

(2) Characteristics of the method

The following are the characteristics of the method:

1) Most of the soil used in this method is low-quality soil from construction sites.
2) Horizontal drains are installed in the process of reclamation, and are quick and easy to install.
3) Preliminary drainage with conduits enhances the effect of the wellpoint.
4) The loading is executed by lowering the water level: it is not necessary to spread geomembranes or perform any extra surface treatment.
5) The multilayered sand mound not only functions as a drainage layer but also reduces the earth pressure on the revetment.
6) The earth brought from another place and placed in a thinner layer on top not only functions as a drainage surface but also as a surcharge and allows traffic to pass over the reclaimed ground in order to bring the earth from the other place for the last time.

5.2 *Design method*

(1) Materials

The drains must be able to convey all the water that drains into them from the soft soil. The examples of drains shown in Table 6 are each composed of a corrugated core enclosed in a filter. Polyethylene resin is used as the core, and nonwoven polyester fabric as the filter. The core is structured so that it is scarcely deformed under the pressure of the soil, and the filter has a high water permeability to prevent clogging. These drains are very flexible and can be wound into rolls, and so can be easily transported and installed continuously over a long distance.

Drains other than these can be used provided that they can let all the water run through themselves, are sufficiently resistant to deformation under the pressure of the soil, and are easy to install.

Table 6. Drains: Examples

Section size B × t (mm)	Area of empty parts of section *S* (cm²)	Diameter obtained by conversion *dw* (mm)	Section form
100 × 3	1	65.5	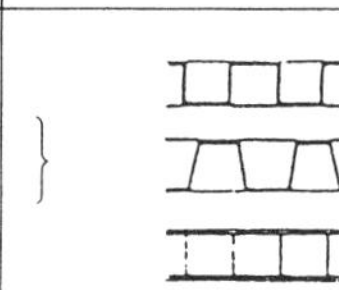
100 × 6	5	67.5	
100 × 10	8	70.0	
150 × 6	8	99.3	
150 × 12	12	103.1	

(Note) The above "diameter obtained by conversion" is calculated using the formula, $d_w = 2(B + t)/\pi$, supposing the external circumference of each drain is circular.

(2) Flowchart of design

The following is a flowchart of the design of soil improvement by this method.

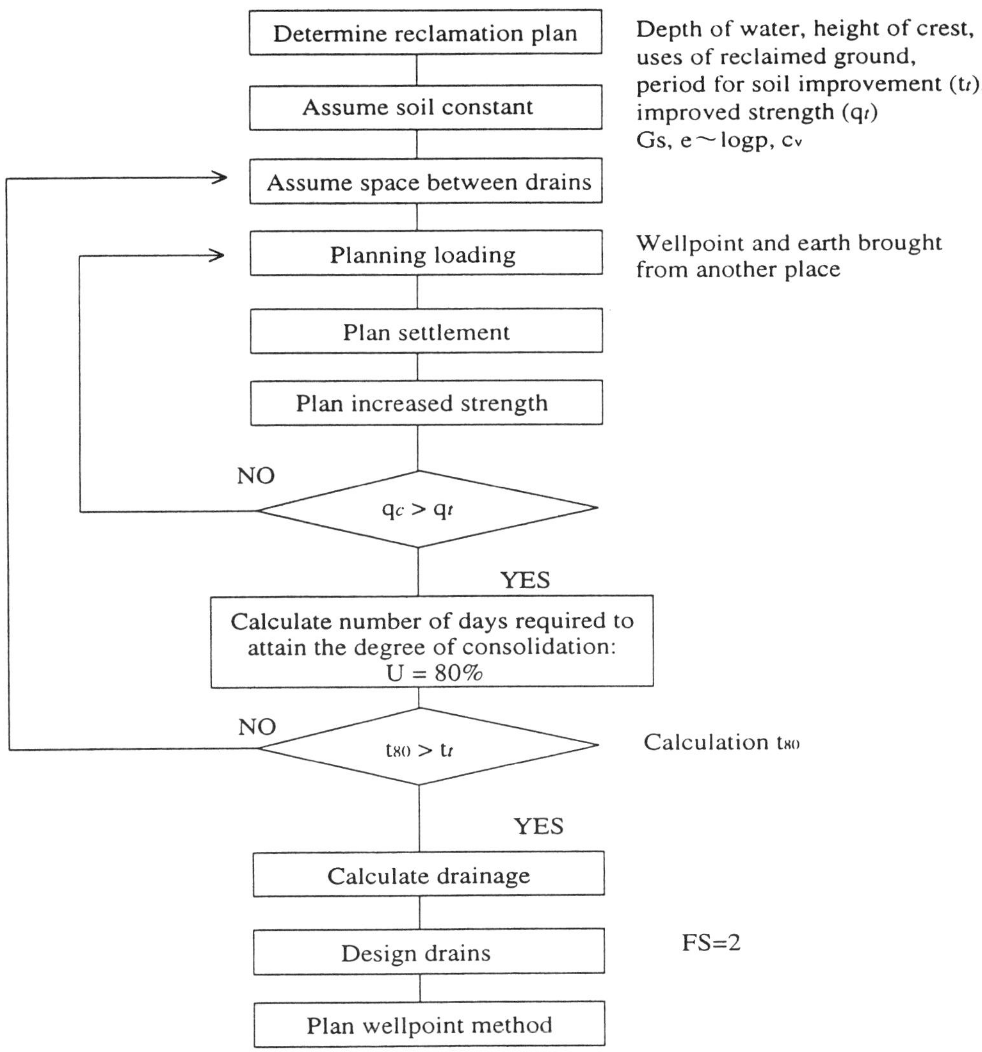

Fig. 26. Flowchart of design

(3) Design steps

1) Calculating the settlement

In calculating the final settlement due to consolidation, use the appropriate equation as follows:

$$S = \frac{e_0 - e}{1 + e_0} H \qquad (5.1)$$

$$S = m_v \Delta p H \qquad (5.2)$$

$$S = H \frac{C_c}{1 + e_0} \log \frac{p_0 + \Delta p}{p_0} \qquad (5.3)$$

where,
- S : settlement due to consolidation
- H : target thickness of layer of soil
- e_0 : initial void ratio
- e : target void ratio
- m_v : volume change rate
- Δp : load change
- C_c : compression index
- p_0 : yield consolidation load

2) Calculating the loading period

Assuming that one-dimensional consolidation applies to the consolidation, calculate the degree of consolidation, which corresponds to the loading period and the time factor, using the following formula:

$$t = \frac{H_d^2}{c_v} T \qquad (5.4)$$

$$U' = \{T^3/(T^3 + 0.5)\}^{1/6} \qquad (5.5)$$

where,
- t : loading period
- H_d : drainage distance; in case of draining to both sides: $H_d = H/2$
- c_v : coefficient of consolidation
- T : time factor
- U : degree of consolidation

We use the degree of consolidation in this formula similar to that of Terzaghi's theoretical formula, because they are precise and simplify the calculation. For reference, when $T_{70} = 0.403$, $T_{80} = 0.567$ and $T_{90} = 0.848$, then the formula gives $U = 0.698$, 0.802 and 0.905, respectively.

3) Selecting the drain material

Calculate the maximum drainage speed per 1 m of drain width from the drainage when the degree of consolidation is 10%, as follows:

$$v = \frac{HL\,\varepsilon_{10}}{t} \qquad (5.6)$$

where;
- v : maximum drainage speed
- H : thickness of soil layer to be improved covered by drains on each level
- L : length of layer covered by the drains
- ε_{10} : volume compression strain at 10% consolidation
- t : loading period till the degree of consolidation of 10%

Figure 27 shows the relationship between the area of inner empty section of each drain and the average drainage speed per drain, which we have obtained by a field experiment. This figure shows that, when selecting the drain, we should fix the area of the inner empty section of the drain so that v, the average drainage speed obtained by the above calculation, is positioned in the lower right of the broken line in the figure. Obtain the required area of the inner empty section of the drain from this figure, and thus select the drain. Note that the drains shown in the figure are only examples; other drains can be used provided that their average drainage speeds corresponding to section size are examined.

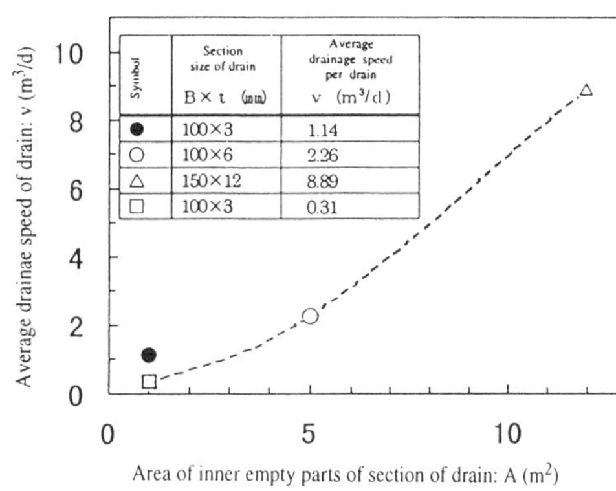

Fig. 27. Relationship between area of inner empty parts of section and average drainage speed of drain

4) Calculating the strength increase

Calculate the increased strength of the improved soil using the following formula:

$$\Delta c = cu/p \times U \times \Delta p \qquad (5.7)$$

where;
- Δc : increased strength
- cu/p : strength increase ratio
- U : degree of consolidation
- Δp : load change

5) Calculating the cone index (q_c)

Assume that the cone index (q_c) and the unconfined compressive strength of cohesive soil (q_u) are related to each other as expressed by Muromachi's formula:

$$q_c = 5\,q_u = 10\,cu \qquad (5.8)$$

where, q_c : cone index (kgf/cm^2)
q_u : unconfined compressive strength (kgf/cm^2)
cu : cohesion (kgf/cm^2)

5.3 *Method of reclamation*

(1) Machines

1) Floating barge for installation of horizontal drains
The floating barge for installation of horizontal drains in the ground should be a catamaran with two floating bodies (made from styrofoam), each with the following dimensions: 1.0 m (w) × 1.0 m (l) × 0.9 m (h) (and with a bottom coated with fiber reinforced plastic (FRP), a friction reducing agent), arranged parallel with each other with a space of 1.2 m in between. There is a mandrel between the two bodies and, by operating the winches, it is possible to wind them up onto the barge and install the drains on the ground. On the barge, there is also a reel unit on which to mount the drains. The barge is structured so that when it is moved back and forth, the reel rotates to automatically pay out the drain onto the ground. The gross weight of this barge is about 4 t and it has a 0.3 m draft. However, the draft can be adjusted by increasing the part of the barge above the water level.

Figure 28 outlines the floating barge for installation of horizontal drains.

2) Wellpoint unit
Figure 29 shows how the wellpoint unit should be installed.

(2) Procedure

In order to securely cut off the water, use steel sheet pile cellular type or steel sheet pile double-wall type when constructing the land-facing side of the revetment. Use soil from construction sites of class 1 or 2 as the soil packing for the side of the revetment. Optional but appropriate sections may be used for the other three sides of the revetment. In the case of the rubble-mound type, however, measures must be taken to prevent reclamation soil from being drawn out, such as by constructing a preventive structure.

a) Building a sand mound
Build a sand mound with a crest of the required width inside the land-facing side of the revetment using a clamshell, etc.

b) Filling soil from construction sites in layers
Fill in soil brought from construction sites systematically layer by layer, because horizontal drains are to be installed on it later. Properly locate each filled in layer, and measure the section of reclaimed ground each time. Use a floating bridge to guide the vehicles to where the soil is to be placed.

c) Installing the horizontal drains
When the depth of the layer of soil reaches the fixed level, install the horizontal drains. The width of the drains should be about 1.5 m. In the case of a band-type drain, mount multiple reels at regular intervals on the shaft aboard the floating barge . First, pull out one end of the drain and attach spacers to ensure adequate space between its neighboring drains. As the spacer material, geogrid, etc. may be used because it is easy to handle and relatively stiff.

From this initial state, operate the mandrel of the base to put the drain below the surface of the water. Then, start installing it in the longitudinal direction by winching the wire line, which has been installed beforehand. Install the horizontal drain so that about 2 m from one end of it lies on the sand mound and the other end is positioned several meters away from the structure to prevent the reclamation soil from being drawn out.

In order to securely position the drain, attach spacers at appropriate intervals to the middle part of it. Drive in bamboo spits through it as needed to prevent it from rising. Having reached the opposite side of the reclaimed ground, cut the drain and finish off the end appropriately. Then, return to the initial side of the ground and repeat the same work until all of the ground is covered by drains.

d) Repeating steps a), b), and c)
Repeat the cycle consisting of building the sand mound, placing the reclamation soil and installing the horizontal drains until the top of the crest is approached.

e) Placing a thinner layer of earth brought from another site
As the top layer, place earth brought from another site in a thinner layer of about 1 m. Perform this process by the floating conveyor method or the like. This layer prevents rainwater from permeating the reclaimed ground and encourages it to promptly drain off from the surface by preventing the surface of the soft reclamation soil from drying.

f) Loading by the wellpoint
Drive the wellpoint down into the multilayered sand mound and install the wellpoint unit on the land-facing side of the revetment. Immediately after completing reclamation of all parts of the land, the water level in the reclaimed ground is fairly high. Therefore, first open the conduits that pass through the land-facing

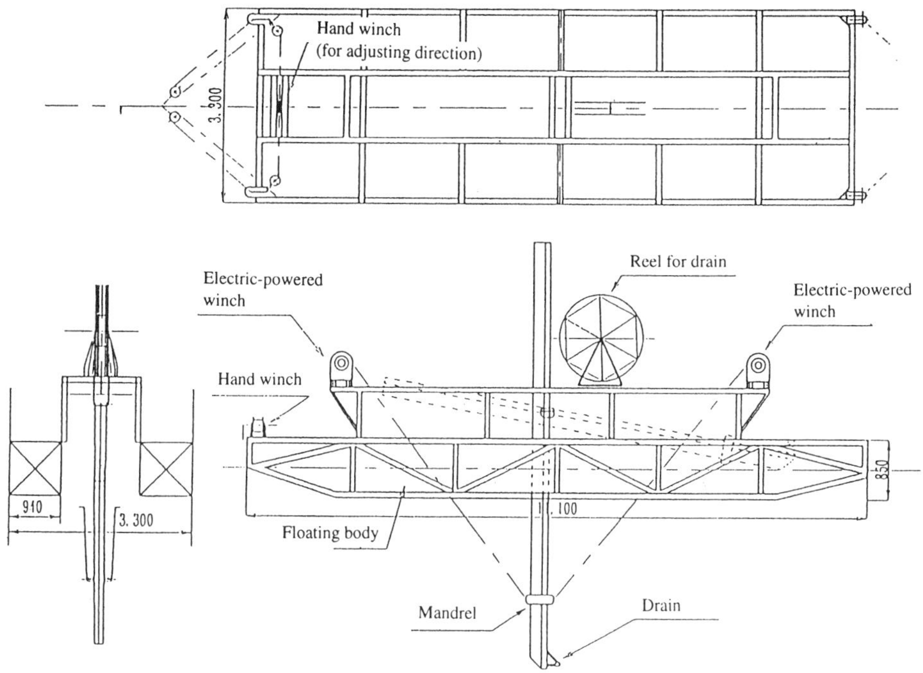

Fig. 28 Floating Barge for Installation of Drains

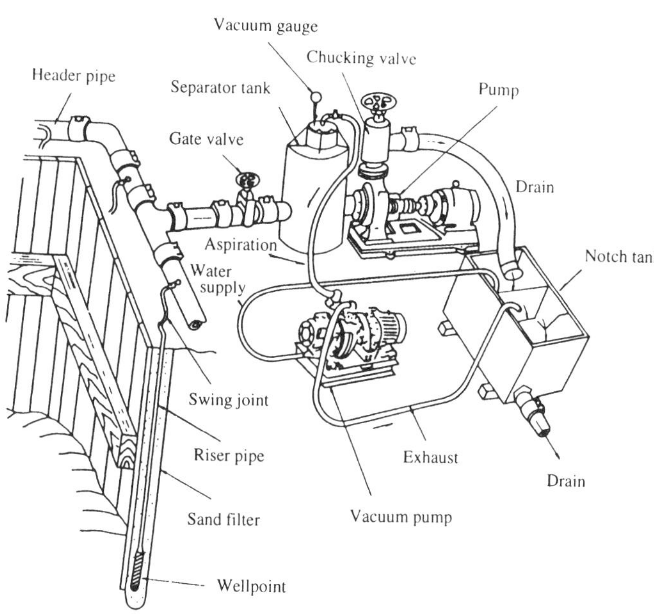

Fig. 29 How to Install the Wellpoint Unit

side of the revetment at appropriate intervals at a suitable time when the tide is right to lower the water level within the multilayered sand mound as much as possible. This is the preliminary loading. In order to make the wellpoint lower the water level as much as possible later, do not operate the wellpoint at this stage.

When the degree of consolidation by this preliminary loading has reached about 80%, operate the wellpoint to lower the water level further.

g) Placing earth brought from another place for the final time

When the degree of consolidation due to loading by the wellpoint reaches about 80%, the reclamation soil should have settled by several meters. Then put earth brought from another place (class 1 or 2) for the final time to offset the settlement. The earth thus placed acts as a load and also as the surface ground of the reclaimed land for parks, housing land and so on.

6 AFTERWORD

This paper explained the techniques of applying geosynthetics to embankments on soft ground mainly by presenting studies in which I have been involved. The results of many other valuable studies made in Japan and overseas have thus hardly been mentioned.

In conclusion, the following tasks remain to be addressed in this field.

In the case of reinforcement of soft ground surface for earth cover, a variety of design concepts that take the bearing capacity mechanism into consideration have been proposed. However, because the method of work execution has a great effect, the design tensile strengths of reinforcing materials have actually been fixed so far based on past experience. Therefore, a more rational design method should be established through an analysis of works executed to date.

On the other hand, in the case of reinforcement of soft ground surface for embankments, when the soft ground is too deep or the embankment too high to secure the stability against sliding failure with sand drains alone, a feasibility study that we conducted has shown that it is advantageous to use synthetic geogrid as a supplement to the sand drains, as another possible way of applying geosynthetics. In order to extend the use of geosynthetics, a limited number of specific uses that can fully compete with existing methods, such as this one, should be developed.

Concerning the combined use of geogrid and the deep mixing method, various methods have been proposed based on concepts similar to that of the pilenet method: conducting piling improvement up to a low improvement rate (about 10 ~ 30%) for all the surface below the embankment through the deep mixing method and then compensating for the insufficient stability against sliding with geogrid, and using geotextile to prevent differential settlement of the improved and non-improved parts. The combined use of the deep mixing method with geogrid to realize a low improvement rate has not yet fully been established as a design method. However, this new and economical attempt at improving soft ground is likely to find practical use.

Further, as a measure to resist earthquakes, we shall study the application of the reinforced earth method using geosynthetics to control settlement during an earthquake of embankments on loose sandy ground that could be liquefied.

Concerning the effective use of low-quality soils generated at construction sites, a design method has been proposed based on the theory for saturated soil, so the current method of reinforcing the embankment using cohesive soil with high water content by draining it can only be applied to saturated soil. It is thus necessary to develop and establish a theoretical method of reinforcement that can be applied not only to saturated soil but also to partially saturated soil. In addition, a new execution method should be developed that enables this method to be used not only for embankments on land but also for those that have foundations under water such as for reclamation works.

REFERENCES

1) Editorial Committee for Comprehensive Bibliography on Reinforced Earth Technology for Slopes/Embankments: *Comprehensive Bibliography on Reinforced Earth Technology for Slopes/Embankments*. Industrial Technology Service Center, pp. 456-471, 1995. (in Japanese)
2) Committee for Generalization of Geo-Textile-Reinforced Earth Method: *Design and Execution Manual for Geo-Textile-Reinforced Earth*. Civil Engineering Research Center, pp. 209-215, 1994. (in Japanese)
3) Kudo, N. and Maruyama, Y.: On Foundation Improvement for Soft Peat Ground. 20th Meeting for Announcement of Results of Technical Studies, Hokkaido Development Bureau, pp. 588-597, 1977. (in Japanese)
4) Nara, A. and Kudo, N.: Improving Very Soft Peat Ground by Pile Net Method. *Execution of Civil Engineering Works*, Vol. 23, No. 6, pp. 27-34, 1982. (in Japanese)
5) Iseda, T., Tanahashi, Y. and Nagamatsu, M.: Analysis of Examples of Pile Net Method and Proposal of Design Guidelines. Proceedings from Civil Engineering Society, Vol. 379, VI-6, pp. 36-44, 1987. (in Japanese)

Earth Reinforcement, Ochiai, Yasufuku & Omine (eds) © 1997 Balkema, Rotterdam, ISBN 90 5410 833 9

Design of reinforced excavations and natural slopes using new European Codes

G.Gässler
FH München, Department of Civil Engineering, Germany

ABSTRACT: In 1994 the European Committee for Standardization (CEN) edited the new European Standard ENV 1997-1 "Eurocode 7: Geotechnical design - Part 1: General rules". For future design of reinforced earth structures the EC 7 will be very important. The essential difference to the conventional design methods is the application of several partial safety factors instead of only one overall safety factor. Design examples of nailed walls are presented showing the application of the various partial safety factors for loads and acting forces as well as for the properties of the materials, including soil friction and cohesion, and the proprties of steel.

1. INTRODUCTION

Over 15 years ago, work started to create common rules for geotechnical design in the European countries. Since the end of the 1980s, this work has been carried out under the direction of the European Committee for Standardization (CEN). The result is published now as ENpr 1997-1 "Eurocode 7: Geotechnical design - Part 1: General rules".

This lecture introduces the general features of the Eurocodes, and deals with the application of the geotechnical design rules of Eurocode 7 - Part 1 (EC 7-1) to insitu reinforced earth structures in particular.

2. THE NEW EUROPEAN CODES IN GENERAL

2.1 *History*

In 1980, the Commission of the European Communities and the International Society of Soil Mechanics and Foundation Engineering (ISSMFE) made an agreement that the ISSMFE should produce a draft for the European Code 7 - Part 1. A group of geotechnical experts from several European countries presented the draft in 1988. The Commission of the European Community then set up a project group which revised the draft and adapted it to the other nearly completed Eurocodes. In 1990, responsibility for the issue of the Eurocodes was transferred to CEN (Comité Européen des Normalisation) in Paris. Part 1 of Eurocode 7 was published as ENpr 1997-1 in English in November, 1994. The letters "pr" denote this stage as a European prestandard.

2.2 *Scope of the present European Codes in civil engineering*

There are nine Eurocodes in total, each containing a set of common unified design rules for some aspect of building and civil engineering works. Each Eurocode is subdivided into various parts. The ENpr 1991-1, or Eurocode 1: *Basis of design and actions on structures* - Part 1: *Basis of design*, contains the principles of design and defines the design values of actions. In addition to Eurocode 1, a further four codes which are of interest to the geotechnical engineer are mentioned here. They are titled as follows:

- EC 2 Design of concrete structures
- EC 3 Design of steel structures
- EC 7 Geotechnical design
- EC 8 Design provisions for earthquake resistance of structures.

A further set of European standards for the execution of special geotechnical work should be mentioned. These include ENpr 1536 *Bored piles*, and ENpr 1537 *Ground anchors*. The latter codes represent the best European knowledge on the execution of ground engineering works at present. However, one has to distinguish codes on design (ENpr 1991 to 1999) from codes on engineering products (ENpr 1536 to 1538). This lecture will only deal with the geotechnical design codes.

2.3 *Objectives of European Codes*

From the outset, the political objective of the Eurocodes, not forgetting the European standards for execution, was to harmonize the very different technical rules of the 15 EU-countries in order to create a real European market for the design and execution of civil engineering works.

It is important to note that, at present, the safety level of building and civil engineering works is extremely different in Europe (not to mention other parts of the world). There are many countries where national standards prescribe specific global safety factors, the magnitude of which can differ from country to country in a range of ±50% for the same check. Furthermore, there are some countries in Europe where the designing engineer fixes the safety factor at his own risk, and there are countries where the insurance company determines safety requirements. Lastly, one has to mention the fact that, due to different safety *definitions*, the safety level is not homogeneous, neither in ground engineering nor in comparision with construction engineering.

It follows from the above discussion that one of the principal objectives of the Eurocodes is to harmonize the understanding of safety requirements in two ways: specifically and internationally. Therefore, as a common basis for all EC-countries, the following two fundamentals feature in the design of building and civil engineering works:
- Limit state design format, and
- Partial factors of safety.

Thus, the above mentioned nine Eurocodes represent a set of fairly well harmonized safety standards for both of the above, covering fifteen European countries and a wide range of materials (e.g., concrete, steel, timber, and importantly soil.

In the following sections the lecture will be focussed on Eurocode 7, Part 1, which concentrates on the functional requirements for geotechnical design. Consequently, one will not find specific calculation procedures for soil nailing in this code. In any case, the design of nailed walls or slopes has to fulfill the above mentioned fundamentals: limit state and partial safety factors.

3. CODES, LICENCES AND RECOMMENDATIONS FOR SOIL NAILING

3.1 *History of soil nailing as a new construction method*

Without doubt, the idea of reinforcing new cut slopes was inspired by the new Austrian tunnelling method (NAT), which had been primarily a rock tunnelling system using a combination of shotcrete and fully bonded steel inclusions to provide early, efficient excavation stability (Bruce & Jewell, 1987). In 1972 the first case of a reinforced cutting was realized in France near Versailles, where a 70° cut slope in cemented sand was reinforced temporarily using grouted steel bars (Rabejac & Toudic, 1974). However, apart from some practical cases, the new method had not been scientifically studied in France in the seventies.

At the beginning of the seventies some applications of the NAT in Germany had been proved successful in less competent materials, such as silts, sands and gravels. Thus, in 1975 the idea of insitu reinforced soil was taken up in Germany, when the German research and development project "Bodenvernagelung" started to study and develop the new technique for stabilizing slopes and excavation cuts.

In the meantime the technical world has nearly forgotten that the technical term *soil nailing* derives from the two German words "Boden" (meaning *soil*) and "Vernagelung" (meaning *nailing*). After several publications on the experimental and theoretical results of the new technique in German and English language (Stocker 1976, Gässler 1977, Stocker & Gässler, 1979), the German contribution on *soil nailing* (Stocker, Körber, Gässler, Gudehus,

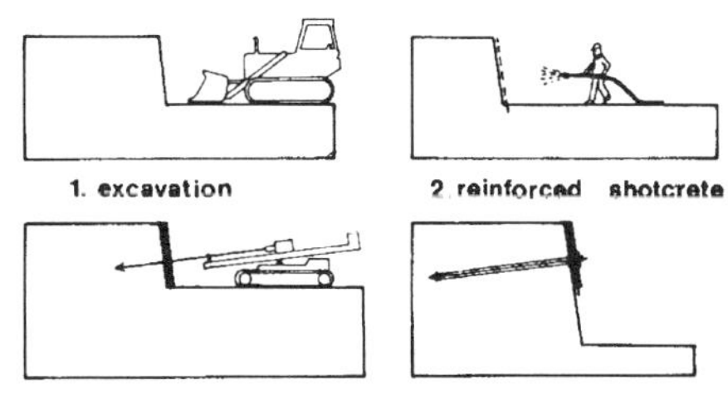

Fig. 1 Construction method of soil nailing (adapted from *soil nailing*, Stocker, Körber, Gässler & Gudehus, Paris, 1979

1979) introduced the new technical term to an international audience at the *International Conference on Soil Reinforcement* in Paris, 1979. Fig. 1 shows the very first sketch of the construction method of soil nailing in its original form from the above mentioned publication.

It should also be mentioned that in the USA a simul-taneous, but independent development of insitu reinforced earth took place in the years from 1976 to 1981, which was called "lateral earth support system" (Shen et al., 1981a). However, today *soil nailing* is also the most common term employed in America.

In France the national project "Clouterre" (clou = *nail*, terre = *soil*) was started in 1986, about ten years after the German research project "Bodenvernagelung" had been initiated. The first results of *Clouterre* were reported to an international forum in 1989 (Plumelle, 1989).

3.2 *The German research and developement project soil nailing (1975-1980)*

As mentioned above, the German research and development project "Bodenvernagelung" started in 1975. The special contractor Bauer GmbH & Co KG and the Institute of Soil Mechanics and Rock Mechanics of the University Karlsruhe in joint venture established a five year research and development project. This research programme covered the following items:

(a) Theoretical stability analyses to study the relevant failure mechanisms of nailed cuttings.

(b) Model tests to study the behaviour of nailed cuttings or nailed walls at limit equilibrium.

(c) Execution of seven full scale nailed walls to develop drilling and construction techniques in cohesive and non-cohesive soils.

(d) Instrumentation of these test walls to observe behaviour during construction, under safe loads and at limit equilibrium.

Concerning item (a), the results of the theoretical stability analyses were presented briefly in the above mentioned contribution in Paris, 1979, and in detail in a German doctoral thesis (Gässler, 1987). The reader is also referred to a comprehensive paper on theory and practical design of soil nailing, presented by the author at IS Kyushu 88 (Gässler, 1988).

The author's philosophy, followed in all his publications from the very beginning, was to design nailed cuttings against the occurrence of a limit state. In the Paris, 1979, paper circular surfaces, two-part wedge mechanisms and the Coulomb wedge were anticipated as potential failure mechanisms for nailed walls (Fig. 2), and a procedure was

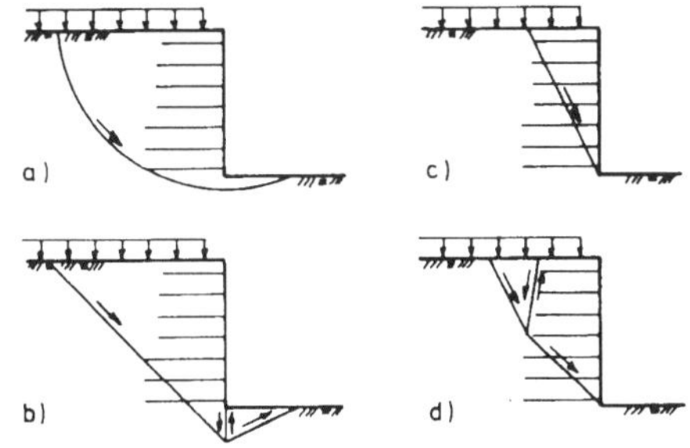

Fig. 2 Potential failure mechanisms of nailed walls (adapted from *Soil nailing*, Stocker, Körber, Gässler & Gudehus, 1979)

presented for the determination of the most critical failure surfaces of a two-part wedge mechanism (Fig. 3).

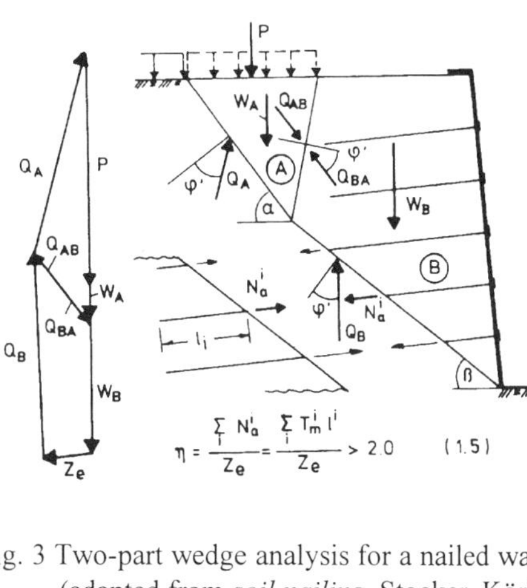

Fig. 3 Two-part wedge analysis for a nailed wall (adapted from *soil nailing*, Stocker, Körber, Gässler & Gudehus, 1979)

Since in the seventies and eighties overall factors of safety were common in geotechnical engineering in nearly all countries, the author used the overall factor η applied to the nail forces at that time (see Fig. 3). This factor $\eta = 2.0$ was estimated to provide for a safe distance from the occurence of failure.

Referring to (b), model tests (scale 1:25) on nailed walls were first presented in Paris, 1979 (Stocker et al.), and then in 1983 (Gässler & Gudehus, 1983). In all tests the observed failure mode was in agreement with theoretical predictions. This has been documented by photographs and by analyses published in the doctoral thesis mentioned above.

Referring to (c) and (d), one can state that all seven fully instrumented test walls up to 6.9 m height yielded many results of theoretical and practical importance. As it is not the object of this paper to deal with the experimental results in detail, the reader is referred to Gässler (1992a & 1993)

Additionally, in the beginning of the eighties the author studied the statistic-probabilistical safety theory after Hasofer & Lind (1974) to deduce partial safety factors for the design of nailed walls following the Eurocode 7, which was in drafting stage at that time. Thus in 1983, for the first time, the set of partial safety factors in Table 1, applied to characteristic values (see ch. 4.3), was proposed to assure a certain safety level (probability of failure $p_f \approx 10^{-6}$) for nailed slopes and walls (Gässler, 1982; Gässler & Gudehus, 1983).

Table 1: Partial safety factors proposed by Gässler (1982), and Gässler & Gudehus (1983)

	Partial safety factors
Nature of forces	
Permanent forces:	
Soil unit weight	$\gamma_\gamma = 1.0$
Variable forces:	
Live loads	$\gamma_Q = 1.1$
Material properties:	
Soil friction	$\gamma_\varphi = 1.1$
Effective cohesion	$\gamma_c = 1.4$
Soil-nail-interaction	$\gamma_N = 1.3$

3.3 *German licences for soil nailing (1984)*

Presenting a new technology soil nailing was not acknowledged as a so-called "common construction method" at its very beginning in Germany. Therefore, until 1983, special permission for each soil nailing project was required by the construction administration of the communities. In all cases this permission was given, as the reliability of design and performance could be proved by the model and field tests of the research programme *soil nailing* and, last but not least, by various well instrumented practical projects.

As the special contractor Bauer had pioneered soil nailing in Germany, the first general licence for soil nailing was issued to the contractor Bauer by the German Federal Institute for Construction Techniques, Berlin, in 1984. Naturally, the design and calculation methods obtained by the research project *soil nailing* were applied in this general licence of *Soil Nailing System "Bauer"*, which is still valid at present. The following items of the *Bauer-licence* shall be given:

(1) General requirements of application
(2) Field of application
(3) Soil exploration
(4) Requirements for the construction of temporary nails and permanent nails and for the facings
(5) Requirements for predrilling the boreholes and installing the nails
(6) Check of internal and external stability
(7) Pull-out tests of nails
(8) Quality control (construction supervision)
(9) Monitoring under service

In contrast to anchorage licences dealing with the construction and installation of single anchors, the licence for soil nailing covers both the nail individually and the nail-reinforced soil system as a whole, including the design and calculation procedure. This is due to the particular history of development of soil nailing: not only the nail as such, but the construction method was entirely new.

Returning to item (6) *check of internal and external stability*, it is interesting to mention that all checks for external stability (e.g., circular surfaces), and internal stability (e,g., two-part wedge mechanism, see Fig. 4) of the system as a whole, as well as the checks for the nail as an element against bond failure with the ground (external stability of a nail) or rupture of the steel (internal stability of a nail) are to be undertaken in limit state using overall safety factors.

The licence of *Soil Nailing System "Bauer"* was the first "code of practice" in the history of soil nailing. However, as this licence has never been

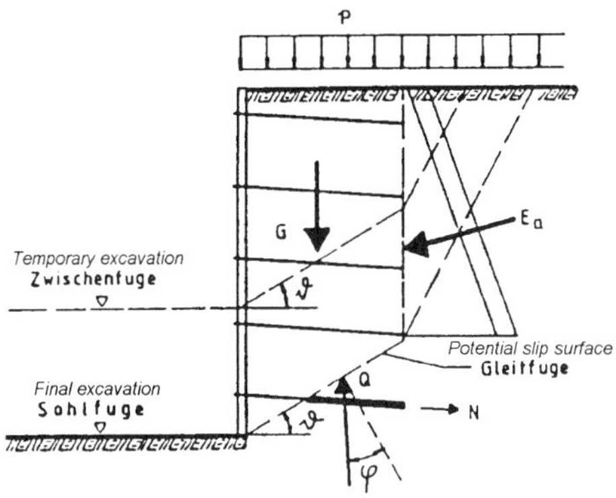

Fig. 4 Two-part wedge analysis for internal stability of a near vertical nailed wall (adapted from the German general licence of *Soil Nailing System "Bauer"*, 1984)

translated into English, no one outside Germany, Austria and Switzerland has taken any notice of it. In the meantime licences for several other German special contractors followed, so that soil nailing became definitely popular as an advantageous and reliable construction method in Germany (Gässler, 1989 & 1995). These licences helped to ensure a high standard of execution of soil nailing.

It should be remarked that as soil nailing has been a steadily developing method in both construction and design in the eighties, it would not have been advisable to produce a German Standard (DIN) for soil nailing. This would have certainly restricted its progress in both practice and theory.

3.4 *French soil nailing recommendations (1991)*

The Soil Nailing Recommendations (original French title: *Recommandation Clouterre 1991*) represent the five years of research, studies and tests (including three field tests on walls, 6 m and 7 m high) of the French national project *Clouterre*, which was conducted from 1986 to 1990. It summarizes the whole design and construction process, from geotechnical investigation to field quality control. In the *Recommendations 91* the reader will find many references to French publications and documents, but relatively few references to research and practical experience outside France. Many important results (e.g. the limit state as optimal design approach, the safety relevant failure mechanisms, or the observed displacements on the top of nail facings) are presented without any background information on results yielded by the research programme *Bodenvernagelung* ten years before.

Nevertheless, the French *Soil Nailing Recommendations 91* represent a remarkable compilation from the studies on nailed walls and slopes carried out by a large group of contracting authorities, prime contractors, research centres and laboratories, consulting firms, and construction companies. It is a very extensive work, from which only the main chapters can be cited here as follows:

(1) The technique used for soil nailed structures: Description and developements
(2) Soil nailing in retaining structures: Mechanisms and behaviour
(3) Conception and design
(4) Investigation and tests
(5) Wall structures construction
(6) Durability of structures
(7) Specifications and inspections

Table 2: Partial safety factors applied to characteristic values, recommended in the *Recommandation Clouterre 1991* (excerpt)

	Partial safety factors
Nature of forces	
Permanent forces:	
Soil unit weight	
de-stabilizing force	$\Gamma_{S1} = 1.05$
stabilizing force	$\Gamma'_{S1} = 0.95$
Other *permanent* forces	
unfavourable forces	$\Gamma_{S1} = 1.20$
favourable forces	$\Gamma'_{S1} = 0.90$
Variable forces:	
Live loads	$\Gamma_{Q} = 1.33$
Material properties:	
Soil:	
tangent of the friction angle	$\Gamma_{m,\varphi}{}^{1} = 1.20$
effective cohesion	$\Gamma_{m,c} = 1.50$
undrained cohesion	$\Gamma_{m,cu} = 1.30$
Mild steel	$\Gamma_{m,\sigma c} = 1.15$
Soil-nail-interaction (tests)	$\Gamma_{m,qs} = 1.40$

With respect to the subject of this lecture, chapter (3) is the most interesting. For the justification of the stability of the soil nailed wall at ultimate limit state, the reader will find there a set of partial safety factors for loads and for material properties, i.e., soil, steel and soil-nail interaction. Table 2 gives an excerpt of the main values from the *Recommandation Clouterre 1991.*

Whereas in early French publications on soil nailing (e. g., Schlosser, 1982) the limit state design with one overall factor had been proposed, the French *Soil Nailing Recommendations 91* follow the safety concept of Eurocode 7 in principle. However, there is no derivation or any explanation of where the values of the various partial safety factors were taken from.

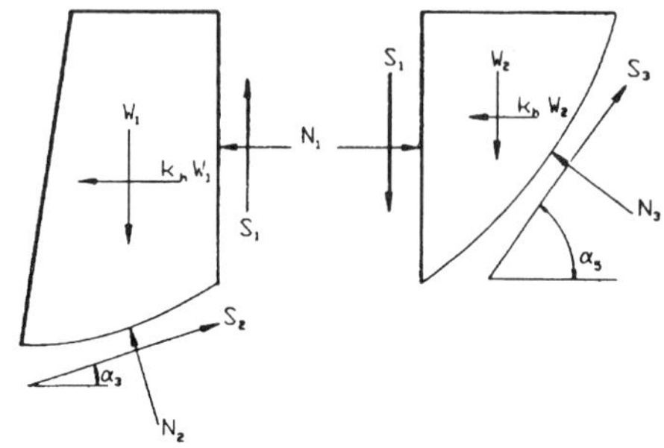

Fig. 5: Design method for nailed walls in the USA after Bang et al. (1992)

3.5 *American research and developement projects*

At the University of California at Davis, a limit state design method was developed in the late seventies, absolutely independently from European activities in soil nailing (Shen et al. 1981b).

The potential failure line is assumed to be a parabola, passing either only through the nailed zone, or through both the nailed zone and the natural soil in the rear. In the latter case, two blocks are separated by a vertical line through the lower end of the nails, where the intersection forces have to be estimated. The global factor FS of safety is calculated by comparing the component of the total resisting force along the direction of driving force with the magnitude of total driving force (Bang, Kroetch & Shen, 1992). Using the same global factor for both nails and soil, the minimum safety factor determines the most critical parabola (Fig.5). For permanent structures, a minimum global factor of 1.5 is proposed.

However, in the author's opinion, it is not expedient in homogeneous soils to seperate the potentially sliding earth block into two blocks, if there is no necessity for it. Therefore, the German design method on the basis of *one* rigid body in rotation (circular slip surface) avoids intersection lines with intersection forces that cannot be determined without questionable assumptions (Gässler, 1988).

Another quite different design method after Juran (1990) has to be mentioned which divides the nail-reinforced block into slices *parallel* to the nails. The potential failure line is assumed to be a logarithmic spiral.

Despite of the fact that this lecture deals with design of nailed slopes, a recent practical recommendation for monitoring performance of nailed structures should be mentioned. It is the *FHWA Soil Nailing Field Inspector's Manual*, edited by the Federal Highway Administration of the U.S. Department of Transportation in 1994. In the author's opinion, it represents the best manual of its kind for soil nailing practice in the world at present.

3.6 *British Standard BS 8006 (1996)*

This new British Standard containes guidelines and recommendations for the application of soil reinforcement techniques to soils, as fill or in situ. In the scope of this lecture it is not possible to report on this extensive standard in detail. However, two main principles of this standard should be noted:

- it is written in a limit state format, and
- guidelines are provided of safety margins in terms of partial load factors and material factors.

The major part of the BS 8006 (1996) deals with reinforced masses of fill using metallic strips or polymeric strips, sheets or grids. However, soil nailing and in situ reinforcing is considered as well, and background information is provided by a substantial bibliography to help the user gain a deeper understanding.

One should also mention the new DOT Advice Note, HA 68/94, which forms part of the Design Manual for Roads and Bridges in the U.K. The Document, entitled *Design methods for the reinforcement of highway slopes by reinforced soil and soil nailing techniques*, is based on two-part wedge analysis, where friction is introduced along the interwedge boundary (Love, 1995). This procedure is in conformity with BS 8006 and will be discussed later.

3.7 *Design methods in Japan*

The author regrets his lack of information on design procedures for soil nailing in Japan. What can be seen from international publications by Japanese authors is that a limit state format with an overall factor F_S seems to be more common (e.g., Maramatsu et al., 1992; Taga et al., 1992; Teramoto et al., 1992; Tsubouchi et al., 1992) than with partial safety factors.

4. THE AUTHOR'S DESIGN METHOD USING PARTIAL SAFETY FACTORS

4.1 Method of kinematics of rigid earth blocks

The kind of plastic limit state, which is reached by earth retaining structures after sufficiently large deformations, depends on the statical and kinematical boundary conditions. It was assumed for nailed walls and slopes that, in the nailed zone as well as in the unnailed zone of the soil, the plastic shear deformations are located in thin shear planes, whereas the greater part of the soil remains rigid in the limit state. Thus, the failure model of nailed walls was based on the kinematics of rigid earth blocks containing two principles (Gudehus, 1981):

(1) Principle of the kinematic compatibility of the failure mechanism

This means the displacements of rigid earth blocks have to be correctly described by a hodograph.

(2) Principle of the minimum of safety by means of the variation of slip planes.

This means, one has to vary the inclinations (and radii) of the slip planes of a potential failure mechanism (e.g., translation mechanism or rotation mechanism) until the most unstable configuration of the slip planes is found.

As a provisional safety definition

$$\eta_N = \frac{Z_a}{Z_g} \qquad (1)$$

can be used. Herein Z_a denotes the available axial nail forces (determined from pull-out tests) and Z_g the axial nail forces in equilibrium or limit state. The principle of the minimum of safety states that failure will occur with that configuration of slip planes, which gives a value of 1 in the safety definition (1). This is analogous to the well known principle of Coulomb's earth pressure theory (c.f. Gudehus, 1981).

Using the regular cross section of a nailed wall, the safety of different potential failure mechanisms consisting of rigid blocks was investigated and compared for various boundary conditions. The procedure systematically develops in three steps:

1. Compiling of potential modes of failure mechanisms,
2. Variation of the slip planes to find the least safe configuration for each failure mode,
3. Determination of the absolutely least safe failure mechanism by means of comparing the least safe configurations of the different failure modes.

The investigated failure modes included:
- translation of a rigid block (simple wedge, TRA-I)
- translation of two rigid blocks (two-part wedge, TRA-II)
- rotation of one rigid block (slip circle, ROT-I)
- rotation of two rigid blocks (ROT-II).

In its simplest form the method of kinematics of rigid earth blocks can be best described by means of example as follows:

In Fig. 6a a cross section of a near vertical slope is shown. The soil has an angle of internal friction $\varphi' = 30°$, cohesion $c' = 7.5$ kN/m², and unit soil weight 20 kN/m³. The vertical nail spacing s_v is 1.1 m, and the horizontal spacing s_h is 1.25 m. As mean value, the *available* nail shear force $T_{m,a} = 30$ kN/m has been anticipated by in situ pull-out tests of several nails.

The assumed failure mechanism is a slip circle (or in other words: a rotation mechanism of one rigid block), determined by the chord inclination angle $\vartheta = 50°$ and the radius $r = 1.5h$. This slip circle is arbitrary and not yet the unsafest configuration. The external and internal forces acting on the sliding earth block are shown in Fig 6b.

The internal force N_i of a nail in one of the lower rows i, which is intersected by the slip circle, is determined by the nearly constant mean shear force T_m (N.B., this shear force acts *parallel* to the nail axis) and the section of the nail beyond the slip plane l_i:

$$N_i = T_m \cdot l_i \qquad (2)$$

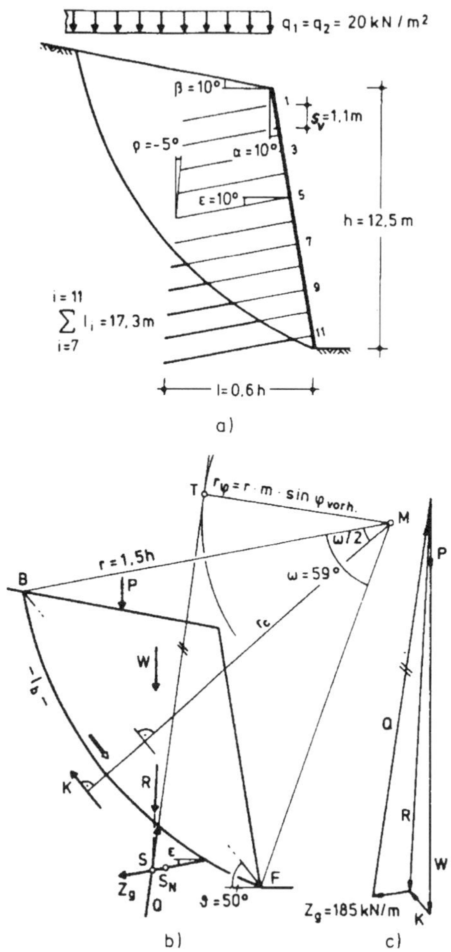

Fig. 6 Stability calculation of a nailed wall: a) cross section with slip circle, b) free body diagram, c) force polygon

The sum of the axial forces, referring to a unit width of the wall, can be expressed by:

$$Z_g = 1/s_h \cdot \sum_{i=j}^{i=n} N_i = T_{m,g}/s_h \cdot \sum_{i=j}^{i=n} l_i \qquad (3)$$

Herein j denotes the upmost row, and n the lowest row that is intersected by the slip plane.

Z_g is located in the centre of gravity S_N of the nailed zone beyond the slip circle. The forces R and Z_g are balanced in point S by the resultant slip circle force Q, the latter determined with the friction circle assumption after Krey (1932). Completing the force polygon in Fig. 6c yields the sum of the axial nail forces in the limit state Z_g = 185 kN/m. From Equ.(3) follows $T_{m,g} = 185 \cdot 1.25/17.3 = 13.4$ kN/m.

The normalized term, equivalent to $T_{m,g}$, is the specific nailing density μ_g, which is defined as follows:

$$\mu_g = T_{m,g} / (\gamma \cdot s_v \cdot s_h) \qquad (4)$$

where γ = unit weight of the soil [kN/m³]
s_v = vertikal nail spacing [m]
s_h = horizontal nail spacing [m]

As with $T_{m,g}$, the index g in μ_g denotes the equilibrium state. The specific nailing density was first introduced by Gässler & Gudehus (1981) for use in design charts (see ch. 6). Thus, one obtains from Equ.(4) $\mu_g = 13.4/(20 \cdot 1.1 \cdot 1.25) = 0.487$.

Instead of the safety factor η_N (see Equ.(1)), it is now more expedient to use the equivalent safety definitions:

$$\eta_T = \frac{T_{m,a}}{T_{m,g}} \quad \text{or} \quad \eta_\mu = \frac{\mu_a}{\mu_g} \qquad (5), (6)$$

which can be obtained from Equ.(1), (3), and (5).

Consistently one yields the same minimal safety factors η_N, η_T, or η_μ for the slip circle in Fig. 6:

$\eta_T = 30/13.4 = 2.24$, or $\eta_\mu = 1.091/0.487 = 2.24$

As second step, one has to vary the parameters ϑ and r of the slip circle, until the *maximum* of $T_{m,g}$, or μ_g, is found. By means of a computer program, the solution of the arithmetic variation is found:

$\mu_{g,max} = 0.636$ with $\vartheta^* = 49.6°$ and $r^* = 3.4 \cdot h$.

The equivalent *minimum* safety factor is then:

$$\eta_{\mu,min} = \mu_a/\mu_{g,max} = 1.091/0.636 = 1.72$$

It is interesting to mention here, that, for the given wall geometry and soil parameters, practically the same result will be obtained on the basis of the two-part wedge mechanism (TRΛ II).

Emphasis is placed here on the very important hypothesis for the validity of limit state design, namely that the maximal resistance of the soil and of the nails in the various rows is simultaneously mobilized at failure. For vertical or near to the vertical slopes this hypothesis is only valid for the tensile resistance (in other words: the pulling-out resistance) of the nails, but *not* for shear forces due to bending. In sandy or clayey soils, the latter are mobilized too late. This was measured by instrumented nails during failure of several full scale test walls carried out in the research project *Bodenvernagelung* (Gässler 1987, 1990a & 1993). Maybe, at *post* failure the bending resistance of the nails can prevent nailed walls from undergoing total collapse. However, this is not relevant for the conditions immediately antecedent to failure. As a conclusion from all the measurements and analyses of several full scale tests, the shear forces of nails of diameters of about 20 to 28 mm are of second order. This is opposite to the French recommendations *Clouterre*, but in agreement with measurements in large shear boxes, carried out by Pedley, M.J., Jewell, R.A. & Milligan G.W.E. 1990.

Therefore, the calculation method shown in Fig. 6 is based only on the *axial* nail forces. The shear forces are not ignored in order to obtain the most simple calculation method; they are ignored for safety reasons. This results in a slightly conservative, but very expedient design procedure.

4.2 *General rules for finding the unsafest failure mechanism*

The third step should now be made to find the least safe failure mechanism in the calculation example of Fig. 6. In principle, there are four possible failure modes to check (see ch. 4.1). The least safe mechanism is the one that leads to the *absolute* maximum $T_{m,g,max}$, or, $\mu_{g,max}$. For various soil properties and boundary conditions the absolutely maximum specific nail densities from the four potential failure mechanisms have been analysed.

Considering the calculation example of Fig. 6, the result is that the ROT-I-mechanism (slip circle) and the TRA-II-mechanism are similarly unsafe, and that both ROT-I and TRA-II are less safe than TRA-I (simple wedge) (cf. Gässler, 1988).

In general, the results of numerous calculations performed by the author (Gässler, 1987) can be summarized into the following rules for practical use:

- In *cohesionless* soils the slip circle and two-part wedge are the two most critical mechanisms for vertical or near vertical nailed walls (the slight difference is here only of academic interest). For less steep walls and/ or longer nails in the upper rows, the slip circle is the least safe mechanism. Especially for high loads beyond the reinforced zone, the two-part wedge is the least safe mechanism.

- In soils with *little cohesion*, both the slip circle and the two-part wedge are, as far as safety is concerned, nearly equivalent for steeper walls.

- In soils with *medium or high cohesion* the slip circle distinctly is the least safe failure mode.

The rotation of two rigid blocks may occur in the case of extremely high surface load beyond the reinforced zone. This has been observed in a model test with sand (Gässler & Gudehus, 1983). Rotation of two rigid earth blocks has also been found as the least safe failure mode for seismic loading (Tufenkjian & Vucetic, 1992). The principal graphic solution for this type of a combined failure mechanism is given in the discussion leader's report *slopes and excavations* of the IS Kyushu '92 (Gässler, 1992b). More background information on this most interesting failure mode is given by Gässler (1987).

4.3 *Deduction of partial safety factors on the basis of the statistic-probabilistic safety theory*

Instead of only one global factor η_N (see Equ. (1)), or η_μ (see Equ. (6)), it is proposed to make use of various partial safety factors. For the derivation of partial safety factors a limit state equation is necessary. For this a limit state equation has been formulated based on the two-part wedge mechanism (Gässler & Gudehus, 1983).

In the limit state equations for every failure mode of nailed slopes, the magnitudes q (e. g., live surface load, dimension kN/m²), φ (friction angle, dimensionless), cohesion c (dimension kN/m²), and T_m (mean nail shear force, dimension kN/m) are scattering, i. e. they are basic variables in the sense of the new statistic-probabalistic safety theory in EC 7. Concerning the statistical distribution of the basic variables, assumptions have to be made, as the available quantity of statistic geotechnical data is very poor at present.

Nevertheless, partial safety factors can be developed on the basis of numerous so-called Level II approach calculations after Hasofer & Lind (1974). The Level II approach applied on soil nailing is appoximately described by Gässler and Gudehus (1983).

Readers interested in the principal procedures of such calculations, are referred to the structural diagram of the computer program that has been developed by the author to derive partial safety factors (Gässler, 1987). As result the following set of factors was achieved, providing a sufficiently high and homogeneous safety level for steep nailed walls (safety index $\beta \approx 4.7$, or in other words: probability of failure $p_f \approx 10^{-6}$):

$$\gamma_Q = \frac{q_d}{q_m} = 1.3 \qquad (7)$$

with q_m : mean value of surcharge q,

$$\gamma_\varphi = \frac{\varphi'_k}{\varphi'_d} = 1.1 \qquad (8)$$

with φ'_k: characteristic value of φ' (=10%-fractile of a log normal distribution, truncated at 20°),

$$\gamma_T = \frac{T_{m,k}}{T_{m,d}} = 1.3 \qquad (9)$$

with $T_{m,k}$: characteristic value of T_m (= 10%-fractile of a log normal distribution).

The index "d" denotes the so-called design point, i. e. the assumed limit state equlibrium with the most unfavourable combination of all basic variables. The values of the partial safety factors in (7), (8) and (9) have already been compiled in Table 1 (see ch. 3.2).

In the Level II approach calculations, it has been found by the author that the scattering of unit soil weight γ[kN/m³] is small compared to the shear parameters φ' or c, so that the safety level is not as much influenced by γ as by φ' or c. Therefore, very early, the author has proposed the partial safety factor $\gamma_G = 1.00$ for soil weight as a permanent load (Gässler & Gudehus, 1983). At this point, it does not matter wether the soil weight is an unfavourable or favourable load. As soil nailed walls are not embedded in the ground, soil weight mostly acts unfavourably. The convention to set $\gamma_G = 1.00$ is very expedient and eases conventional and computer calculation.

However, in the meantime, the partial safety values given in Equ. (7), (8) and (9) are obsolete. In future, the partial safety factors for loads will be prescribed by the Eurocode 1, and the partial safety factors for the material properties of the ground, say φ', c', T_m, will be recommended by Eurocode 7.

5. THE NEW EUROPEAN CODES (EC 1, EC 7)

It is not the object of this lecture to present the content of both codes in detail. The lecture only aims to give a brief overview and to set out what is important for the design of nailed slopes.

5.1 *Eurocode 1, Part 1 (ENV 1991-1:1994)*

The EC 1-1 describes the principles and requirements for safety, serviceability and durability of structures. It is based on the limit state concept used in conjunction with a partial factor method. For the design of new structures, the EC 1 is intended to be used for direct application, together with the design Eurocodes (EC 2 to EC 9; cf. ch. 2.2). The EC 1 is divided into a main text and a series of annexes. The main text includes the principles and most of the application rules necessary for direct application for designs in the field. The annexes are informative only. In order to give an idea of the content of the main text, the titles of the main sections are given as follows:

1 General
2 Reqirements
3 Limit states
4 Actions and environmental influences
5 Material properties
6 Geometrical data
7 Modelling for structural analysis and resistance
8 Design assisted by testing
9 Verification by the partial factor method.

The principles comprise (cf. EC 1-1 sec. 1.4):

(a) general statements and definitions for which there is no alternative

(b) requirements and analytical models for which no alternative is permitted unless specifically stated.

The application rules are generally recognized rules which follow the principles and satisfy their requirements. It is permissible to use alternative rules to the application rules given in the Eurocode.

In sec. 3 of EC 1-1 it is formulated, as a principle, how limit state design shall be carried out, namely by:

- setting up structural and load models for relevant ultimate and serviceability limit states to be considered in the various design situations and load cases;
- verifying that the limit states are not exceeded when *design* values for actions, material properties and geometrical data are used in the models.

Design values are generally obtained by using the *characteristic* values in combination with partial safety factors.

A characteristic value F_k of an *action* is the principal representative value of an action. If this characteristic value can be fixed on a statistical basis, it is chosen so as to correspond to a prescribed probability of not being exceeded on the unfavourable side during a reference period, e.g., design working life (EC 1-1, sec. 1.5.3.14)

A characteristic value X_k of a *material property* (or *resistant* force) is a value having a prescribed probability of not being attained in a test series. This value generally corresponds to a specified fractile of the statistical distribution of the particular property of the material (EC 1-1, sec. 1.5.4.1).

The design value F_d of an action is expressed in general terms as:

$$F_d = \gamma_F \cdot F_k$$

with γ_F: partial factor for actions (see Table 3)

The design value X_d of a material property (e.g., soil friction or yield point of steel) is generally defined as:

$$X_d = X_k / \gamma_M$$

with γ_M: partial factor for material property, given in EC 2 to EC 9 (for ground properties see Table 4 in ch. 5.2 of this lecture).

Table 3: Partial safety factors for actions in limit states (excerpt from EC 1-1, Table 9.2)

	Partial safety factors
Case B (Failure of structure or structural element governed by strength of material)	
Permanent actions	
- unfavourable	γ_{Gsup} = [1.35]
- favourable	γ_{Ginf} = [1.00]
Variable actions	
- unfavourable	γ_Q = [1.50]
Case C (Failure in the ground)	
Permanent actions	
- unfavourable	γ_{Gsup} = [1.00]
- favourable	γ_{Ginf} = [1.00]
Variable actions	
- unfavourable	γ_Q = [1.30]

One has to state that in the currently available set of Eurocodes, the partial safety factors are partly based on probabalistic considerations (very few on Level II approach calculations; cf. ch. 4.3), partly on a historical or empirical derivation (the latter seems to be predominant in geotechnical design).

Concerning partial factor for actions, one has to differentiate three cases of ultimate limit states:

Case 1: Loss of static equilibrium of a structure as a whole (e. g., hydrostatic uplift)

Case 2: Failure of structure or structural elements, including those of the footing, piles, basement walls etc., governed by the strength of structural material

Case 3: Failure in the ground

As can be seen in Table 3, the partial safety factors are different in these cases. (As Case 1 is less relevant for soil nailed structures, the respective partial factors are not given in the excerpted Table 3.) There are good reasons for dividing limit states into three cases. The main reason for introducing Case 3 is the fact that the partial safety factor for permanent unfavourable actions $\gamma_{Gsup} = 1.35$, used in structural engineering (EC 2, EC 3), is absolutely unacceptable in geotechnical design. The factor 1.35 is too high and not justified, because the unit weight of soil does not scatter to this extent. Hence, it would not make any sense to apply this factor 1.35 to the unit soil weight (in other words: to the dead weight of destabilizing earth blocks).

The essential point of the new safety concept in the ECs is the following (cf. EC 1-1, sec. 9.4):
When considering a limit state of static equilibrium of the structure as a rigid body, it shall be verified that:

$$E_{d,dst} \leq E_{d,stb}$$

where: $E_{d,dst}$: design value of the total effect of destabilizing actions (or forces)
$E_{d,stb}$: design value of the total effect of stabilizing (or resistant) forces.

This equation is a symbolic expression, which is to be replaced by an interaction formula or limit state equation.

When considering a limit state of rupture or excessive deformation of a section, member or connection, it shall be verified that:

$$E_d \leq R_d$$

where: E_d: design value of the effect of actions, such as internal force or moment
R_d: design value of the resistance associating all relevant structural properties.

5.2 Eurocode 7, Part 1 (ENV 1997-1:1994)

This prestandard applies to the geotechnical aspects of the design of buildings and civil engineering works. It is subdivided into the following sections:

1 General
2 Basis of geotechnical design
3 Geotechnical data
4 Supervision of construction and maintenance
5 Fill, ground improvement and reinforcement
6 Spread foundations
7 Pile foundations
8 Retaining structures
9 Embankments and slopes.

This prestandard is intended for experimental practical application, and this is why certain safety elements in this code have been assigned indicative values which are identified by [ ]. Thus, the partial safety factors in Table 4 are given as so-called boxed values. The authorities in each member country of the European Community are expected to assign definitive values to these safety elements.

Hence each member state has issued a National Application Document (NAD) giving definitive values for safety elements and referencing compatible supporting standards.

The sections 2 and 8 are the most important for the design examples in the next chapter. However, this lecture can only refer to one essential point, namely the selection of characteristic values of soils. EC 7 says that this selection shall take account the geological and other background information (data from previous projects), the variabilities of the property values and the extent of the zone of the ground governing the behaviour of the geotechnical structure at limit state. This is rather general and it seems that, as with the conventional *calculation* value, the assessment of the *characteristic* value will depend on individual engineering judgement and less on statistics.

Table 4: Partial safety factors for ground properties (excerpt from EC 7-1, Table 2.1)

Case C Failure in the ground)	Partial safety factors
Ground properties:	
Soil friction $\tan\varphi'$	γ_φ = [1.25]
effective cohesion c'	γ_c = [1.60]
undrained cohesion cu	γ_{cu} = [1.40]

6. PRACTICAL DESIGN AFTER EC 7

6.1 *Design example No. 1*

Consider a 10 m high, near vertical slope of an urban excavation pit in sand. The groundwater table is situated very deep beneath the bottom of the excavation. The owner of the neighbouring property will not allow any construction elements to be placed in his ground. Under these conditions a nailed wall will undoubtedly represent an expedient solution. (Note that an anchored wall would require too long anchor rods!)

Thus, a nailed wall could be designed as shown in Fig. 7.

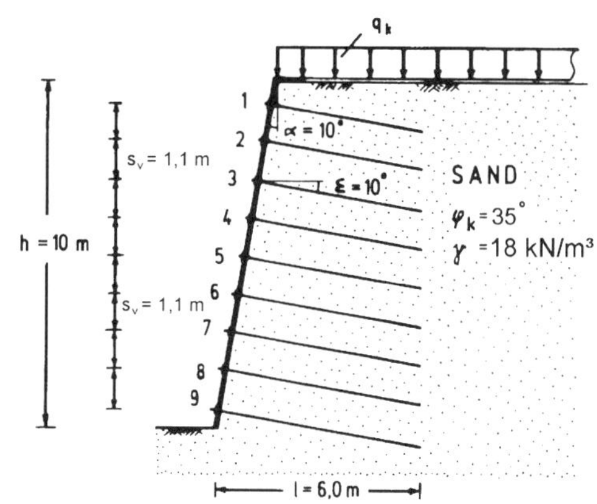

Fig. 7 Design example for a near vertical nailed wall in sand charged with a live load

The following data are given:

characteristic soil parameters: $\varphi'_k = 35°$
$\gamma_k = 18$ kN/m³
characteristic live load: $q_k = 23$ kN/m²
characteristic value of pull-out resistance of a set of test nails: $T^{\#}_{m,k} = 30$ kN/m
(The symbol # detotes $T_{m,k}$ as a test result)

predesigned wall dimensions:
wall inclination $\alpha = 10°$
nail length at the bottom: $l = 6{,}0$ m
vertical nail spacing: $s_v = 1.1$ m
line through lower nail ends: vertical

The required *horizontal* nail spacing s_h [m] and the required diameter of the steel bar in the nails are to be determined for the nailed wall in complete state.

1st Calculation step: Derivation of design values:

Design values of soil parameters:
$\varphi'_d = \arctan(\tan\varphi'_k/\gamma_\varphi) = \arctan(\tan 35°/1.25) = 25°$

Comment: For limit state case 1C: $\gamma_\varphi = 1.25$ (see Table 2.1 in EC 7-1, or Table 4 in ch. 5 of this lecture)
$\gamma_k = \gamma_d = 18$ kN/m³

Comment: For permanent actions in case 1C, the partial factors γ_{Gsup} and γ_{Ginf} are 1.00. (see Table 9.2 in EC 1-1, or Table 3 in ch. 5 of this lecture). Hence, for unit soil weight, the design value is identical with the characteristic value.

Design value of load:
$q_d = \gamma_Q \cdot q_k = 1.30 \cdot 23 = 30$ kN/m²

Comment: In case 1C, $\gamma_Q = 1.30$ for unfavourable variable actions (see Table 9.2 in EC 1-1)

2nd Calculation step: Failure mode

Now, one has to select the failure mechanism which the calculation should be based on. In this example the two-part wedge mechanism (TRA-II) will be the basis of the limit state design.

Comment: Considering the boundary conditions and soil parameters, a solution based on the slip circle (see Fig. 6) would be practically equivalent and yield the same result (see also ch. 4.2)

3rd Calculation step: Identifying the unsafest two-part wedge mechanism

One has to identify the two-part mechanism which requires the maximum reinforcement force, i. e. the maximum mean shear force per nail meter $T_{m,max}$. This can be found by searching systematically, varying the inclination ϑ of the slip surface of the reinforced earth block.

Comment: Opposite to DOT Advice Note, HA 68/94, U. K. (see ch. 3.6), it is not advisable to vary any other parameters. It is justified by model tests (Gässler & Gudehus, 1983) that the interwedge boundary coincides with the line of lower nail ends, if the nails are not too long, say l/h < 0.7. Thus, for soil nailing, it is not recommended by the author to assume an interwedge boundary within the reinforced zone.

In this case it is expedient to start the variation of the slip surface with $\vartheta^{(1)} = 35°$ and $\Delta\vartheta = 5°$.

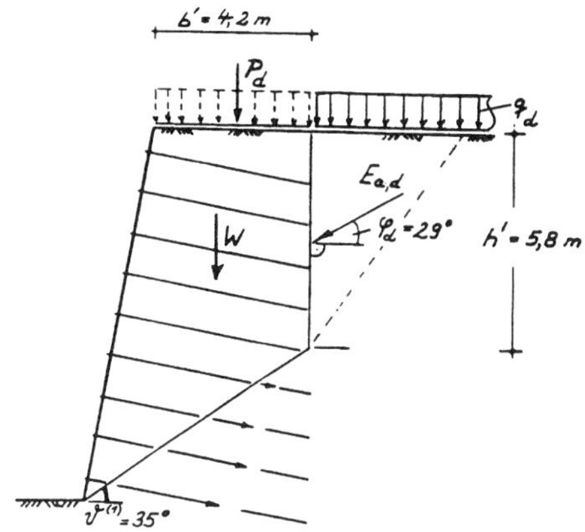

Fig. 8 Two-part wedge mechanism (not reinforced wedge replaced by active earth pressure $E_{a,d}$)

1st variation $\vartheta^{(1)} = 35°$ (see Fig. 8):
Driving forces:

$P_d = q_d \cdot b' = 30 \cdot 4.2 = 126$ kN/m
Comment: P_d is the resultant load on the reinforced wedge
$W_d = V \cdot \gamma_d = 38.55 \cdot 18 = 694$ kN/m
Comment: V is the Volume of the reinforced wedge per unit width of the wall
$E_{a,d} = \frac{1}{2} h' \cdot (\gamma_d \cdot h' + 2q_d) \cdot K_a(\varphi'_d, \delta_{a,d})$
Comment: In limit state case 1C the design ground properties γ_d, φ_d and the design live load q_d are used to calculate the design earth pressure force (see EC 7-1, ch. 2.4.2, (16)P).
The coefficient of active earth pressure K_a is also a function of the interwedge friction $\delta_{a,d}$, which should, in the author's opinion, be set at φ'_d. The *DOT Advice Note, HA 68/94, U. K.*, safely sets $\delta_{a,d}$ at ½ φ'_d. Of course, details of that kind are not dealt in the EC 7. The coefficient of active earth pressure K_a can be calculated using the well-known formula.

With $K_a(\varphi'_d, \delta_{a,d}) = 0.308$ the design value of the active earth pressure acting on the reinforced wedge is yielded as follows:
$E_{a,d} = \frac{1}{2} \cdot 5.8 \cdot (18 \cdot 5.8 + 2 \cdot 30) \cdot 0.308 = 147$ kN/m.

Resistant forces:

The free body diagram in Fig. 9a shows both driving forces and resistant forces acting on the nailed wedge.

The resultant nail force Z_d (= sum of the intersection nail forces from row No. 6 down to row No. 9) that is required for the equilibrium in the design state is obtained from the force polygon in Fig. 9b.

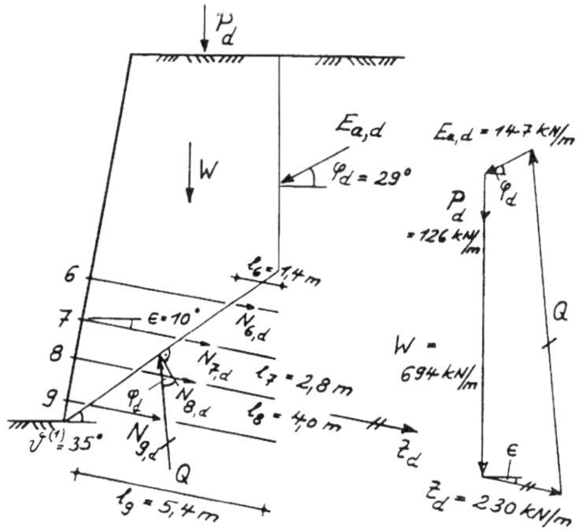

a) b)

Fig. 9 a) Free body diagram with $\vartheta^{(1)} = 35°$,
b) Force polygon

Referring to Equ. (3) in ch. 4.1, the resultant nail force Z_d = 230 kN/m is equal to the following expression:

$$Z_d = 1/s_h \cdot \sum_{i=6}^{i=9} N_{i,d} = T_{m,d}/s_h \cdot \sum_{i=6}^{i=9} l_i$$

with: s_h = 1.0 m (= unit width).

$$230 = T_{m,d} \cdot (1.4 + 2.8 + 4.0 + 5.4)$$

Now, one obtains the mean shear force per nail meter $T_{m,d}$ required for equilibrium at $\vartheta^{(1)} = 35°$:

$$T_{m,d} = 230/\ 13.6 = 16.9 \text{ kN/m/m}.$$

2nd variation $\vartheta^{(2)} = 40°$ (see Fig. 10):

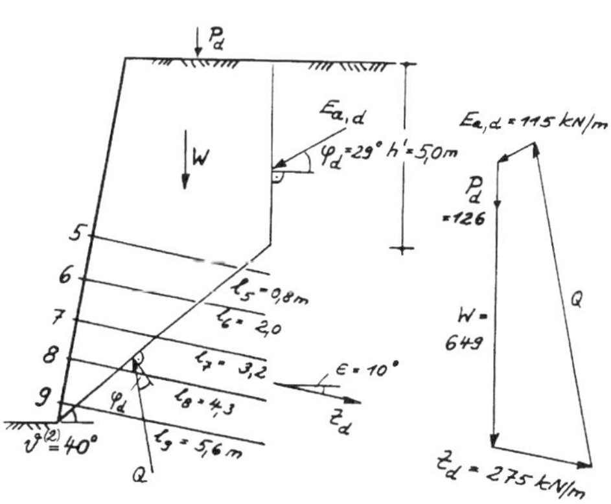

a) b)

Fig. 10 a) Free body diagram with fwith $\vartheta^{(2)} = 40°$,
b) Force polygon

3rd variation $\vartheta^{(3)} = 45°$ (see Fig. 11).

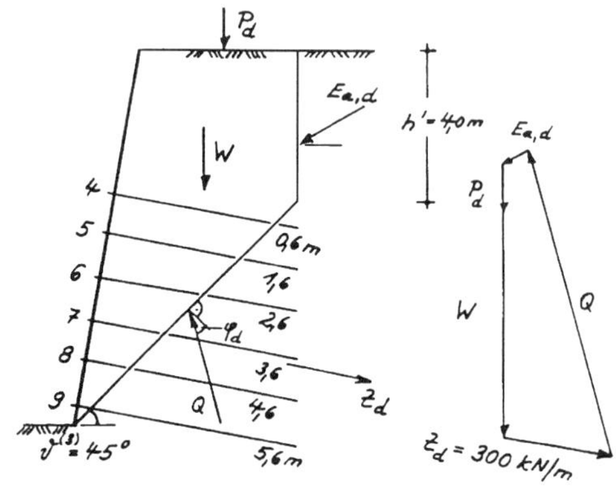

a) b)

Fig. 11 a) Free body diagram with $\vartheta^{(3)} = 45°$,
b) Force polygon

Now, the obtained results from variation are compiled in the following table:

No. of variation	ϑ [°]	Z_d [kN/m]	$\sum l_i$ [m]	$T_{m,d} = Z_d/\sum l_i$ [kN/m/m]
1	35	230	13.6	230/13.6=16.9
2	40	275	15.9	275/15.9=17.3
3	45	300	18.6	300/18.6=16.1

Thus, at $\vartheta^{(2)} = 40°$, the maximum value of the mean shear force per nail meter required for equilibrium in the least safe mechanism is found:

$$T_{m,d,max} = 17.3 \text{ kN/m/m}.$$

4th Calculation step: Determination of $T_{m,k,max}$

As the EC-7-1 does not contain a partial safety factor for nailed walls, the factor γ_N = 1.3 is recommended by the author. This factor is prescribed in the German DIN V 1054-100, which was issued in April, 1996, as a preliminary national standard (cf. ch. 5.2). The factor γ_N = 1.3 is slightly below the value recommended in the *Recommandation Clouterre 1991* (cf. Table 2 in ch. 3.4).
The characteristic mean shear force per nail meter is obtained using Equ. (9):

$$T_{m,k,max} = \gamma_N \cdot T_{m,d,max} = 1.3 \cdot 17.3 = 22.5 \text{ kN/m/m}$$

5th and final calculation step: Evaluation of the required *horizontal* nail spacing

Finally, one obtains the required *horizontal* nail spacing s_h [m] by the following equation:

$$s_h = \frac{T^{\#}_{m,k}\ [\text{kN/m}]}{T_{m,k,max}\ [\text{kN/m/m}]} = \frac{30}{22.5} = 1.35\ [\text{m}] \quad (10)$$

It should be mentioned here, that this $T_{m,d,max}$-method, used by the author since 1982, leads the designer to a unique and optimized solution. It is not a trial-by-error method based on a complete but guessed initial lay-out as many other methods available. In this sense, the British DOT Advice Note, HA 68/94, (Love, 1995) is very similar to the author's method.

The same solution, s_h = 1.35 m, can be easily obtained from design charts (see Appendix) using the specific nailing density μ. The design charts in the Appendix are valid for the limit equilibrium or design state. This means the input *and* output values are design values. All parameters in the design charts are normalized. The use of normalized, or dimensionless, values, say l/h (l: horizontal nail length at the bottom, h: slope height), or $q/(\gamma \cdot h)$ (q: load, γ: unit soil weight), helps considerably to reduce the number of required design charts.

The geometric parameters of the nailed wall, normalized nail length l/h = 0.6, inclination α = 10° and line through lower nail ends vertical (i.e. $\rho = 0$) lead the designer to the chart in the left column and second row (see Appendix).

With the design values, soil friction $\varphi'_d = 29°$, and normalized design load

$$q^*_d = q_d/(\gamma_d \cdot h) = 30/(18\ 10) = 0.17,$$

one obtains the design value of the specific nailing density μ_d = 0.87 from the selected chart. (The asterisk symbolizes q^*_d as a normalized magnitude). It is obvious that μ_d represents the maximum value required for equilibrium in the design state. Thus:

$$\mu_{d,max} \leftarrow \mu_d$$

Now, using the partial safety factor $\gamma_N = 1.3$ yields the respective characteristic value:

$$\mu_{k,max} = \gamma_N \cdot \mu_{d,max} = 1.3 \cdot 0.87 = 1.13$$

Finally, application of Equ. (4) leads to the required *horizontal* nail spacing s_h:

$$s_h = T^{\#}_{m,k}/(\gamma_d \cdot s_v \cdot \mu_{k,max}) = 30/(18 \cdot 1.1 \cdot 1.13) = 1.34\ \text{m}$$

One can see that the solution is identical to the one in Equ. (10). The difference is meaningless and comes from a slight inaccuracy due to the use of the design chart.

Design charts can be very expedient in practice. Very often a quick pre-design of a nailed slope project is needed to estimate the costs. A complete set of 36 design charts can be found in Gässler (1987) covering frequent geometric and static boundary conditions of nailed walls.

The next check required is the check of the steel bar in the nails. It is a check of a material property or dimension and is, according to EC 1, to associate with case 1B. The check shall fulfill the general requirement of all ECs (see ch. 5.1 of this lecture):

$$E_d \leq R_d$$

Generally, a set of partial safety factors different from Case 1C shall be used now for the *design* action in Case 1B (cf. Table 3, ch. 5). However, in this example the *design* nail force is directly given from the stability calculation in Case 1C, so that the partial safety factors in Table 3 need not be applied.

The maximum design value of the intersection nail force $N_{d,max}$ is found in the free body diagram of the least safe failure mechanism ($T_{m,d,max}$ mechanism) in Fig. 10a. One obtains $N_{d,max}$ according to Equ. (2) as follows:

$$N_{d,max} = T_{m,d,max} \cdot s_h \cdot l_{max} \quad (2a)$$

with l_{max}: longest nail section beyond the slip plane.

Normally, as it is here, the longest nail section is to be found in the row at the bottom. Thus, one yields:

$$N_{d,max} = 17.3 \cdot 1.35 \cdot 5.6 = 131\ \text{kN} \quad (11)$$

It is important to see now, that the nail force, having been a *resistant* force, becomes an *acting* force, when the material property (or the cross area) of the steel is checked:

$$E_d \leftarrow N_{d,max} \quad (12)$$

The resistant force is determined by the material property, i. e. the tensile strength of the steel, and the cross section area of the bar. Commonly used nail diameters are given by Bruce & Jewell (1986/87), by Gässler (1990b) or by the French recommendations *Clouterre*.

In this case a steel bar of diameter d = 20 mm, and of normal quality is chosen with the characteristic tensile strength at the yield point:

$$f_{yk} = 500\ \text{N/mm}^2.$$

The design value is derived from:

$$f_{yd} = \frac{f_{yk}}{\gamma_M} \qquad (13)$$

with γ_M: partial safety factor for the material steel (here: concrete steel), given in EC 2

With γ_M = 1.15 the design value of the tensile strength is calculated:

$$f_{yd} = \frac{500}{1.15} = 435\ \text{N/mm}^2 \qquad (14)$$

The material resistance R_d is obtained by:

$$R_d = f_{yd} \cdot A_S \qquad (15)$$

with A_S: cross section area of the steel bar (A_s = 314 mm² for d = 20 mm)

In the given case R_d is calculated:

$$R_d = 435 \cdot 314 = 137000\ \text{N} = 137\ \text{kN}$$

Finally, the check of the steel bar in the nails can be verified as follows:

$$E_d = 131\ \text{kN} \leq R_d = 137\ ✓$$

6.2 *Design example No. 2*

Now, consider a 10 m high nailed cutting slope in cohesive soil. There is no surcharge acting on the crest of the slope. On the uphill side, the natural slope rises with the inclination $\beta = 10°$ (Fig. 12). The following data are given:

characteristic soil parameters: $\varphi'_k = 29°$
$\gamma_k = 19\ \text{kN/m}^3$
cohesion $c'_k = 15\ \text{kN/m}^2$

predesigned wall dimensions:

wall inclination	$\alpha = 10°$
constant nail length	$l = 6.0$ m
vertical nail spacing:	$s_v = 1.10$ m
horizontal nail spacing:	$s_h = 1.25$ m
line through lower nail ends:	$\rho = 10°$

The required characteristic value of the mean shear force $T_{m,k}$ [30 kN/m] is to be determined for a quick pre-design (estimated design working life of the structure: > 50 years; cf. EC 1, Table 2.1: Design working life classification). The solution is to be achieved by means of design charts.

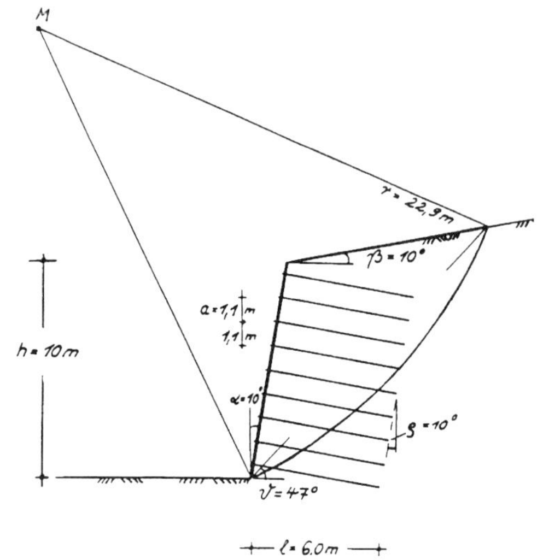

Fig. 12 Design example for a nailed cutting slope in cohesive soil (with unsafest slip circle)

1st Calculation step: Derivation of design values:

Design values of soil parameters:
$\varphi'_d = \arctan(\tan\varphi'_k/\gamma_\varphi) = \arctan(\tan 29°/1.25) = 24°$
$\gamma_k = \gamma_d = 19\ \text{kN/m}^3$
$c'_d = c'_k / \gamma_{c'} = 15/1.6 = 9.4\ \text{kN/m}^2$
Comment: For limit state case 1C: $\gamma_{c'} = 1.6$ (see Table 2.1 in EC 7-1, or Table 4 in ch. 5.2 of this lecture)

2nd Calculation step: Normalizing of data for the design chart

Normalized cohesion is to be found by dividing c' by the unit soil weight and by the height of the slope (cf. Gässler, 1987):

$$c'^*_d = c'_d / (\gamma_d \cdot h) = 9.4/(19 \cdot 10) = 0.05\ [-]$$

3rd Calculation step: Application of the design chart

For the given geometric parameters the adequate design chart in Fig. 13 was selected from a set of charts developed by the author for cohesive soils (Gässler, 1987).

The basis of the limit state design chart in Fig. 13 is the slip circle (rotation mechanism of one rigid block, ROT-I) as shown in Fig. 6b. Very quickly the specific nailing density $\mu_d = \mu_{d,max} = 0.58$ is obtained from the chart.

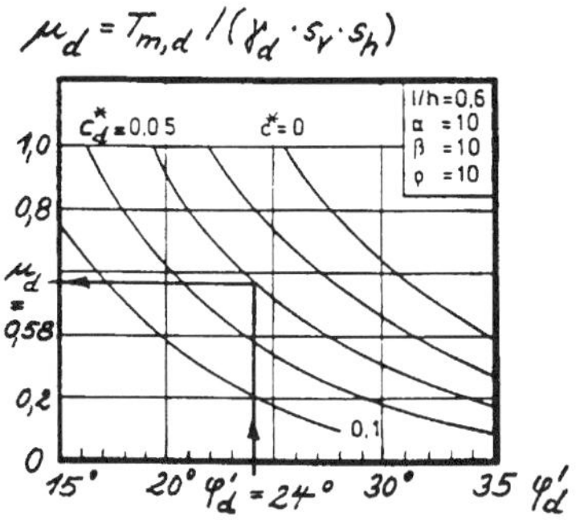

Fig. 13 Design chart for a nailed slope with l/h=0.6, α= 10°, β= 10°, ρ=10° in cohesive soil

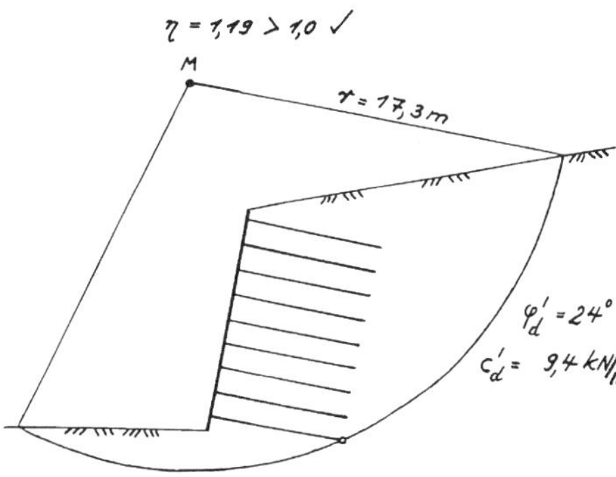

Fig. 14 External stability check: unsafest slip circle

Now, using the partial safety factor $\gamma_N = 1.3$ yields the respective characteristic value:

$$\mu_{k,max} = \gamma_N \cdot \mu_{d,max} = 1.3 \cdot 0.58 = 0.75$$

4th Calculation step: Determination of $T_{m,k,max}$

The characteristic mean shear force per nail meter is obtained using Equ. (4):

$$T_{m,k,max} = \mu_{k,max} \cdot \gamma_k \cdot s_v \cdot s_h$$

$$T_{m,k,max} = 0.75 \cdot 19 \cdot 1.1 \cdot 1.25 = 20 \text{ kN/m/m}.$$

Herewith, the solution is found. However, the design chart does not give any information on the unsafest slip circle. Its geometric parameters can only be achieved from the computer program which the design charts for cohesive soils have been based on (cf. ch. 4.1 and Fig. 6b). Fig. 12 shows the unsafest clip circle with the chord angle $\vartheta = 47°$ and the radius r = 22.9 m.

In the given case it is also recommended to check the external stability. Fig. 14 shows the least safe external slip circle that was found using a conventional computer program. Naturally, following EC 7, the input data were *design* values, say, $\varphi'_d = 24°$ and $c'_d = 9.4$ kN/m². (As no nails are intersected by the check of external stability, no design value for the nails was put in the program.)
Fulfilling the requirements of EC 1 and EC 7, the difference between the resisting moments M_{Rd} and the driving moments M_{Fd} has to be greater or equal zero in the *design* state. The same reqirement can be formulated as the following ratio:

$$\frac{M_{Rd}}{M_{Fd}} \geq 1 \qquad (12)$$

The ratio, resisting moments over driving moments, is 1.19 in Fig. 14. Consequently the respective slip circle in Fig. 14 is safer than the slip circle in Fig. 12, of which the ratio M_{Rd}/M_{Fd} is exactly 1. As the external stability is checked using *design* values, the minimal ratio in Equ. (12) does not need to reach values in the range of 1.5, as it is usually required for *overall* safety factors applied to conventional *calculation* values φ' and c'.

Further stability checks need not be carried out. For example, the external stability checks of sliding or bearing capacity do not govern the design in normal cases. Of course, the justification of the facing is required. However, this cannot not be considered in the scope of this lecture.

7. CONCLUSIONS

The use of the new European Codes, *EC 1 and EC 7* in particular, yields a safe and practicable design of nailed walls and cutting slopes. Thinking in terms of the design limit state and application of several partial safety factors instead of only one global safety factor will be new and unusual for many practising engineers, but in the author's opinion, the Eurocodes can be easily applied to soil nailing without major problems. However, one must not forget that *EC 7- Part 1* in particular is in a pre-liminary state and is not yet completely applicable in practice to all types of retaining structures. Concerning soil nailing, two examples have been given which demonstrate that the partial safety factors approach adopted by the Eurocodes provides an easy and expedient design for soil nailed structures.

REFERENCES

Bang, S., Kroetch, P.P. & Shen, C. K. 1992. Analysis of soil nailing system. *Proc. of the Symp. on Earth Reinforcement Practice*, Fukuoka, Japan, Nov., Vol. 1: 457-462

Bruce, D.A. & Jewell, R.A. 1987. Soil nailing: Application and practice, part 1 and 2. *Ground Engineering.* Nov. 1986, Jan. 1987.

Gässler, G. 1977. Large scale dynamic test of insitu reinforced earth. *Proc. Dynamical Methods in Soil and Rock Mechanics, Karlsruhe*. Vol. 2, ed. by G. Gudehus, Rotterdam: Balkema: 333-342

Gässler, G. 1982. Anwendung des statistischen Sicherheitskonzeptes auf verankerte Wände und vernagelte Wände. Vorträge Baugrundtagung in Braunschweig. *German Society for Geotechniques (Deutsche Gesellschaft für Geotechnik e.V.)*, Essen: 49-82

Gässler, G. 1987. Vernagelte Geländesprünge - Tragverhalten und Standsicherheit. (Doctoral Thesis). *Veröffentlichung d. Inst. f. Bodenmech. u. Felsmech., Univ. Karlsruhe, Germany,* H. 108.

Gässler, G. 1988. Soil nailing - Theoretical basis and practical design. *Proc. Int. Geot. Symp. on Theory and Practice of Earth Reinforcement*, Fukuoka, Japan. A.A. Balkema: 283-288.

Gässler, G. 1989. Planung, Ausschreibung und Überwachung von Vernagelungsprojekten. *Tiefbau, Ingenieurbau, Straßenbau.* Bertelsmann-Verlag. Sept. 31. Jahrgang, Heft 10: 626-640

Gässler, G. 1990a. Contribution to the discussion of session 3. *Proc. Int. Reinforced Soil Conference, Glasgow, Scotland,* Sept. 1990, ed. by A. McGown et al., Th. Telford London: 283-284

Gässler, G. 1990b. In-situ techniques of reinforced soil. State-of-the-Art Lecture, *Proc. Int. Reinforced Soil Conf., Glasgow, Scotland*, Sept. 1990, ed. by A. McGown et al., Th. Telford London: 185-196.

Gässler, G. 1992a. Full scale test on a nailed wall in consolidated clay. *Proc. of the Symp. on Earth Reinforcement Practice*, Fukuoka, Japan, Nov., Vol. 1: 475-480

Gässler, G. 1992b. Report and Discussion of Session 5. *Proc. of the Symp. on Earth Reinforcement Practice, Fukuoka*, Japan, Nov., Vol 2: 955-960.

Gässler, G. 1993. The first two field tests in the history of soil nailing on nailed walls pushed to failure. *Renforcement des sols: Expérimentation en vraie grandeur des années 80.* Int. Symposium Paris, Nov., Presses de l'école nationale des Ponts et Chaussées, Paris: 7-34.

Gässler, G. 1995. Stabilisation of a cutting slope along a high-speed railway line using extremely long nails. *The practice of soil nailing in Europe.* ed. by T.S. Ingold. London: Thomas Telford: 214-226

Gässler, G. & Gudehus, G. 1981. Soil nailing - some aspects of a new technique. *Proc. 10th Int. Conf. Soil Mech. and Found. Eng.*, Stockholm, 1981, Vol. 3: 665-670.

Gässler, G. & Gudehus G. 1983. Soil nailing - statistical design. *Proc. 8th Eur. Conf. Soil Mech. and Found. Eng.*, Helsinki, Vol. 2: 491-494.

Gudehus, G. 1981. *Bodenmechanik.* Enke Verlag, Stuttgart

Hasofer, A.M. & Lind, N.C. 1974: Exact and invariant second-moment code format. *Journ.. Eng. Mech. Div. ASCE*, Vol. 100, EM1: 111.

Krey, H. 1932. *Erddruck, Erdwiderstand und Tragfähigkeit des Baugrundes.* 4th ed., Berlin: Verlag von Wilhelm Ernst & Sohn.

Love, J.P. 1995. Reinforced soil and soil nailing for slopes - Advice Note HA68/94. *The practice of soil nailing in Europe.* ed. by T.S. Ingold. London: Thomas Telford: 44-54.

Maramatsu, K., Nagura, Sueoka,T., Suami, K. & Kitamura, T. 1992. Stability analysis for reinforced cut slopes with a rigid facing. *Proc. of the Symp. on Earth Reinforcement Practice*, Fukuoka, Japan, Nov., Vol. 1: 503-508

Pedley, M.J., Jewell, R.A. & Milligan G.W.E. 1990. A large scale experiment study of soil reinforcement interaction. *Ground Engineering.* July/Aug. & Sept.

Plumelle, C. et al. 1989. Expérimentations du project national français Clouterre sur le clouage des sols. *Proc. 12th Int. Conf. on Soil Mech. and Found. Engg.*, Rio de Janeiro. Vol. 2: 1303-1306.

Rabejac, S., Toudic, P. 1974. Construction d'un mur de soutenement entre Versaille-Chantier et Versailles-Matelots. *Revue Generale des Chemins de Fer.* 93e année, Avril: 232-237.

Schlosser, F. 1982. Behaviour and design of soil nailing. *Proc. of Symposium on soil and rock improvement techniques, including geotextiles, reinforced earth and modern piling methods.* Bangkok, Nov./Dec.

Shen, C.K., Bang, S. & Herrmann, L.R. 1981a. Field measurement of an earth support system. *ASCE, J. of the Geot. Eng. Div.*, 1981, Vol. 107, No. GT 12: 1609-1624.

Shen, C.K., Bang, S., Romstad, K.M., Kulchin, L. & De Natale, J.S. 1981b. Ground movement analysis of an earth support system. *ASCE, J. of the Geot. Eng. Div.*, 1981, Vol. 107, No. GT 12: 1625-1642.

Stocker, M. 1976. Bodenvernagelung. Vorträge Baugrundtagung in Nürnberg, *German Society for Geotechniques (Deutsche Gesellschaft für Geotechnik e.V.)*, Essen: 639-652

Stocker, M., Körber, G., Gässler, G. & Gudehus, G. 1979. Soil nailing. *Int. Conf. in Soil Reinforcement. Paris.* March, Vol. 2: 469-474.

Stocker, M. & Gässler, G. 1979. Ergebnisse von Großversuchen über eine neuartige Baugruben-wand-Vernagelung. *Tiefbau, Ingenieurbau, Strassenbau.* Bertelsmann-Verlag. Sept. 21. Jahrgang: 746-751

Taga, N., Tayama, S., Uehara, S. & Doi, Y. 1992. Stability nomograms for reinforced earth with steel bars against shallow sliding of steep slopes. *Proc. of the Symp. on Earth Reinforcement Practice*, Fukuoka, Japan, Nov., Vol. 1: 549-554

Teramoto, K., Taga, N., Naruse, T. & Tayama, S. 1992. Model loading tests of reinforced slope with steel bars. *Proc. of the Symp. on Earth Reinforcement Practice*, Fukuoka, Japan, Nov., Vol. 1: 561-566

Tsubouchi, T., Hiroi, K. & Onodera, S. 1992. Search for critical failure lines of models reinforced with rockbolts and ground anchors. *Proc. of the Symp. on Earth Reinforcement Practice*, Fukuoka, Japan, Nov., Vol. 1: 567-572.

Tufenkjian, M.R. & Vucetic, M. 1992. Seismic stability of soil nailed excavations. *Proc. of the Symp. on Earth Reinforcement Practice*, Fukuoka, Japan, Nov., Vol. 1: 573-578.

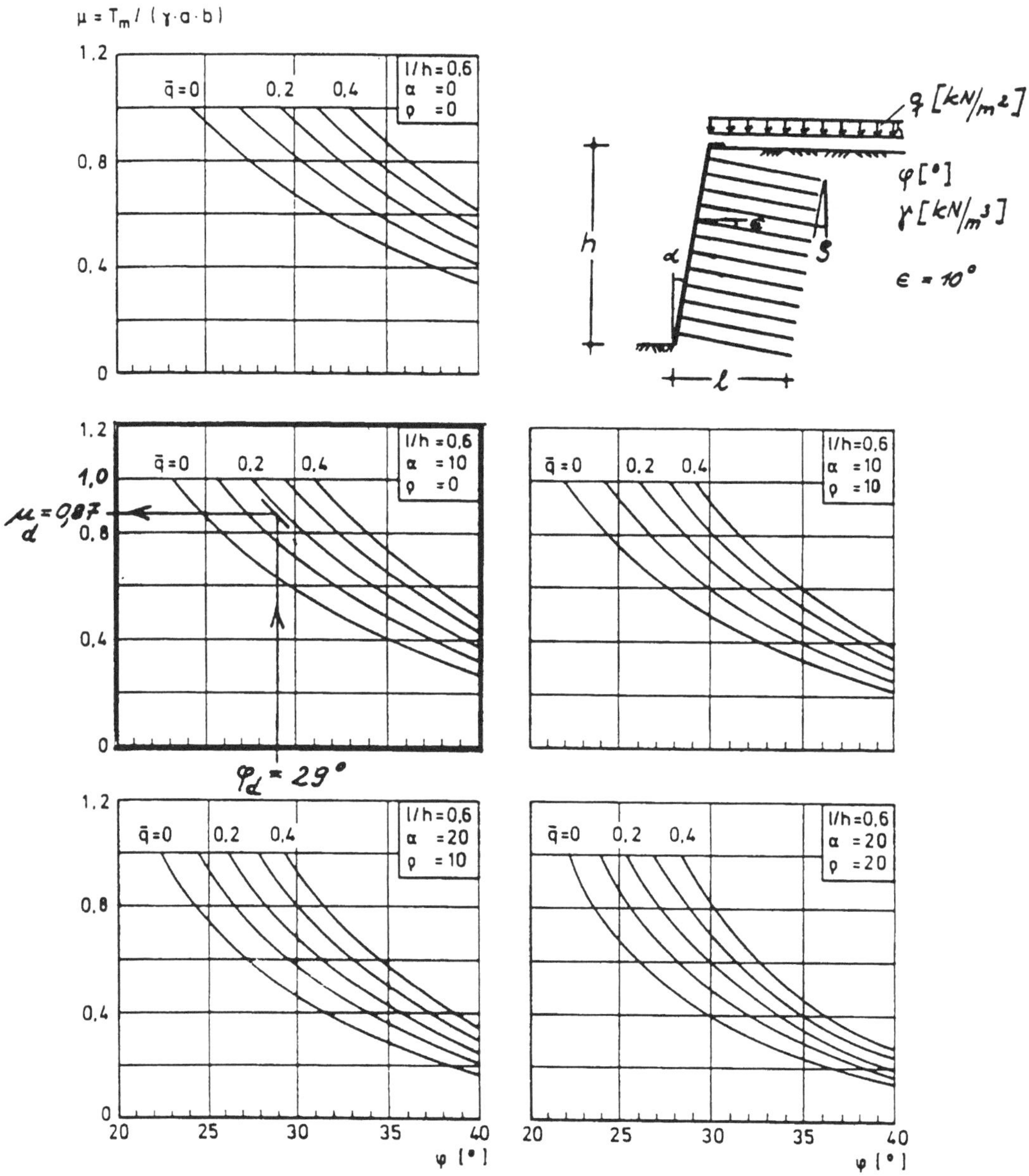

Design charts for nailed walls and slopes in non-cohesive soil with nail length l/h = 0.6 and with constant surcharge $q^* = q/(\gamma \cdot h)$ acting on the horizontal crest

Special reports

Earth Reinforcement, Ochiai, Yasufuku & Omine (eds) © 1997 Balkema, Rotterdam, ISBN 90 5410 833 9

Performance of geosynthetic-reinforced slopes and walls during the Northridge, California Earthquake of January 17, 1994

D.M.White
Applied Geotechnology, Inc., Bellevue, Wash., USA

R.D.Holtz
University of Washington, Department of Civil Engineering, Seattle, Wash., USA

ABSTRACT

This paper reports on the performance of seven geosynthetic-reinforced slopes and walls, all of which experienced significant to moderate shaking during the Northridge, California earthquake of January 17, 1994. No adverse performance of geosynthetic-reinforced structures was observed, even in areas where other types of structures experienced failures and/or other poor performance.

1. INTRODUCTION

The January 17, 1994 Northridge, California earthquake, felt widely throughout Southern California, demonstrated several shortcomings in the design and construction of many civil engineering structures. On the other hand, the earthquake also provided an opportunity to study the successful (non-failure) performance of newer technologies such as geosynthetic-reinforced structures. We have examined and documented the performance of three walls and four slopes, with varying types and methods of reinforcement and all of which experienced shaking in the Northridge event. In no case was a failure of the reinforced slope or wall observed, even in areas where other nearby structures failed.

The field observations took place during the Spring of 1994; note that this was four and six months after the event. The locations of the seven reinforced soil structures investigated are shown on Fig. 1. At some locations, it was possible to obtain the plans and/or design methodology.

A discussion of each case is presented below.

2. OBSERVATIONS AND DISCUSSION

2.1 Valencia Water Reclamation Plant

The Los Angeles County Sanitation District No. 32 constructed an approximately 80m long modular block faced reinforced wall as part of the expansion of the Valencia Water Reclamation Plant (Fig. 2). The wall was completed less than 1 year prior to the Northridge event. The wall is located approximately 23 km north of the epicenter, 5 km northwest of the I-5 Gavin Canyon Undercrossing, and 7 km northwest of the failed overcrossing at the intersection of the I-5 and Antelope Valley Freeway (Highway 14). Shaking of a Modified Mercalli Intensity (MMI)=VII was experienced in the area.

The wall was constructed with Keystone blocks and Miragrid reinforcement and was used to establish additional level ground in the area immediately west of some clarifier tanks. The wall was constructed across and filled in a portion of a former streambed. Face height varied from 1 to 5 m over its length, although for most of the length its height was near 5 m. Two or three courses of modular blocks (each approximately 0.6 m high) were constructed below grade, reportedly for increased passive resistance, for a total wall height of approximately 6 m. The wall supported an improved roadway and lighting poles that were founded immediately below the surface along its entire length. The wall transitioned to a conventional reinforced concrete wall at the north end of the project, where a steel tank structure (approx. 4 m high and 15 m diameter) was supported on the backfill (Fig. 3). Apparently, even

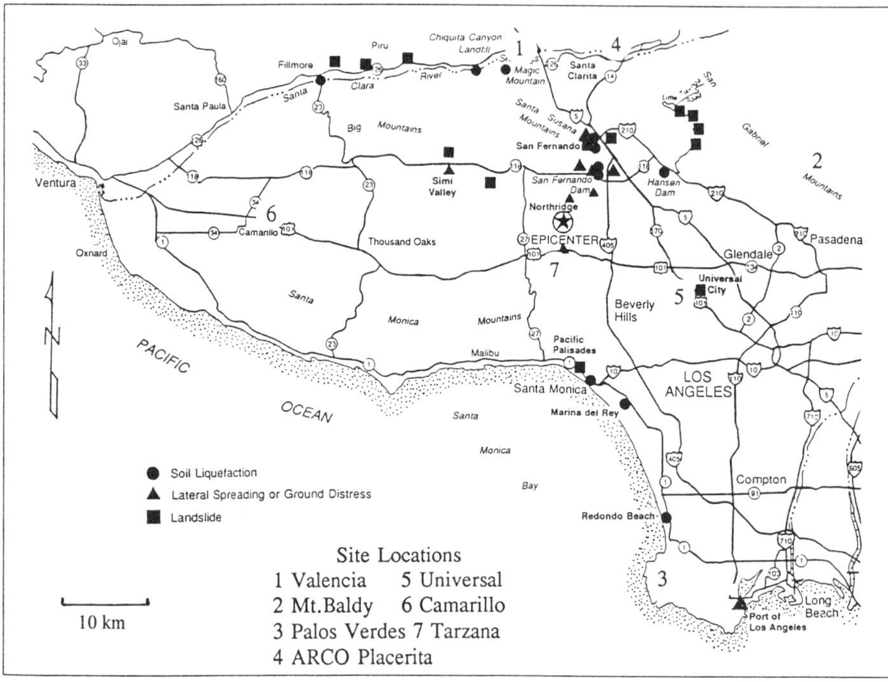

Figure 1 - Map of Southern California showing sites of interest (after UC Berkeley, 1994)

Figure 2 - View of Valencia plant wall; modular wall is centered below tank; I-5 frontage is top right; Santa Clarita River is at base

in 1993 when the reinforced wall was constructed, confidence in geosynthetic technology was limited.

Performance of both the modular block and reinforced concrete portions of the Valencia wall was adequate. Detailed examination of the area indicated that backfill grade was maintained, and no cracking was visible in the soil at the surface above either the block or reinforced concrete wall at the time of our inspection. On the other hand, Bathurst and Cai (1995) reported some cracks at the top of the backfill. Slight (estimated less than 50 mm) bulging of the wall for three to four courses of block at midheight was observed, and limited amounts of pea gravel (backfill) were resting along the exposed portion of several courses of blocks and at the wall base. Both of these imperfections were reported to have occurred during construction. Another construction difficulty was observed; sometime after the wall was completed, approximately a 1 x 1 m square of the uppermost layer of geogrids was cut and removed to allow for the excavation for the foundations of the light poles (Fig. 4). Although it is believed that this removal of geosynthetic reinforcement at several locations did not adversely affect wall performance, this observation provides an important lesson to geotechnical engineers responsible for the design of reinforced soil systems.

Further observations at the Valencia wall included poor detailing and inadequate construction of erosion control measures at the south end of the block wall. This resulted in a partial washout of the fill material at that location. In the treatment plant itself, significant cracking occurred in the concrete walls of two clarifier tanks within 100 m of the retaining wall. The tanks had to be taken out of service for repairs.

This wall was also analyzed by Bathurst and Cai (1995). Their pseudo-static analysis indicated marginal dynamic factors of safety.

2.2 Mount Baldy Highway

Caltrans constructed a Reinforced Earth wall on Highway 2, the Mount Baldy Highway, at a location approximately 35 km east of the epicenter. The wall is approximately 100 m long and 10 m high (Fig. 5). MMI V shaking was reported in this area. Inspection of the pavements and guard rail above the wall indicated no signs of sliding and/or cracking. A maintenance concern was noted in that vegetation was growing from the cracks between several wall panels.

2.3 Palos Verdes Estates

A 1992 slope repair to Via Zurita in Palos Verdes Estates was constructed of an unknown type of geosynthetic reinforcement. The roughly 70 m high 2:1 slope is located approximately 55 km south of the Northridge epicenter on a switchback or U-turn in the roadway. The slope had reportedly been failing by sliding into the adjacent canyon prior to the repair activity (Fig. 6). Estimated MMI at this site was VI.

Post-earthquake reconnaissance showed no displacements in the soil mass and no cracks in the asphalt pavements above the slope. Some separations on the order of 5 to 7 mm were observed between the concrete curb and the pavement both within and outside the repair area.

2.4 ARCO Placerita Field

A geosynthetic-reinforced slope was constructed to establish a grade separation between two adjacent drilling pads in the oil well field near Newhall, California (Fig. 7). This site is approximately 20 km northeast of the Northridge epicenter, and estimated MMI was VII. The slope is approximately 10 m high on a 1.5:1 inclination to accommodate the desired working space on the soil pads.

Performance of the slope during the earthquake was adequate, with no displacements or cracking observed in the earth mass or in the pads. Nearby temporary job trailers were shaken from their foundations, resulting in some damage, but no earthquake-related damage to permanent well head equipment, pipelines or structures was reported. Unreinforced natural cut slopes in a weakly cemented sand and gravel at a similar or slightly steeper inclinations approximately 200 m south of this location failed during the earthquake, closing an oil field road (Fig. 8).

2.5 Universal Overcrossing

Three modular block faced retaining walls at an overcrossing in the Universal Studios area in Los Angeles were examined as they were reportedly constructed with geosynthetic reinforcement in the backfill. However, inspection verified that these were crib walls and not reinforced with geosynthetics. Performance on all but one wall was adequate. The southernmost of the walls experienced cracking with a relative displacement

Figure 3 - Transition from modular block to reinforced concrete wall at Valencia.

Figure 4 - Foundation formworkfor light pole at top of modular block wall at Valentia. Note the top layer of geogrid removed during excavation ofthe foundation.

Figure 5 - Reinforced Earth wall on Mount Baldy Highway (Highway 2).

Figure 6 - View across reinforced slope repair at Via Zurita. Slope steepens and repair begins above vegetation change at lower part of photograph.

Figure 7 - Reinforced soil slope at ARCO Placerita Field separating adjacent drilling pads.

Figure 8 - Natural slope failure approximately 200 m south of reinforced slope in Figure 7.

Figure 9 - Camrillo slope reinforced to mitigate deformations due to potential fault rupture.

Figure 10 - Reinforced slope in Tarzana. Residence at top reportedly experienced over $150,000 damage to foundation system in the Northridge event

Figure 11 - Cracks in concrete block splash walls at Tarzana reinforced slope

of approximately 20 mm at a distance of 8 to 10 m behind the wall face, probably the result of backfill settlement and/or outward movement of the wall face during the Northridge event. This structure was also described by the UC Berkeley team (Stewart et al., 1994a and b).

2.6 Camarillo

The instrumented reinforced slope at the Spanish Hills residential development in Camarillo, discussed by Bray et al. (1993), was also observed (Fig. 9). This slope was west of Northridge at an epicentral distance of approximately 45 km, and experienced proportionately lower levels of shaking (MMI=VI). However, it was examined due to its previous documentation and instrumentation. The Tensar UX1500 reinforcement was installed for both reinforcement of slopes and mitigation of potential displacements due to rupture of a fault underlying the site. Slopes (10 m high at 2:1) in the area of the surface fault rupture were examined and found to be crack-free, but they did contain some erosional gullies. At the time of the Northridge earthquake and our inspection, no construction on the building pads had begun and it was easy to observe the performance of the earth fill. Detailed observation revealed no cracking or earthquake-related displacement of the reinforced and well-vegetated slopes. The only adverse performance observed was the erosion of several approximately 100 mm deep gullies into the relatively flat, exposed (unvegetated) earth at the top of the slope.

2.7 Tarzana Slope

In 1985, a 60m long, 15m high, 1.5:1 slope separating the building pads for the residences at the top of the slope (19538 and 19542 Redwing St. in Tarzana) from the adjacent and lower properties to the east was constructed using Tensar SS1 geogrids for reinforcement (Fig. 10). This construction was to repair a landslide. This slope is located only 6 km southwest of the Northridge epicenter and received strong shaking (MMI VIII). This slope repair was one of the earliest (1985) uses of geosynthetics for repairs to residential slopes in the region. The use of this technology was due in part to the unsuccessful performance of five previous conventional attempts (overexcavate, backfill and revegetate) to repair the same slope. Failures occurred during each of the wetter years between construction in the mid-1960's and most recent failure during the 1984-85 wet season. The slope supports the foundations of the east deck to the residence at 19542 Redwing and further failure would have threatened the structure itself. A single, mid-height, 1.5 mm wide bench and concrete lined v-ditch with two downdrains are present on the face of the slope to convey stormwater from the slope in accordance with local practice.

The slope performed well during the earthquake, as evidenced by the lack of displacement and earth cracking in the face. The only damage to the slope following the quake was to the concrete drainage ditches on its face, which showed signs of minor cracking (up to 2 mm width) at joints and shape changes (Fig. 11). These cracks could not be directly attributed to the earthquake by the owner and may be shrinkage-related. Some additional cracking was evident in the tennis court surface and in the low (0.5 m) stucco-faced, concrete block retaining wall on its uphill side. The new cracks in the court joined pre-existing cracks of a similar nature and were a maximum of 3 mm wide. The retaining wall cracks approached 10 mm width with 1 to 2 mm vertical displacement. A swimming pool and deck approximately 10 m beyond and 1.5m downhill of the tennis court showed no new cracking or displacement from the earthquake.

Greatest damage in the area of the slope was at the top of the slope, where the residence at 19542 Redwing experienced a reported $150,000 damage to the structure and foundation. Casual observation of the interior and exterior walls showed minimal cracking and the house was habitable, but foundations were damaged by the shaking and will require replacement.

3. COMMENTS AND CONCLUSION

Performance of the seven geosynthetic-reinforced slopes and walls shaken in the Northridge earthquake of January 17, 1994, was adequate, particularly when compared to the performance of other structures in the immediate vicinity. While this performance is desirable, it may also be the result of overly conservative designs, possibly because soil reinforcement is relatively new technology. As an understanding of geosynthetic reinforcement performance during seismic shaking develops and as the design process achieves greater certainty, reinforcement designs are likely to become more economical. Research into seismic design processes should also evaluate whether stiffening of slopes by geosynthetics has the potential to increase the damage to brittle structures founded upon these slopes and walls.

ACKNOWLEDGMENTS

The authors gratefully acknowledge the support of the National Science Foundation (Grant No. CMS-9412803) for this research. The cooperation of the various property owners in providing access to the privately owned sites is also greatly appreciated. Information provided by engineers at American Geotechnical, Caltrans, Mirafi, Reinforced Earth, and Tensar was also very helpful.

REFERENCES

Bathurst, R. J. And Cai, Z. (1995) "Pseudo-Static Seismic Analysis of Geosynthetic-Reinforced Segmental Retaining Walls", *Geosytnethics International*, Vol. 2, No. 5, pp. 787-830

Bray, J. D., Ashmawy, A., Mukhopadhyay, G. and Gath, E.M. (1993), "Use of Geosynthetics to Mitigate Earthquake Fault Rupture Propagation Through Compacted Fill", *Proceedings,* Geosynthetics '93, Vancouver, B. C., Vol. 1, pp. 379-392.

Stewart, J.P., Seed, R.B., Riemer, M. and Zornberg, J.G., (1994a), "Northridge Earthquake, Geotechnical Structures", *Geotechnical News*, Vol. 12, No. 2, pp 59-62.

Stewart, J.P., Bray, J. D., Seed, R.B., Sitar, N. (1994b) "Preliminary Report on the Principal Geotechnical Aspects of the January 17, 1994 Northridge Earthquake", Earthquake Engineering Research Center, University of California at Berkeley, Report No. UCB/EERC-94/09, 238pp.", *Geotechnical News*, Vol. 12, No. 2, pp 59-62.

UC Berkeley, (1994) "Preliminary Report of the Geotechnical Aspects of the January 17, 1994, Northridge Earthquake", Geotechnical Engineering Group, Department of Civil Engineering, *Geotechnical News*, Vol. 12, No. 1, March 1994.

Earth Reinforcement, Ochiai, Yasufuku & Omine (eds) © 1997 Balkema, Rotterdam, ISBN 90 5410 833 9

Performance of reinforced soil structures during the 1995 Hyogo-ken Nanbu Earthquake

F.Tatsuoka
University of Tokyo, Japan

J.Koseki
Institute of Industrial Science, University of Tokyo, Japan

M.Tateyama
Railway Technical Research Institute, Japan

ABSTRACT: The performance of a number of mechanically stabilized earth retaining walls and nailed slopes during the 1995 Hyogo-ken Nanbu earthquake is reviewed. The behaviour of the reinforced soil retaining walls is compared with that of the conventional retaining walls. Some lessons from the case histories and limitations of the current seismic design methods are discussed. Reconstruction of damaged structures by soil reinforcing techniques is also presented.

1 INTRODUCTION

This paper will discuss the seismic stability of reinforced soil structures, including reinforced sloped embankments, soil retaining walls and natural slopes. This topic has always been a major concern of geotechnical engineers in high seismicity regions including Japan. The major issues may include:
a) stability for seismic loads relative to the stability for static loads;
b) seismic stability relative to other types of civil engineering structures (e.g., elevated RC frame structures); and
c) relative seismic stability among different types of soil structures (e.g., unreinforced versus reinforced soil structures).

Issue a) is related to the aseismic design methodologies for soil structures. Compared with ordinary steel and RC structures, soil structures in secondary applications are usually not aseismic-designed. Important soil structures are usually aseismic-designed, but they are by the limit equilibrium-based pseudo-static method using a relatively a low seismic coefficient k_h such as 0.20. Probably, this situation results from such a consideration that in case soil structures in secondary applications are damaged, the influence of the damage would not be vast and serious and they could be easily repaired. Also, the use of higher k_h values could result in higher construction cost. However, to be consistent with the recent trend of aseismic design methodologies for steel and RC structures which employ much higher seismic loads than before, we need rationales for the above-mentioned conventional aseismic design methodologies or new aseismic design methodologies for soil structures including reinforced soil structures.

For "reinforced soil retaining walls (RWs) to retain embankment (Fig. 1d)", for example, Issues b)

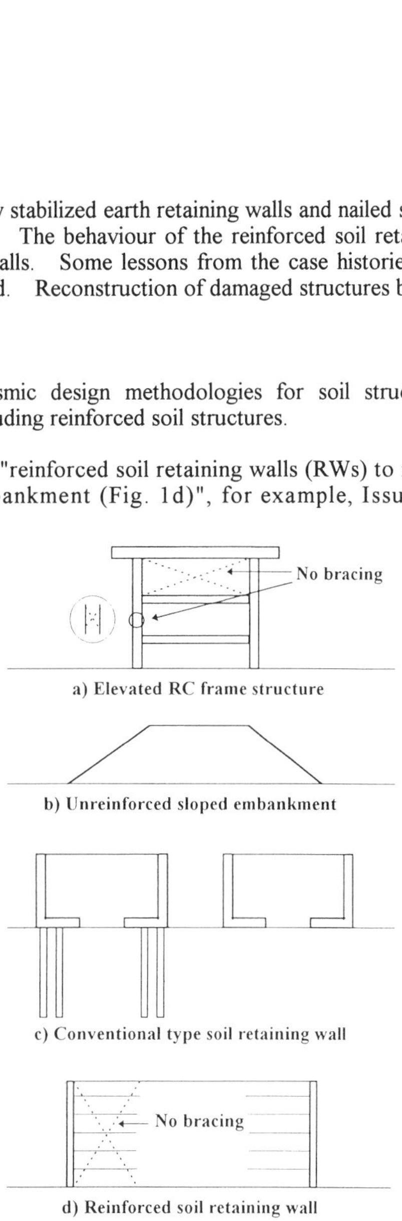

Fig. 1 Schematic figures of a) elevated RC frame structures, b) sloped embankment, c) RC RWs, and d) Reinforced soil RWs.

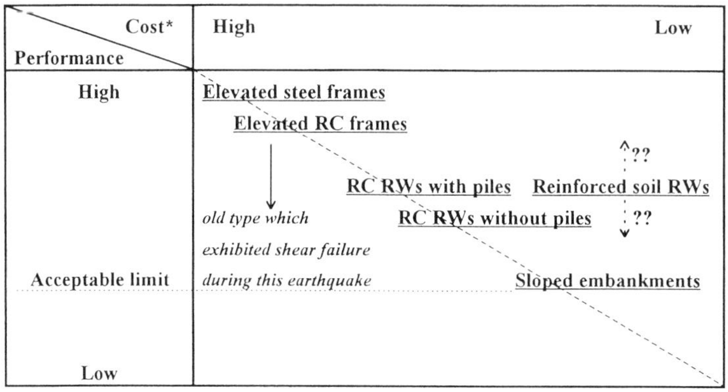

* at the time of construction

Fig. 2 Approximated relative cost and performance of typical different types of elevated structures

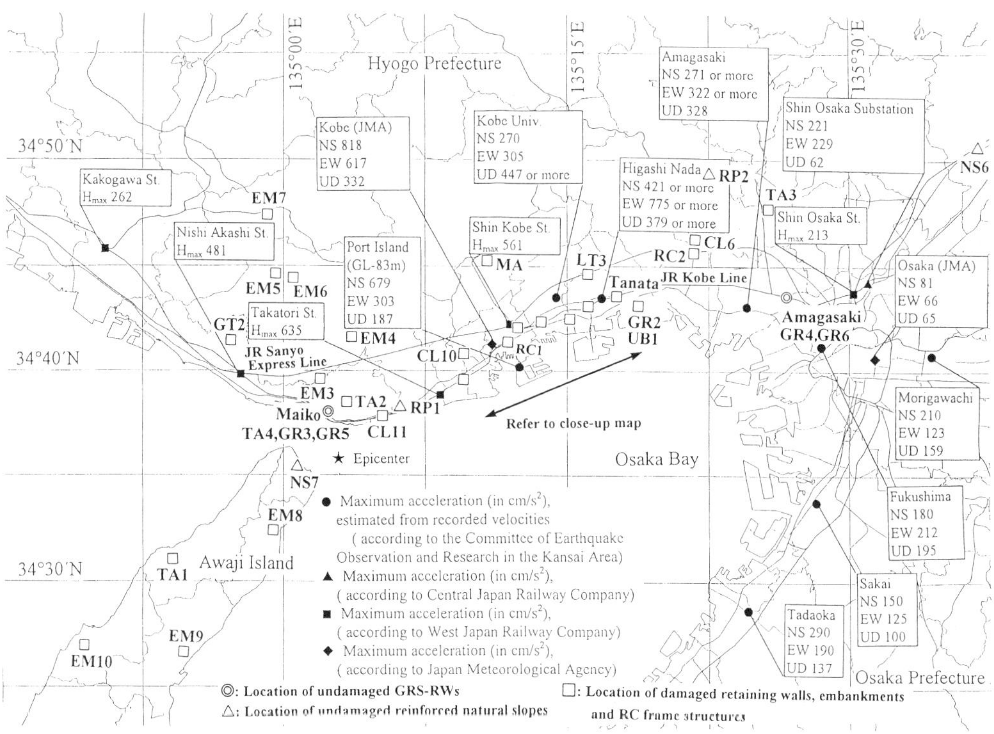

Fig. 3 Recorded peak ground acceleration and locations of typical structures described in the paper.

and c) are linked to their advantages over unreinforced sloped embankments (Fig. 1b), conventional type soil RWs (Fig. 1c), and elevated RC frame structures (Fig. 1a). It is common to try to select a structure type which is the most cost-effective while satisfying the required level of performance. In Fig. 2, the structures having the same cost-effectiveness are located at the same distance from the diagonal line extending from the top left corner to the bottom right corner. Reinforced soil RWs are usually more cost-effective than the conventional soil RWs, located more toward the top right corner. Even when a higher cost-effectiveness is assured, however, it is not certain whether or not the performance (i.e., seismic stability in this case) of a given reinforced soil structure could be equivalent to, or higher than, that of the conventional structures (e.g., conventional type RC RWs and elevated RC frame structures).

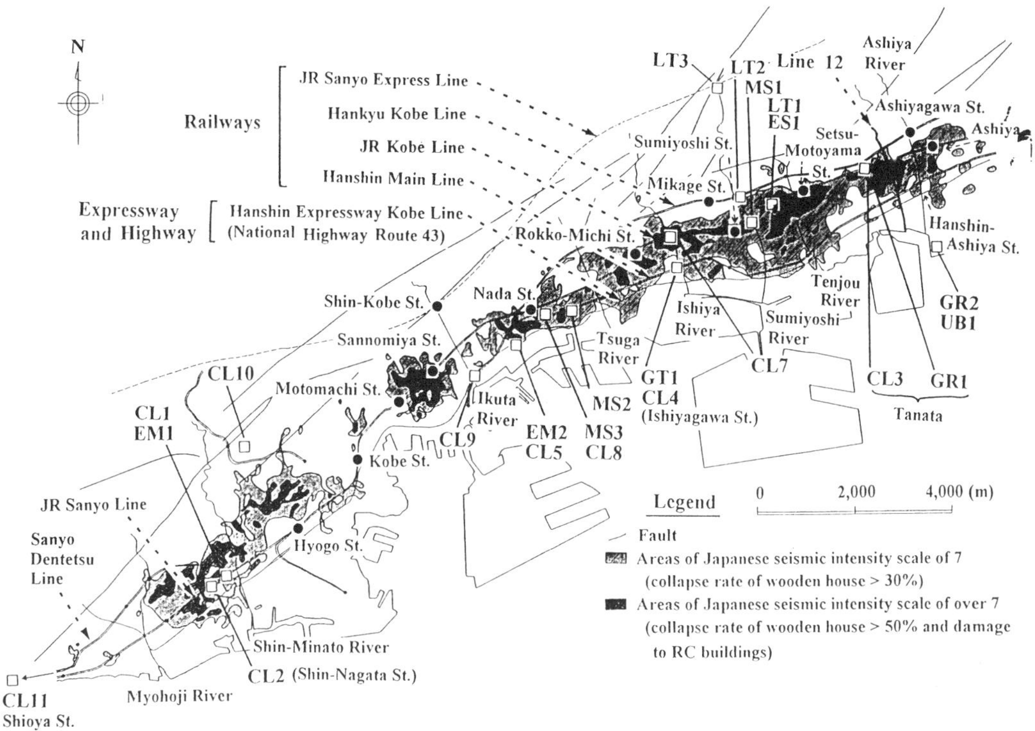

Fig.4 Areas of JMA scale equal to seventh or higher and locations of typical structures described in the paper.

Varying degrees of damage to a great number of various type old and recent civil engineering structures caused by the 1995 Hyogo-ken Nanbu earthquake provided us a great opportunity to examine the above issues. It will be attempted to review the performance of reinforced soil structures, as much as possible, in view of the points described above.

2 OVERVIEW OF THE EARTHQUAKE AND DAMAGE TO CIVIL ENGINEERING STRUCTURES

2.1 *The Earthquake*

At 5:46 a.m. 17th January 1995, a devastating earthquake measuring 7.2 on the Richter scale hit the southern part (i.e., "nanbu" in Japanese) of Hyogo Prefecture, including Kobe City. The representative recorded peak horizontal ground accelerations (PHGAs) and vertical ones are shown in Fig.3. Fig.4 shows the areas where the Japanese Meteorological Agency seismic intensity scale (JMA scale) was seventh or higher, estimated from the collapse ratio of wooden houses (Chuo Kaihatsu Corp., 1995),

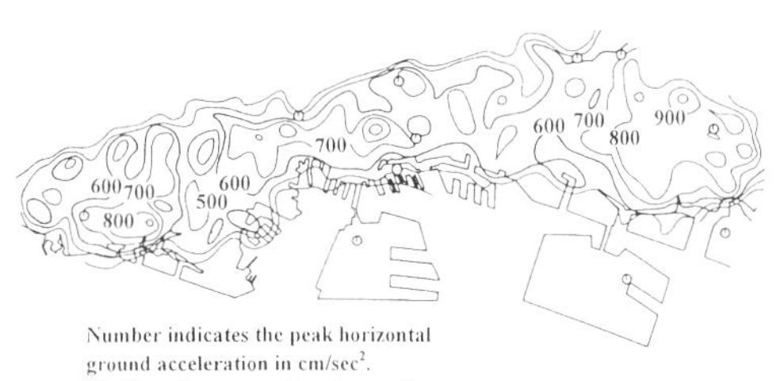

Fig. 5 Estimated contours of PHGA (after Fig. 13 of Sato, 1996).

which is consistent with the distri-bution of PHGA estimated by Sato (1996) (see Fig. 5).

Fig. 6 shows the response spectrum curves for a damping ratio equal to 5 % for the ground horizontal acceleration recorded at Kobe (JMA) and others (see Fig. 3). The predominant period was around 0.3 - 1.0 seconds. This and other similar records suggest that structures having natural period of around 0.3 - 0.5 seconds while having low damping properties should have proven to be most seriously damaged,

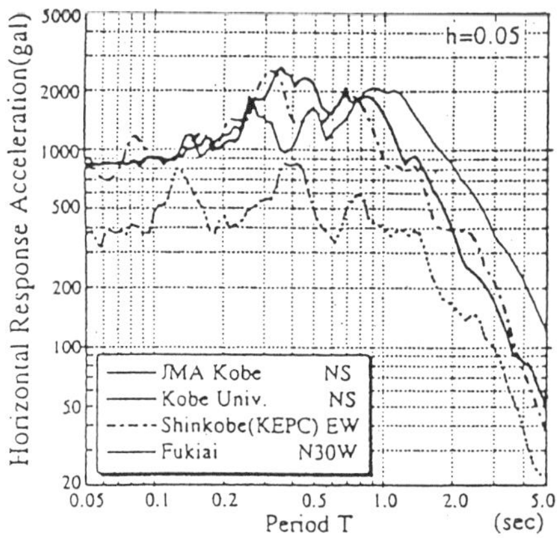

Fig. 6 Response spectrum curves of some typical recorded earthquake records (after Fig. 3 of Sato, 1996)

because the response could be large at the initial stage of loading and it could not decrease much with the increase in the natural period T_n although the damping increased as the damage became more severe.

2.2 *Damage to civil engineering structures*

Perhaps one of the best case histories to examine relative seismic stability of typical civil engineering structures is the performance of elevated RC frame structures, embankment slopes and conventional type soil RWs constructed for and adjacent to the JR Kobe and Sanyo Lines located in the affected areas (see Figs. 3 and 4 for the locations). The performance of reinforced soil structures will be reviewed later. The railway was extended from Osaka to Kobe by 1874, constructed mainly on the ground surface, partly on embankments with vertical:horizontal slopes of 1:1.5. The railway was extended to the west of Kobe Station as the Sanyo Line from 1894.

Elevated RC frame structures: For grade separation in the urban area, with a length of about 0.74 km between San-no-miya and Motomachi Stations (see Fig. 4), elevated RC frame structures were constructed by 1938. They were 8-9 m high on average. It is considered that the design was rather conservative as evident from a relatively short center-to-center spacing of 5.5 m between RC columns in the longitudinal direction. A large number of RC columns were damaged, but none of them totally collapsed, and after necessary repair, the railway was reopened only 34 days after the earthquake.

After World War II, the railway for a length of about 3.3 Km between Sumiyoshi and Nada Stations (see Fig. 4) on the ground surface and embankments was reconstructed by 1976 on elevated RC frame structures, including RC bridge girders and their RC frame abutments. Based on the design code specified in 1970, the design was more economical than before, as seen from a wider center-to-center spacing of 8 m between RC columns. Many of the columns were much more seriously damaged and some totally collapsed mainly due to shear failure as typically seen from Figs. 7a and b. Seventy one columns, which were nearly a half of the total columns, were reconstructed. The undamaged columus were only 20 % of the total columns (RTRI, 1996). The railway was reopened on 1st April 1995, 74 days after the earthquake.

The likely reasons for such serious damage to the RC columns are as follows:

1) Also outside the areas where Japanese wooden houses were seriously damaged (as shown in Fig. 4), a large number of RC columns were seriously damaged. Typically, about 30 columns of the elevated RC frame structures for the Bullet Train Line (Shin-kansen) located north of Ashiya (RC2; see Fig. 3) were totally collapsed and reconstructed. The behaviour of the elevated RC frame structures of the Kobe Line and the Bullet Train Line would be due partly to highly amplified responses, caused by the following two factors; i) most of the damaged elevated RC frame structures had predominant period T_n about 0.5-0.8 second, and ii) the RC columns had low damping ratios when compared with soil structures, which exhibited high damping ratios when undergoing large deformation before reaching ultimate failure.

2) The elevated RC frame structures had no overall bracing and no diagonal steel reinforcement in each column (Fig. 1a). They were, therefore, basically not well prepared for large tensile loads acting in the diagonal direction caused by lateral seismic loads. The damaged RC columns were prone to shear failure, mainly due to over-estimation of the shear strength of concrete in the design and an insufficient amount of stirrups. Shear failure of RC columns occurs in a rather brittle manner, easily resulting in total collapse. Recent RC columns are, therefore, designed not to fail by shear failure but by bending failure, based on the design code specified in 1983. Note that many of the independent RC piers supporting steel or RC girders for elevated highways located inside and outside the most severely shaken areas were also seriously damaged, mainly by shear failure, and some of them were totally collapsed.

Reinforced soil RWs are reinforced with horizontal reinforcement with or without a vertical full-height rigid facing, while no inextensible members are

a) b) d)

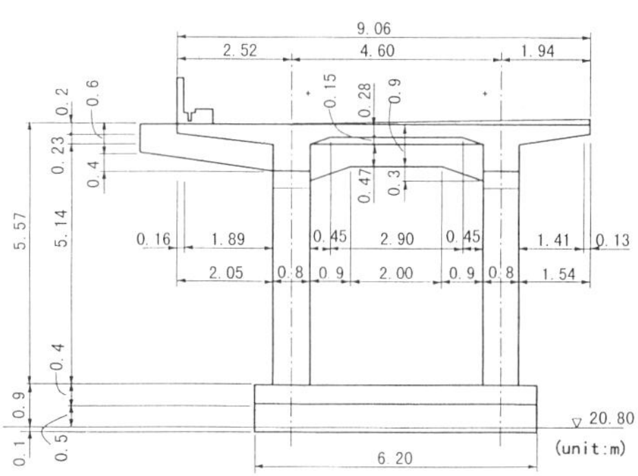

c)

Fig. 7 a) Damaged RC frame bridge abutment and b) damaged Rokko-michi Station consisting of elevated RC frame structures, c) typical dimensions of elevated RC frame structure, and d) inverted T-shaped RC RW adjacent to Rokko-michi Station.

ranged in the diagonal direction where tensile strains become the largest (Fig. 1d). This feature is similar to that of elevated RC frame structures. This point will be discussed later.

Soil structures: Many of the sloped embankments constructed for grade separation of the JR Kobe and Sanyo Lines suffered from large settlement and lateral movement, accompanied by longitudinal tensile cracks on the top near the slope and along the slope surface, resulting in large settlement and lateral movement of the railway tracks. At some locations, the toes of slopes were supported by low gravity type RWs, which were 1.0 - 1.5 m in the total height with a buried depth of 0.5 m (Fig. 8). Despite their small heights, most of them completely overturned to the ground surface. Most of the slope surface had been covered with a cast-in-place concrete lattice and precast concrete plates to protect the slope against against heavy rainfall, but they were found to be ineffective in preventing the deformation of the embankment due to seismic loads. This damage was induced very likely by the high seismic force applied to the facing itself.

The oldest type of RW is masonry RWs. They were constructed more than 70 years ago without aseismic design. They were most seriously damaged among different types of RWs. Most of the masonry RWs located in the areas where the JMA scale was equal to seventh or higher were more-or-less damaged as typically shown in Fig. 9. Many masonry RWs supporting the backfill behind the RC bridge abutments moved outward, resulting in large settlement of the backfill relative to the pile-supported RC bridge abutments.

For a length of about 600 m between Setsu-motoyama and Sumiyoshi Stations (see Fig. 4), the north side slope of embankment was retained by leaning-type unreinforced concrete soil RWs constructed around 1938. The RWs were either broken at the ground surface level and the upper part overturned completely to the ground surface or overturned about the bottom, resulting in the back face facing up (Fig. 10a). This damage was perhaps triggered by large horizontal seismic loads acting on the RW structure itself and large seismic earth pressure imposed on its back face from the backfill. It seems, however, that the former factor should be the major factor for complete overturning of the RW.

Shin-Nagata Station of the JR Sanyo Line was constructed around 1965 on the crest of a sloped embankment. Both sides was supported by cantilever-type (or inverted T-shaped type) RC RWs for a total length of about 800 m (Fig. 11a). The walls were aseismic-designed using k_h= 0.2, but it is

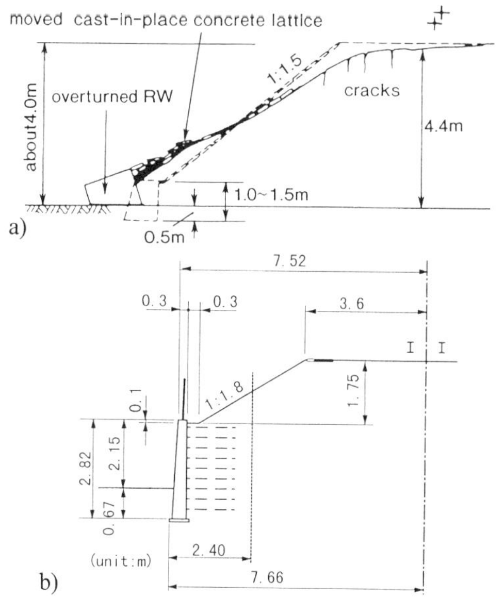

Fig. 8 Cross-sections of a) a typical damaged south slope of the embankment and b) a GRS-RW by reconstruction, between Setsu-motoyama and Sumiyoshi Stations, JR Kobe Line (ES1 in Fig. 4).

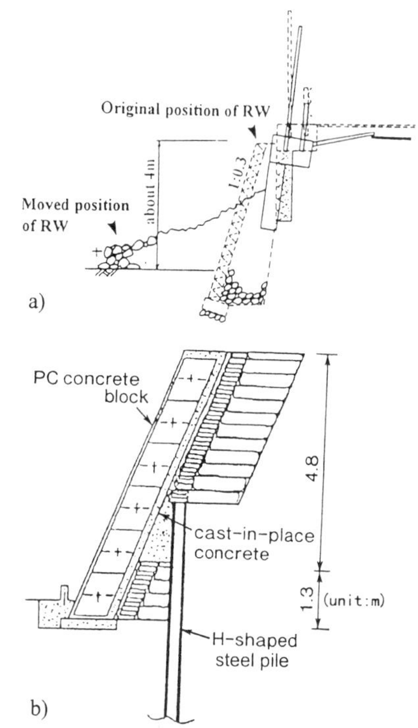

Fig. 9 Cross-sections of a) a typical damaged masonry RW constructed in around 1925 (MS1 in Fig. 4) and b) a GRS-RW by reconstruction (the facing consists of PC concrete blocks, which were fixed afterwards to each other and to the backfill), between Setsu-motoyama and Sumiyoshi Stations, JR Kobe Line.

very likely that they were not supported by piles despite a weak subsoil. Most of these RWs considerably tilted and slid outward at the bottom inducing large settlement at the top of the railway embankment. It seems that the damage was due to insufficient values of both the bearing capacity in the subsoil beneath the wall and the sliding resistance at the wall base. The most serious damage was cracking in the facing, probably due to extra-ordinary large seismic earth pressure from the sloped backfill (Fig. 11a).

Inverted T-shaped and buttressed RC RWs which was constructed around 1938 at Shioya Station of the Sanyo Dentetsu Line, running in parallel to the JR Sanyo Line (see Fig. 4), were also seriously damaged (Fig. 12, CL11 in Fig. 3). There were no piles. Although the wall did not totally collapse, most part tilted largely for a length of about 360 m, and some part of the wall structure ruptured. Similar to the RWs at Shin-nagata, the strength of the wall structure itself and the bearing capacity of the supporting subsoil below the wall were seemingly not sufficient to prevent, respectively, the failure of the wall structure and the large tilting of the wall. The walls were removed after the earthquake.

The seismic behaviour of a RC RW which was constructed most recently (in 1992) to retain the south face of the embankment for the JR Kobe Line

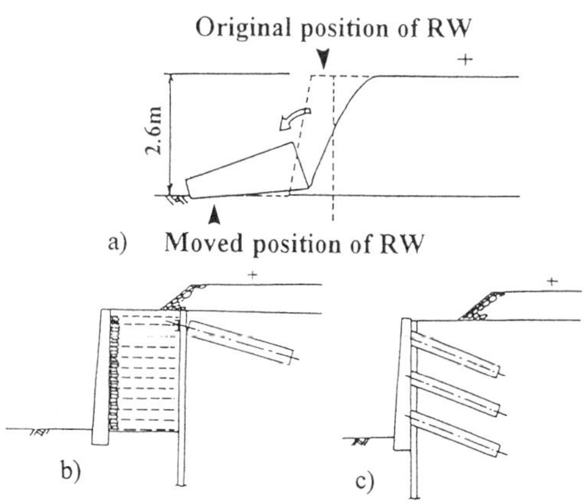

Fig. 10 Cross-sections of a) a typical damaged leaning-type RW constructed in 1937 (LT1 in Fig. 4), and b) a GRS-RW and c) a nailed RW by reconstruction, the north side slope of the embankment between Setsu-motoyama and Sumiyoshi Stations, JR Kobe Line.

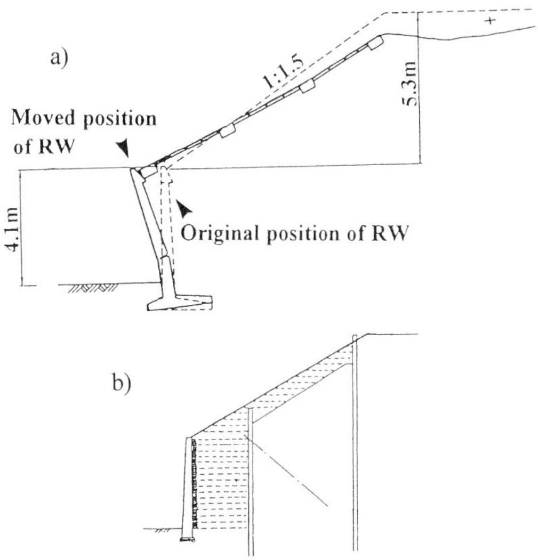

Fig. 11 Cross-sections of a) a typical damaged cantilever-type RC RW without a pile foundation (CL2 in Fig. 4) and b) a GRS-RW by reconstruction, at Shin-nagata Station, JR Sanyo Line.

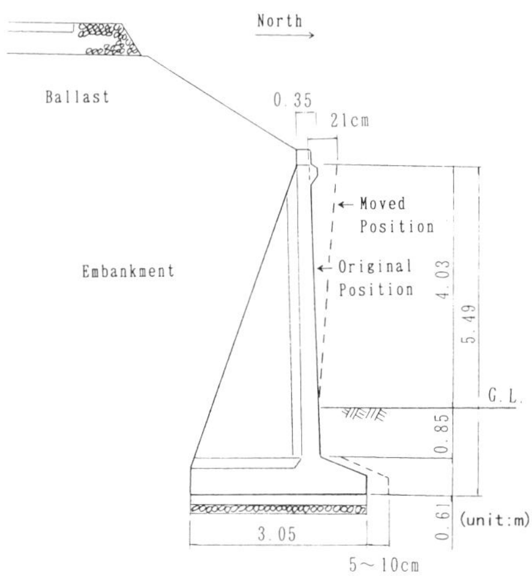

Fig. 12 Typical cross-section of damaged buttressed RC RW without a pile foundation (CL11 in Fig. 4), Shioya Station, Sanyo Dentetsu Line.

at Tanata site (CL3 in Fig. 2) will be discussed later in this paper.

Gravity-type unreinforced concrete RWs is another old type. They were constructed more than 60 years ago. The walls at the Ishiyagawa Station of the Hanshin Line (Fig. 13, GT1 in Fig. 4) were most seriously damaged, although these walls were aseismic-designed by using k_h= 0.2.

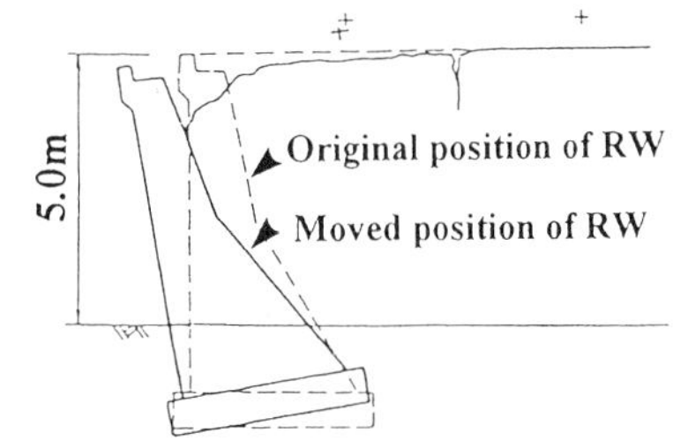

Fig. 13 Cross-section of the seriously damaged gravity type unreinforced concrete RW at Ishiyagawa Station (GT1 in FIg. 4), Hanshin Railway Main Line.

In Tables 1 and 2, damaged embankments and conventional type soil retaining structures constructed for railways and roads, including those described above, are summarized. The features of the damage could be summarized as follows:

1) In contrast to the damaged RC columns and piers, most seriously damaged soil structures were located in the zones where Japanese wooden houses were seriously damaged (Fig. 4). The major reason is perhaps their low amplified responses due to high damping ratios at large strains and presumably the undamaged natural periods T_n that were much shorter than the range of T_P (Fig. 6).

2) A number of embankments and gravity type soil RWs were seriously damaged. Their failure was triggered by i) shear failure within the slope or overturning of the wall induced by high seismic loads applied to the embankments and wall structures, and/or ii) failure in the subsoil. The former cases were observed mostly in the most severely shaken areas. Among the latter cases, catastrophic failure was triggered by soil liquefaction in the sub-soil, which was observed outside the most severely shaken areas shown in Fig. 4 (e.g., RWs; UB1 and GR2 and embankments; EM3 through EM9). A similar type of failure was observed also for a large portion of river dikes.

3) Among the conventional RWs, which are masonry RWs, leaning-type (supported type) unreinforced concrete RWs, gravity-type unreinforced concrete RWs, and cantilever-type or inverted T-shaped type RC RWs (mostly without piles), the first three types of RW were most seriously damaged, and the failure was catastrophic, despite that except for the masonry RWs, they had been aseismic designed. It is likely that even without seismic earth pressure applied to the back face, they would have ruptured or tilted or even completely overturned due to seismic loads exerted on the wall structure itself. A very wide gravity-type unreinforced concrete RW supported by piles would have been stable even during this level of earthquake, but this type of RW is never cost-effective.

4) For many of the damaged RWs, the bearing

Table 1 Damaged retaining walls for railways and roads.

Facility	Site	Location	Type of wall	Height/ length of wall	Subsoil condition	Brief description of damage/ permanent restoration method
Railway	MS1	Between Setsu-Motoyama & Sumiyoshi Stations of JR Kobe Line	Masonry	4 m/ 50 m	-	Total collapse/ GRS-RW
Railway	MS2	Adjacent to Nishi-Nada Station of Hanshin Main Line	Masonry	3.4 to 3.8 m/ 80×2 m	-	Tilting of upper wall, settlement of embankment/ GRS-RW
Road	MS3	City Road Nishi-Nada-Harada and Rokko-Sannomiya Lines at Iwayakita 3 & 4-chome, Nada-ku, Kobe City	Masonry	max. about 5 m/ 70 m	-	Vertical and horizontal cracking of wall, lateral deformation and settlement of embankment/ ?
Railway	LT1	Between Setsu-Motoyama & Sumiyoshi Stations of JR Kobe Line	Leaning-type	2.6 m/ 500 m	Pleistocene gravel (N_{SPT}= 15~50 up)	Complete overturning, partial breakage at the level of subsoil surface/ GRS-RW, RW with embankment reinforced by large diameter nailing
Railway	LT2	Between Okamoto and Mikage Stations of Hankyu Kobe Line	Leaning-type	8.0 m/ 500×2 m	-	Tilting on both sides, cracking near the bottom, settlement of embankment/ U-shaped RW filled with cement-treated soil
Road	LT3	City Road Higashi-Nada-sato No.143 Line at Higashi-Nada-ku, Kobe City	Leaning-type	max. about 5 m/ 160 m	-	Vertical opening and horizontal sliding at construction joint/ partial reconstruction to increase wall height
Railway	GT1	Adjacent to Ishiyagawa Station of Hanshin Main Line	Gravity-type	5.0 m/ 200×2 m	Holocene sand (N_{SPT}=10~30)	Tilting on both sides, partial breakage at construction joint and overturning/ viaduct
Road	GT2	City Road Ookubo No.18 Line at Ookubo-cho, Akashi City	Gravity-type	max. 3.0 m/ 160 m	-	Tilting, longitudinal cracking of embankment/ reconstruction of original RW
Railway	CL1	Between Hyogo & Shin-Nagata Stations of JR Sanyo Line	Cantilever-type	4.0 m/ 400×2 m	-	Tilting on both sides, settlement of embankment, deformation of footpath/ reinforcement by anchoring and tie-rods
Railway	CL2	At Shin-Nagata Station of JR Sanyo Line	Cantilever-type	4.1 m (+5.3 m for overlying embankment)/ 200 m	Holocene clay (N_{SPT}= 5)	Tilting and sliding, cracking at the middle height, settlement of embankment/ GRS-RW, cantilever-type RW with pile foundation
Railway	CL3	At Tanata between Ashiya & Setsu-motoyama Stations of JR Kobe Line	Cantilever-type with pile foundation	5.4 m/ 50 m	Holocene sand and clay (N_{SPT}=25~50 & 10~25)	Tilting and sliding, settlement of embankment/ reinforcement by horizontal tie-rods connected to upper RW adjacent to RC box
Railway	CL4	Adjacent to Ishiyagawa Station of Hanshin Main Line	Cantilever- type	5.0 m/ 30 m	Holocene sand (N_{SPT}=10~30)	Tilting, cracking at the middle height of a section without counterforts/ viaduct
Railway	CL5	Between Higashi-Nada & Kobe-kou Stations of JR Kobe Line (freight branch)	Cantilever-type	4.5 m (+1.8 m for overlying embankment)/ 50 m	-	Tilting, cracking near the bottom, settlement of embankment/ cut-off sheet piles with tie-rods & outer backfill reinforced by geogrid
Road	CL6	Prefectural Highway Shioze-Mondosou Line at Koma-no-machi, Takarazuka City	Cantilever-type	about 4 m/ 80 m	(Adjoining waterway)	Subsidence, tilting and sliding of wall, longitudinal fissuring of embankment/ reconstruction of original RW
Road	CL7	City Road Yamate Main Line at Yuminoki-1-chome, Nada-ku, Kobe City	Cantilever-type	about 3 m/ 60 m	-	Tilting of wall, longitudinal fissuring of embankment/ Terre Armee
Road	CL8	City Road Nishi-Nada-Harada and Rokko-Sannomiya Lines at Iwaya-Kita 3 & 4-chome, Nada-ku, Kobe City	Cantilever-type ? (not confirmed)	max. about 5 m/ 250 m	-	Tilting and vertical cracking of wall, opening of vertical joint, settlement of embankment/ Terre Armee
Road	CL9	Regional Highway Baiko-Hamabedori-Wakihama Line at Masago-dori 1 & 2-chome, Chuo-ku, Kobe City	Cantilever-type ? (not confirmed)	max. 3.7 m/ 310 m	-	Tilting of wall (cracking of side wall of RC box culvert adjoining the wall)/ reconstruction of original RW ?
Road	CL10	Regional Highway Sanroku Line at Hiyodorigoe, Hyogo-ku, Kobe City	Cantilever-type ? (not confirmed)	max. 4.5 m/ 40 m	Holocene sand (N_{SPT}=3~50)	Sliding of wall/ reinforcement by earth anchor fixed to firm subsoil
Railway	CL11	Shioya Station of Sanyo Dentetsu Line	Cantilever-type	about 4 m/ 360 m	-	Tilting of wall/ viaduct
Road	UB1	City Road Ashiya-hama Line at Midori-cho, Ashiya City	U-shaped with inner backfill	about 2 m/ 560 m	-	Uneven settlement and tilting/ reconstruction of original RW, correction of wall height
Railway	GR1	At ***Tanata*** between Ashiya & Setsu-motoyama Stations of JR Kobe Line	GRS with rigid facing	max. 4.5 m/ 300 m	Holocene sand and clay (N_{SPT}= 5~50 & 5)	Tilting and sliding, partial cracking of facing at the middle height, settlement of embankment/ reinforcement by horizontal tie-rods connected to upper RW adjacent to RC box
Park	GR2	In Maihama Park at Ashiya-hama, Ashiya City	GRS with concrete-block facing	5.3 m (+1.0 m for overlying embankment)/ 90 m	-	Uneven settlement and opening of facing blocks (+extensive sand boil and fissures due to liquefaction at the subsoil surface)/ ?
Road	GR3	Near approach road to Akashi Kaikyo Bridge (under construction) at ***Maiko***, Tarumi-ku, Kobe City	GRS with flex. metal-mesh facing	max. 4.0 m/ 95 m	-	Slight differential horizontal movement of sound barrier foundation at the top of embankment (uneven settlement & fissures at the subsoil surface)
Railway	GR4	Near ***Amagasaki*** Station of JR Kobe Line	GRS with rigid facing	max. 8 m (av. 5 m)/ 1000 m (including two abutments by GRS-RW)	-	No damage

Table 1 Damaged retaining walls for railways and roads. (continued)

Facility	Site	Location	Type of wall	Height/ length of wall	Subsoil condition	Brief description of damage/ permanent restoration method
Railway	GR5	Near approach road to Akashi Kaikyo Bridge (under construction) at ***Maiko***, Tarumi-ku, Kobe City	GRS with rigid facing	max. 6.7 m/ 150 m	-	No damage
Railway	GR6	Near ***Amagasaki*** Station of JR Kobe Line	GRS with rigid facing	max. 8 m/ 400 m	-	No damage
Road	MA	Kita-Kobe Line of Hanshin Express way at Kami-Tanigami, Yamada-cho, Kita-ku, Kobe City	Multiple-anchor-reinforced soil with RC facing	max. 4.6 m/ 24 m	-	Tilting of cap concrete without anchors at the top of RW/ ?
Road	TA1	National Highway Route No.28 at Hokutan-cho, Awaji Island	Terre Armee	max. 8.9 m/ ? m	-	Tilting of facing, partial cracking of facing at the corner/ reconstruction using lengthened metal-strip
Park	TA2	In Hoshi-ga-oka Park at Tarumi-ku, Kobe City	Terre Armee	max. 5.1 m/ 43 m	-	Tilting and sliding of facing, partial cracking of facing at the corner/ ?
Others	TA3	In Midori-ga-oka Pool at Itami City	Terre Armee	max. 3.3 m/ 36 m	-	Tilting of facing, partial cracking of facing at the corner, settlement of embankment/ ?
Road	TA4	Approach Road to Akashi Kaikyo Bridge (under construction) at ***Maiko***, Tarumi-ku, Kobe City	Terre Armee	max. 6-7 m/ ? m	-	Noticeably compressing at the bottom of facing / ?

RW: retaining wall, GRS: geogrid-reinforced soil, Terre Armee: metal-strip reinforced soil with discrete RC facing

Table 2 Damaged embankments for railways and roads.

Facility	Site	Location	Slope angle/ facing condition	Height/ length of embankment	Subsoil condition	Brief description of damage/ permanent restoration method
Railway	EM1	Between Hyogo & Shin-Nagata Stations of JR Sanyo Line	1:1.5/ cast-in-place concrete lattice and precast concrete panel	4.4 m/ 450 m (full embankment)	-	Settlement and lateral deformation, sliding of concrete facing/ embankment partially reinforced by geogrid and covered with geomembrane (artificial lawn type)
Railway	EM2	Between Higashi-Nada & Kobe-kou Stations of JR Kobe Line (freight branch)	1:1.5/ vegetation	6.0 m/ 320 m (full embankment)	-	Settlement and lateral deformation, longitudinal crack/ embankment reinforced by large diameter nailing
Road	EM3	Ookura-dani Interchange of the Second Shinmei Expressway	-	15 m/ 30 m (full embankment)	Soft clay and partial sand (N_{SPT}<5, former pond)	Collapse of embankment/ ?
Road	EM4	Regional Highway Kobe-Kakogawa-Himeji Line at Zenkai, Ikawadani-cho, Nishi-ku, Kobe City	1:2.0/ precast concrete panel	7 m/ ? m (half bank)	(existing pond at slope toe)	Sliding of slope and facing/ reinforcement by sheet pile driven at the middle of slope
Road	EM5	Regional Highway Kobe-Kakogawa-Himeji Line at Minami, Kande-cho, Nishi-ku, Kobe City	-	max. 8 m/ ? m (half bank)	(existing pond at slope toe)	Collapse of embankment/ ?
Road	EM6	Regional Highway Kobe-Kakogawa-Himeji Line at Wada, Oshibedani-cho, Nishi-ku, Kobe City	-	? m/ ? m (half bank)	(existing pond at slope toe)	Sliding of slope/ ?
Road	EM7	City Road Fukui Line at Fukui, Miki City	-	max. 13 m/ 66 m (full embankment)	(existing pond at slope toe)	Collapse of embankment/ protection work against seepage of water from existing pond?
Road	EM8	Town Road Ura No.105 Line at Kusumoto, Higashiura Town, Tsuna District	-	? m/ 116 m (half bank)	(swamp)	Collapse of embankment/ reconstruction of original embankment
Road	EM9	Town Road Takataki Line at Ikuho, Tsuna Town, Tsuna District	-	max. 7.5 m/ 34 m (half bank)	(swamp)	Collapse of embankment/ cast-in-place concrete lattice and partial drainage mat in embankment
Road	EM10	Town Road Hirata-Shitamichi Line at Takayama, Ichinomiya Town, Tsuna District	-	? m/ 30 m (half bank)	-	Collapse of embankment/ reconstruction of original embankment

capacity of the subsoil and shear resistance along the wall base was not enough to prevent the failure by overturning and base sliding. Although the use of a strong pile foundations could have effectively prevented this type of failure, the construction cost would have become high.

5) A number of RC columns and piers were seriously damaged or even totally collapsed. Compared with them, a number of modern RC RWs supported by piles and reinforced soil RWs performed much better.

Brittle versus ductile behaviour: The ductility of a structure means the capability for large energy absorption after yielding, which means a large area between the post-yield load-displacement curve and the horizontal axis (Fig. 14). In this schematic figure, according to concrete engineering, the ductility μ for a given structure is defined as the ratio δ_u/δ_y, where δ_u and δ_y are the deformation at the ultimate failure where the strength drops rapidly to zero or a near-zero residual strength

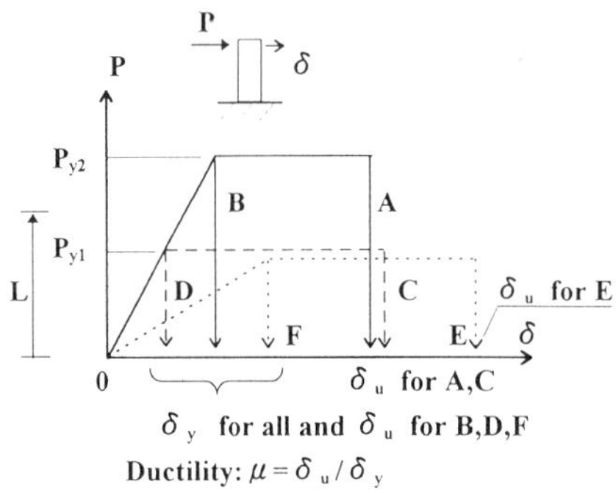

Type	Preyield stiffness	Yield strength	Ductility
A	High	High	High
B	High	High	Low
C	High	Low	High
D	High	Low	Low
E	Low	Low	High
F	Low	Low	Low

Fig. 14 Seismic structure characterization by pre-yield stiffness, yield strength and ductility (a schematic figure).

and the deformation at yielding, respectively.

In Fig.14, structures A, B, C and D have the same high pre-yield stiffness, while structures A and B have a high yield strength P_{y2}, and structures C and D have a low yield strength P_{y1}. Structures E and F have a low pre-yield stiffness and a low yield strength P_{y1} (n.b., for simplicity, structures C through F are assumed to have the same yield strength). Structures A and C have the same high ductility, and structure E has the highest ductility, while structures B, D and F have zero ductility.

For load L lower than the yield strength P_{y1}, structures A, B, C and D exhibit the same and small deformation δ, while structures E and F exhibit the same but larger δ. The behaviour of structures E and F is of ordinary soil structures in comparison with steel and RC structures (i.e., structures A, B, C and D).

In dynamic behaviour, when the load L exceeds the resistance P available at a given δ, the difference L - P is resisted by the inertia force of the structure. After yielding, a low value of P results in a large value of L - P with large deformation as obtained by double integration of the acceleration.

As the load L exceeds P_{y1}, δ of structure D becomes larger than that of structure C, and δ of structure F becomes larger than that of structure E. The difference increases with the difference L - P_{y1}. As the load L exceeds P_{y2}, δ of structure B becomes larger than that of structure A. It is specified in the current design codes for RC structures that the shear behaviour of columns and piers should be, comparatively, like that of structure B and the bending behaviour should be like that of structure C so that shear failure does not occur before bending failure occurs; in other words, if failure occurs, it should be bending failure, exhibiting sufficiently ductile behaviour.

At and near JR Rokko-michi Station (see Fig. 4), most of the RC columns of the elevated RC frame structures were seriously damaged or totally collapsed by shear failure. Adjacent to these structures, both sides of the railway embankment were supported by inverted T-shaped RC RWs without a pile foundation. A part of the RWs moved at the base only about 10 cm and/or tilted outward slightly for an angle of about 2 degree (Fig. 7d). When L exceeds P_{y2}, δ of structure B could become larger than that of structure E. If such RC columns which exhibited shear failure can be modeled by structure B, these RC RWs could be modeled by structure E.

The illustrations described above suggest that, for loads exceeding the yield strength, large ductility could be more important than large pre-yield stiffness. It is also understood that it is not enough for a structure to have a reasonably high yield strength and pre-yield stiffness so that it does not exhibit too large displacements against moderate seismic loads. The structure should also be ductile enough as not to totally collapse by very large seismic loads.

On the other hand, failure is brittle for soil structures exhibiting flow failure due to soil liquefaction, for structures resting on liquefying sub-soil deposits and for gravity type RWs exhibiting overturning failure. These structures could be modeled by structure F. This type of structure has the lowest seismic stability for any level of load.

2.3 *Natural slopes*

The earthquake did not seriously damage natural slopes except one landslide at Nikawa, Nishinomiya City, where 11 houses were destroyed and 34 persons were killed. Sassa et al. (1996) summarized locations of the landslides (Fig. 15) and reported that their distribution overlapped that of aftershocks, which were not inversely proportional to the

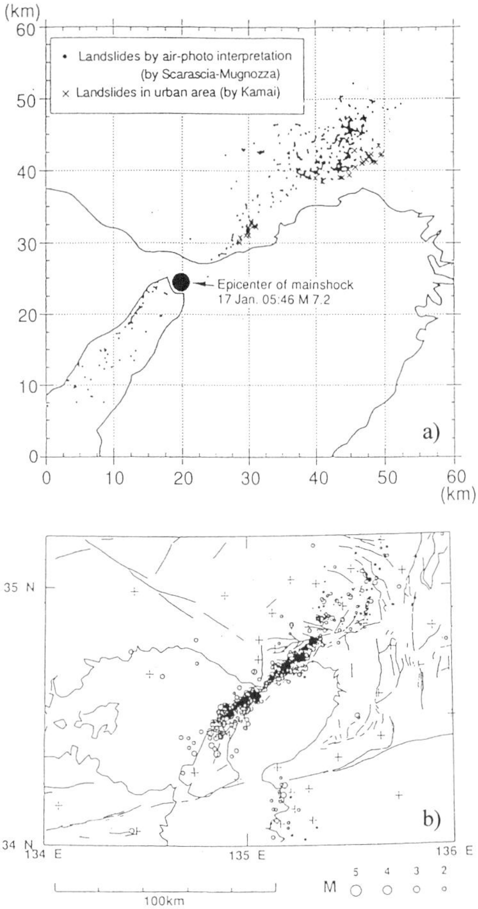

Fig. 15 Distributions of a) slope failures and land slides and b) aftershocks (after Sassa, et al., 1996).

Fig. 16 a) Cross-section and b) front views of the GRS-RW at Maihama Park, Ashiya City (GR2 in Fig. 4) (after Naemura and Miyatake, 1995).

epicentral distance, but rather distributed linearly along active faults. Many of them occurred in the region of Mt. Rokko, which was outside the most severely shaken areas shown in Fig. 4, and the size of each case was not extensive. Based on these observations, it seems that slopes which had been nearly at critical condition at the time of the earthquake suffered from failure by relatively low earthquake motion in the mountain areas. Relatively low ground water level due to the driest season of the year may have contributed to reducing both the number of slope failures and landslides and their extent as pointed out by Sassa et al. (1996).

3 PERFORMANCE OF GEOTEXTILE-REINFORCED SOIL RETAINING WALLS

3.1 *GRS-RWs without full-height rigid facings*

Around the periphery of the most severely shaken areas (see Fig. 4), several types of geotextile-reinforced soil retaining walls (GRS-RWs) without full-height rigid (FHR) facings was constructed as secondary applications.

Fig. 16 (GR2 in Fig. 4) shows a GRS-RW that was constructed in May 1993 at Maihama Park in a reclaimed land (Naemura and Miyatake, 1995). The wall was 5.0 m high and 90 m long. The facing was inclined at 0.5:1.0 in horizontal:vertical. The reinforcement was a high density polymer grid (the tradename is Tensar SR-55) having a nominal tensile strength of 49 kN/m (5 tonf/m). The reinforcement was on average 4.5 m in length with a vertical spacing of 1.4 m. The facing was a stack of discrete concrete blocks having a hollow core with vegetation inside. The facing was not connected with the reinforced backfill, but separated by a layer of gravel.

Noticeable soil liquefaction occurred in the subsoil in front of the wall. The ground settled unevenly with cracks, from which sand and water erupted. One large crack reached to the subsoil below the facing blocks, which resulted in a separation between adjacent blocks of about 10 cm and noticeable unequal settlement of the facing. Yet, the deformation of the facing was smaller than that in the subsoil, and the integrity of the facing and backfill was maintained. In fact, the wall has been in service since the earthquake.

Fig. 17 (GR3 in Fig. 3) shows another GRS-RW, similar to the above, which was constructed in September 1993, along the west-side slope of an approach road to Akashi Kaikyo Bridge (MKS, 1995, Nishimura et al., 1996). The wall was 4 m high and 95 m long. The reinforcement was Tensar SR-55, as for the aforementioned GRS-RW, and the facing consisted of L-shaped steel wire frame. On the ground surface in front of the wall, cracks with a maximum opening of 20 cm appeared, and an unequal settlement of about 20 cm was observed. One of the construction joints of the foundation for noise barrier walls on top of the GRS-RW exhibited a lateral shear displacement of about 2 cm. Despite the above, no problematic deformation of the GRS-RW was noticed.

These two case histories show some ductility of GRS-RWs in comparison with the catastrophic brittle failure of several unreinforced embankments triggered by the soil liquefaction in the subsoil (as listed in Table 2).

3.2. *GRS-RWs having full-height rigid facings*

GRS-RWs having full-height rigid (FHR) facings were constructed at four locations in 1990 - 1994 in the affected area (marked by symbols ◎ in Fig. 3). The walls had been aseismic-designed using $k_h = 0.2$. These GRS-RWs were constructed by following the staged construction procedures shown in Fig. 18 (Tatsuoka et al., 1992, 1996c). The reinforcement was a grid made of fibers of polyvinyl alcohol (the tradename is Vinylon) coated with soft PVC for protection, with a nearly rectangular cross-section of 2 mm times 1 mm and an aperture of 20 mm. The nominal tensile rupture strength is 29 kN/m (3tonf/m). The following three walls were located in the areas where the JMA scale was fifth or sixth:

a) Amagasaki, No. 1 (Fig. 19; wall height H= 5 m on average and total length L= about 1 km): The wall was completed in April 1992 to support two new tracks added on both sides of an existing railway embankment of the JR Kobe Line (GR4 in Fig. 3). The backfill was as basically a cohesionless soil including a small amount of fines. A number of foundations for a steel frame structure for electricity supply had been constructed inside the reinforced backfill. Four pairs of bridge abutments of GRS-RW had been constructed. No deformation of the wall by the earthquake was observed.

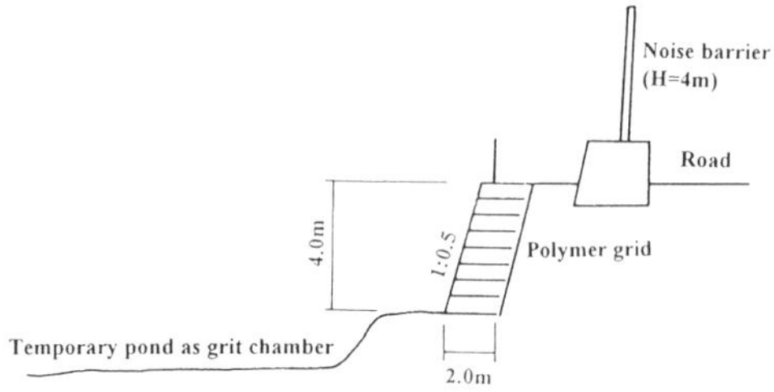

Fig. 17 Cross-section of the GRS-RW along the west-side slope of one of the approach roads to Akashi Kaikyo Bridge (GR3 in Fig. 3; after Mitsubishi Yuka Sanshi Products Co.,Ltd., 1995).

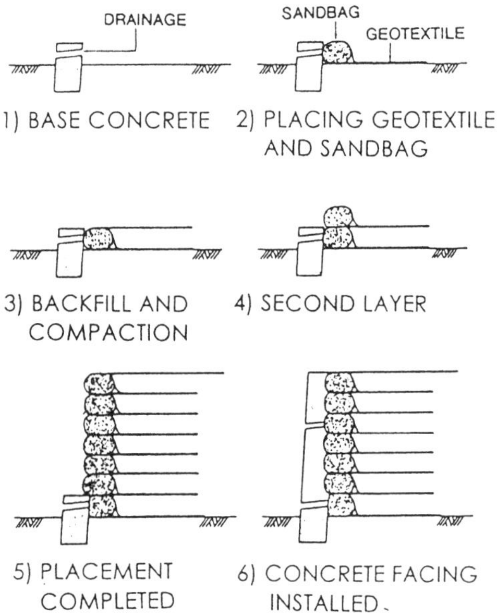

Fig. 18 Standard staged construction procedures for GRS-RWs having full-height rigid facings.

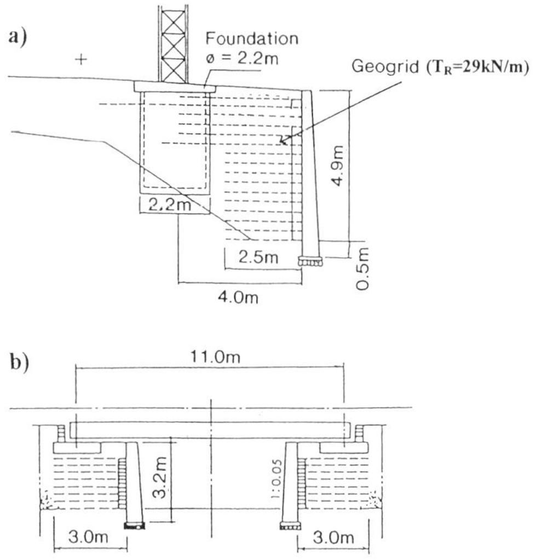

Fig. 19 a) Typical cross-section of Amagasaki No. 1 GRS-RW and b) typical GRS bridge abutments (GR4 in Fig. 3).

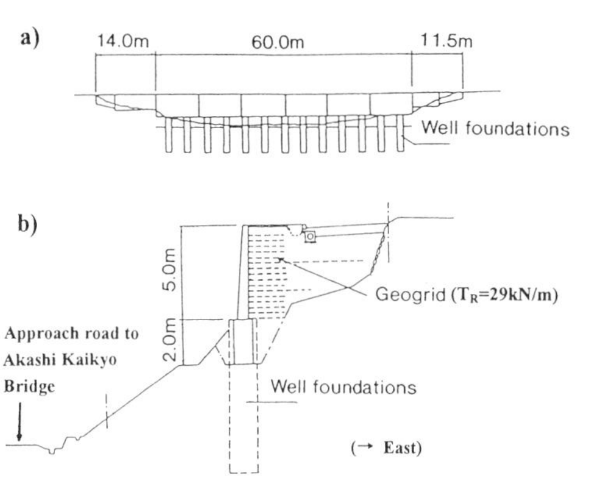

Fig. 20 a) Front view and b) cross-section of Tarumi GRS-RW (Block B) (GR5 in Fig. 3).

b) Maiko at Tarumi-ku, Kobe City (GR5 in Fig. 3): The wall having Blocks A and B, constructed in 1993 to expand the road immediately east of one of the approach roads to Akashi Kaikyo (Strait) Bridge. The backfill was a well-graded gravel. On the opposite side of the approach road, there exists the GRS-RW without a FHR facing described in section 3.1. Block B (Fig. 20; H= 4.9 - 5.1 m and L= 60 m) was located on a steep slope facing the approach road. The facing was supported with well foundations. The wall moved outward about 2 cm at a maximum at the top of the facing. Block A (Fig. 21; H= 3.7 - 6.6 m and L= 55 m), facing the direction opposite to the approach road, moved outward about 1 cm relative to the low-height gravity type RW immediately to the right of the GRS-RW in Fig.21a. On the left of the GRS-RW in Fig. 21a, a RC wall had been constructed, supported by steel pipe piles (Fig. 21c). Although the height of the RC wall was shorter than that of the GRS-RW, it moved outward about 2 cm relative to the GRS-RW showing a maximum depression of about 10 cm at the road face behind the RC wall. This fact indicates that RC RWs supported by piles could be less cost-effective in terms of seismic stability than GRS-RWs having FHR facings. A similar case was observed on a larger scale at Tanata, to be described later. The displacement of the masonry walls on the left end and the gravity type RW on the right end was not observed, probably due to their small height (about 2.2 m). No further displacement was observed after the earthquake in all these walls.

c) Amagasaki, No. 2 (Fig. 22) (H= 2.0 - 8.0 m and L= 500 m; GR6 in Fig. 3): The site is west of the Amagasaki No. 1 GRS-RW, adjacent to Amagasaki Station (Fig. 19). The wall was completed in March 1994 for the JR Fukuchiyama Line. No displacement of the wall by the earthquake was observed.

In the areas adjacent to these GRS-RWs having FHR facings, a number of wooden houses, railway and highway embankments and conventional types of RWs were seriously damaged (see Fig. 3). The degree of damage was, however, not as serious as that observed in the areas where the JMA scale was equal to seventh or higher (Fig. 4). In one of these areas, at Mori-Minami-cho 1-chome in Higashi-Nada-ku, Kobe City (the local name is Tanata), GRS-RWs having FHR facings (H= 1.5 m - 6 m and L= 305 m) were completed in February 1992 on the south slope of the embankment of JR Kobe Line (Fig. 23; GR1 in Fig. 4). The wall construction was to increase the number of railway tracks from four to five (Figs. 23a and b). The subsoil condition is relatively good (see Figs. 23d and e). In sections where the wall was higher than 1.5 m, H-shaped steel piles at a spacing of 1.5 m with temporary anchors were provided at two elevations to retain the embankment before the slope was excavated for the construction of the wall (Fig. 23d). The anchors were removed as the GRS-RWs were being constructed.

The bottom of the wall moved outward on average about 5 cm relative to the supporting foundation subsoil, pushing the subsoil in front of the wall laterally. At the highest part of the wall, which was in contact with a RC box culvert structure crossing the railway embankment, the largest outward displacement occurred, which was 26 cm at the top of the wall and 10 cm at the ground surface level (Fig. 24a).

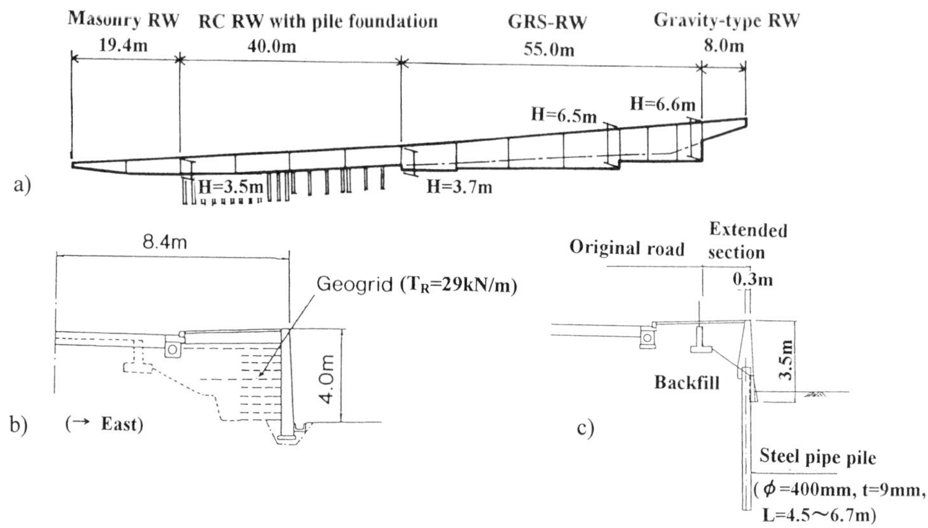

Fig. 21 a) Front view and b) cross-section of Tarumi GRS-RW (Block A) (GR5 in Fig. 3) and c) cross-section of the adjacent RC RW with piles.

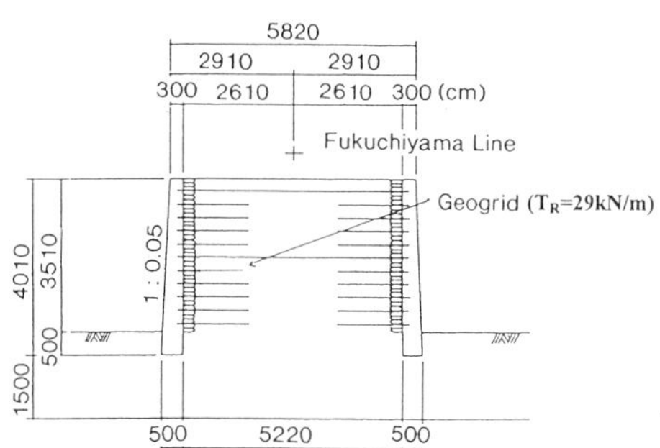

Fig. 22 Typical cross-section of Amagasaki No. 2 GRS-RW (GR6 in Fig. 3).

Despite the noticeable movement of the wall, the performance of the GRS-RW walls were considered quite satisfactory for the following reasons;

a) At the site and in the adjacent areas, the seismic intensity was extra-ordinarily high (Fig. 4). More than 80 % of the wooden houses in front of the wall totally collapsed, but most of them were not particularly old (Fig. 25). In the areas surrounding the wall, many RC buildings and RC piers for highways were seriously damaged or totally collapsed. No doubt, this GRS-RW experienced the highest seismic load among the modern GRS-RWs.

b) On the opposite side of the RC box structure, a RC RW with a largest height of about 6 m was constructed concurrently with the GRS-RW (Fig. 23e). The wall was supported by a row of bored piles, despite the sub-surface condition was similar to that of the GRS-RW. Consequently, the construction cost per wall length for the RC RW was about double to triple that for the GRS-RW. The RC RW displaced similarly to the GRS-RW; i.e., at the interface with the side of the RC box structure, the outward lateral displacement was 21.5 cm at the top and 10 cm at the ground surface level (Fig. 24b). This comparable behaviour of these two different types of RWs constructed adjacent to each other was similar to that at Maiko described above. Moreover, the GRS-RW having a FHR facing had been constructed adjacent to and to the west of the RC RW (see Fig. 23a), which was 3 m high and 45 m long. The wall did not move. In the design of RC RWs, such as the one at Tanata, seismic earth pressure calculated by the Mononobe-Okabe method using k_h= 0.2 is resisted by the lateral and rotational resistance of the pile foundation, which results mainly from passive earth pressure in the subsoil in front of the pile foundation. Therefore, some lateral displacement of the pile was inevitable to mobilize the large passive earth pressure.

c) The design standard for railway earth structures (Ministry of Transport, 1992) specifies the minimum allowable length of grid reinforcement for the GRS-RW system to be the larger value of either 35 % of the wall height or 1.5 m. For most of the GRS-RWs constructed so far, to be conservative, several top reinforcement layers were made longer than the others at lower levels (Figs. 19, 20 and 21). For the Tanata GRS-RW, however, all the reinforcement layers were truncated to nearly the same length (Fig. 23d). That was due to a construction constraint to allow trains to run in the area where the top several reinforcement layers should have been extended. This arrangement may have reduced the seismic

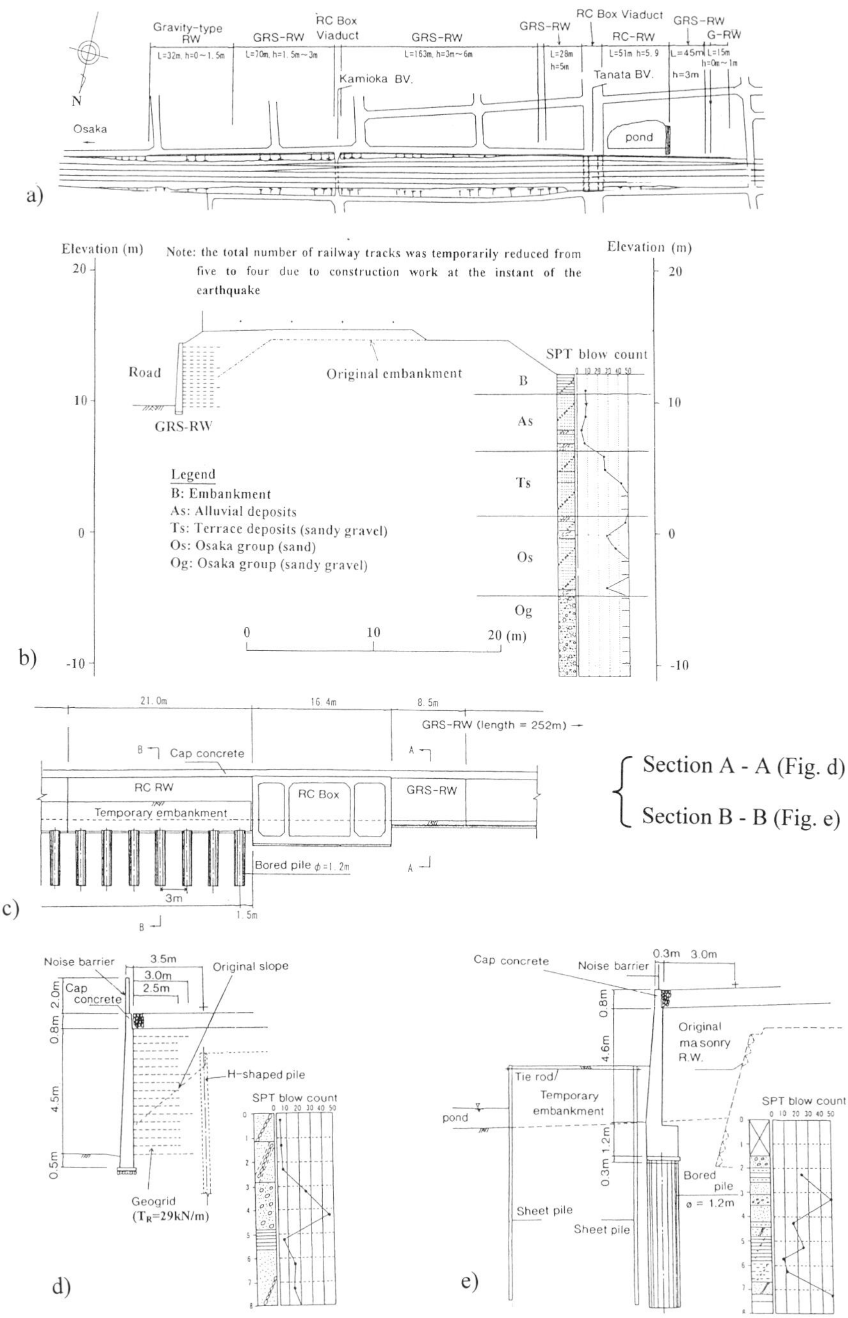

Fig. 23 a) Plan, b) cross-section of the embankment and c) front view from the south and cross-sections of d) GRS-RW and e) RC RW with pile (GR1 in Fig. 4), at Tanata between Ashiya and Setsu-motoyama Stations, JR Kobe Line.

stability of the wall, particularly in terms of overturning.

It is very likely that GRS-RWs having FHR facings would not have collapsed even if they had been constructed on liquefying sub-soil. Murata et al. (1994) performed a shaking table test under plane strain conditions of a large model GRS-RW con-

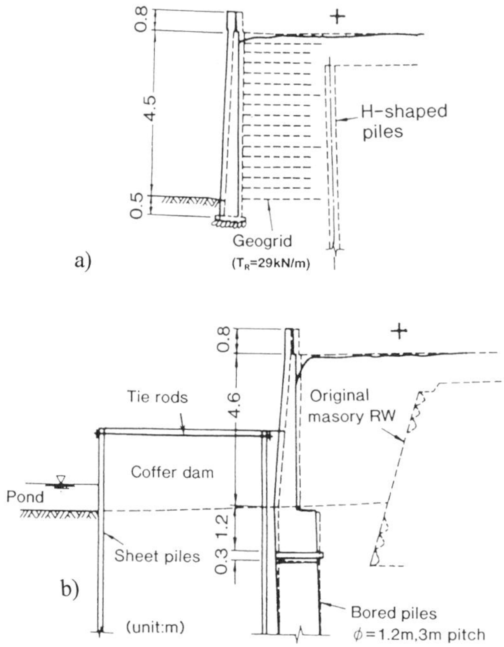

Fig. 24 Deformation due to the earthquake of a) GRS-RW and b) RC RW with piles at Tanata.

Fig. 25 View in front of the GRS-RW immediately after the earthquake

structed on a saturated sand layer (Fig. 26). The sand zones on both the sides of the wall liquefied by the application of 360 cycles of sinusoidal loads at a frequency of 2 Hz with a peak acceleration of 400 cm/sec^2. It seems that the sand zone beneath the center of the wall did not liquefy due to the weight of the wall, as seen from the excess pore water pressure recorded beneath the center of the wall. As a result, the facings settled about 2cm relative to the reinfor-

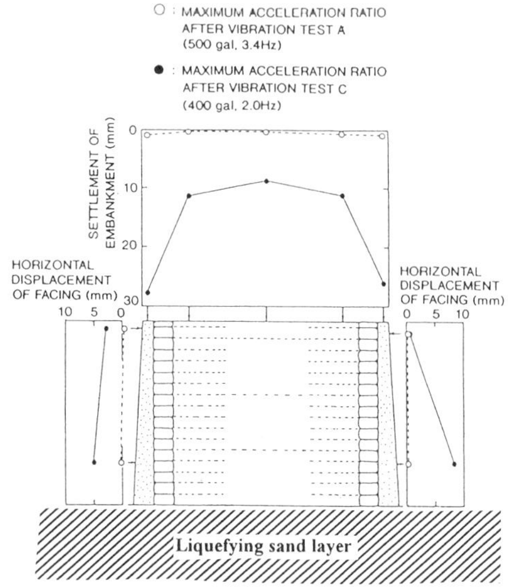

Fig. 26 Deformation of a large GRS-RW model constructed on a liquefied sand deposit (after Fig. 5 of Murata et al., 1994).

-ced backfill, but the wall maintained its integrity. This result suggests that when the foundation subsoil liquefies, a GRS-RW having a FHR facing will deform and displace more than when the subsoil does not liquefy, but may not totally collapse.

4 PERFORMANCE OF STEEL-REINFORCED SOIL RETAINING WALLS

4.1 *Terre Armee walls*

Among other types of permanent reinforced soil RWs that had been constructed in the areas affected by the earthquake, Terre Armee walls had the largest number; 66 walls had been constructed within a radius of 70 km from the epicenter. Although they were located outside the most severely shaken areas where the JMA scale was equal to seventh or higher, several walls had been constructed in the next most severely shaken areas with a JMA scale equal to sixth. They exhibited high seismic resistance without any total collapse. However, three walls exhibited relatively large deformation as the GRS-RW at Tanata, and one of them was partly reconstructed after the earthquake.

Fig. 27 shows a relatively large scale Terre Armee RW that was completed in January 1994 in Awaji Island (TA1 in Fig. 3) as a part of the approach road

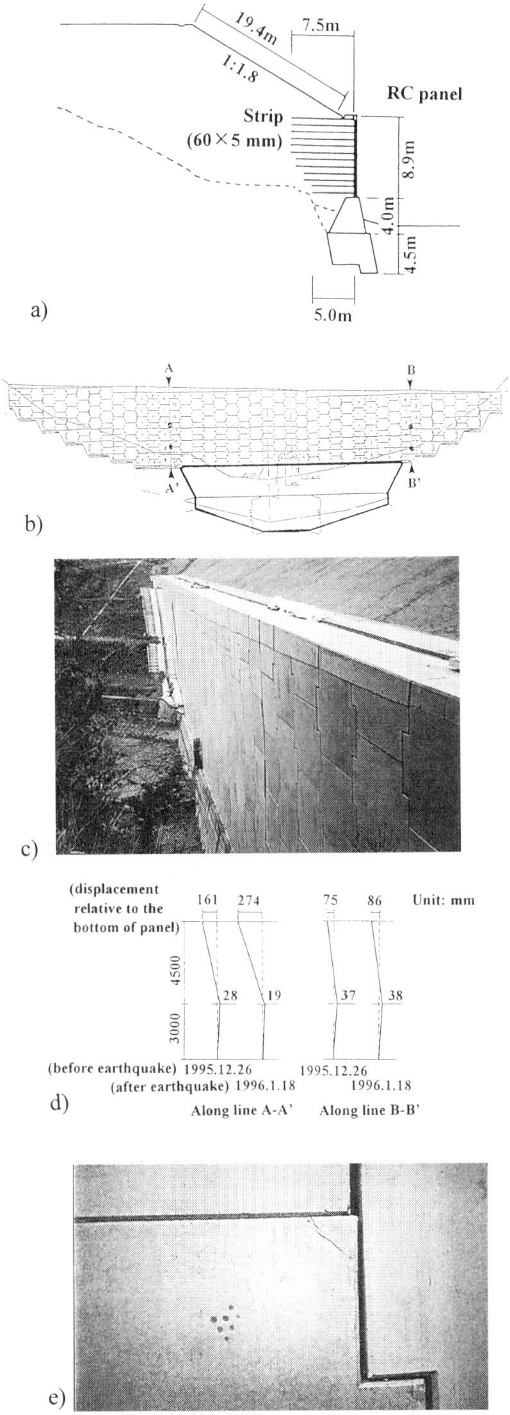

Fig. 27 a) Cross-section, b) front view, c) general view, d) wall deformation and e) cracking at the corners of RC panel, Terre Armee wall in Awaji Island (TA1 in Fig. 3) (JTAA, 1995).

(National road No. 28) to Akashi Kaikyo bridge (JTAA, 1995). The largest wall height was 9 m and the total wall area was 464 m^2. The backfill soil is a medium-grained sand including fines less than 20%. The facing consisted of discrete precast RC panels. The wall had been aseismic-designed by the pseudo-static stability analysis using k_h= 0.15. The largest overhanging of 27.4 cm at the top of the facing relative to a level of 4.5 m down from the top was found after the earthquake (Fig. 27d). About a half of that deformation was considered due to the earthquake. Corner portions of some panels cracked (Fig. 27e), probably due to relative movements between the adjacent panels during the earthquake.

According to the information obtained by the authors, for conservatism, the top 4.5 m height for the entire wall length was reconstructed by increasing the length of the top six reinforcement layers from 7 m to 9 m aiming at increasing the pull-out resistance. This re-arrangement was based on another pseudo-static stability analysis using k_h= 0.4.

Fig. 28 (TA2 in Fig. 3) shows another Terre Armee RW, which has two continuous wall faces bending at a right angle at the center (JTAA, 1995). The RW was retaining an earthfill for Hoshiga-oka Park of Kobe City, completed in November 1989. The largest wall height and total wall area were, respectively, 4.5 m and 162.7 m^2. The backfill was a gravelly soil. The wall exhibited relatively large deformation and displacement. Note that a masonry wall located 100 m from the site collapsed totally (Fig.29). The largest overall outward movement of the facing relative to the facing bottom was reported to be about 9 cm. However, from the fact that the largest lateral opening between adjacent panels was about 20 cm (Fig. 28e) and a number of openings with noticeable widths appeared on the crest on the backfill (Fig. 28f), it is very likely that the whole wall moved outward much more than 9 cm. Also for this wall, corner portions of some RC panels cracked.

Fig. 30 (TA3 in Fig. 3) shows another permanent Terre Armee RW which exhibited relatively large deformation (JTAA, 1995, Matsui et al., 1996). This wall was completed in June 1980, having a largest wall height of 3.27 m and a total wall area of 120 m^2. The top of the wall moved outward 94 mm at the largest. However, considering the fact that the top surface of the backfill settled 20 cm to 100 cm, it is very likely that the base of the facing also moved outward to some extent. Also for this wall, corner portions of some RC panels cracked (Fig. 30g).

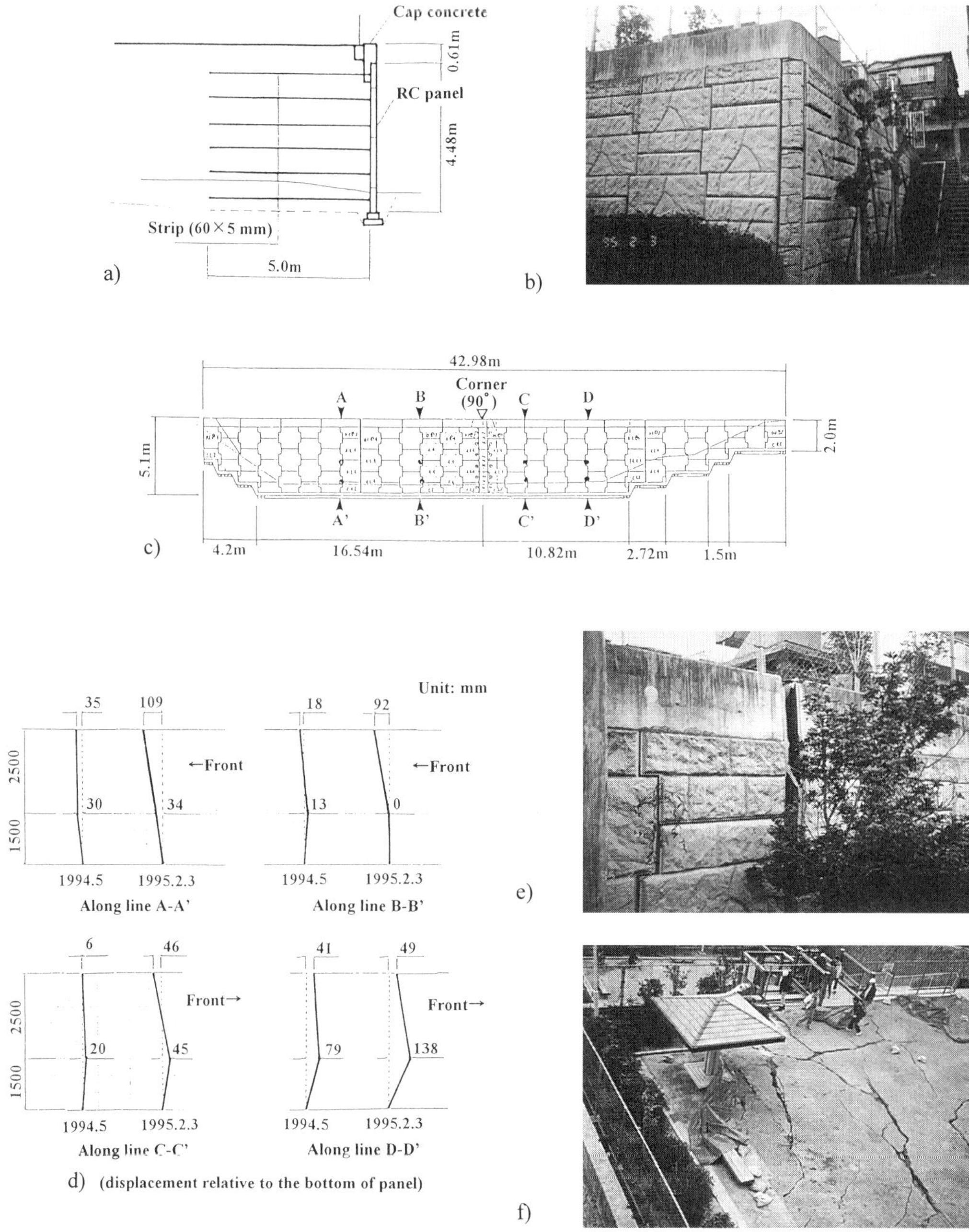

Fig. 28 a) Cross-section, b) general view, c) expanded front view, d) wall deformation, e) opening at a vertical construction joint and f) cracking on the embankment surface, Terre Armee wall for Hosiga-oka Park, Kobe City (TA2 in Fig. 3) (JTAA, 1995).

Fig. 29 A collapsed masonry wall located near the Terre Armee wall TA2 (JTAA, 1995)

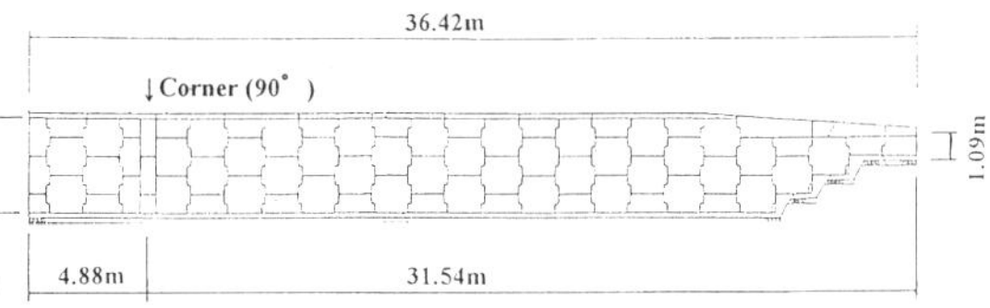

c)

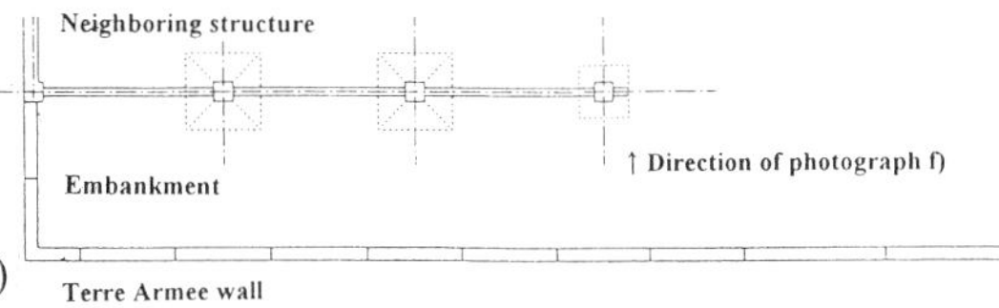

d)

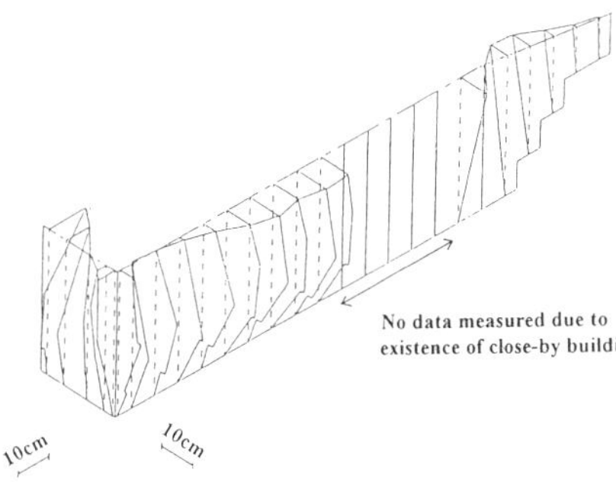

e)

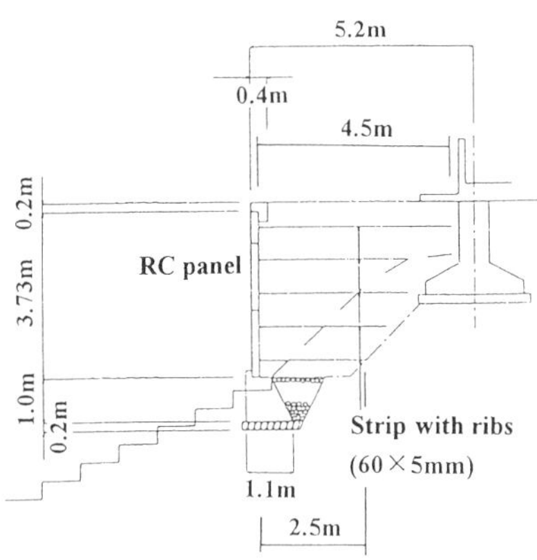

a)

f)

b)

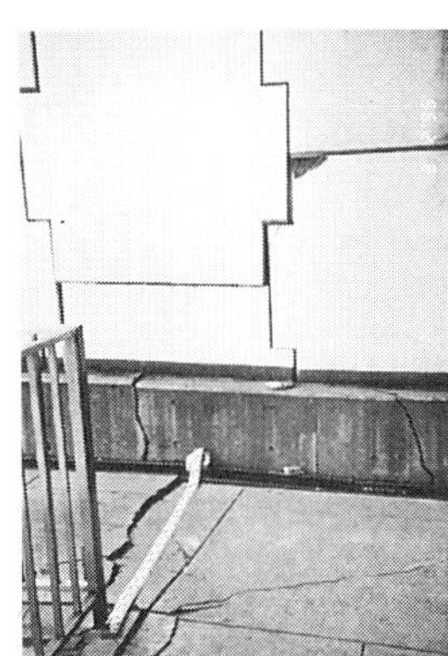

g)

Fig. 30 a) Cross-section, b) general view, c) expanded front view, d) plan, e) wall deformation, f) settlement of embankment relative to the neighboring structure, and g) cracking of the corner of RC panel, Terre Armee wall at Itami City (TA3 in Fig. 3) (JTAA, 1995, Matsui et al., 1996).

Fig. 31 Temporary Terre Armee wall at Tarumi-ku (by the courtesy of Prof. Akagi, T. of Toyo University).

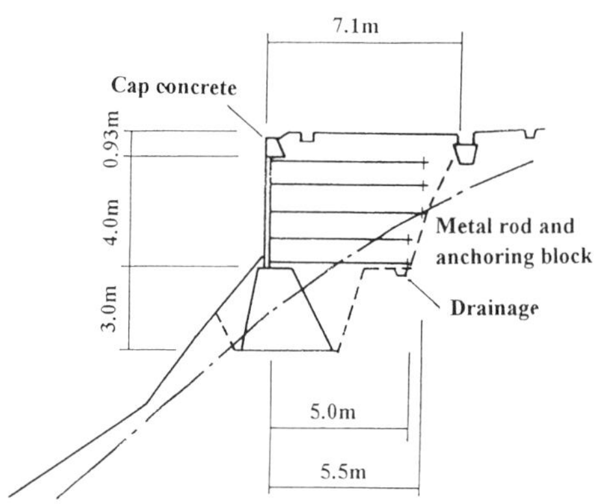

Fig. 32 Cross-section of a multiple-anchor soil RW (MA in Fig. 3).

There was a 6 -7 m high temporary Terre Armee wall with a metal skin facing at Tarumi-ku for the construction of an approach road to Akashi Kaikyo Bridge. Some part of the facing, particularly near the bottom, was compressed noticeably (Fig. 31, TA4 in Fig. 3), inducing some outward displacement and settlement at the entire wall. As the wall did not show any sign of total collapse, only some remedy work was performed.

In summary, the Terre Armee RWs generally behaved very well. The RWs are made relatively deformable by using a discrete panel facing to accommodate possible deformation of the subsoil and backfill. It seems that because of the above, the walls deformed to some extent during the earthquake to a higher level than the conventional RC walls that were stable during the earthquake and the GRS-RWs having FHR facings. Cracks observed at the corner potions of RC panels suggest the above. The ductile behaviour of the wall may have contributed to avoiding total collapse by reducing the seismic response presumably due to large energy absorption by the backfill and the dynamic earth pressure exerting on the back face of the reinforced zone. It seems that as a result, the residual deformation has become larger.

4.2 *Other types of walls*

In the mountain areas, where the seismic intensity was lower than in the areas shown in Fig. 4, some multiple-anchor soil RWs had been constructed. Precast RC panels were anchored with steel tie rods having small anchor concrete blocks at the deepest ends. The wall shown in Fig. 32 (MA in Fig. 3) had a largest wall height of 4.6 m and a total wall area of 87 m^2 (JMAA, 1995). No structural damage and deformation was observed at the wall face, but the cap concrete, that was not anchored, moved backward slightly.

5 PERFORMANCE OF REINFORCED NATURAL SLOPES

5.1 *Nailed slopes*

Felio et al. (1990) reported good performance of eight nailed slopes during the 1989 Loma Prieta earthquake. Torii (1996) summarized the performance of seven nailed slopes during this earthquake (Table 3 and Fig. 33a). The PHGAs listed in Table 3 were estimated from the PHGA contours shown in Fig. 33b. It is reported that no noticeable deformation of the nailed slopes was observed despite relatively high PHGAs, which was in contrast to serious damage to adjacent civil engineering structures and unreinforced slopes (except the one case described below).

Among them, perhaps the behaviour of the nailed wall NS6 (Fig.3) was most carefully monitored at the time of the earthquake. The slope was in the process of excavation while having exhibited noticeable deformation (Fig. 34) (Fujii et al., 1996). The wall was nearly vertical while the soil was not very stiff. At the time of the earthquake, the slope had been excavated to a depth about 4.0 - 5.0 m. The wall displaced outward 1.9 mm and 3.2 mm at the largest at the measuring sections Nos. 1 and 2, showing an overturning mode as observed before and after the earthquake. No damage to the houses and other structures in the adjacent areas was observed,

Table 3 List of the nailed slopes located in the affected areas (see Fig. 33)
(after Torii,T., Research Committee on Reinforcement Technology for Natural Slopes, 1996).

No.	Slope height and wall area	Slope (V:H)	Soil type	Reinforcement Material				Facing	Permanent Spacing temporary	Esti-mated PHGA
					(mm)	(m)	(m)			
NS1	20 (m) ---* (m^2)	1:0.5	Weathered granite	Threaded deformed steel bar (TD bar)	25	V: 1.25 H: -	5.0	Shotcrete thichness t= 50 (mm)	Permanent	350 (gals)
NS2	4	1:0.4	Decomposed granaite	TD bar	29	V: 1.5 H: -	4.0	Precast RC panels t= 300	Permanent	300
NS3	3	1:0.4	Holocene sand	TD bar	-	V: 1.5 H: -	5.5	Precast RC panels t= 300	Permanent	700
NS4	10	1:0.3	Terrace gravel/ weathered granite	Deformed steel bar (D bar)	22	V: 1.5 H: 1.5	6.8	Shotcrete t= 100	Temporary	500
NS5	7	1:0.6	Terrace gravel/ weathered granite	D bar	22	V: 1.5 H: 1.5	4.0	Shotcrete t= 100	Permanent	500
NS6	5m at the time of earthquake (final 14.3m)	1:0.1	Pleisto-cene sand and clay	D bar	22	V: 1.5 H: 1.5	5.5-6.2	Shotcrete t= 100	Temporary	200
NS7	Total 16m (backfilled by concrete to a height of 6.4m)	1:0.2	Granite covered with debris	D bar	25	V: 1.5 H: 1.4	3-7	Shotcrete t= 150 with lattice RC frame as leaning type RW	Permanent	600 (PVGA =300)

*) not reported

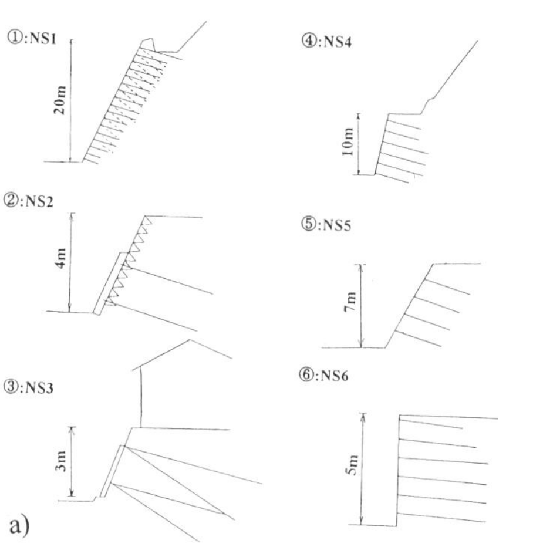

a)

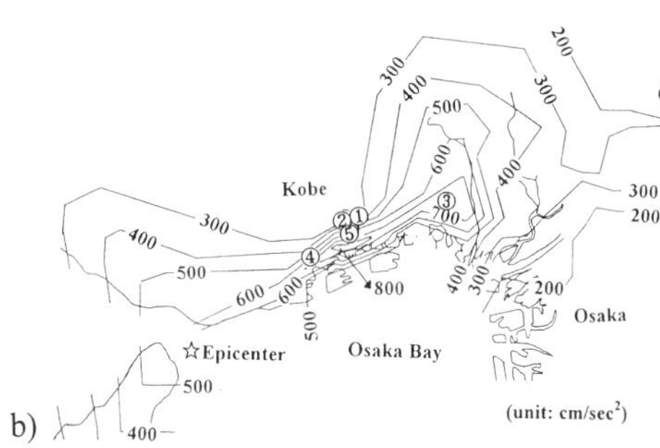

b)

Fig. 33 a) Cross-sections of the nailed slopes listed in Table 3 and b) locations of the nailed slopes and contours of estimated PHGA (according to Ohbayashi Corp., 1995) (after Torii,T., Research Committee on Reinforcement Technology for Natural Slopes, 1996).

because the site was a bit far away from the epicenter and the estimated PHGA was relatively low. It seems that the slope was already near the critical state at the time of the earthquake, and the slope would have not been stable without nailing during the earthquake. Despite higher PHGAs, the other slopes were more stable probably due to better soil conditions and/or larger inclination of the slopes.

A nailed slope located in Awaji Island (NS7 in Fig.3), on an approach road to Akashi Kaikyo Bridge, was located closest to the epicenter (Fig. 35). Before the earthquake, some deformation of the excavated slope face had been noticed after having excavated to the bottom of the slope. Therefore, two layers of ground anchor had been installed at depths of 8.7 m and 11.6 m from the slope crest. After well foundations were constructed in front of the slope, the space between the slope face and the back face of the well foundation had been backfilled with

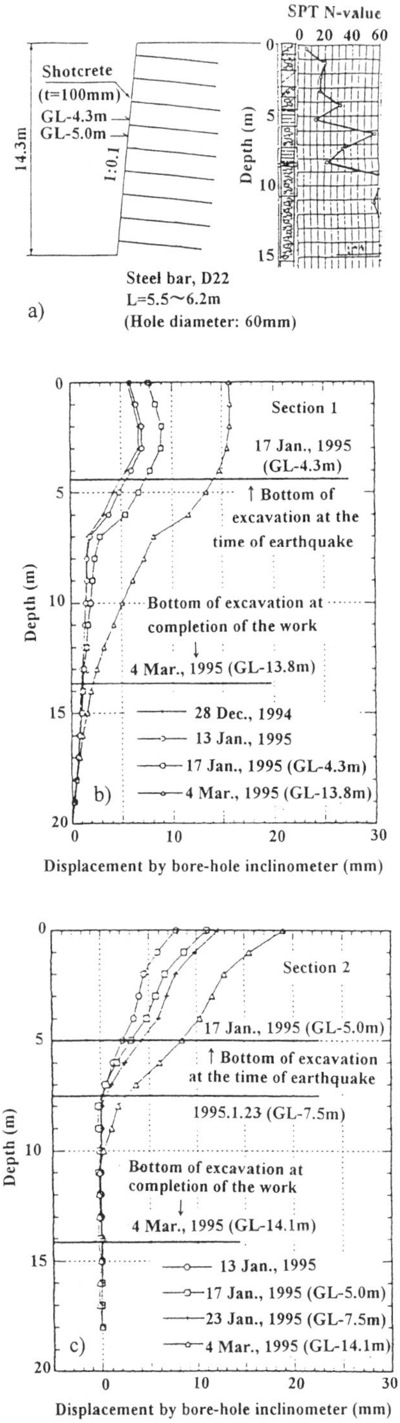

Fig. 34 a) Cross-section at the end of excavation and b) displacements at the slope face of the nailed slope NS6 in Fig. 3 (Fujii et al., 1996).

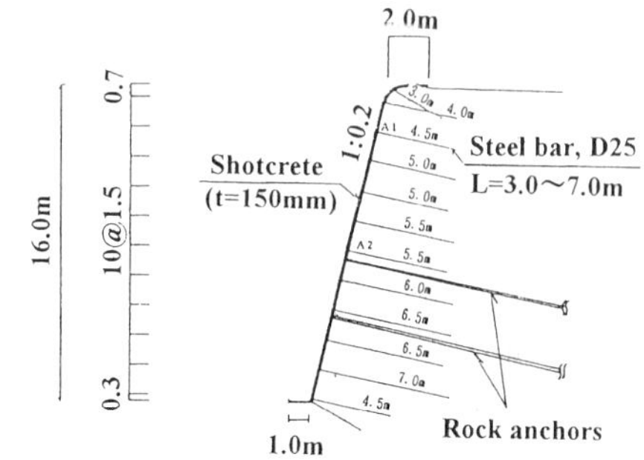

Fig. 35 Cross-section of a nailed slope in Awaji Island (NS7 in Fig. 3; Tsukuda et al., 1996).

concrete up to a depth of 9.6 m from the slope crest by the time of the earthquake. The results of pseudo-static seismic analysis performed after the earthquake for the slope with and without nailing showed that the slope had become more stable by nailing (Tsukuda et al., 1996). Although the reinforced slope was not damaged by the earthquake, a safety factor equal to 0.8 was obtained when applying k_h= 0.6, which is equal to the estimated PHGA/g, normal to the slope face. They considered that this discrepancy was due to the fact that the direction of the PHGA was largely inclined (more than 60 degrees) from the normal to the slope surface. The authors consider, however, that many other factors should be considered as well.

Root-piled slopes: Ten slopes had been stabilized by the root pile technique in the areas where the estimated PHGA was 100 gals or more (Fujii et al., 1996). Among them, the following two cases are important.

A slope located north-east of Suma-koen Station of the Sanyo Dentetsu railway (RP1; Fig. 36) (see Fig. 3) experienced an estimated PHGA of about 500 gals. A 6 - 7 m high slope had been stabilized by root piling, adjacent to a 1 - 3 m high slope that had been retained only by a masonry RW. The slope was of a residual soil, basically a gravelly sand with inter-bedded clay. The root-piled slope had two different facings, masonry and shotcrete. A number of hair cracks were observed on the shotcrete facing and on the road face behind the wall face, and an opening of 5 cm appeared between the masonry and shotcrete facings. However, the damage was not serious at all. The deformation of the masonry facing was smaller. Compared with the above, the masonry facing of the unstabilized slope collapsed, and several nearby wooden houses were seriously damaged.

In the second case located south of JR Takarazuka Station on the right bank of Mukogawa river (RP2 in Fig. 3; Fig. 37), the estimated PHGA was about 500

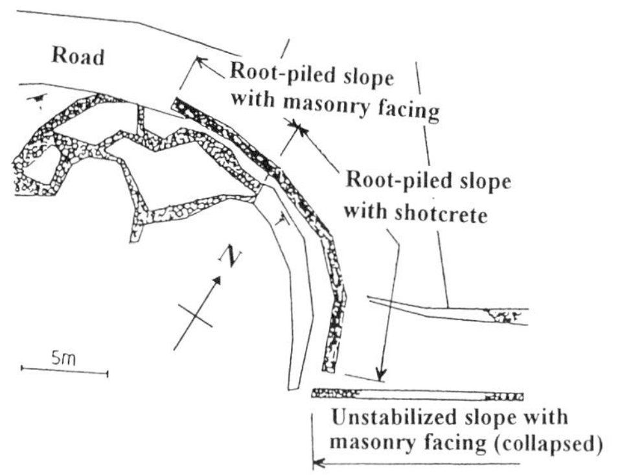

Fig. 36 Plan of the root-piled slopes and adjacent damaged unreinforced slope near Suma-Koen (RP1 in Fig. 3; after Fujii et al., 1996).

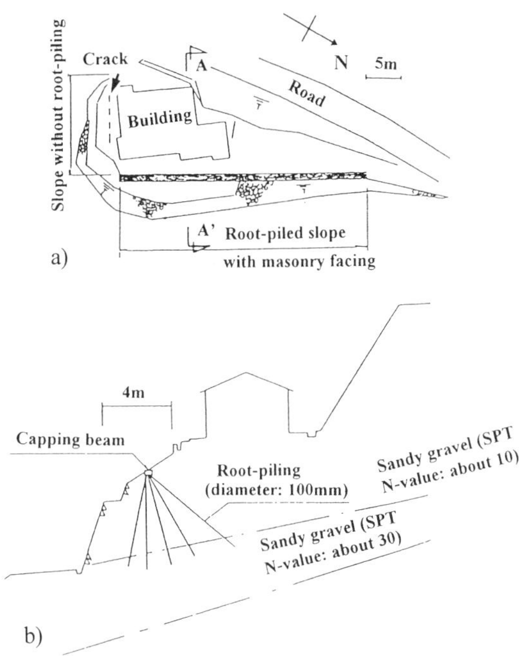

Fig. 37 a) Plan of the root-piled slope and adjacent damaged unreinforced slopes and b) cross-section A-A' of the root-piled slope near Takarazuka (RP2 in Fig. 3; after Fujii et al., 1996).

gals. An about 7 m high slope having a masonry RW had been stabilized by root piling, while the adjacent slope with a height of 5.0 m at the maximum had been retained only by a masonry RW. The slope soil was basically a sandy gravel. In the root-piled slope, the soil mass between the wall face and the root piles seems to have moved slightly outward while the root-piled zone and the soil behind it did not, judging from some small cracks and no cracks on the ground surfaces, respectively, in front of and behind the capping beam. Compared with the good performance of the root-piled slope, on the crest of the slope without root-piling, a crack with a maximum width of about 20 cm appeared immediately behind the masonry RW, associated with a 7 cm deep depression at a distance of 20 cm from the masonry RW. This fact shows that the unstabilized slope moved outward to a some extent.

These two case histories show relatively high seismic stability of the root-piled slopes compared with the unstabilized slopes. It seems, however, that the stability of the zone in front of root piles would be prone to larger outward displacement when subjected to higher seismic loads. This zone can be stabilized by nailing from the wall face.

6 LESSONS FROM THE CASE HISTORIES AND SOME CONSIDERATIONS ON SEISMIC DESIGN

6.1 *Summary of the seismic behaviour of reinforced RWs*

Fig. 38 is the authors' evaluation of the relative construction cost and seismic performance of several different types of RWs and sloped embankments. From the case histories described in this report, the following conclusions are reached:

1) Among the structures located in the severely shaken areas, all the geosynthetic- and steel-reinforced soil RWs performed better than not only most of the masonry RWs and the gravity and leaning type unreinforced concrete RWs, but also a number of cantilever or inverted T-shaped RC RWs which exhibited large displacements (most of them were not supported by piles). The performance of the reinforced soil RWs was even better than a number of RC columns and piers which collapsed by shear failure. When following the aseismic characterization of structure illustrated in Fig. 14, the masonry RWs and the gravity and leaning type unreinforced concrete RWs are classified as "structure F", the cantilever or inverted T-shaped RC RWs are classified as between "structures E and F", and the RC columns are classified as "structure B".

2) The good seismic performance of the reinforced soil RWs is comparable to that of the cantilever or inverted T-shaped RC RWs which were supported by piles and performed very well. Therefore, both of them could be classified as "structure E". Between the two, the reinforced soil RWs are more cost-effective.

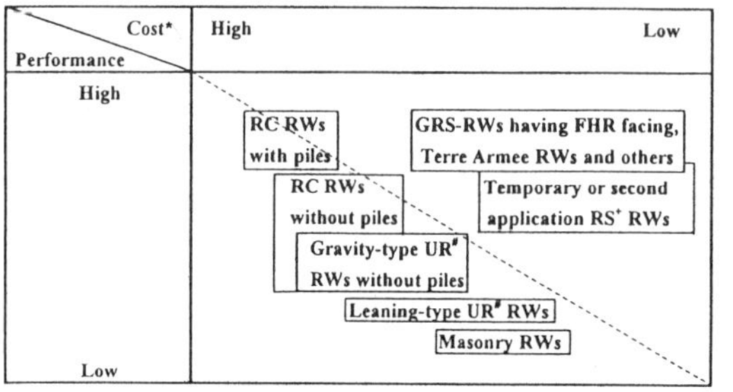

* at the time of construction # unreinforced concrete + reinforced soil

Fig. 38 Approximated evaluation of cost and seismic performance of different types of soil retaining structures.

6.2 *Analysis by current seismic design methods*

The seismic stability of the damaged conventional soil RWs (LT1, GT1, CL2 and CL4 in Fig. 4) and the GRS-RW having a FHR facing at Tanata (GR1 in Fig. 4) was analyzed by limit equilibrium-based pseudo-static stability analysis in the framework of the current seismic design procedure (Koseki et al., 1996). In the analysis, the effects of vertical seismic force was found to be much smaller than those of horizontal seismic force, and therefore will not be discussed herein.

The seismic stability of the GRS-RW was evaluated by the two-wedge method (Fig. 39; Horii et al., 1994). In this method, the same horizontal seismic coefficient is applied to the facing and the reinforced and unreinforced zones of backfill, while no seismic load is applied to the subsoil. The seismic loads are resisted mainly by the tensile force in the reinforcement and partly by the reaction force at the bottom of the facing. Sliding at the base of the facing and reinforced backfill and the overturning of the facing together with the active zone F are also examined. To prevent sliding at the base, several bottom reinforcement layers are made long enough as in the case of RC RWs without a pile foundation. When the safety factor for bearing capacity at the facing base is becoming less than unity, the direction of the earth pressure acting on the back face of the facing is inclined more horizontally so that the safety factor does not become less than unity. The increased tensile force resulting from this procedure is accommodated by rearranging the reinforcement. This methodology was supported by the results of a large-scale shaking test on a GRS-RW model constructed on a liquefiable sand deposit (Fig. 26). In the test, the wall was able to maintain its stability even when the load acting at the bottom of the facing had reached the bearing capacity of the subsoil, which was decreased substantially by liquefaction.

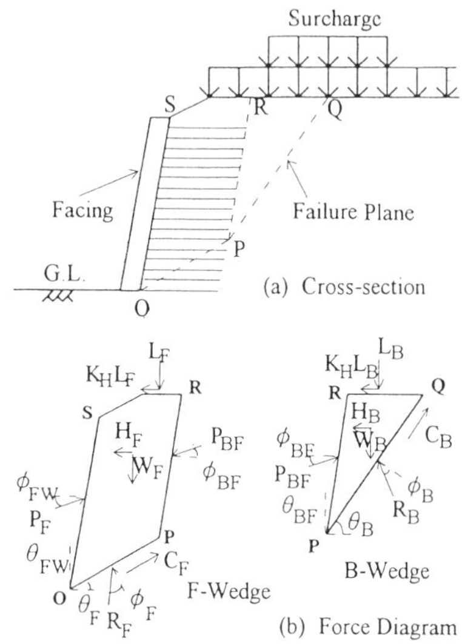

Fig. 39 Two-wedge method to evaluate the seismic stability of GRS-RWs (after Horii et al., 1994).

The design parameters employed in the analysis against external instability are shown in Table 4. The results are summarized in Fig. 40 in terms of a ratio C_r defined as;

$$C_r = k_{h,crit}/(PHGA/g) \qquad (1)$$

where $k_{h,crit}$ is the critical horizontal seismic coefficient yielding a safety factor of unity. The value of PHGA was estimated as approximately 600 gal to 800 gal based on Fig. 5. For the RWs which actually failed, the values of $k_h = C_r \cdot (PHGA/g)$ as obtained would have under-estimated the correct values to be used in the aseismic design, because the critical PHGA at which the structures were about to fail should be equal to or smaller than the actual PHGA. The reverse is true for the RWs which did not fail.

In Fig. 40, the potential failure modes against external instability are also indicated. These failure modes were found consistent with the observed behaviour. The value of C_r against external instability for the GRS-RW at Tanata was 0.6 - 0.8, which was similar to those for the other conventional RWs excluding the cantilever-type RC RW at Shin-Nagata (CL2 in Fig. 4; see also Fig. 11). A very low value of C_r for the wall at Shin-Nagata would be due to the fact that the backfill was considerably sloped. Based on the results shown in Fig. 40 (excluding the wall at Shin-Nagata), the following conclusions could be drawn:

1) With respect to the external stability of the conventional RWs which had failed, the correction

Table 4 Design parameters for external seismic stability analysis of RWs (after Koseki et al., 1996).

Site		Sumiyoshi (Leaning type)	Ishiyagawa (Gravity type)	Ishiyagwa (Cantilever type)	Shin-Nagata (Cantilever type)	Tanata (GRS-RW)
Schematic diagram		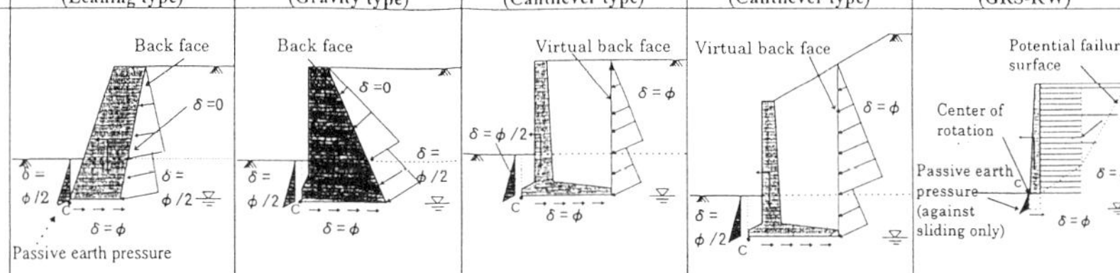				
Concrete wall	γ (kN/m^3)	23.2	22.7	22.5	23.2	24.5
Soil property used to evaluate earth pressure	γ (kN/m^3)	18.1	17.5	17.5	16.1	16.7
	c (kN/m^2)	0.0	0.0	0.0	0.0	0.0
	ϕ (°)	45	42	42	36	41
Soil property used to evaluate bearing capacity	γ (kN/m^3)	19.7	18.9	18.9	17.3	17.4
	c (kN/m^2)	0.0	0.0	0.0	49.1	0.0
	ϕ (°)	41.8	35.8	35.8	0	33.4
Coefficient of bearing capacity	N_c	91	37	37	5.1	26
	N_γ	150	32	32	0	17
	N_q	82	24	24	1	15

* Existence of wooden piles in the subsoil beneath RW was assumed.

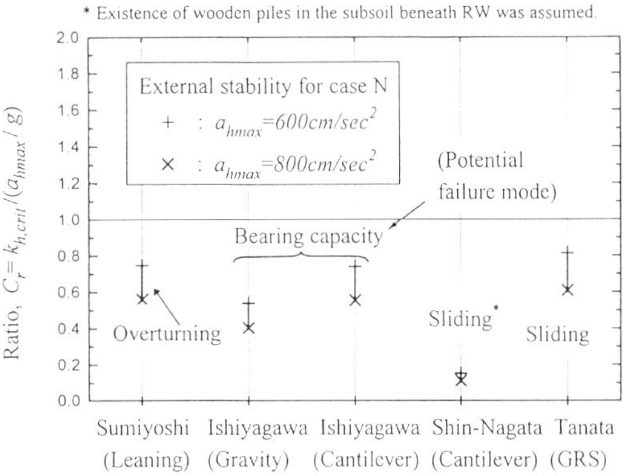

Fig. 40 Ratio of critical horizontal seismic coefficient to estimated PHGA for external stability of RWs (after Koseki et al., 1996).

factor C_r to be used to convert the design PHGA/g into the design seismic coefficient k_h in such that $k_h = C_r \cdot$(PHGA/g) may be larger than 0.6 (but very likely smaller than 1.0).

2) For the GRS-RW at Tanata, the above correction factor for external instability may be smaller than that for the conventional RWs, considering the fact that the GRS-RW was only slightly damaged and has been reused while the other conventional RWs had to be removed for reconstruction.

3)As a whole, when the same pseudo-static seismic stability analysis method is used for both conventional RWs and GRS-RWs, the design would become more conservative for GRS-RWs.

In the current design procedure for the GRS-RW as employed in the aforementioned analysis, the safety factor against overturning is defined as:

$$F_s = \left\{ \frac{M_T + M_{WW} + M_{PV}}{M_{PH} + M_{WH}} \right\}_{min} \quad (2)$$

where M_T, M_{WW} and M_{PV} are the resisting moments due to, respectively, the tensile force of reinforcement acting at the back face of the facing, the self weight of facing, and the vertical components of both the earth pressure acting on the boundary between the front (F) and the back (B) wedges in the backfill ($P_{BF} \cdot \sin\phi_{BF}$ in Fig. 39) and the resultant earth pressure acting on the back face of the facing ($P_F \cdot \sin\phi_{FW} - P_{BF} \cdot \sin\phi_{BF}$ in Fig. 39). M_{PH} and M_{WH} are the overturning moments due to horizontal components of the earth pressure acting on the back face of the facing ($P_F \cdot \cos\phi_{FW}$ in Fig. 39) and the inertia force of the facing, respectively. This safety factor is principally based on the equilibrium of moments acting on the facing about its toe, but it is somewhat ambiguous because the vertical component of earth pressure acting on the boundary between F-wedge and B-wedge was expediently introduced in order to reflect the effect of the length of reinforcements (or the width of F-wedge).

On the other hand, when the equilibrium of moments about the facing toe acting on the facing and the F-wedge are considered, the safety factor can be defined alternatively as:

$$F_s = \left\{ \frac{M_{FT} + M_W}{M_P + M_R + M_H} \right\}_{min} \quad (3)$$

where M_{FT} and M_W are the resisting moments due to the tensile force of reinforcement mobilized at the boundary between F-wedge and B-wedge (PR in Fig. 39) and the bottom of F-wedge (OP in Fig. 39), and the self weight of both the facing and F-wedge, respectively. M_P, M_R and M_H are the overturning

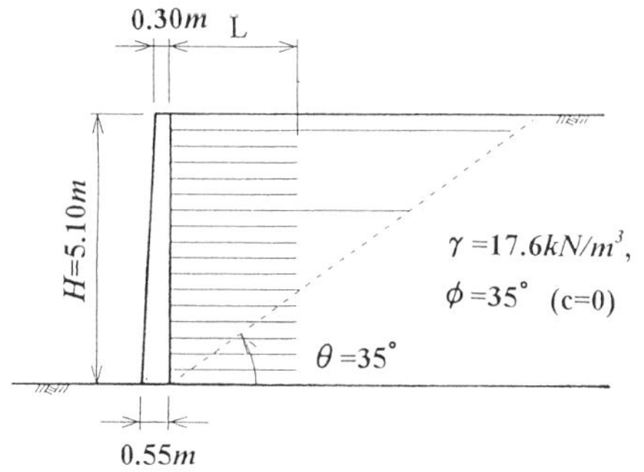

Fig. 41 A model GRS-RW having a FHR facing to estimate the effects of different stability analysis methods (after Horii et al., 1994).

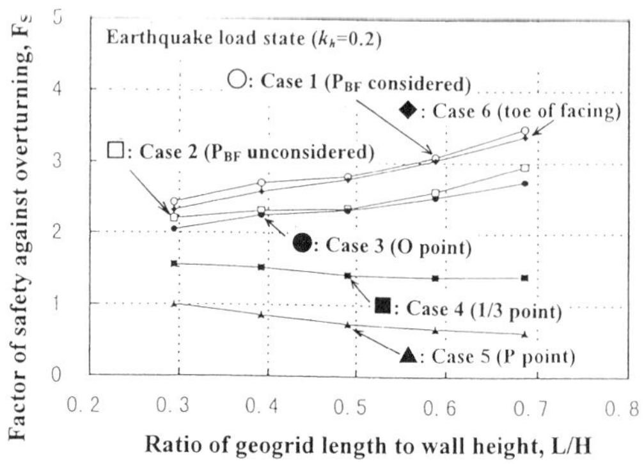

Fig. 42 Effects of different assumptions on the results obtained by two different seismic stability analysis methods for a model GRS-RW having FHR a facing.

moments due to, respectively, the earth pressure acting on the boundary between F-wedge and B-wedge (PBF in Fig. 39), the reaction force at the bottom of F-wedge (R_F in Fig. 39), and the inertia forces of both the facing and F-wedge.

The issues of differences in the safety factor between the two methods and the effects of the assumed location of the reaction force at the bottom of F-wedge or the facing have been discussed by Horii et al. (1994). Safety factors against overturning of a model wall shown Fig. 41 were evaluated under the following six conditions. Case 6 was added on by the authors.

Based on Eq.2

Case 1: Based on the equilibrium around its toe of moments acting on the facing about the facing toe, considering the effects of the vertical component of the earth pressure acting on the boundary between F-wedge and B-wedge in the backfill (P_{BF} in Fig. 39) (the current design procedure).

Case 2: The same equilibrium as above is considered, but not considering the effects of the component P_{BF}.

The alternative design procedure based on the equilibrium moments acting on the facing and the F-wedge about the facing toe (i.e., based on Eq.3)

Case 3: The reaction force at the bottom of F-wedge (R_F in Fig. 39) is assumed to act at point O in Fig. 39.

Case 4: The reaction force R_F is assumed to act at the one third point of the length OP from the point O.

Case 5: The reaction force R_F is assumed to act at point P in Fig. 39.

Case 6: The reaction force R_F is assumed to act at the facing toe (M_R is set to be zero).

Fig. 42 shows the change of safety factor against overturning by the change of the reinforcement length (L) normalized by the wall height (H) under seismic conditions when k_h = 0.2. The results can be summarized as follows;

1) In the alternative design procedure (Eq. 3), the results of cases 4 and 5 are not realistic, because the safety factor decreases as L/H increases. On the other hand, the stabilization effect by widening the reinforcement zone is reflected in the results of cases 3 and 6. The assumptions made in cases 3 and 6 may be acceptable when a FHR facing is employed with the subsoil having a sufficient bearing capacity against the somehow concentrated reaction force R_F at the facing base.

2) When based on Eq. 2, cases 1 and 2 yield similar results to cases 6 and 3, respectively. Therefore, the current design procedure considering the effect of the vertical component of the earth pressure acting on the boundary between F-wedge and B-wedge in the backfill (i.e. case 1) is judged to be reasonable as long as the momentary concentration of the reaction force at the facing toe due to the inertia force during an earthquake is allowed.

6.3 *Limitation of current seismic design*

Felio et al. (1990) reported that the safety factors of eight nailed slopes which did not fail during the 1989 Loma Prieta earthquake obtained based on the current limit equilibrium-based seismic stability analysis using the value of k_h equal to the estimated PHGA divided by g are considerably lower than unity. This was also the case with the nailed slopes during this earthquakes. They considered that the main reason is the hidden conservatism by neglecting effects of the bending and shear resistance of the reinforcement and the rigidity of facing.

The Tanata GRS-RW survived despite the use of k_h = 0.2 in the design, which is much lower than the PHGA/g at the site (about 0.6 - 0.8). Some hidden conservatism may be one of the reasons. Other reasons may include a) under-estimated shear strength of backfill soil, and b) no consideration of passive earth pressure on the front face of the buried part of facing. The effects of H-shaped steel piles remaining behind the reinforced zone at the time of the earthquake would be small. Collins et al. (1992) also reported that during the 1989 Loma Prieta earthquake, two geogrid-reinforced RWs having wrapped-around wall face behaved very well for estimated PHGA/g of around 0.4, despite the design k_h was 0.1 - 0.2. Fukuda et al. (1994) also reported that a 6.7 m high geogrid-reinforced RW having a facing of steel wire mesh with a wall face slope of 1:0.3 in V:H, which was not aseismic-designed, was stable despite that the estimated PHGA was 310 gals.

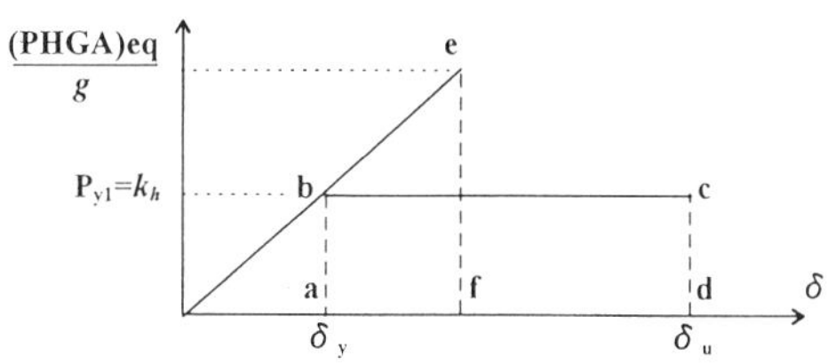

Fig. 43 Method to obtain the equivalent PHGA of a ductile structure.

It seems to the authors that the effect of the flexibility and ductility of reinforced soil RWs is yet another reasons. It is conceivable that as a RW becomes more flexible, the seismic earth pressure acting on its back face becomes smaller; and as the structure becomes more ductile, the design procedure where PHGA/g is used as k_h and the structure is judged to fail when the limit equilibrium is lost for the first time is too conservative. On the other hand, as the current seismic design procedures for soil structures are mostly limit equilibrium-based stability analysis, they cannot evaluate the deformation and displacement of soil structure caused by seismic loads, therefore they cannot evaluate the effects of flexibility and ductility on the seismic stability of a soil structure.

Effects of ductility: When some post-yield deformation without exhibiting complete collapse is allowed to occur, it could be judged that ductile structures C and E illustrated in Fig. 14 can sustain seismic load L exceeding the yield strength P_{y1}. In concrete engineering, the value of PHGA/g which is larger than P_{y1}, is obtained in such a way that the areas *abcd* and *abef* of the two load-deformation curves for linear elastic and ductile elasto-perfectly plastic materials are the same (Fig. 43). The above procedure is equivalent to the use of k_h

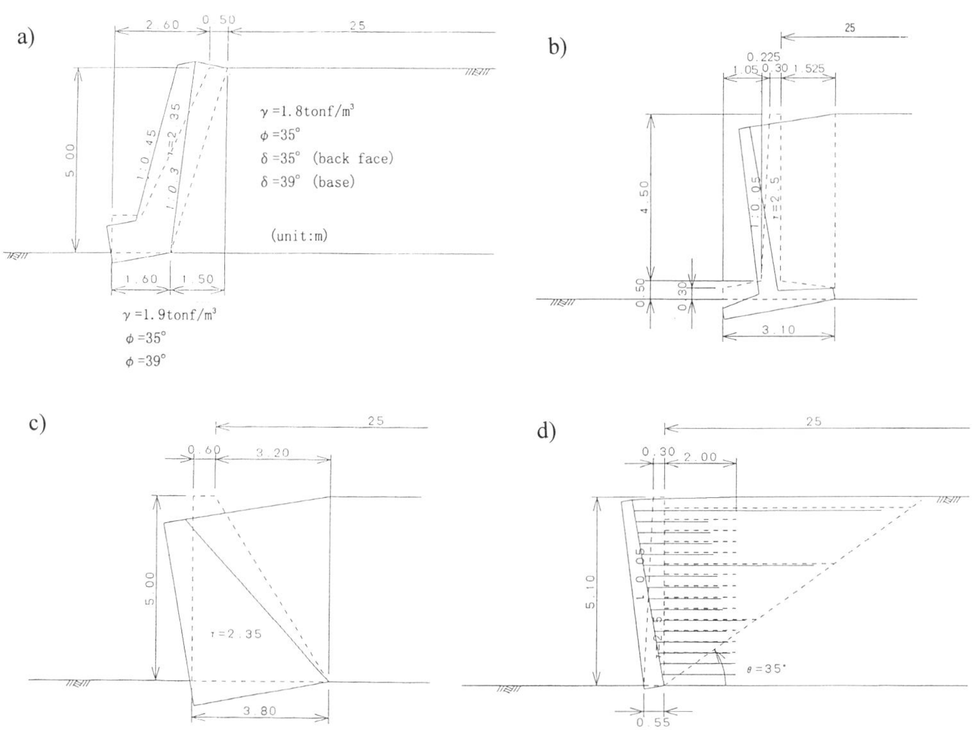

Fig. 44 Model RWs to evaluate the ductility in overturning failure.

smaller than PHGA/g (= L) in the limit equilibrium-based stability analysis. According to this methodology, the design k_h value for ductile soil structures could be set to be smaller than that for brittle soil structures.

Although in a very approximated way, one aspect of the structural ductility of RW could be evaluated by the rate of decrease in the safety factor for overturning with the increase in the overall overturning displacement as shown in Fig. 44. It was assumed in this analysis that the three conventional types of RWs overturn about the heel of the base, while the GRS-RW about the heel of the facing base. The backfill was assumed not to move, while the space produced by the displacement of the wall was filled up by additional soil. It was assumed that the bearing capacity failure and wall structure failure do not take place. Fig. 45 shows the reduction rate of safety factor for overturning when k_h= 0.2, plotted against the rotational angle θ of the wall. F_{s0} and F_s mean the safety factors before and after the occurrence of overturning. It is seen that the safety factor at seismic conditions (with k_h=0.2) decreases as the wall overturns, and the reduction rate is larger, following the order: the leaning type RW, gravity type RW, inverted T-shaped RC RW and GRS-RW. As the area between each curve and the horizontal axis becomes larger, the RW could be considered to be more ductile.

Effects of flexibility: It may be needed to classify RWs into dynamically rigid and flexible RWs and intermediate ones. "Dynamically rigid RWs" could be defined as those of which the dynamic displacement is much smaller than that of the backfill, and therefore the seismic earth pressure becomes larger than that at static states. The most conventional type soil RWs (i.e., leaning type and gravity type unreinforced concrete RWs and cantilever and inverted T-shaped RC RWs) which are stable during earthquakes are categorized as "rigid RWs". On the other hand, "dynamically flexible RWs" could be defined as those of which the displacement is similar to, or larger than, that of the backfill, and the seismic earth pressure is not larger, or smaller, than that at static states. Reinforced soil RWs may be categorized as "flexible RWs" or "intermediate ones".

Rigid walls should be designed to displace smaller than the backfill when subjected to seismic loads. The seismic earth pressure is evaluated usually by the Mononobe-Okabe method. Against the seismic earth pressure and the gravitational and inertia force applied to the wall structure, it is evaluated whether the wall structure, the bearing capacity of subsoil (or piles when used) and the shear strength at the interface between the wall base and the subsoil are

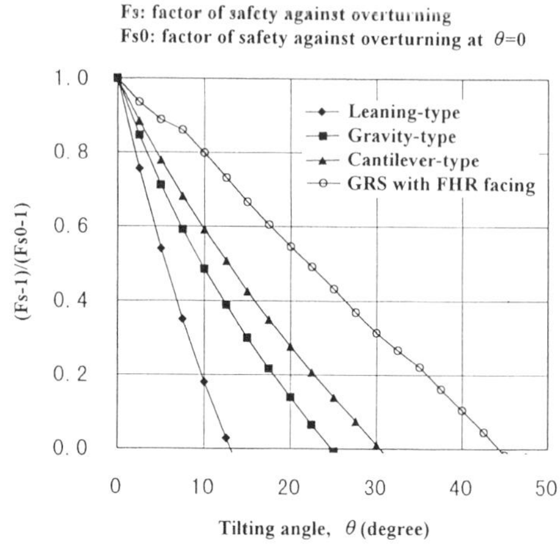

Fig. 45 Change in the safety factor due to overturning evaluated by the limit equilibrium pseudo-static method (k_h= 0.2).

all large enough, and whether the wall does not overturn. The deformation of the wall by seismic force is usually not examined.

In the current design methodologies, reinforced soil RWs are treated as rigid RWs. To the authors, it would be more reasonable to consider reinforced soil RWs as "flexible RWs" or "intermediate ones". In that case, the design seismic earth pressure could be made smaller than that for rigid RWs. However, in the following two types of analysis, it should be ensured that a given reinforced soil RW displaces and deforms in a ductile manner as "structure E", behaving as a wide flexible monolith;

1) the overall stability analysis for overturning and sliding at the base as that for rigid RWs; and
2) the internal deformation of the reinforced zone, including the issue of internal stability,

6.4 *Deformation of reinforced backfill as flexible RWs*

In case reinforced soil RWs are designed as "flexible RWs" or "intermediate ones", the following three types of internal deformation of the reinforced backfill should be evaluated (Fig. 46):
a) shear deformation along the potential failure plane;
b) simple shear deformation along horizontal planes; and
c) deformation due to overturning moment.

a) The horizontal outward seismic loads increase the disturbing shear stress τ_w acting along the potential failure plane (Fig. 46a). In an unreinforced backfill,

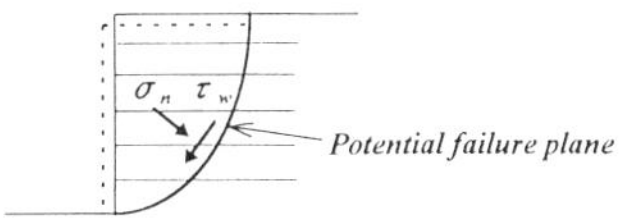

a) shear deformation along potential failure plane

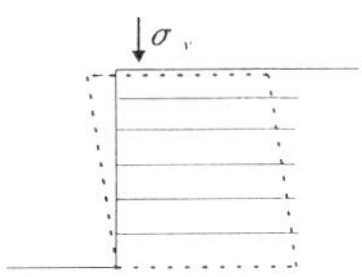

b) simple shear deformation along horizontal planes

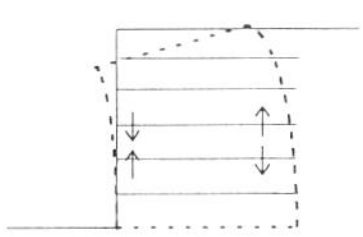

c) deformation due to overturning moment

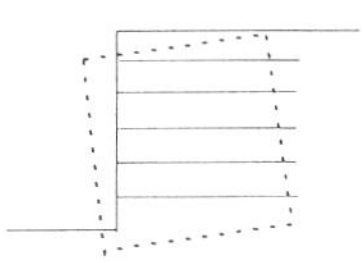

d) overturning due to deformation of subsoil layer

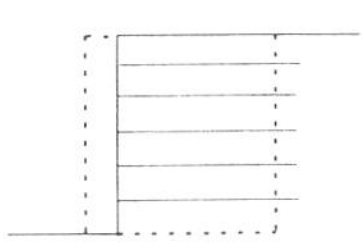

e) sliding at the boundary of reinforced zone and subsoil layer

Fig. 46 Different modes of deformation and displacements of GRS-RWs subjected to seismic loads.

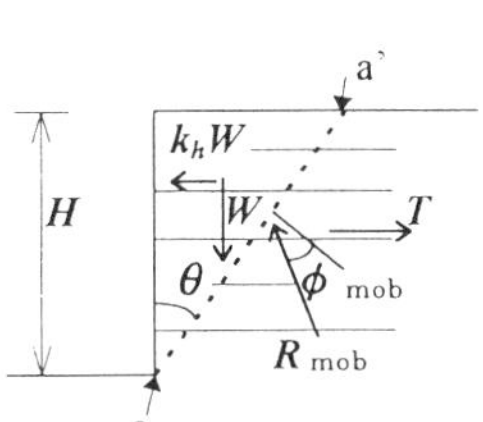

$W=(\gamma \cdot H^2 \cdot \tan\theta)/2$

k_h: **horizontal seismic coefficient**

H: **wall height**

T: **total reinforcement force**

Fig. 47 Simplified force equilibrium in reinforced soil RWs.

the normal stress σ_n acting on the potential failure surface decreases at the same time, which leads to a reduction in the soil shear strength $\tau_f=\sigma_n \cdot \tan\phi$. In a reinforced backfill, tensile force induced in the reinforcement resists directly against the shear stress τ_w and restrains the reduction in σ_n. This mechanism can be explained also by using a simple model shown in Fig. 47. For a backfill of cohesionless soil, the local safety factor $(F_s)_{rf}$ along a given plane a-a' in the direction at the angle θ from the vertical is defined as:

$$(F_s)_{rf}=\tan\phi / \tan\phi_{mob} \tag{4}$$

where ϕ and ϕ_{mob} are the angle of internal friction (peak shear strength) and the mobilized angle of friction, referring to the mobilized reaction force R_{mob} acting on the plane. From the force equilibrium among the weight of the active zone W, the outward horizontal seismic force $k_h \cdot W$ and the total reinforcement force T, it follows:

$$(F_s)_{rf}=\tan\phi \cdot \frac{T/W+\tan\theta-k_h}{1-(T/W)\cdot\tan\theta+k_h\cdot\tan\theta} \tag{5}$$

The safety factor increases as T increases. The safety factor when the backfill is not reinforced is:

$$(F_s)_{unrf}=\tan\phi \cdot \frac{\tan\theta-k_h}{1+k_h\cdot\tan\theta} \tag{6}$$

The ratio between the above two is;

$$\frac{(F_s)_{rf}}{(F_s)_{unrf}}=\frac{1+(T/W)/(\tan\theta-k_h)}{1-\{(T/W)\cdot\tan\theta\}/(1+k_h\cdot\tan\theta)} \tag{7}$$

Since the value of θ at which the value of Fs becomes the minimum is around $45^\circ - \phi/2$ and T/W is much smaller than 1.0, the ratio for the critical plane increases with k_h. That is, with the increase in k_h, a reinforced RW becomes more stable relative to the corresponding unreinforced RW. This means that the advantages of reinforced RWs become more apparent in the aseismic-design. It is also the case with the relative stability between unreinforced and reinforced slopes.

Note also that the use of a FHR facing is effective to restrain the development of a failure plane which intersects the facing at intermediate heights.

b) In case the deformation of reinforced backfill is restrained to simple shear deformation along horizontal planes, the stress condition becomes uniform along each horizontal plane and the length of each horizontal plane is kept constant. To realize these conditions, the following two requirements should be satisfied:

i) Relatively inextensible reinforcement is arranged horizontal to achieve zero or nearly zero normal strains in the horizontal directions.

ii) A facing with sufficient axial rigidity while having sufficient vertical bearing capacity at the facing base is arranged to transmit the axial force so that the upward shear stress τ_{hv} can be activated along the vertical face of the backfill in contact with the back face of the facing (Fig. 48). Without this complementary shear stress τ_{hv}, the shear stress τ_{vh} cannot be exerted on the horizontal planes. As the back face of the reinforced zone is inside a soil mass, the complementary shear stress τ_{hv} is exerted automatically on this plane.

In addition, the shear strength $(\tau_{vh})_{max}=(\tau_{hv})_{max}$ is controlled by the vertical and horizontal stresses σ_v and σ_h as;

$$(\tau_{vh})_{max}=(\tau_{hv})_{max}=(1/2)\cdot\{(\sigma_v+\sigma_h)^2\cdot\sin^2\phi-(\sigma_v-\sigma_h)^2\}^{1/2} \qquad (8)$$

It is seen from the above equation that under a constant σ_v, the shear strength $(\tau_{vh})_{max}$ (and shear rigidity) increases as σ_h increases, resulting in smaller shear deformation of the reinforced backfill. Therefore, the simple shear deformation becomes smaller by maintaining larger horizontal stress σ_h. The use of a FHR facing to which reinforcement layers are connected contributes to the above, although it is in an indirect way. It is to be noted that even under the same seismic conditions, the stresses in the FHR facing of GRS-RW become much smaller than those in the cantilever-type RC RWs, since the FHR facing is supported at many elevations by reinforcement layers, behaving as a continuous vertical beam with a rather short span (Tatsuoka, 1993).

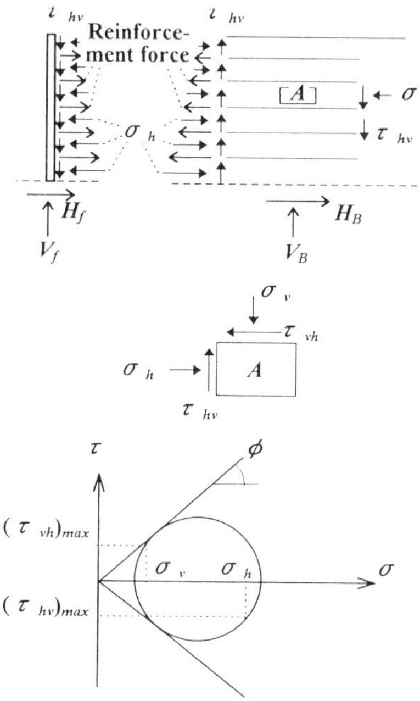

Fig. 48 Stress condition on the back of a full-height rigid facing.

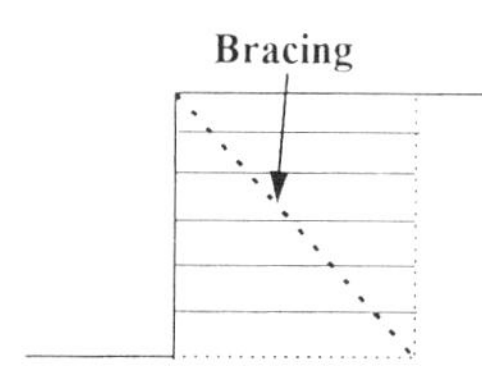

Fig. 49 Bracing in a reinforced soil RW.

Moreover, as the shear deformation of the reinforced backfill increases, a constant length of facing results in a reduction in the height of the reinforced backfill zone adjacent to the facing, while the height of the backfill trends to increase due to dilatancy effects. Then, the vertical stress σ_v may increase in that zone. This factor would be an additional contribution of a FHR facing. It is noted, however, that this behaviour is realized only after large deformation.

Note that all the above-mentioned possible contributions of a FHR facing to the seismic stability of GRS-RW are indirect. The use of a bracing as shown in Fig. 49 would be much more direct and efficient. Nevertheless, it would not be very practical in actual construction. Indeed, the low capability for resisting against simple shear deformation is one of the largest problems in the seismic stability of reinforced soil RWs. This situation is similar to that of elevated RC frame structures (Fig. 1a).

When the horizontal displacement is uniform throughout the reinforced and unreinforced zones of an embankment, the amount of simple shear deformation is independent of the aspect ratio of the reinforced zone, or the reinforcement length. In case the displacement decreases away from the wall face, then the use of longer reinforcement, particularly at higher levels, would be effective to restrain the simple shear deformation (Fig. 50).

c) When using a FHR facing supported by a subsoil having sufficiently large bearing capacity, the overturning moment due to seismic force $k_h\cdot W$ and some additional seismic earth pressure exerting at the

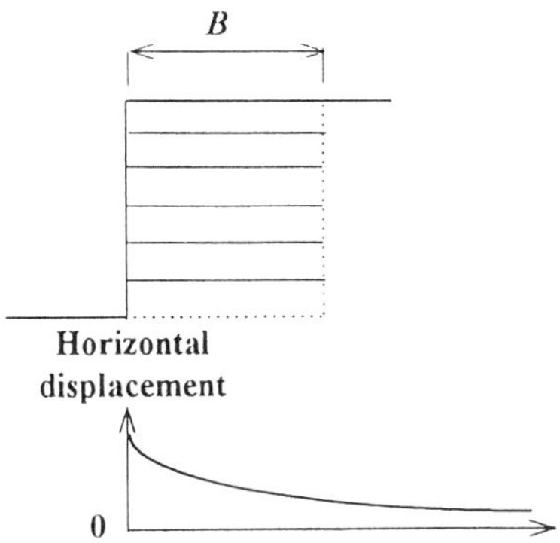

Fig. 50 A possible distribution of outward lateral displacement in reinforced soil RWs.

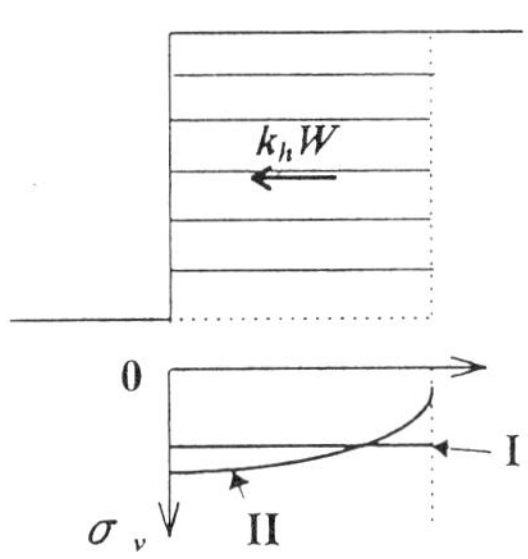

I: before earthquake

II: during earthquake

Fig. 51 Change in the vertical pressure at the base of reinforced soil RW without a FHR facing.

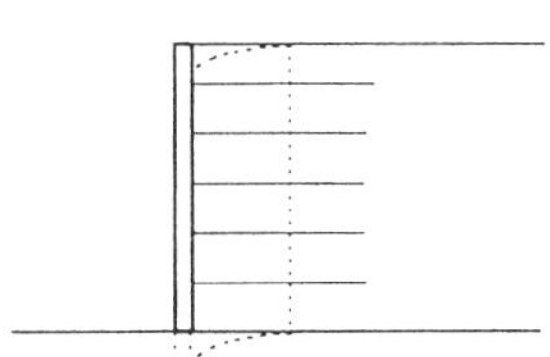

Fig. 52 Vertical shear deformation near the FHR facing due to the loss of bearing capacity of the subsoil

back face of the reinforced zone is resisted by the shear stress τ_{hv} acting upwards along the front face of the backfill (Fig. 48). The vertical load V_f at the facing base is the weight of the facing plus the integration of τ_{hv}. With this, the overall over-turning of the reinforced backfill is restrained. Without a FHR facing, on the other hand, the vertical stress σ_v within and at the base of the reinforced backfill increases near the wall face and decreases near the back of the reinforced zone (see Fig. 51). This results in the deformation of the backfill as shown in Fig. 46c and the deformation of the subsoil as shown in Fig. 46d, both of which increase the overall tilting of the reinforced backfill.

Although the bearing capacity at the base for a FHR facing is important, it could be still secondary for the ultimate failure of a reinforced soil RW as was observed in a large model during shaking table tests (Fig. 26). When the bearing capacity at the facing base is lost, the flexible reinforced backfill soil would exhibit vertical shear deformation to accommodate it as illustrated in Fig. 52, while increasing the horizontal shear deformation and overall overturning deformation. On the other hand, for conventional RWs (e.g., gravity type unreinforced concrete RWs and cantilever RC RWs) without piles, stress tends to be concentrated at and near the toe of the base as the wall tends to overturn, which may lead to the loss of the vertical bearing capacity, resulting in the overall overturning displacement. To avoid this behaviour, most of recent conventional type RC RWs are supported with piles except when constructed on a very firm subsoil. This arrangement makes the conventional RWs less cost-effective than reinforced soil RWs.

6.5 *Reinforced soil bridge abutment subjected to seismic loads on the crest*

Bridge abutments should be able to support a RC block on which a bridge girder is directly placed, exhibiting vertical and horizontal displacements that are within the allowable limits. This was also one of the major issues when designing the three GRS bridge abutments having FHR facings shown in Fig. 53 (Tatsuoka et al., 1996c). Soon after the wall completion, cyclic lateral loads were applied to the front face of RC block denoted by the letter A in Fig. 53. The width of the reinforced bridge abutment is about 9 m. As seen from Fig. 54, for a maximum load of 98 kN (10 tonf), the displacement at the facing having a width of about 9 m was very small. Importantly, the FHR facing displaced not only near the top but near the bottom, which indicates that the FHR facing as a whole resisted against the lateral load. This is another important possible contribution of FHR facing to the seismic stability of reinforced soil RW. The lateral load in this test was limited to 98 kN (10 tonf) to avoid any possible damage to the structure. In similar lateral loading tests, a lateral load of 196 kN (20 tonf) was applied to a 3 m-wide test section of a 5 m-high full-scale model GRS-RW having a FHR facing (Tamura et al., 1994, Tateyama et al., 1994). The wall and facing did not show any sign of failure.

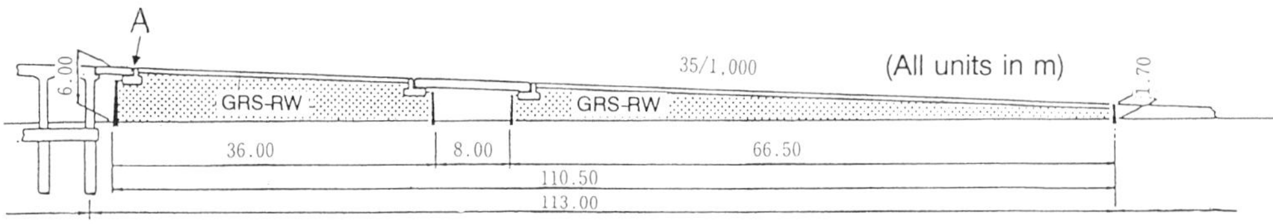

Fig. 53 Side view of GRS-RW and GRS bridge abutments for Seibu Line, at Nerima; the letter A shows the location of the lateral loading test (Tatsuoka et al., 1996c).

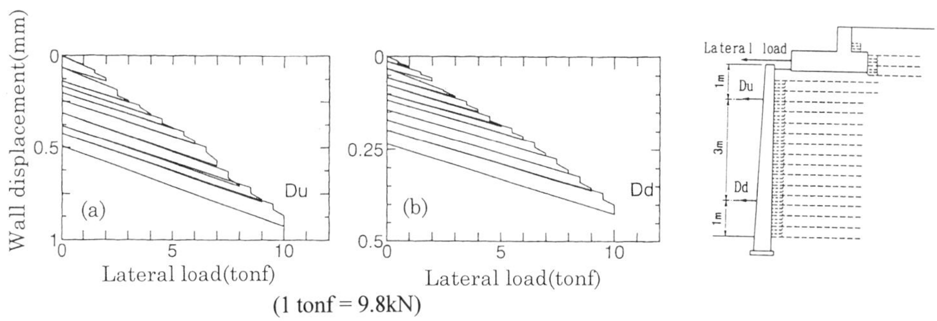

Fig. 54 Results of cyclic lateral loading test on a GRS bridge abutment for Seibu Line.

6.6 *Possible future trends of aseismic design*

So far, for most slopes, whether reinforced or not, the seismic stability is not evaluated in ordinary design procedures. It is also the case with usual highway and railway embankments having a height lower than, say, 5 m. As discussed earlier, reinforced soil structures become more stable relative to unreinforced ones as seismic loads increase. Therefore, if seismic stability analysis becomes a standard design procedure, more often than not reinforced soil structures will be adopted.

After this earthquake, the following two stage aseismic-design procedure has been accepted by Japanese civil engineers to be applied to civil engineering structures to a greater extent than before the earthquake. That is, the following two level seismic loads are used, associated with each respective objective of aseismic design (JSCE, 1996):
1) Level one seismic load, which is expected to strike the structure once or twice during its service life. Structures should not be damaged under Level one earthquake motions.
2) Level two seismic load, which is very rare in its occurrence but has a very high impact in its intensity. It includes ground motions generated in epicentral regions of large scale inter-plate earthquakes and those generated in near-field area of causative inter-plate faults. Structures may be damaged under Level two earthquake motions, where design should be practiced to control their damage processes. It is considered that the seismic loads in the most severely shaken areas shown in Fig. 4 are equivalent to Level two seismic load. It is apparent that if the above procedure is to be adopted also for soil structures, the following two problems should be solved:

1) When the pseudo-static seismic stability method is used, the ratio "k_h/(design PHGA/g)" should be less than unity. Then, what value should be adopted for the ratio ? The ductility should be taken into account in a proper way when designed for Level two seismic load. Then, how could the ductility be defined ? And how could the ductility be reflected in the ratio "k_h/(the design PHGA/g)" ?
2) The dynamic flexibility of structure should be defined properly. Then, what would be a proper method to evaluate the effects of flexibility on the seismic earth pressure acting on the back of a soil retaining structure ?

Even if sophisticated seismic design methods considering the above two factor are developed and can be applied to reinforced soil RWs, it is not likely that they are applied equally to unreinforced and reinforced slopes. This is because the soil conditions in natural slopes are much more difficult to identify and much more heterogeneous than in backfills. A much simpler but rational method need to be developed.

6.7 *Seismic stability of PL/PS reinforced soil*

One of the advantages of reinforced soil structures in reducing the effects of seismic loads may be its flexibility. However, if it is too flexible, the deformation may exceed the allowable limit. The preloading and prestressing technique has been proposed originally to increase the vertical stiffness of reinforced soil structures (Tatsuoka et al., 1996a,b,c, Uchimura et al., 1996). As shown in Fig. 55, a vertical load is applied as preload to a reinforced backfill by using a pair of reaction blocks and a set of tie rods (or by using a top reaction block and a set of tie rods with the bottom ends anchored in the subsoil). After some creep deformation of the backfill has occurred at the preloading stage, the load is reduced to a lower level, which is then maintained as prestress. When sufficient amount of prestress has survived, a high pressure level (thereby high stiffness) is maintained in the backfill, which results in smaller deformation when subjected to seismic loads. Furthermore, when a PL/PS reinforced backfill is to be subjected to large shear deformation, it would exhibit highly ductile behaviour. This is because, as the length of the tie rods are maintained nearly constant, the height of backfill is forced to reduce with shear deformation, which results in an increase in the pressure level and thereby the increase in the stiffness. This mechanism will be enhanced if the backfill is well compacted to induce considerable dilative characteristics upon shear deformation. This behaviour is similar to that of saturated dense sand subjected to shear deformation under undrained conditions.

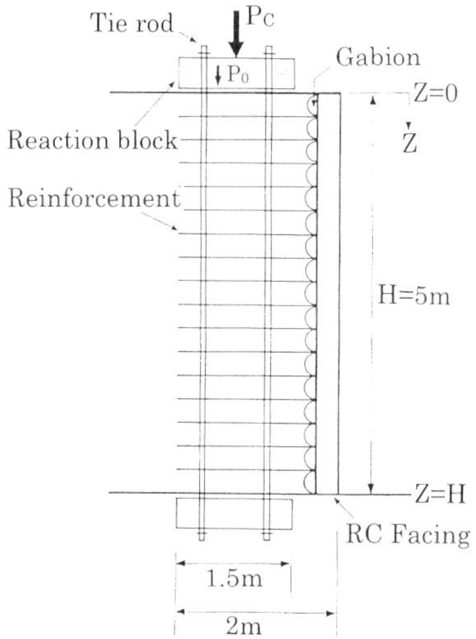

Fig. 55 Schematic diagram of PL/PS geosynthetic-reinforced soil RW (Tatsuoka et al., 1996a, b and c, Uchimura et al., 1996).

7. REMEDY WORKS AND RECONSTRUCTIONS BY SOIL REINFORCEMENT TECHNIQUES

After the earthquake, a large amount of remedial and preventive work of soil structures (i.e., sloped embankments, RWs and slopes) was performed by means of a variety of soil reinforcement techniques. This is another good evidence how these techniques have been accepted by civil engineers because of their cost-effectiveness and their performance.

7.1 *GRS-RWs with FHR facings*

Most of the embankment slopes and conventional type RWs for railways that were damaged seriously were replaced with GRS-RWs with FHR facings. Typical cases are shown in Figs. 8b, 9b, 10b and 11b. The total length of the walls constructed for remedial work is 1.1 km with the wall height ranging from 2 m to 5 m.

7.2 *Terre Armee walls*

Immediately to the south of JR Nada Station (see Fig. 4), the south slope of a road had been retained by RC RWs for a total length of 260 m at two elevations (see Fig. 56a) (Kobayashi and Ohtani, 1996). These RC RWs and the backfill moved significantly. The walls were replaced by Terre Armee walls, which were designed by using k_h= 0.15 (see Fig. 56b). The backfill soil consisted of 76 % gravel, 15 % sand and 9 % fines.

7.3 *Large diameter nailing in embankment slopes*

Many embankment slopes for railways were reinforced by using 40 cm-diameter nails as shown in Fig. 57. Each nail consists of a tendon of fiber-reinforced plastic (FRP) with a diameter of 37 mm (Tateyama et al., 1992). Each nail is produced by auguring the embankment slope, and then extruding the auguring device, supplying cement slurry from the blades at the end of the augur rod and mixing it with the surrounding soil while leaving the FRP rod at the center of the cylindrical zone of augured soil.

To construct a vertical RW, first vertical cement slurry-mixed soil columns are produced in the slope as pre-propping. Then, while excavating the slope, these large diameter nails are installed in stages. Finally, a thin RC facing is cast-in-place on the excavated slope. This method has been used to excavate the railway embankment slopes while allowing trains to run on the embankment for a total

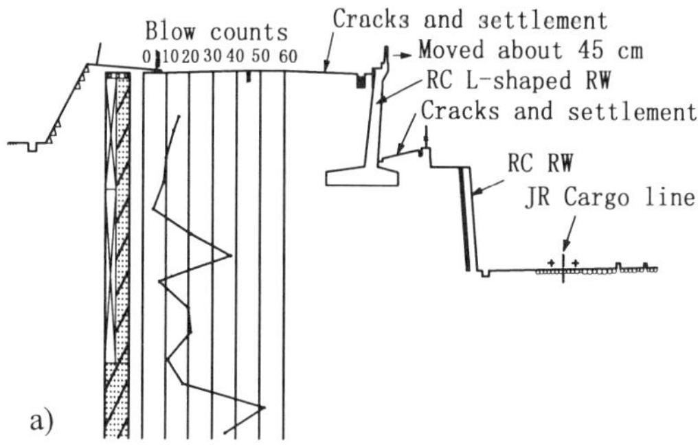

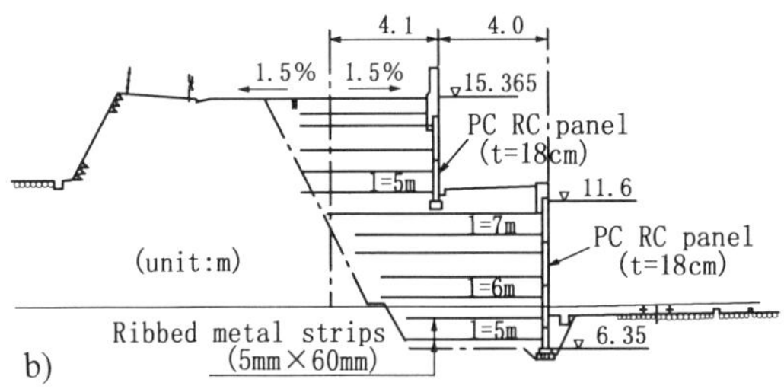

Fig. 56 a) Damaged RC cantilever RW for road embankment and b) Terre Armee wall by reconstruction, south of Nada Station, JR Kobe Line (Kobayashi and Ohtani, 1996).

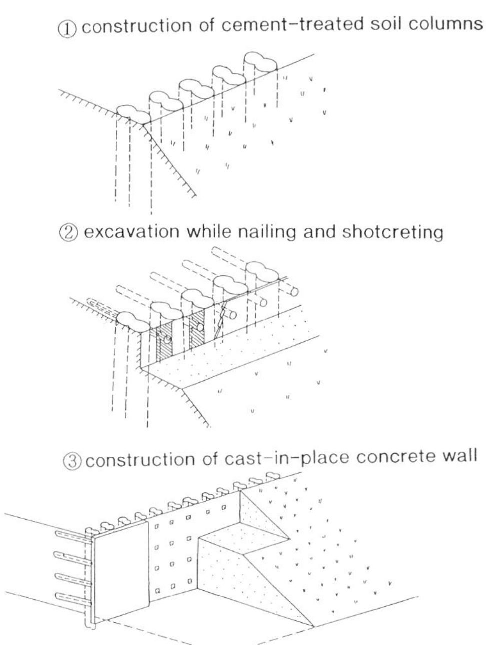

Fig. 57 Illustration of the large diameter nailing technique for vertical cutting of embankment slope (Tateyama et al., 1992, Tatsuoka, 1993).

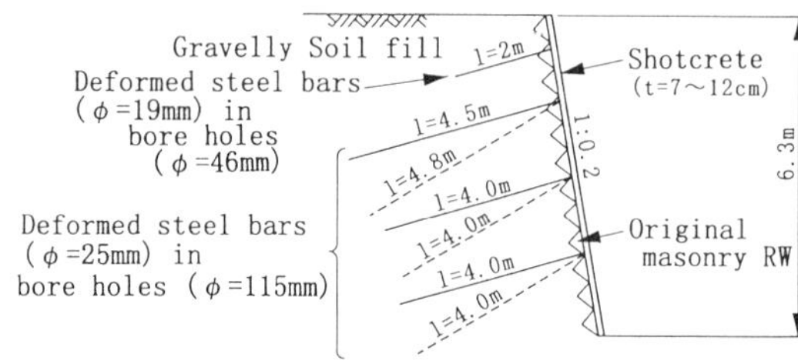

Fig. 58 Typical preventive work of slope by the root piling technique (Kobayashi and Ohtani, 1996).

length of more than 2 km. A completed structure is a sort of reinforced soil RW with a FHR facing. As the remedial work after this earthquake, damaged RWs and embankment slopes for railways were reconstructed by this technique for a total length of 530 m with the wall height ranging up to 6 m (Fig. 10c).

7.4 *Root-piling in natural slopes*

More than forty slopes were remedied or reinforced preventively for possible earthquakes in the future by using the root piling technique from either the crest of slope or the slope face or wall face. The case shown in Fig. 58 is typical of the walls stabilized by root piling from the wall face (Kobayashi and Ohtani, 1996). This masonry wall, located in Chuo-ku, Kobe City, was moved by the earthquake, while cracks appeared on the wall face and on the ground surface on the crest. The backfill soil was a sand gravel showing a SPT blow count of about five.

8 CONCLUSIONS

Different types of soil retaining walls (RWs) located in the severely shaken areas during the 1995 Hyogoken-Nanbu earthquake performed differently. In general, older RWs were damaged more seriously, while masonry, leaning-type and gravity-type unreinforced concrete RWs showed a very low stability against the strong seismic shaking. In addition, many cantilever-type or inverted T-shaped RC RWs, mostly without piles, performed poorly.

A great number of RC columns and piers collapsed by shear failure in a brittle manner.

In comparison with the above, a number of geogrid- or metal-reinforced soil RWs performed very well. In particular, the geogrid-reinforced soil RWs with FHR facings that were constructed in 1992 at Tanata did not collapse despite the fact that the site was located in one of the most severely shaken and seriously damaged areas. Based on these experiences, many damaged embankment slopes and conventional RWs were replaced with GRS-RWs with FHR facings.

A number of natural slopes which had been stabilized by the nailing and root piling techniques also performed very well. A large amount of natural slopes and backfills of RWs were reinforced by these techniques after the earthquake.

For better understanding of the seismic behaviour and for developing more rational seismic design methodology of soil structures, the effects of flexibility on the seismic earth pressure and the effects of ductility on the design seismic coefficient to be used in limit equilibrium-based pseudo-static stability analysis should be studied. The internal deformation of reinforced backfill when subject to seismic loads and the effects of the facing rigidity and reinforcement length should also be studied.

It is not easy to restrain the simple shear deformation of reinforced soil RWs because of the lack of a bracing component, as in elevated RC frame structures. Preloading and prestressing the reinforced backfill may be effective to prevent such deformation.

ACKNOWLEDGMENTS

The authors wish to thank Messrs. Miyatake,H. and Tsutsumi,T. of Public Works Research Institute, Ministry of Construction, Horii,K. of Chuo Kaihatsu Corporation, Ohtani,Y. of Hirose Co.,Ltd., Torii,T. of Construction Project Consultants Inc., and Shintani,H. of Mitsubishi Yuka Industrial Products Co.,Ltd. and Mr. Sato,T., Ms. Torimitsu,M. and Dr. Kodaka,T. of IIS, Univ. of Tokyo for their assistance in the preparation of this paper. They also appreciate a careful review of the manuscript by Prof. Wu, J.T.H, the University of Colorado at Denver.

REFERENCES

Chuo Kaihatsu Corp. 1995. Report of damage investigation for the Great Hanshin earthquake, No.2- damage ratio investigation and geology and ground conditions. (in Japanese)

Collins,J.G., Chouery-Curtis,V.E. and Berg,R.R. 1992. Field observation of reinforced soil structures under seismic loading, Int. Symp. Earth Reinforcement Practice (Ochiai et al., eds), IS Kyushu '92, Balkema, Vol.1, pp.223-228.

Felio,G.Y., Vucetic,M., Hudson,M. Barar,P. and Chapman,R. 1990. Performance of soil nailed walls during the October 17, 1989 Loma Prieta earthquake, 43rd Canadian Geotechnical Conference, Quebec, pp.165-173.

Fujii,T., Sueoka,T., Muramatsu,M., Mega,M. and Ohtani,M. 1996. Behaviour of soil nailed walls and root pile structures caused by the 1995 Hyogoken-Nanbu earthquake, Symposium on Reinforcement of Natural Slopes, the Japanese Geotechnical Society, pp.1-191. (in Japanese)

Fukuda,N., Tajiri,N., Yamanouchi,T., Sakai,N. and Shintani,H. 1994. Applicability of seismic design methods to geogrid reinforced embankment, Proc. 4th Int. Conf. on Geotextiles, Geomembranes and Related Products, Singapore, pp.533-536.

Horii,K.,Kishida,H., Tateyama,M. and Tatsuoka,F. 1994. Computerized design method for geosynthetic-reinforced soil retaining walls for railway embankments, Recent Case Histories of Permanent Geosynthetic-Reinforced Soil Retaining Walls, Tatsuoka & Leshchinsky (eds.), Balkema, pp.205-218.

Japan Multiple Anchor Association (JMAA), 1995. Report on the behaviour of multiple-anchor-reinforced soil retaining walls during the Hyogoken-Nanbu earthquake. (in Japanese)

Japan Society of Civil Engineers, 1996. Proposal on earthquake resistance for civil engineering structures (Special Task Committee of Earthquake Resistance of Civil Engineering Structures), The 1995 Hyogoken-nanbu Earthquake -Investigation into Damage to Civil Engineering Structures-, Committee of Earthquake Engineering, Japan Society of Civil Engineers, pp.297-306.

Japan Terre Armee Association (JTAA), 1995. Report on the behaviour of Terre Armee walls during the Hyogoken-Nanbu earthquake. (in Japanese)

Kobayashi,K and Ohtani,Y. 1996. Damage and restoration of reinforced earth structures in the 1995 Hyogoken-Nambu Earthquake, The Foundation Engineering & Equipment, Vol.24, No.10. (in Japanese)

Koseki,J., Tateyama,M., Tatsuoka,F. and Horii,K. 1996. Back analyses of soil retaining walls for railway embankments damaged by the 1995 Hyogoken-Nanbu Earthquake, The 1995 Hyogoken-nanbu Earthquake -Investigation into Damage to Civil Engineering Structures-, Committee of Earthquake Engineering, Japan Society of Civil Engineers, pp.101-114.

Matsui,T., Kobayashi,K., Kumada,T. and Ohtani,Y. 1996. Damage of reinforced earth structures by steel reinforcements in the 1995 Hyogoken-Nanbu earthquake, Tsuchi-to-Kiso, Vol.44, No.2, pp.76-78. (in Japanese)

Ministry of Transport, 1992. Design standards for railway structures - soil structures-. (in Japanese)

Mitsubishi Yuka Industrial Products Co.,Ltd., 1995. Survey report on the damage induced by the 1995 Hyogoken-Nanbu earthquake. (in Japanese)

Murata,O., Tateyama,M. and Tatsuoka,F. 1994. Shaking table tests on a large geosynthetic-reinforced soil retaining wall model, Recent Case Histories of Permanent Geosynthetic-Reinforced Soil Retaining Walls, Tatsuoka & Leshchinsky (eds.), Balkema, pp.259-264.

Naemura,S. and Miyatake,H. 1995. Damage to earth structures for road facilities in the Hanshin-Awaji earthquake disaster, Technical information on geosynthetics, Vol.11, No.2, Japan Chapter of the Int. Geosynthetics Society, pp.3-8. (in Japanese).

Nishimura,J., Hirai,T., Iwasaki,K., Saitoh,Y., Morishima,M., Shintani,H., Yoshikawa,S. and Yamamoto,H. 1996. Earthquake resistance of geogrid-reinforced soil walls based on a study conducted following the southern Hyogo earthquake, Prof. Int. Symp. on Earth Reinforcement, IS Kyushu '96, Balkema Vol.1, pp.439-444.

Ohbayashi Corp., 1995. Preliminary report on the 1995 Southern Hyogo Prefecture Earthquake, Ohbayashi Corporation Technical Research Institute, p.16.

Research Committee on Reinforcement Technology for Natural Slopes, 1996. Activity report, Symposium on Reinforcement of Natural Slopes, Japanese Geotechnical Society, pp.1-191. (in Japanese)

Railway Technical Research Institute (RTRI), 1996. Reconnaissance report on damage of railway facilities in Hyogo-ken Nanbu earthquake, Special No.4, pp.56-62. (in Japanese)

Sassa,K., Fukuoka,H., Scarascia-Mugnozza,G. and Evans,S. 1996. Earthquake-induced-landslides: distribution, motion and mechanisms, Special Issue of Soils and Foundations on Geotechnical Aspects of the January 17 1995 Hyogoken-Nambu Earthquake, pp.53-64.

Sato,T. 1996. Estimation of peak acceleration in the severely damaged area during the 1995 Hyogo-ken Nanbu earthquake, The 1995 Hyogoken-Nanbu Earthquake -Investigation into Damage to Civil Engineering Structures-, Committee of Earthquake Engineering, Japan Society of Civil Engineers, pp.35-44.

Tamura,Y., Nakamura,K., Tateyama,M., Murata,O., Tatsuoka,F. and Nakaya,T. 1994. Full-scale lateral loading tests of column foundations in geotextile-reinforced soil retaining walls, Recent Case Histories of Permanent Geosynthetic-Reinforced Soil Retaining Walls, Tatsuoka & Leshchinsky (eds.), Balkema, pp.277-285.

Tateyama,M., Murata,O., Tamura,Y., Nakamura,K., Tatsuoka,F. and Nakaya,T. 1994. Lateral loading tests of columns on the facing of geotextile-reinforced soil retaining wall, Recent Case Histories of Permanent Geosynthetic-Reinforced Soil Retaining Walls, Tatsuoka & Leshchinsky (eds.), Balkema, pp.287-294.

Tateyama,O., Tarumi,H., Tamura,Y. and Tatsuoka, F. 1992. Permanent cut of an embankment slope by soil nailing allowing very small deformation, Earth Reinforcement Practice (Ochiai et al., eds), IS Kyushu, Balkema, Vol.1, pp.555-560.

Tatsuoka,F., Murata,O. and Tateyama,M. 1992. Permanent geosynthetic-reinforced soil retaining walls used for railway embankment in Japan, Geosynthetic-Reinforced Soil Retaining Walls, Wu (ed.), Balkema, pp.101-130.

Tatsuoka,F. 1993. Roles of facing rigidity in soil reinforcing, Keynote Lecture, Proc. Int. Sym. on Earth Reinforcement Practice, IS Kyushu '92, Balkema (Ochiai et al., eds), Vol. 2, pp.831-870.

Tatsuoka,F., Tateyama,M. and Koseki,J. 1996. Performance of soil retaining walls for railway embankments, Soils and Foundations, Special Issue of Soils and Foundations on Geotechnical Aspects of the January 17 1995 Hyogoken-Nambu Earthquake, pp.311-324.

Tatsuoka,F., Uchimura,T. and Tateyama,M. 1996a. Preloaded and prestressed reinforced soil, Soils and Foundations (accepted for publication)

Tatsuoka,F., Uchimura,T., Tateyama,M. and Muramoto,K. 1996b. Creep Deformation and Stress Relaxation in Preloaded/Prestressed Geosynthetic-Reinforced Soil Retaining Walls, Session "Measuring and Modeling Time-Dependent Soil Behavior", ASCE Washington Convention, pp.258-272.

Tatsuoka,F., Tateyama,M, Uchimura,T. and Koseki,J. 1996c. Geosynthetic-Reinforced Soil Retaining Walls as Important Permanent Structures, The 1996-1997 Mercer Lecture, Proc. of Euro-Geo, First European Geosynthetic Conference and Exhibition, Maastricht, the Netherlands, pp.3-24.

Tsukuda,C., Yamamoto,A. and Fukuda,N. 1996. Stability analysis of cut slope reinforced with steel bars at north area of Awaji Island during Hyogoken-Nambu earthquake, Symposium on Reinforcement of Natural Slopes, the Japanese Geotechnical Society, pp.295-298. (in Japanese).

Uchimura,T., Tatsuoka,F., Sato,T., Tateyama,M. and Tamura,Y. 1996. Performance of preloaded and prestressed geosynthetic-reinforced soil, Proc. Int. Symp. on Earth Reinforcement, Fukuoka, Balkema (Ochiai et al., eds), Vol. 1, pp.537-542.

Embankments

Earth Reinforcement, Ochiai, Yasufuku & Omine (eds) © 1997 Balkema, Rotterdam, ISBN 90 5410 833 9

Chairman's report: Embankments

C.K.Shen
Hong Kong University of Science and Technology, People's Republic of China

In Session 1 - EMBANKMENTS, there are altogether 20 papers written by authors from 7 countries in 3 different continents. A thorough and constructive review of each individual paper was presented by Professor D. T. Bergado, the able Discussion Leader of this Session. It is therefore not the intention of this Reporter to give further detailed account of the papers. Rather, a few thoughts pertaining to a broader picture of the subject area will be outlined.

The titles and authors of the 20 papers are tabulated in Table 1; furthermore, an attempt has been made to group the papers into a number of subsets as shown in Table 2. Since many of the papers deal with more than on subset listed, as such, the grouping is indeed subjective, serving only as a reference.

The papers presented in this Session cover a broad range of approaches and considerations in studying reinforced embankments and other earth retaining structures, which include theoretrial formulation, numerical simulation, model testing, laboratory testing and design methodology. All of the above are essential elements forming the wealth of knowledge upon which safe and economic design of reinforced earth structures can be achieved. One of the most encouraging indications gleaned from the papers is the increasing application of reinforced earth technology to "problem soils"; e.g. very soft clay foundation and claying fill.

As the practice of ground improvement moves into areas of problem soils and hostile environments, it is conceivable that the complex soil behavior and the uncertain site conditions will not allow us to carry out a design or analysis with a high degree of confidence. Thus field performance monitoring with extensive instrumentation remains a viable means to provide valuable and meaningful information to identify the prototype behaviour and to guard the safety of a structure. When properly documented, field records are the best information for case studies from which lessons can be learned and experiences can be shared. We therefore must report cases of both success and failure. Fortunately or perhaps unfortunately, of the 20 papers, none was dedicated to address failure.

The advancement and practice of soil reinforcement technology is somewhat unique in the filed of geotechnical engineering, because its activities are largely market driven, e.g. by the various reinforcing elements available on the market and the promotion of the manufacturers of their respective products. This is quite different from the normal progression of research, development and application in geotechnical engineering practice. In the field of reinforced soil the availability and the types of

Table 1. List of Session 1 Papers

1. Chilean experiences with geosynthetics in embankments
P.M. Acevedo and C.A. Martinez
2. Effects of reinforcement rigidity on the behavior of reinforced soil wall-embankment system on soft ground
M.C. Alfaro, S. Hayashi and N. Miura
3. Limit design of earth reinforcement methods considering displacement field
K. Arai and K. Kasahara
4. Dynamic response analysis of geogrid reinforced steep embankment subjected to an earthquake
T. Fujii, N. Fukuda and N. Tajiri
5. Creep behavior of laboratory embankment reinforced with geogrid
H. Furuya, M. Toriihara and K. Hirama
6. Geogrid reinforced railway embankment on piles - monitoring
E. Gartung and J. Verspohl
7. In situ failure test of high water content soft clay embankments reinforced by GHDs
M. Kamon, T Akai, M. Fukuda and Y. Nanbu
8. Direction and magnitude of reinforcement force in embankments on soft soils
S. R. Kaniraji
9. Geosynthetic-reinforced soil retaining wall using clay on a very soft ground for Hokuriku bullet train yard in Nagano
K. Kojima, N. Sakamoto, M. Tateyama and O. Maruyama
10. Some factors affecting the structural performance of reinforced fills spanning voids
C. R. Lawson, C.J.F.P. Jones and G.T. Kempton
11. Importance of strong motion in the design of earth reinforcement
S.P.G. Madabhushi
12. Effect of geogrid stiffness on the resistance behavior of reinforced embankment under loading
Y. Miyata, H. Ochiai and K. Kogure
13. Evaluation of effects of soil reinforcement using a new prototype shear apparatus
H. Ochiai, N. Yasufuku, G.L. Xu, T. Yamaji and T. Hirai
14. Numerical simulation of beam-shaped soil structure reinforced by geosynthetics
H. Ohta, S. Goren, A. Iizuka, T. Yamakami, K. Yamagishi and N. Moroto
15. A case history of the construction of a reinforced high embankment on an extra soft ground
H. Oikawa, S. Sasaki and N. Fujii
16. Centrifugal and finite element modelling of reinforced embankments on soft clay
J.S. Sharma and M.D. Bolton
17. Behavior of reinforced earth embankments on liquefiable sandy ground
R. Uzuoka, M. Mihara, F. Oka and A. Yashima
18. A partial factor approach for reinforced fill slope design in Hong Kong
H.N. Wong
19. Application of the velocity field method to stability analysis of earth reinforcement
J.J. Yang, N. Moroto, H Ochiai and A. Suzuki
20. Review of design temperature for reinforced fill slopes in Hong Kong
K.C. Yeo and P.L. R. Pang

Table 2. Grouping of Papers

EMBANKMENT FOUNDED ON SOFT & VERY SOFT CLAY FOUNDATIONS	ANALYTICAL SOLUTION & PREDICTION	SPECIAL CONSIDERATIONS IN DESIGN	MODELS NOT FOUNDED ON SOFT CLAY FOUNDATION
* Field performance and case study #6, #9, #15 * Parametric study using numerical solutions #2, #10 * Centrifuge model study #16 * Design considerations #1, #6, #8, #9	*Analytical formulation #3, #8, #19 * Numerical prediction #4, #5, #14, #17	* Reinforcement #2, #8, #9, #20 * Interface #11, #13, #14 * Loading #5 * Foundation soil #1, #9 * General #18	#12, #14

merchandise appear to dictate the state of prevalent practice and application, while academic pursuit and fundamental research seems to lag behind. For the profession as a whole, I think it is important that a balance be maintained for research, development and marketing to warrant a healthy environment to nurture the continuing growth of this technology.

The performance of reinforced earth structures under seismic loading has been widely discussed, particularly after the 1995 Kobe earthquake. The fact that reinforced earth walls performed better than other types of earth retaining structures is reassuring, but our knowledge on the interactive behavior of a soil-reinforcement system during earthquake is still lacking. More fundamental and experimental research are indeed needed. Dynamic centrifuge model testing is one of the more promising research areas to understand the prototype behavior.

In short, I believe the use of reinforcing elements for ground improvement is still a young and emerging field in geotechnical engineering. Close cooperation and collaboration among the manufacturers, the users, and the researchers will give the impetus to advance the state of knowledge in reinforced soil technology.

Earth Reinforcement, Ochiai, Yasufuku & Omine (eds) © 1997 Balkema, Rotterdam, ISBN 90 5410 833 9

Discussion leader's report: Embankments

D.T. Bergado & P.V. Long
School of Civil Engineering, Asian Institute of Technology, Bangkok, Thailand

1 INTRODUCTION

In this Session 1 on Embankments, a total of 20 papers has been submitted. The papers can be roughly classified into 4 groups as follows:

1) Design and analyses (7 papers)
2) Modeling and full scale tests (7 papers)
3) Reinforced soil structures spanning pile foundation or voids (3 papers)
4) Dynamic response of geosynthetic reinforced structures (3 papers)

2 DESIGN AND ANALYSES

In this category, the papers of the following authors are included, namely: Ochiai et al, Kaniraj, Arai and Kasahara, Acevedo and Martinez, Yang et al, Yeo et al, and Wong. For geotextile reinforced embankment on soft ground, three issues are involved, namely: (a) Inclination of the reinforcing force, (b) Mobilized strains in the geotextile reinforcement, (c) Modification of shear strength at the failure plane, and (d) Effect of confinement on the properties of the geotextile reinforcement. Accordingly the research directions are mostly directed into these issues. The first three issues have been addressed by the submitted papers and further elaborated herein.

The effect of reinforcement on soil strength at the failure plane has been presented by Long et al (1996) and illustrated in Fig. 1. The tensile force mobilized in the reinforcement, T_{mob}, can be decomposed into 2 components:

$$Tangential: \quad T_p = T_{mob}(\cos\beta) \tag{1}$$

$$Normal: \quad T_n = T_{mob}(\sin\beta) \tag{2}$$

The projection of the resistant force on the tangent direction is given as follows:

$$T_r = T_p + T_n(\tan\phi) = T_{mob}(\cos\beta + \sin\beta\tan\phi) \tag{3}$$

The resisting moment caused by the reinforcement is given as:

$$M_{rr} = T_r\,(R) \tag{4}$$

The geotextile reinforce embankment can be assumed to be equivalent to an unreinforced embankment with the following modified parameters:

$$\phi_G = \phi \tag{5}$$

$$C_G = C + \Delta C_G \tag{6}$$

Then, the following expression can be written:

$$M_{rr} = \Delta C_G(\overline{AB})R \tag{7}$$

where AB is the arc length in Fig. 1.

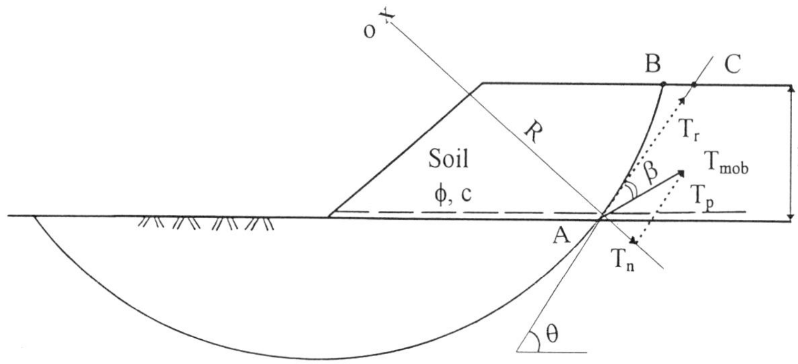

Fig. 1 Effect of Geotextile Reinforcement on Soil Strength.

Equating Equas. 4 and 7 yields:

$$\Delta C_G = \left(\frac{T_r}{\overline{AB}}\right) \quad \textbf{(8)}$$

Approximating the arc length AB by the straight line AC and substituting to Equa. 8:

$$\Delta C_G = \left(\frac{T_r(\sin\theta)}{H}\right) \quad \textbf{(9)}$$

From Equas. 3 and 9, the following expression can be obtained:

$$\Delta C_G = \left(\frac{(\alpha)T_{mob}}{H}\right) \quad \textbf{(10)}$$

where:

$$\alpha = (\cos\beta + \sin\beta\tan\phi)\sin\theta \quad \textbf{(11)}$$

The orientation and magnitude of geotextile reinforcement force with the slip failure are also the key parameters affecting on the stability analyses of reinforced embankments on soft ground. Presently, these parameters have been selected arbitrarily and even independently. Bergado et al (1996) presented the deformation behavior of geotextile reinforcement at the vicinity of the shear surface which addressed these parameters. Both the large direct shear tests with inclined geotextile reinforcements as well as as the finite element modeling for reinforced soil mass that simulate the actual conditions of slip failure in the field have been conducted.

The mobilized strain in geotextile reinforcement near the shear surface can be calculated from the displacement of the reinforcement measured at locations G_1 and G_2 in Fig. 2.

$$\varepsilon = \left(\frac{d_1 - d_2}{l_{12}}\right) \quad \textbf{(12)}$$

where d_1, d_2 are the measured displacements of the geotextile at points G_1 and G_2, respectively, and l_{12} is the original distance between these points. The value of l_{12} of 50 mm was set for all tests. The inclination factor, I_f, of geotextile reinforcement is expressed by the following equation:

$$I_f = 1 - \alpha/\alpha_0 \quad \textbf{(13)}$$

where α_0 and α, respectively, are the original and current inclination angles of the geotextiles as seen in Fig. 3. As shown in Fig. 3, for shear displacement, u, the interface surface between geotextile and backfill soil OBA moved to O'B'A'. Assuming that the current interface surface O'B'A' can be approximated by a parabola, the following expressions can be obtained: (Bergado et al, 1996):

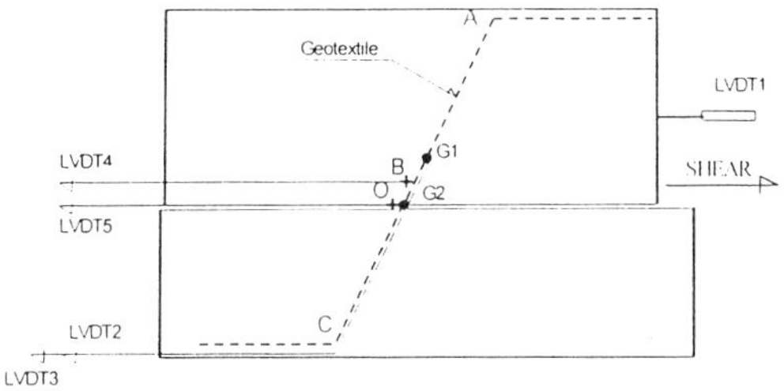

Fig. 2. Instrumentation for Large Direct Shear Test

$$\tan \alpha = \frac{Y_A X_{B'}^2 - Y_B X_{A'}^2}{X_{A'} X_{B'}^2 - X_{B'} X_{A'}^2} \qquad \textbf{(14)}$$

where Y_A and Y_B are the vertical coordinates of point A and B as illustrated in Fig. 3.

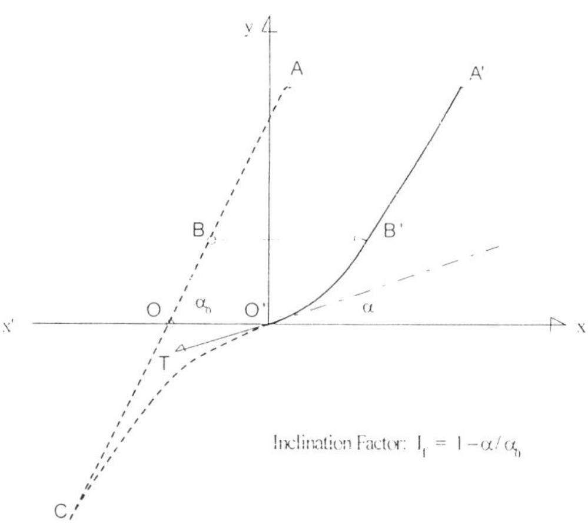

Fig. 3. Schematic diagram for calculating the inclination factor.

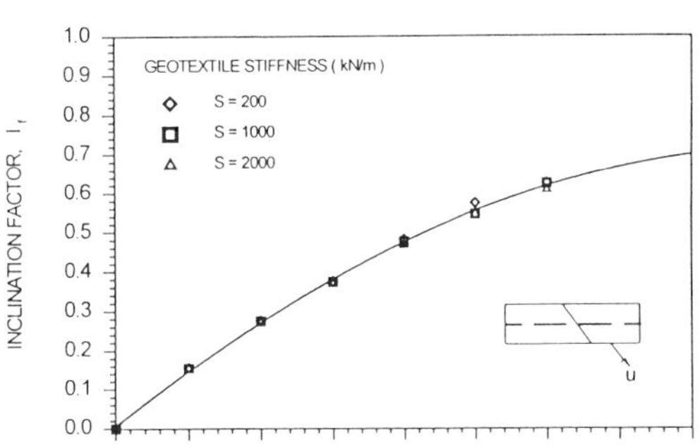

Fig. 4. Inclination factor-shear displacement plots from FEM.

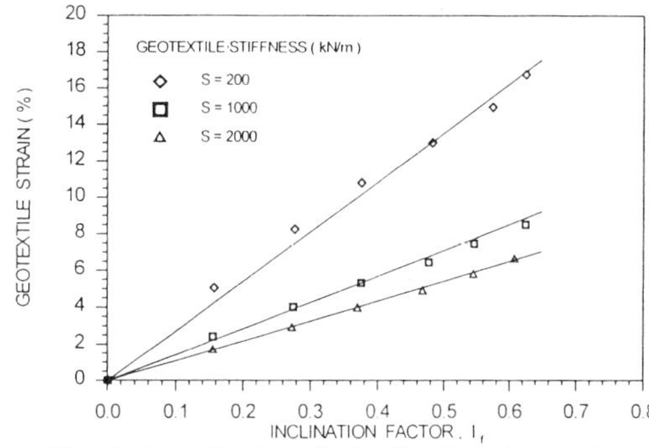

Fig. 5. Localized strain-Inclination factor plots from FEM.

The PLAXIS finite element program (Vermeer and Brinkgreve, 1995) was employed in the validation of the direct shear test results across geotextile reinforcement by conducting the back-analyses of the test data. The good agreement between the calculated and measured data has been presented by Bergado et al (1996) indicating the success of utilizing the PLAXIS software. The calculated values of inclination factor, I_f, are plotted in Fig. 4. indicated that the inclination factor is dependent on the shear displacement but not on the geotextile stiffness. As shown in Fig. 4, the value of I_f is less than 1.0 even at large displacement of 0.8 m. The values of the geotextile strain near the intersection of the reinforcement and shear surface, denoted as localized strain ε_{lc}, are presented in Fig. 5 as a function of the inclination factor. Subsequently, the relationship between localized strain, ε_{lc}, and inclination factor, I_f, can be approximated by straight lines at different geotextile stiffness. The curve fitting is plotted in Fig. 6 and the following expression can be obtained:

$$\varepsilon_{lc} = 225(I_f) S^{-0.4} \qquad \textbf{(15)}$$

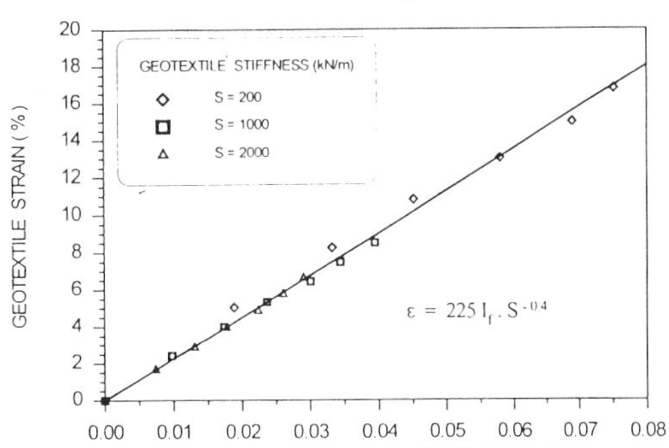

Fig. 6. Localized strain-normalized inclination factor from FEM.

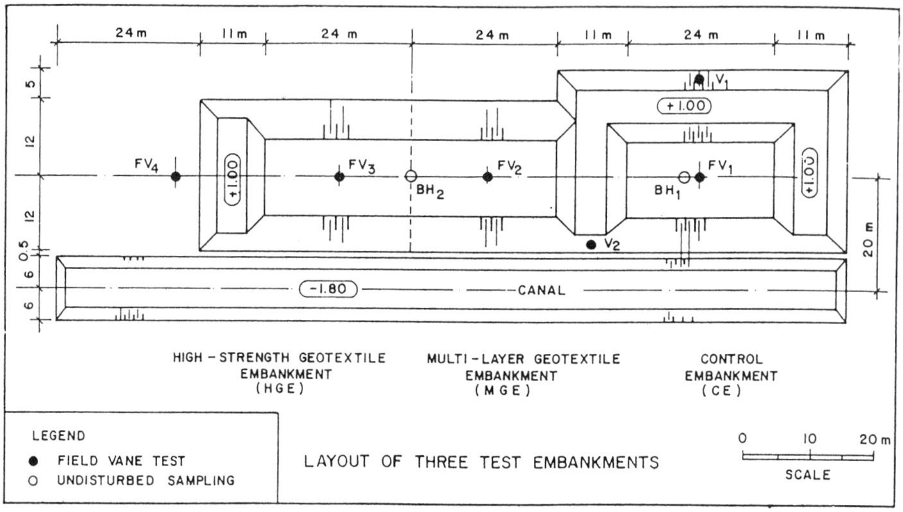

Fig. 7. Layout of the Test Embankment together with Locations of Field Investigation.

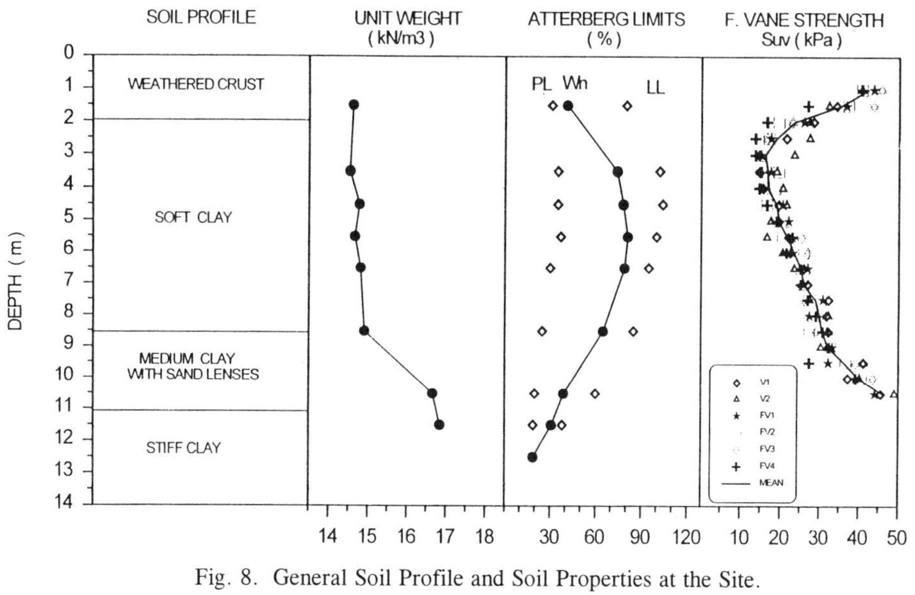

Fig. 8. General Soil Profile and Soil Properties at the Site.

Thus, the localized strain, ε_{lc}, in geotextile reinforcement associated with the slip failure has been found to be a function of geotextile stiffness and the corresponding inclination factor.

3 FULL SCALE TESTS AND MODELING

The papers in this group include those by Marolo et al, Furuya et al, Miyata et al, Sharma and Bolton, Kojima et al, Kamon et al, and Oikawa et al. The topics addressed in the submitted papers dealt with effects of reinforcement rigidity or stiffness, creep behavior of reinforcement, foundation behavior beneath reinforced structures reinforcement by geosynthetic horizontal drains (GHD), and behavior of geotextile-reinforced embankment on soft ground from centrifuge model tests and finite element modeling. The utilization of GHD reinforcement accelerated pore pressure dissipation and increased the soil shear strength for high water content cohesive embankment fills.

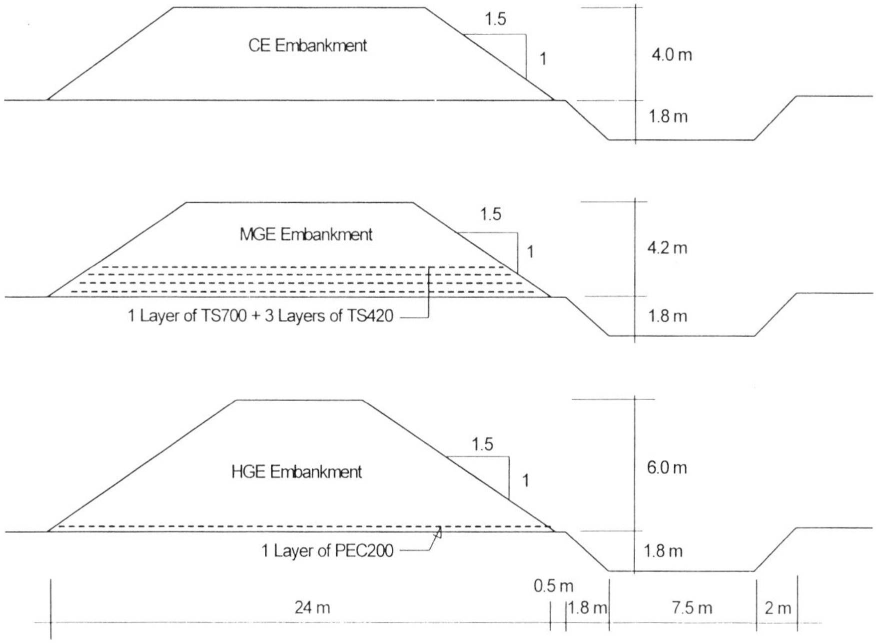

Fig. 9. Cross Sections of Three Test Embankments.

Concerning the failure mechanism and performance of geotextile reinforced embankments on soft Bangkok clay, three full scale instrumented test embankments with and without geotextile reinforcements have been constructed to failure (Bergado et al 1994a). The layout of the test embankments is shown in Fig. 7. The generalized soil profile and soil properties are presented in Fig. 8. One embankment was reinforced with multiple layers of low strength, nonwoven, needle-punched geotextile. The other was reinforced by a single layer of high strength composite woven/nonwoven geotextile. The unreinforced embankment was built

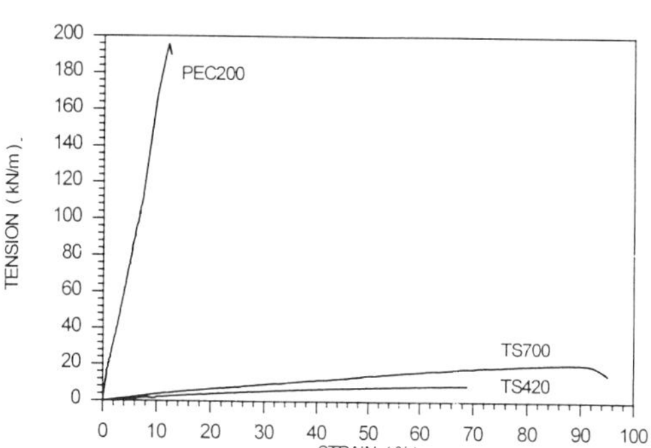

Fig. 10 In-Air Tension-Strain of Geotextile Reinforcements (ASTM D4595).

nearby as a control embankment.

The cross-sections of the three embankments are presented in Fig. 9. All embankments were designed with 1.5 to 1.0 side slopes. A canal of 2.0 m deep and 7.5 m wide was excavated along the toe of the embankments in order to reduce the fill required for the embankment loading to failure and to ensure that the failure will occur in the proposed side. The multi-layer reinforced embankment (MGE) contained four layers of reinforcements as follows: one layer of nonwoven, needle-punched geotextile (TS-700) and three layers of nonwoven, needle-punched geotextile (TS-420). The single layer reinforced embankment (HGE) was reinforced with high strength geotextile (PEC-200) placed at the embankment base. The wide-width tensile strength versus strain relationships of the geotextile reinforcements are given in Fig. 10. The silty sand materials with maximum dry density of 18 kN/m^3 which are widely used for highway fills, was utilized as backfill materials. Standard instrumentations were installed in the foundation soils, namely: settlement plates, piezometers, and inclinometers. The displacements and strains in the geotextile reinforcements were measured by wire extensometers, Glotzl extensometers, and strain gages. Details of the foundation and geotextile instrumentations are given by Bergado et al (1994b). The construction sequences of the three test embankments are presented in Fig. 11. Detailed description of the embankment constructions were given by Bergado et al (1994b). The embankment loading was calculated from the fill thickness and its total unit weight with adjustments due to rainfall during construction. The net embankment height is defined as the difference between the current elevation of embankment crest and the original elevation of the embankment base. In the reinforced embankment, a two step failure occurred. The primary failure corresponded to the bearing capacity failure of the foundation. The secondary failure occurred during rupture of the reinforcements. The failure of MGE embankment compared to that of the CE embankment is shown in Fig. 12.

The limit equilibrium method, assuming circular slip failure and using the PC-STABL6 software, was utilized to back-analyze the stability of the three embankments. The free body reinforcement force was utilized for the single layer reinforced (HGE) embankment. The modified soil strength was employed in the multi-layer reinforced (MGE) embankment. In this case, the resultant force in the direction tangent to the slip surface, T_r, can be calculated as in Equa. 3. The direction of the reinforcement force at onset of failure can be represented by the inclination factor, I_f, as given in Equa. 13. The limiting strain, ε_{lm}, at the onset of

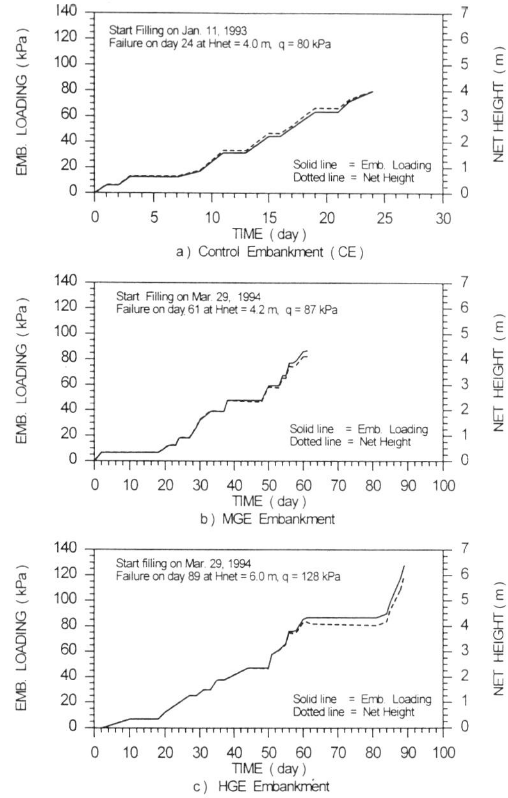

Fig. 11. Construction Sequences of Three Test Embankments.

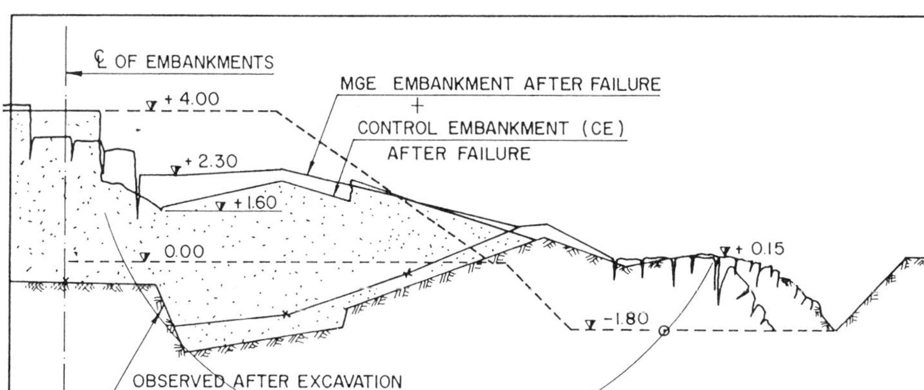

Fig. 12 Cross Section of MGE Embankment after Failure Measured on May 31, 1993.

failure can be obtained as follows:

$$\varepsilon_{lm} = \varepsilon_c + \varepsilon_{lc} \tag{16}$$

where ε_c is the critical strain corresponding to the critical height just prior to the bearing capacity failure of foundation soil, and ε_{lc} is the localized strain associated with slip failure as given Equa. 15. The results of the back-analysis are tabulated in Table 1. The interpreted and predicted values of ε_{lc} are plotted in Fig. 13 showing good comparison for a large range of geotextile stiffness including the Sackville Embankment (Rowe et al, 1995) and Guiche Embankment (Delmas et al, 1992).

For the soft Bangkok clay, the multiple layer reinforced embankment height was 1.15 times the unreinforced embankment height using lower strength geotextile reinforcement. The single layer, high-strength geotextile reinforcement increased the embankment height to 1.5 times the unreinforced embankment height. The critical strain at primary failure can be taken as 2.5 to 3.0 %. The limiting strain at secondary failure can be taken as 12 to 14%. At primary failure the reinforcement force can be assumed in the horizontal direction, (I_f =0.0). During the secondary failure, the reinforcement force can be assumed at bisectional direction (I_f =0.50).

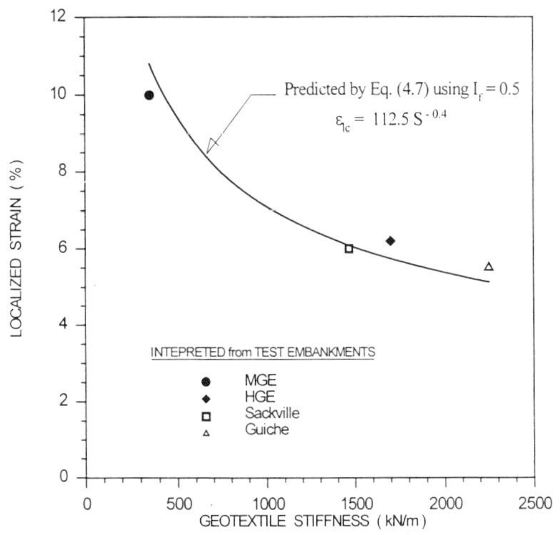

Fig. 13. Interpreted and Predicted Localized Strain in Geotextile Reinforcement.

4 REINFORCED-SOIL STRUCTURE SPANNING PILE FOUNDATIONS OR VOIDS

Three papers were submitted in this category, by Lawson et al, Gartung and Verspohl, and Ohta et al. Based on the measured sagging of embankment base between pile caps and above voids and the reinforcement strain measurements, good

Table 1. Back-Analyses Results for Three Test Embankments Using Actual Properties of Embankment Fill.

Net Embankment Height (m)	FEM Consolidation Analysis	LEM Using Circular Slip Surface: Reinforcement ε (%)	Tmob (kN/m)	If	Tr (kN/m)	Embankment Fill γ (kN/m3)	φ (deg)	c (kPa)	FS	Notes
CE Embankment										
4.0 m - Case 1						18.5	40	22	1.02	
4.0 m - Case 2	Collapse					18.5	30	15	0.98	
MGE Embankment										
4.2 m - Case 1	Critical	3.0	10.5	0	12.5	19.2	38	17	0.98	
4.2 m - Case 2	Collapse	13.9 *	26.4	0.5	30.5	19.2	30	10	1.00	
HGE Embankment										
4.2 m - Case 1	Critical	2.3	39.1	0		19.2	38	17	1.00	Before induced failure
4.2 m - Case 2		8.5	145	0.5		19.2	30	10	1.25	After induced failure
6.0 m - Case 3	Collapse	13.0	200	0.5		19.2	30	10	1.01	Previous failure surfac
6.0 m - Case 4	Collapse	13	200	0.5		19.2	30	10	1.03	New failure surface

Notes:
1) * Geotextile strain obtained from back calculated value of tensile force for FS = 1
2) All the other values of geotextile strain are taken from the measurements
3) Case 1 and 2 corresponding to strength parameters of embankment fill at peak and critical state, respectively
4) Case 3 using a reduction factor of 0.75 for undrained shear strength of sub-soils along the previous sliding surface

performances of geosynthetic reinforced structure were observed. Furthermore, a unique combination concerning the void space or span, fill height, reinforcement strength, and reinforcement stiffness was obtained to satisfy serviceability criteria. The use of reinforced structure for fills spanning voids or pile is now an increasingly accepted practice.

5 DYNAMIC RESPONSE IN GEOSYNTHETIC REINFORCED STRUCTURES

There were three papers submitted in this category, namely: Fujii et al, Uzuoka et al, and Madabhushi. It was demonstrated that the reinforced embankment maintained its stability under maximum earthquake horizontal accelerations of 310 gals. Moreover, a phase difference was observed between the horizontal inertia force in the reinforced area and the area behind the reinforced zone preventing their simultaneous occurrence. The geotextile reinforcements can control the occurrence of tensile stresses within the reinforced embankment. Increasing the rigidity of the reinforced embankment by increasing the reinforcement stiffness may reduce the amount of deformation and subsidence. Further study is needed to incorporate the results of dynamic response analyses to the seismic design methods of geotextile reinforced embankment. In addition, further studies are needed for the effectiveness of the reinforcement in the prevention of embankment failure and reducing the embankment deformation and subsidence.

6 CONCLUDING REMARKS

The technique of geotextile reinforced embankment has been increasingly used for construction on soft ground foundation. For geotextile reinforced embankment on soft ground,three relevant issues have been dealt with and discussed in this report, namely: the direction of the reinforcing force, the mobilized strains in the geotextile reinforcement, and the modification of soil shear strength at the slip plane. Moreover, a new alternative to geotextile reinforcement, the geosynthetic horizontal drain (GHD), has been developed for utilizing saturated clayey embankment fill. Furthermore, the results of three full scale test embankments with and without geotextile reinforcements constructed to failure are presented in this report. These results confirmed and validated the aforementioned three issues of reinforced embankments on soft ground. Finally, our basic tools to understand and interpret the behavior of reinforced embankment may include theoretical formulation, numerical simulations, full scale tests, laboratory model tests, and design methodology. Considering these tools, the importance of conducting full scale tests together with subsequent data monitoring should be emphasized in order to yield the most valuable and reliable information.

7 REFERENCES

Bergado, D. T., Long, P.V., Lee, C.H. and Werner, G. (1994a), Performance of reinforced embankment of soft Bangkok clay with high strength geotextile reinforcement, Geotextile and Geomembranes, Vol. 13, pp. 403-420.

Bergado, D. T., Long, P.V., Loke, K.H. Christopher, B.R. and Delmas, P. (1994b), Geotextile reinforcement in full scale test embankment on soft ground, Proc. Fifth Int'l. Conf. on Geotextiles, Geomembranes, and Related Products, Singapore, Vol, 1, pp 21-24.

Bergado, D.T., Long, P.V. and Werner, G. (1996), Deformation behavior of geotextile reinforcement at vicinity of shear surface, Proc. First European Geosynthetics Conf. (EuroGeo 1), Maastrict, The Netherlands, pp. 177-182.

Long, P.V., Bergado, D.T. and Balasubramaniam, A.S. (1996), Stability analysis of reinforced and unreinforced embankments on soft ground, Geosynthetics International, Vol. 3, No. 5.

Delmas, P., Queroy,D. Quaresma, M. and De Saint A. (1992), Failure of an experimental embankment: Guiche, Geotextiles, Geomembranes, and Products, den Hoedt (ed), Balkema Printers, Rotterdam, The Netherlands.

Rowe, R. K., Granendran, C. T., Landva, A.O. and Valsangkar, A. J. (1995) Construction and performance of a full scale reinforced test embankment, Sackville, New Brunswick, Canadian Geotech. Journal, Vol. 110, No. 3, pp. 512-534.

Vermeer, P.A. and Brinkgreve R. B. J. (1995), Finite element code for soil and rock analysis, Balkema Printers, Rotterdam, The Netherlands.

Wall structures

Earth Reinforcement, Ochiai, Yasufuku & Omine (eds) © 1997 Balkema, Rotterdam, ISBN 90 5410 833 9

Chairman's report: Wall structures

L.E.Wichter
Brandenburg Technical University, Germany

ABSTRACT: For technical session (2) „Wall Structures" a total of 48 papers (including the poster session) were submitted. Twelve papers were selected for oral presentation. The topics of the papers were: model tests, field tests, case histories, analysis, and dimensioning. The session was lead by L. E. Wichter (Germany), co-chairman T.W. Cousens (United Kingdom), discussion leader K.C. Yeo (Hong Kong), and secretary R. Kitamura (Japan). This report gives a summary of the contributions.

1. *Design and Dimensioning*

Anderson, Boyd, Segrestin and Worrall (The Reinforced Earth Company) show in their paper that it is necessary to have more logical consistency and compatibility between national standards. Safety factors and design philosophies should be based on consistent mechanical models.

Zhang Yemin et al. proposed a design method for the construction of reinforced slopes for agriculture storage rooms.

Greenwood and Yeo presented a report on the assessment of geogrids for soil reinforcement in Hong Kong. The use of partial safety factors based on statistical values, and representing the levels of risk and uncertainty for the different influences, seems them to be the best way of dimensioning.

Lesniewska proposes a general classification system of reinforced soil retaining walls regarding the possible failure mechanisms.

Mak and Lo presented a paper on a limit state design specification for reinforced soil walls, using partial safety factors. They set off the potential problems of using a statistical definition of characteristical values.

Pun and Man describe the use of reinforced fill structures in Hong Kong. They give an interesting overview on the design and construction standards, the geotechnical control, performance, research, and development undertaken by the Hong Kong Geotechnical Engineering Office (GEO).

Xueqiang and Zude presented a paper about the designing methods for reinforced vertical slopes using the theories of soil plasticity and limit state analysis

2. *Case histories*

A paper about a 7 km long reinforced earth seawall for a new runway of Sidney airport was presented by M. Boyd. The author describes the construction of the 5 m high sea walls and gives some informations about the considerations on durability of the steel reinforcing strips in aggressive seawater environment, and on corrosion monitoring.

Raybould, Hadley, and Boyd describe the design, construction and performance of a 40 m high and 210 m long reinforced fill wall in Hong Kong. The wall became necessary, together with many other earthworks, to construct the Route 3 expressway linking Hong Kong Island to the chinese border. The construction was equipped with numerous monitoring devices such as inclinometers, piezometers, etc.. During construction time (16 months) a washout failure of a section of the wall occurred showing the importance of temporary drainages for structures built across valleys in areas subjected to heavy rainfalls.

Frankenberger et al. presented a report on the experiences with reinforced earth walls during Northridge (California) earthquake in 1994. Within the affected area a number of 23 reinforced earth structures withstood the earthquake without damage though in the neighbourhood many collapses of highways and commercial structures occurred.

Guo's and Luo's report deals with the application of reinforced earth technologies in the loess areas of China. The authors give an overview of the facing and reinforcing materials. It is very interesting that, even for very high retaining structures (≥ 20 m) polypropylene plastic strips are used successfully, together with loess as backfill material. In a second report they describe in detail the construction of a 50 m high wall using these materials.

Rao et al. describe the construction of a panel-faced reinforced soil wall near the city of New Delhi. The use of fly ash as fill material makes this report very interesting in an environmental aspect.

Another very important contribution was presented by Kobayashi, Tabata and Boyd. They studied the behaviour of reinforced earth structures in the vicinity of Kobe during the Great Hanshin Earthquake in 1995. More than 120 reinforced earth structures were inspected after the earthquake, and they showed only minimal damages. No structure failed. Pseudo-static analyses of some selected structures indicated that present design methods may be very conservative, and the structure's bearing capacities are considerably higher than estimated.

The findings of Kobayashi et al. were confirmed by the observations of Nishimura et al., who investigated ten geogrid-reinforced soil walls in the area of the Southern Hyogo earthquake which happened in 1995. Though having undergone seismic activities of 6 to 7, these structures were not destroyed or even damaged seriously. The stability of the walls was, in all cases, higher than the results of calculations using the conventional design methods.

Otani, Mega and Matsui made the same observations on steel-reinforced earth structures after the 1995 Hyogoken-Nambu earthquake. They checked about 40 such structures after the earthquake and found, that reinforcement protected them from severely being damaged. They found that reinforcement also increased the resistance against failures of the natural slopes in the earthquake area.

Lin Xin describes the performance of a reinforced embankment based on a pile construction. The piles were necessary to stabilize the embankment against sliding on a deeper slip plane.

Marchal, Bastick and Belblidia presented a report on a 24 m high Terre Armee wall near Pont de Normandie in France. Great attention was payed to a good design of the facing which forms a part of the landscape in the neighbourhood of characteristic chalk cliffs.

Nakajima et al. describe the construction of a 8 m high reinforced soil wall with a concrete block facing. They observed a maximum reinforcement strain of about 1 % and a horizontal displacement of the facing of 6.5 cm, both measured at the time of completion of construction, and no further increase.

Sakajo et al. describe the performance of a steel strip reinforced wall of 60 m length and 12 m maximum height. They observed a wall base settlement of 100 mm and a lateral displacement of 15 mm at the top of the wall after the end of the measures. Maximum tension force in the reinforcement strips was measured near the wall facing, but no clear distinction between active and passive zone in the reinforced soil mass was found.

Singh et al. analysed the failure of a fully anchored conventionally designed breast wall constructed in an unstable slope area. The construction became unstable after a period of heavy monsoon rainfalls. The authors make a proposal for a better „combined system", with the generous provision of chimney and slot drains to ensure good drainage conditions.

Smith describes the design and construction of the very impressive Terratrel Ro-Ro ramp for second Severn Crossing. The 21.000 kN caissons for the foundation of the construction piers were transported from the precast yard on a 5.000 kN transporter, so that the overall weight of the transport was 26.000 kN. The transports used a 13 m high reinforced embankment in difficult marine environment.

Stewart, Truong and Segrestin describe another three very high reinforced earth constructions in Tennessee, USA. The overall height of their structures ranged from 17.5 to 30.5 m, and they were carried out without the slightest problem. No post construction movement or settlement was observed.

Sumanaratne and Mallawaratchie presented a paper on the use of treated bamboo strips as reinforcement and used tyres as facings in Sri Lanka. They remark that these construction methods could be adopted ideally in countries like Sri Lanka were advanced technologies and, above all, capital intensive equipment is lacking.

Won, Hull and De Ambrosis described the performance of a geosynthetic-reinforced segmental block wall structure in New South Wales, Australia. They found that geogrid strain data and numerical modelling had suggested that load bolts may overestimate geogrid tension in this type of construction.

3. *Model Tests*

M. Hyodo et al. presented a paper on model tests in a pull-out aparatus. They used rubber strips and PEHD strips embedded in Aio sand. The strips of different tensile stiffnesses were equipped with strain gauges. The authors found that walls reinforced with strips of high stiffness need only small displacements to activate the tensile forces along the whole lengths of the strips, while along strips of lower stiffness the friction predominates near the front of the wall and decreases rapidly with increasing distance from the wall facing.

Yogarajah and Saad describe the performance of two large scale tests to determine the behaviour of polymeric-reinforced single and multi segmented walls. They found significant differences between these wall types when considering the horizontal earth pressures and strain distributions along the reinforcement layers.

C.C. Huang and B. Huang performed an infiltration test on the crest of a wrap-around model wall. The 2.77 m high, 6 m wide and 3 m long reinforced construction consisted of an alluvial clay and was exposed to water infiltration over a period of 50 days. The long-term percolation had some influence on the outward deformation of the wall facing. The test shall be used to evaluate the feasability of using clay as backfill material, and is not finished.

Tsukamoto et al. presented a report on a large-scale model test to study the lateral earth pressures on a rigid facing reinforced retaining wall. They used granular backfill material and found, that in their tests neither collapse states nor fully mobilized states for active earth pressure could be observed under the surcharge pressures they used.

Uchimura et al. describe the performance of preloaded and prestressed geosynthetic-reinforced soil. They found that the rigidity of a reinforced soil mass can be increased largely by proper preloading and prestressing.

Arab, Villard and Gourc presented the results of two large scale loading tests. They used woven and nonwoven geotextile and observed only very small displacements until 2/3 of the failure loads were reached. Finite element analysis gives rise to the assumption that a compaction effect causes this bearing characteristic.

Benigni, Bosco, Cazzuffi and De Col describe the construction of a 5 m high geocomposite-reinforced wall in Northern Italy. The wall was loaded, and measurements showed that the tensile strength of the reinforcement was mobilized only minimally.

Chang and Milligan carried out a series of small-scaled pullout tests on soil nails. They found that pullout tests conducted under conditions in which no overall failure is occurring, and where the active and transition zones are not fully developed, may be unsafe for pullout under ultimate failure conditions.

Okabayashi, Kawamura and Okada carried out two series of centrifugal tests on three-dimensional models of reinforced retaining walls, and verified the test results with three-dimensional FEM analysis. They found that the shape of the reinforcement mainly influenced the stability of the walls under their test conditions.

Centrifuge model tests were also carried out by Porbaha using different foundation conditions, nonwoven fabric, and cohesive soil as backfill. Overturning and rotational sliding were observed as failure modes.

Da Silva and Abramento propose a method for verifying the performance of reinforced soil walls under working conditions, using an analytical model based on shear-lag formulation.

Sreekantiah and Kishan describe geogrid-reinforced earth model retaining walls and found a remarkable decrease in lateral deformation and vertical settlement when increasing the number of reinforcement layers.

Tajiri et al. described a series of very interesting full scale loading tests with different types of facings. They found that the displacement of the wall facings could be reduced by the gravity effect of it's facing, and they have studied the shapes of the failure planes.

4. *Analyses*

Rowe and Ho carried out a FEM study on the influence of some parameters on the deformation behaviour of reinforced walls. The found out that the deformation of walls constructed with wrap-back type facings is smaller when the intermediate layers that form part of the facings are not to far from the anticipated failure plane, because they share the load with the main layers. Location of panel connections and the panel continuity had no significant influence on the force in reinforcement and deformation. The effect of higher backfill soil stiffness on the deformation behaviour was relatively insignificant. Rigid facings on a rigid base lead to a clear decrease in horizontal facing deformation.

M. Bastick and P. Segrestin presented a study on the use of double wedge equilibrium method for reinforced earth structures design. They found that the method is suitable together with the use of partial safety factors for loads and material properties. The attempts to define global factors of safety using double-wedge failure mechanisms lead to meaningless results.

Kim et al. carried out a theoretical study on the stress-strain behaviour of reinforced walls using a nonlinear elastic soil behaviour constitutive law and nonlinear response of the fabric reinforcement.

Boyle and Holtz used sixteen case histories reported in the literature to assess the accuracy of methods for the prediction of wall deformations. They found that the predicted deformations ranged from 40% to 460% of the actual deformations of walls constructed with cohesionless soils.

Cousens and Pinto studied the effect of compaction on model fabric reinforced brick faced earth retaining walls. In order to avoid or minimise unwanted effects of compaction they propose to compact the backfill in the direction towards the wall, and to use only light compacting device within 1 or 2 m from the wall facing. In a second paper the same authors describe the creep behaviour of the model walls. They found that creep does not seem to be a major problem of reinforced earth constructions, which is in agreement with the observations of many other researchers.

Cardoso and Lopes used FEM analysis to perform a numerical study on the influence of the construction method on the behaviour of geosynthetic reinforced walls.

Gourves, Reiffsteck and Vignon presented two different methods of stability analysis which were developed for geocell retaining structures.

Nakane et al. analized three geotextile-reinforced soil walls using a two-dimensional elasto-plastic FE model. The aim was to get some information about the influence of facing rigidity on the deformation behaviour of such walls.

Rajagopal and Sri Hari propose a method of design for anchored retaining walls. Their proposal is based on the results of model tests in a 0.8 x 0.6.0.8 m test box.

5. *Conclusions*

A main result of the observations described in the session is the high resistance of reinforced earth structures against beeing damaged by earthquakes. Another aspect is that there is obviously a lot of work to do in order to come to a better agreement of calculation results and field observations. The contributions have shown that it is possible to construct high and very high reinforced walls safely and, in general, with lower expenditure than conventional retaining walls.

Earth Reinforcement, Ochiai, Yasufuku & Omine (eds) © 1997 Balkema, Rotterdam, ISBN 90 5410 833 9

Discussion leader's report: Wall structures

K.C.Yeo
Geotechnical Engineering Office, Hong Kong, People's Republic of China

ABSTRACT: This report presents a summary of all the forty-eight papers which are represented by more than seventeen countries, submitted for this technical session. The range of topics is widely spread, and can be categorised into five main groups, namely, Classification/Specification/Standards, Analyses, Laboratory Model Tests, Field Trials/Tests and Case Histories. There is a sharp increase on number of papers by Asian/Pacific Asian countries, which indicates the importance of reinforced soil technology in Asian development. Importance issue as raised in these papers are highlighted by the writer, and in the writer's view, this would be used to gauge the advancement of reinforced soil technology in the future.

1 INTRODUCTION

There are in total forty-eight papers submitted for the technical session on wall structures of IS Kyushu'96 as compared to twenty-two in 1992. A comparison of the breakdown with respect to country/continent is shown in Figure 1. It is noted that there is a marked sharp increase on the number of papers submitted by Asian/Pacific Asian countries.

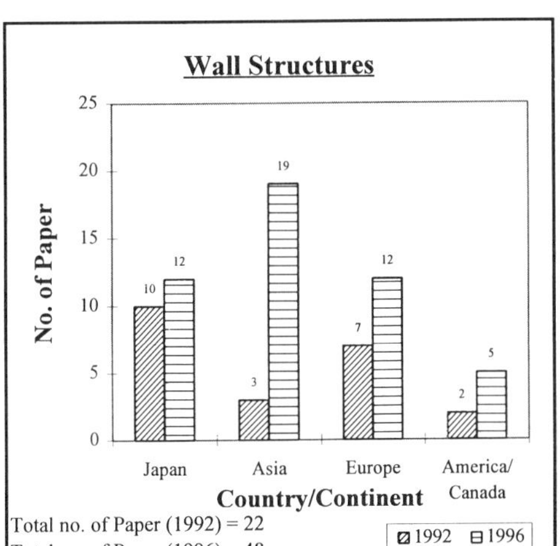

Figure 1 Comparison of Number of Papers Submitted for IS Kyushu' 92 & 96

This indicates the importance on the use of reinforced soil technology in the development of Asia over the next decade.

The submitted papers covered a wide range of topics, and can be categorised into five main groups, namely, Classification/Specification /Standards, Analyses, Laboratory Model Tests, Field Trial/Tests and Case Histories. Table 1 shows the breakdown of the number of papers with respect to each group. The papers included in each group are briefly summarised in the following sections, and readers are drawn attention to important issue as raised in the papers.

Table 1 Number of Papers Submitted for Each Group of Topic(s)

Wall Structures		
Topic		**No. of Papers**
Classification / Specification / Standards		5
Analyses	Numerical	3
	Others	5
Laboratory Model Tests		11
Field Trials / Tests		9
Case Histories (Case Studies)	Seismic	4
	Others	11
	Total :	48

2 CLASSIFICATION/SPECIFICATION/ STANDARDS

Five papers are included in this section on classification/specification/standards, which also includes some discussion on the issue of partial safety factors.

Leśniewska proposed a classification system of reinforced soil retaining walls taking into account of the internal structure and the most probable failure mode. She concluded that much attention should be paid on the lack of mechanical uniformity of many reinforced soil structures, which is important in deriving the appropriate theoretical description.

Pun & Man and Greenwood & Yeo described the use of an endorsement (pre-approval) system, and outlined the importance of taking the appropriate design conditions when endorsing the system.

Mak & Lo described the development of a limit state design specification based on reliability analysis (calibration) of a limited number of data.

Anderson et al demonstrated the use of different design codes may result in quite different safety factor for the steel/geosynthetics reinforcement , and suggested the need for a rational approach.

In addition, Greenwood & Yeo and Anderson et al proposed that different group of partial factors, i.e. related to reduction of strength, related to uncertainty etc, must be treated differently, to achieve compatibility, effectiveness and consistency in design.

The writer would like to draw attention that as different type of reinforced soil wall, i.e. with /without structural facing, extensible/ inextensible reinforcement etc, may render different behaviour, it is important to classify the structure and to develop appropriate design method for the structure. In addition, it is vital to exercise control over the use and design of reinforced soil structures, e.g. by modified (international)design code, by providing specification or by pre-approval(endorsement) system, to avoid failure that may be caused by inappropriate design method or construction procedure applied. Means of handling the partial factors used in design should be considered carefully to produce compatible and consistent design solution.

3 ANALYSES

Eight papers are included in this section on analyses, which cover a wide range of analytical methods.

Rowe & Ho, based on their previous numerical modelling work using finite element analysis, gave an account on the effect of intermediate layers, interface shear, panel continuity, panel location, facing rigidity, backfill stiffness and foundation stiffness, on the behaviour of a modelled reinforced soil wall.

Cardoso & Lopes used a numerical model, again based on finite element analysis, to study the effect of wall construction technique, i.e. use of a fully propped rigid panel & use of incremental panels.

Boyle & Holtz used (performance) charts developed from two independently calibrated (finite element) numerical models to predict and compare the actual deformation of wall in sixteen case histories.

Bastick & Segrestin compared the use of three limit equilibrium methods of analysis for reinforced soil wall design, namely coherent gravity method, slip circle method and two-part wedge method.

Kim et al carried out internal stability analysis of a reinforced soil structure based on limit equilibrium, using a non-linear hyperbolic load-extension relationship for the reinforcement.

da Silva & Abramento used an analytical model based on shear-lag formation, to verify the loads along the length of the reinforcement at various depth of an instrumented reinforced soil wall.

Yang & Liu employed limit analysis to develop upper bound solutions for vertical reinforced soil slope, taking into account the reinforcement with and without bending stiffness.

Zhang et al developed a design procedure for a tied back to back wall based on limit equilibrium.

The writer would like to draw attention to the feasibility of using a calibrated numerical model to develop design (performance) chart that can be used in limit states design, and the continuous effort of researchers in developing other design methods based on limit equilibrium methods of analysis.

4 LABORATORY MODEL TESTS

Eleven papers are included in this section on various laboratory model tests, i.e. soil nails, geocell, soil anchors, centrifuge, laboratory-scaled reinforced soil walls etc, to study the behaviour of different soil reinforced system for wall structures.

Chang & Milligan carried out small-scaled pullout tests to study the effect of transition zone in a nailed wall. They concluded that under the conditions in which no overall failure is occurring and transition zone are not fully developed, results yielded from pullout tests conducted may be unsafe for ultimate failure conditions.

Gourvès et al carried out a study on geocell reinforced soil wall using simple shear and compression tests, and carried out full scale and centrifuge model tests with analyses based on two part wedge limit equilibrium and FLAC (finite differences) methods of analysis.

Rajagopal & Sri Hari carried out pullout tests to study the effect of shape and size of anchor plate for a model soil anchored wall.

Cousens & Pinto and Pinto & Cousens studied the effect of compaction and creep behaviour of a modelled fabric reinforced soil wall with brick facing.

Porbaha used centrifuge model tests to study the failure modes of a non-woven geotextile reinforced soil wall with different length of reinforcement using cohesive backfill on firm/rigid foundation.

Hyodo et al studied the effect of reinforcement stiffness on the deformation of geosynthetic-strip modelled reinforced soil walls with single layer of reinforcement. The strip was pull-out with the wall subjected to a rotational movement at toe.

Yogarajah & Saad studied the effect of facing construction on the development of earth pressure and load-strain performance of the geogrid reinforcements.

Sreekantiah & Tito Kishan applied surcharge load to study the wall deformation of a geogrid reinforced wall varying spacing of the reinforcement.

Tsukamoto et al studied the (active) earth pressure development on the front wall, side wall and back wall of the retained geogrid-reinforced soil, using large-scaled model test.

Okabayashi et al carried out centrifuge model test of a steel strip reinforced soil wall varying the facing thickness, length and width of the reinforcement, and verified some of the test results with a 3-D finite element analysis.

The writer would like to draw attention to the continuing effort of using laboratory model test to study the behaviour of different reinforced soil system, and the validity of the these test results in accounting for the effect of construction technique.

5 FIELD TRIALS/TESTS

Nine papers are included in this section on field trials/tests.

Huang & Huang carried out infiltration test on two 2.8m high clay walls, reinforced with a woven & non-woven composite geotextile and a PVC coated polyester fibre geogrid respectively, to study the permeability of compacted clay and the stability of clay reinforced soil structures.

Uchimura et al developed a new construction method to prestress the reinforced soil vertically by metallic tie rod, thus increasing the shear rigidity of the soil, and carried a field trial on a 5m high block. In addition, they also conducted a laboratory experiment to study the relaxation of the pre-stressed reinforced soil.

Won et al carried out a field study on the behaviour of a 9.5m high geogrid reinforced four terraced (tiered) segmental block wall, instrumented to measure strains in the reinforcements, base pressures, settlement and settlement profile of the wall and also lateral deformation of the soils. Numerical analysis based on FLAC (finite difference analysis) calibrated by (back analysed) pull out test was then carried out to assess the performance of the structures.

Tajiri et al carried out full scale failure experiments of three 5m high geogrid reinforced soil walls with different facing, namely expanded polystyrol (EPS) blocks, full height concrete panel & concrete block. Instrumentation was installed to measure earth pressure on the facing of the walls, base pressures, displacement of the facing and strains in the reinforcement.

Nakane et al carried out finite element numerical analysis to model the same experiment, and to study the effect of facing rigidity.

Nakajima et al also carried out field study on the performance of a 8m high geogrid reinforced soil wall of concrete block facing, with instrumentation to measure strains of the reinforcement, earth pressures on the facing of the wall, base pressure distribution, settlement and displacement of the wall.

Benigni et al conducted field study on the construction and performance of a 5m high geocomposite (non-woven polyester sheet with woven polypropylene thread knitted at constant spacing both directions) reinforced soil wall with strong fill and weak reinforcement. Instrumentation was carried out to measure elongation of the geocomposite and the deflection of the wall.

Arab et al carried out finite element analysis to simulate the behaviour of two instrumented 4.5m high reinforced soil walls, one with woven and one with non woven reinforcement. The walls were loaded to failure and instrumentation was installed to measure the load settlement, wall deflection and strains along the reinforcement sheet with respect to increase of load.

Sakajo et al carried out field study of the performance of a 12m high steel strip reinforced wall with numerical modelling based on finite element analysis, with instrumentation data on both the vertical and horizontal displacements of the wall and settlement of the fill.

The writer would like to draw attention to the importance of quality field studies with high quality instrumentation measurement, to be used to calibrate numerical models, to develop design (performance) charts for limit states design.

6 CASE HISTORIES

Fifteen papers are included in this section of case histories.

Four papers are concern with the effect of earthquake on reinforced soil structures. Frankenberger et al gave an account of the behaviour of Reinforced Earth structures (23 number) in Northridge Earthquake in LA, California on 17.1.1994. Kobayashi also reported on the performance of Reinforced Earth structures (124 number) in the Great Hanshin Earthquake, near the vicinity of Kobe on 17.1.1995. Otani et al reported on the performance of 47 steel reinforced soil structures, including Reinforced Earth walls and pile reinforced slopes, during the same earthquake, and carried out some case studies with analysis on factor of safety. On the other hand, Nishimura et al reported on the performance of 10 geogrid reinforced soil structures during the same earthquake, and compare the safety factor obtained based on some existing methods of analysis for design. All the four papers confirmed the effectiveness of reinforced soil structures under seismic impact. The analysis carried out indicated that under the effect of the seismic force recorded, some of the structures would have already failed. However, most of the reported reinforced soil structures suffered either little or no damage during the earthquake.

Marchal et al reported on the use of an innovative facing for a 24m high Reinforced Earth wall near Pont de Normandie in France.

Boyd reported on the use of a 5m high Reinforced Earth seawall at Sydney Airport in Australia.

Stuart et al reported on the construction of 17m to 30m high three tiered Reinforced Earth walls in Eastern Tennessee of the United States of America.

Smith reported on the construction of a 12m high Terratrel (Reinforced Earth structure with wire-mesh facing) temporary jetty ramp for Severn river in Wales of the United Kingdom.

Raybould et al reported on the construction of a 40m high Reinforced Earth wall in Hong Kong.

Lin reported a case of using piles installed at the toe of a reinforced soil wall on soft foundation (which is used to support a road embankment) to prevent a sliding failure, in Chongqing of the PR China.

Guo & Luo presented an account of case histories on reinforced soil structures using loess as backfill, also in the PR China. The reinforcements used include ribbed and plain steel strips, polypropylene strip & reinforced concrete bar.

In addition, Luo & Guo gave an account on the design of a 50m high reinforced soil wall using oil painted ribbed steel strips as reinforcement, in Shan Xi Province of the PR China.

Rao et al described the design, construction and monitoring of a 6-7m high geogrid reinforced soil wall using fly-ash as backfill, in New Delhi of India.

Singh et al described the construction of an anchored earth wall of 3-4 m high using semi-Z shaped mild steel reinforcement with cohesive backfill, in the sub-Himalayan region of India.

Sumanaratne & Mallawaratchie described the use of used Tyre facing with treated bamboo strips as reinforcement, to construct reinforced soil wall of 2m high (which is used to support a 3m high road embankment), in Sri Lanka.

The writer would like to draw attention to the effectiveness of designing and constructing reinforced soil structures in areas proning to earthquake, and the lack of a possible international design code of reinforced soil structures taking into account of the appropriate earthquake impact. It is also important to share international experience on the use of reinforced soil technology, so that technical transfer can be efficiently transmitted to less privileged countries.

7 DISCUSSION

To ensure the development and advancement of the state-of-the-art in a positive direction, it is important to set some milestones to gauge the progress on the use of the technology. The writer would like to suggest five areas for further discussion and consideration as follows:

a. The need for developing Asian countries to

exercise control over the use of reinforced soils based on modified code, specification or pre-approval system.

b. The use of reliability analysis and treatment of partial safety factors, to ensure consistency and effectiveness in design.

c. The use of calibrated numerical model with quality field study to develop design (performance) charts for limit states.

d. The boundary conditions should be carefully considered in design for different types of reinforced soil structures. Effect of compaction, method of construction etc, clearly affect the behaviour of the reinforced soil wall.

e. The need of an international design code for reinforced soil wall, taking into account of the appropriate effect of earthquake.

REFERENCES

BSI (1995). *British Standard Code of Practice for Strengthened/Reinforced Fills (BS 8006)*. British Standards Institution, London.

Foundations

Earth Reinforcement, Ochiai, Yasufuku & Omine (eds) © 1997 Balkema, Rotterdam, ISBN 90 5410 833 9

Discussion leader's report: Foundations

E.M.Palmeira
University of Brasilia, Brazil

ABSTRACT: This work summarizes and discusses the papers presented to the session on Reinforced Foundations at the IS Kyushu'96. These papers gave a important contribution to the state-of-the-art on the subject and also emphasized that further research is needed in order to a better understanding on the mechanisms involved in foundation reinforcement.

1 INTRODUCTION

The use of soil reinforcement in geotechnical engineering has increased markedly in the last decades. Reinforced slopes, embankments and retaining structures are now common solutions for designers for most of the routine works in the profession. Despite the large amount of research and construction experiences already gained there is still the need for additional research on several topics to a better understanding on failure mechanisms and, in particular, deformation mechanisms and long term performance of reinforced soil structures. The use of reinforcement inclusions to improve foundation conditions is certanly one of these topics. Because of the scale of the problems and consequences of potential failures, the general approach towards the reinforcement of foundations has been the use of conservative solutions in most of the cases. The papers presented to this conference show that researchers and engineers are concerned with a wide range of problems where reinforcement can increase the strength of the foundation soil and reduce its compressibility.

Depending on the main subject studied the works presented at this symposium can be divided in the following categories, as schematically shown in Figures 1 (a) to (h):

. Buried pipes in reinforced ground;
. Reinforced embankments on poor foundation soils;
. Bearing capacity of reinforced granular layers on weak soils;
. Bearing capacity of footings on reinforced slopes;
. Bearing capacity of reinforced foundation soil;
. Reinforced deep and shallow fondation elements.

The types of studies and investigations conducted in these topics were analytical studies, numerical analysis, case histories, field tests and physical modelling. The results obtained show a significant contribution to the state of the art and will be discussed later in this work.

The distribution of papers on each subject is presented in Figure 2. A total of 27 papers were presented in this session. This represents a 23% increase on the number of papers presented in the same session at the last Kyushu conference (IS Kyushu'92), in 1992. Seventy four percent of the present papers dealt with bearing capacity of reinforced granular layers, reinforced foundations or reinforced embankments on poor soil. Compared to the subject of papers presented at the last IS Kyushu'92, it can be observed that there was a substancial reduction on the number of papers on bearing capacity of reinforced foundation and significant increases of papers on bearing capacity of granular layers on weak soils and reinforced embankments on soft subgrade.

The works presented and their results are summarised and discussed in the next sections for each of the subjects listed above.

2 BURIED PIPES IN REINFORCED GROUND

Two papers were presented on this subject and the problems approached are presented schematically in Figure 1(a). These papers involved field and laboratory model tests of pipes subjected to pull-out and vibration with different reinforcement arrangements.

The presence of the reinforcement increases pull-out resistance and reduces pipe settlements, as reported by Babin et

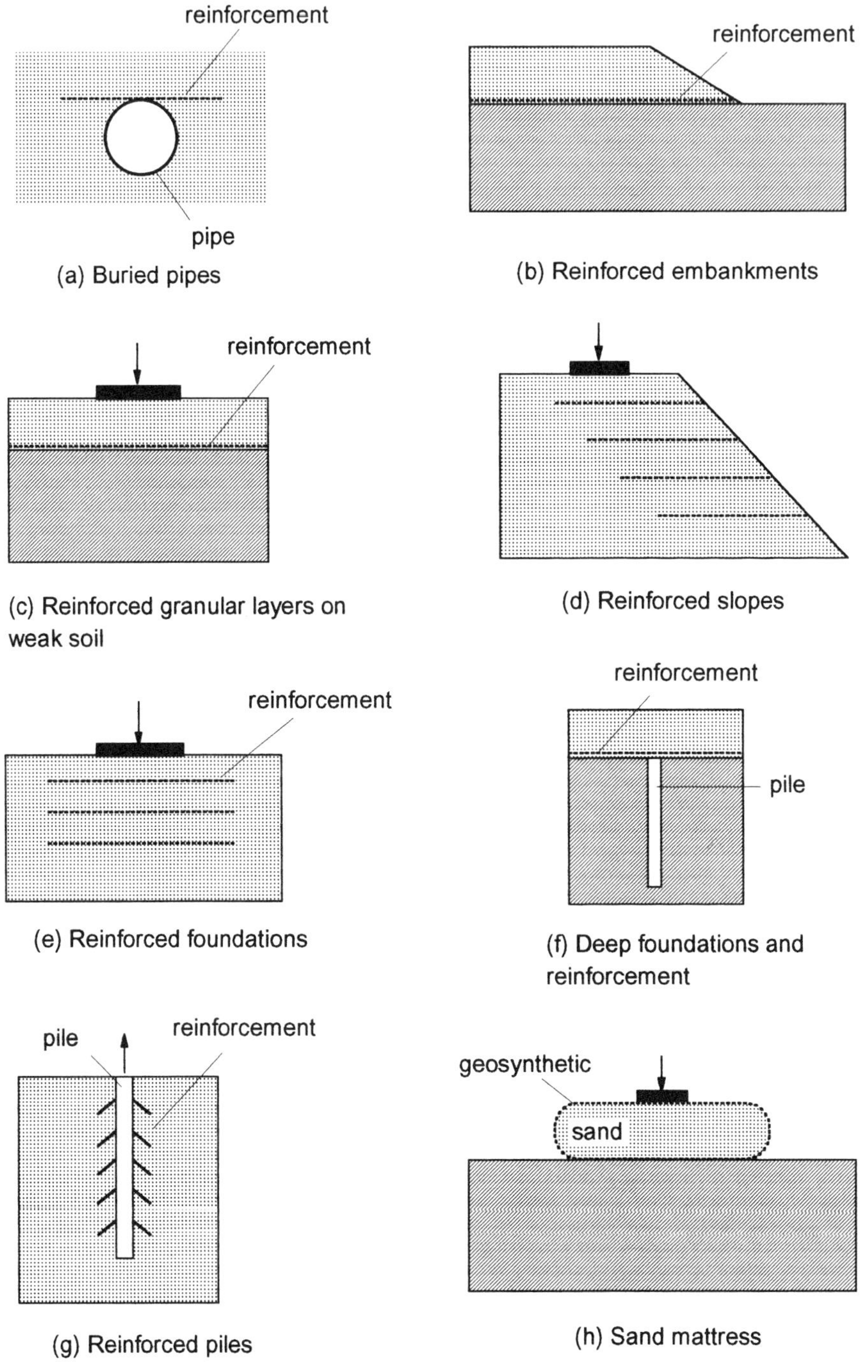

Fig. 1. Types of problems studied.

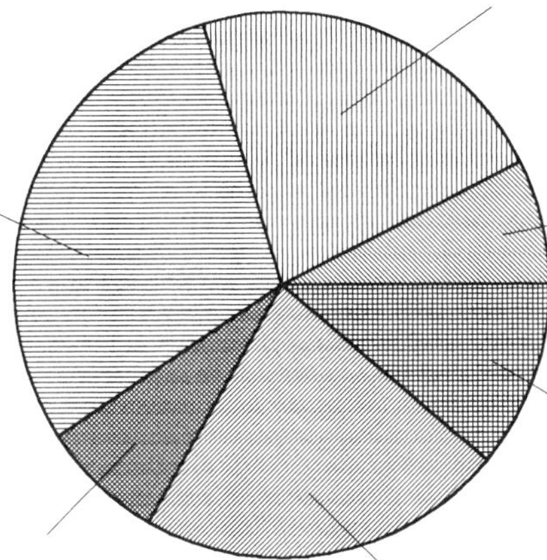

Fig. 2. Distribution of papers in the session.

al (1996). Field pull-out tests showed that pipelines in reinforced soil can resist twice as much pull-out loads than in unreinforced soil. The Authors also presented a empirical expression to estimate the ultimate pull-out load that can be applied to a pipe-line in reinforced soil. The influence of the reinforcement in reducing erosion was also investigated.

Nagase et al (1996) performed model tests of pipes buried in sand on a shaking table and showed that geotextile reinforcement reduced pipe flotation due to vibration. Appropriate reinforcement dimensions were also investigated. Differences between flotation resistance were attributed by the Authors to bending stiffness and geonet mesh opening sizes. However, it may also have been caused by tensile stiffness and/or soil-reinforcement interaction, which depends on net geometry and mesh size. Tests trying to study each factor (bending stiffness, tensile stiffness and soil-reinforcement interaction) independently should be performed to identify the role of each mechanism.

3 REINFORCED EMBANKMENTS ON POOR FOUNDATION SOILS

The geometrical aspects of the studies on reinforced embankments are presented in Figure 1(b) with the traditional arrangement for the reinforcement installed at the base of the embankment. In one of the papers the embankment rested on a reinforced layer in the foundation soil. Six papers were presented on this subject.

As its has been observed in several previous works in the literature the presence of the reinforcement reduced fill deformations and settlements.

Behbahani (1996) described case histories on the use of geosynthetic reinforcement for the stabilization of embankments on soft soils. Reductions in fill deformations and settlements due to the use of reinforcement were observed by the Authors.

Imanishi et al (1996) presented a case history of a geosynthetic reinforced road embankment on very soft soil. The Authors observed a redistribution of strains in the reinforcement during construction as the soft subgrade was partially replaced by the geonet reinforced embankment.

Klosek (1996) addressed the use of geosynthetics as reinforcement of embankments built on areas with mining activity. Theoretical and finite element method predictions showed that the reinforcement brings significant reductions on vertical and horizontal stresses in the foundation.

Poopath (1996) reported smaller pavement deflections for an embankment reinforced by a weak and extensible non woven geotextile. The question in this case is whether the geotextile layer increased the foundation strength by drainage instead of other reinforcing mechanism, because of its low tensile stiffness.

The length of the reinforced granular layer below the embankment was investigated by Shukla and Chandra (1996) assuming the foundation soil composed by Winkler springs. The stress-strain behaviour of the embankment was neglected. A limit length for the contribution of the reinforced granular layer was estabilished and the influence of pre-stressing the geosynthetic reinforcement was also investigated. It should be pointed out that soils behave in a different way, besides the difficulties in determining appropriate values for the soil reaction modulus in this case.

Tanabashi et al (1996) presented results of finite element analysis of embankments on soft soil. The analysis approached the important aspect of construction effects and how reinforcement performs under these circunstances. The results showed that the effect of the embankment weight was greater than the effect of the bulldozer weight for low embankment heights (of the order of 0.3 to 0.5m). In fact, one would expect the other way round in constructions on soft soils. Unfortunately the Authors did not comment on the magnitude of stresses transferred to the foundation and if the finite element formulation accounted for large deformations, which would be appropriate under these circunstances.

4 BEARING CAPACITY OR REINFORCED GRANULAR LAYERS ON WEAK SOILS

The bearing capacity of reinforced granular layers was investigated in 7 papers presented to the symposium. The studies involved the traditional arrangement with the reinforcement placed on the foundation surface (Figure 1(c)) as well as arrangements with multiple layers of reinforcement. In one of the cases the influence of the presence of piles under the reinforced zone was also investigated.

Significant reductions in settlements and rut depths were observed by Manjunath and Dewaikar (1996), Cancelli et al (1996) and Knapton and Austin (1996) in large scale laboratory tests. The ratio between the size of the containers used in the tests and the size of the footings may have caused some boundary effects. The Authors did not comment if this effect was somehow investigated or evaluated.

Cancelli et al (1996) observed a punching failure mechanism in large laboratory tests of unreinforced paved roads and an reduction of 30 to 50% on settlements with the use of geosynthetic reinforcement. The influence of the presence of the reinforcement on the asphalt thickness was also investigated.

Fakher et al (1996) emphasised the importance of dimensional analysis on the interpretation of model loading tests of reinforced ground. The Authors also performed model tests on extremely soft subgrades, with undrainded strength of the order of 0.06 kPa. Special techniques were employed for sample preparation and undrained strength measurements. The Authors comment on the importance of reinforcement flexural rigidity (in plane bending stiffness) as a dominant factor in constructions on super soft clays.

Hirao et al (1996) used model tests of footings on granular fill on soft subgrade and concluded on the importance of the reinforcement bending stiffness on the bearing capacity of the system. Again, it is not clear if reinforcement bending stiffness, different interaction mechanisms or modelling limitations were responsible for different responses of the system.

Knapton and Austin (1996) reported large scale tests of unpaved roads reinforced by a geogrid layer. Tests results showed that the presence of the reinforcements reduced rut depths by 50 to 70%. The Authors did not comment on the possible influence of boundary effects on the test results.

Manjunath and Dewaikar (1996) studied the behaviour of square footings on granular layers on soft soil subjected to inclined loads. The granular layer was reinforced by a non woven, needle punched, geotextile. The influence of the inclination of the load on the footing was investigated for reinforced and unreinforced cases. The bearing capacity for the reinforced case was 30 to 50% greater than for the unreinforced case. Tilting of the footing in reinforced cases was also considerably less than in unreinforced cases (20 to 40% less).

Shivashankar et al (1996) also performed model tests on footings on granular soil ovelying soft soil with and without the presence of piles in the foundation. The test results showed that the presence of piles under the reinforced mass significantly increases the stiffness of the system. Optimum reinforcement length was also investigated. Some modelling limitations due to the use of real reinforcement materials in the model tests can be identifyed.

Tanaka et al (1996) performed tests with geogrid reinforced clay-cement layers. The appropriate grid mesh size was observed to be very important in this type of application. The Authors also provide suggestions in terms of design procedures for this type of application.

5 BEARING CAPACITY OF FOOTINGS ON REINFORCED SLOPES

The bearing capacity of footings on reinforced slopes was also investigated in two papers, as presented in Figure 1(d).

Das at al (1996) investigated the optimum position for the footing on the slope, depth of reinforced zone and spacing between reinforcement layers. A procedure for estimating bearing capacity was also presented. This procedure is limited to a certain range of slope angles and model tests results are required for its use.

Ohtsuka et al (1996) observed that the reinforcement effect depends on the reinforcement stiffness and that there is a limit value of reinforcement strength beyond which the reinforcement effect is constant. This seems difficult to understand. Could it be due to reinforcement location or a different failure mechanism such as reinforcement pull-out, for instance? The same Authors also observed a limit value for the reinforcement length which, on the other hand, is reasonable and has also been observed in other works in the literature.

6 BEARING CAPACITY OF REINFORCED FOUNDATION SOIL

A total of seven papers studied the use of multi-layers of reinforcement in the foundation soil (Figure 1(e)).

Physical and numerical modelling and theoretical studies by Ju et al (1996), Nataraj et al (1996) and Zhao et al (1996) showed that the presence of the reinforcement almost doubled the foundation bearing capacity. Limiting values for the influence of reinforcement length, number of reinforcement layers and reinforced zone depth were identified. Regarding the model tests by Ju et al (1996) it can be argued on the possible influence of side friction along the in internal walls of the container used for the tests and on the limitations on the scaling of the reinforcement material in view of the other dimensions adopted.

Otani and Yamamoto (1996) performed model tests of footings on reinforced and unreinforced foundations using aluminium rods as the foundation layer. The displacements and rotations of some rods were measured and it was observed that the latter was a function of the reinforcement position and length. The Authors attributed to reinforcement length and bending stiffness a major influence on the deformation mechanism of the reinforced foundation. However, perhaps the same conclusion could also be reached based on different mechanisms of interaction having occurred.

Simonini (1996) performed finite element analyses and model tests of circular footings on reinforced sand. A satisfactory agreement between numerical and experimental results was achieved and both studies showed a punching failure mechanism for the foundation. Triaxial tests on specimens reinforced with horizontal reinforcement layers were also conducted as part of the study. These tests were also modelled by finite elements whose predictions copared well with the test results. A similar numerical study was conducted by Sitharam et al (1996) where it was observed that the point of maximum tension in the reinforcement layer depended on the reinforcement stiffness. A similar work was presented by Yamamoto and Otani (1996).

An interesting design method for the estimate of bearing capacity of the multi-layer reinforced foundation was introduced in Zhao et al (1996). The method is based on the slip-line theory and its use is very straightforward. Design charts and a worked example are provided. The Authors observed a limiting value for the depth of the reinforced zone in the foundation beyond which the reinforcement strength has little influence. This was also observed by Ju et al (1996) in model tests. The results of the method are consistent with the results presented by Ju et al (1996) and by Simonini (1996). It would be interesting to see how the method's predictions compare with model or field test results.

7 REINFORCED DEEP AND SHALLOW FOUNDATION ELEMENTS

Figures 1(f) to (h) shows schematically the arrangements studied for the combination of reinforcement and deep foundation elements and for the use of geosynthetic mattresses for load tranference to the foundation layer.

A Finite Element study performed by Nakai and Uno (1996) predicted that the uplift capacity of a pile can increase by 70% with the use of protruded bars orientated downwards, for which case the influence of the bars bending stiffness was negligible (Figure 1(g)). The reinforcement bending stiffness was observed to be important only for horizontal bars. The Authors could have commented on the effects of some reinforcement modelling and construction difficulties.

Ohkubo et al (1996) studied the problem of reinforced embankments founded on piles on soft soil. The Authors observed that the presence of geogrids on top of the piles had a significant effect on the displacements of the pile heads using the Finite Element method. Maximum geogrid strains were observed to occur near the pile head and negative strains were predicted in between piles. That might be expected for the region below the embankment slope due to the mobilization of different horizontal pile displacements. However, the Authors also observed that for the central region of the embankment, which seems unexpected.

Yasufuku et al (1996) conducted tests with geosynthetic (geogrid) mattresses resting on a foundation composed by springs (Fig. 1(h)). The magnitude of the stresses transferred to the foundation depended on the ratio between mattress thickness and footing width. A simple design approach to determine the mattress thickness was presented. Again, it should be pointed out the simplicity of the use of springs to simulate soil and the limitation of this procedure as commented earlier in this work.

8 CONCLUSIONS

This report tried to summarize the works presented in the session on Reinforced Foundations at the IS Kyushu'96. The main conclusions are as follows:

- A significant contribution to the state-of-the-art has been given by high quality papers, which unfortunately the limited amount of space does not allow the Author of this report to elaborate more.
- A range of relevant problems were investigated and design methodologies presented which enhanced the potentials of the use of reinforced foundations.
- Some papers attributed to reinforcement bending stiffness a important influence on the behaviour of the reinforced soil. However, for polymeric and flexible metalic grids this influence is likely to be negligible. Perhaps the role of other factors, such as soil-reinforcement interaction, may have been minimised or obscured in these works.
- Some aspects of the use of reinforcement in foundations have still to be better investigated. One of these aspects is the deformability of the reinforced mass. This is a key factor for the use of reinforcement in foundations of structures. The prediction of settlements of footings on unreinforced granular media is already very complex. Efforts should be concentrated in developing accurate methodologies for this type of prediction when reinforcements are present.
- One should to bear in mind the limitations of numerical and physical modelling and should try to evaluate them. Problems such as reinforcement modelling and approach used to determine stresses along the soil-reinforcement interface in the Finite Element Method may influence the results obtained. The transformation of three dimensional problems in two dimensional equivalents is also subject to criticisms. The same applies to physical models if enough care is not given to proper scalling of materials and boundary effects.

REFERENCES

All the references presented in the text can be found in the first volume of the International Symposium on Earth Reinforcement - IS Kyushu'96.

Slopes and excavations

Earth Reinforcement, Ochiai, Yasufuku & Omine (eds) © 1997 Balkema, Rotterdam, ISBN 90 5410 833 9

Discussion leader's report: Slopes and excavations

B. Myles
Soil Nailing Limited, UK

Ladies and gentlemen, to open this discussion session I will very briefly introduce the 24 papers forming the basis of the discussion and classify these into subject areas to give structure to the discussion. The papers are numbered 1 to 24 as follows:

1) A Asaoka et al, Stability analysis of reinforced soil structures introducing some constraint conditions upon the 3-D velocity field

2) D Asbey-Palmer et al, An innovative solution to the stabilisation of the Cairmuir landslide incorporating reinforced earth

3) S Bang et al, Deep seated stability of soil nailing walls

4) B Berardi, A design method for the reticulated pile structure for the stabilisation of slopes and excavations

5) E C Drumm et al, Instrumentation and monitoring of a nailed mine waste slope

6) M Ehrlich et al, Parametric numerical analysis of soil nailing systems

7) S Ghalib et al, Earth reinforcement applications for hazardous waste containment

8) E Guler, Comparison of different design concepts in slope stability analysis of geosynthetic reinforced slopes

9) T S Ingold and B Myles, Ballistic soil nailing

10) M J Kenny and Y Kawai, The effect of bending stiffness of soil nails on wall deformation

11) J S Kim et al, A large-scale experimental study of soil-nailed structures

12) R Kitamura et al, In-situ test of reinforced volcanic ash with steel bars and panel facings

13) M Kulczykowski, Stability analysis of reinforced slopes based on apparent cohesion method

14) D Y Lin et al, Predicting seismic performance of geogrid reinforced slopes

15) H I Ling et al, A new concept of seismic design of geosynthetic reinforced soil structures: permanent-displacement limit

16) H Matsuoka and Y Sugiyama, Failure mechanism and effective reinforcement of granular soil slope

17) R L Michalowski, Instability patterns of reinforced soil structures

18) K Nishida and T Nishigata, Reaction of reinforcing force and restrainment effect on soil nailing

19) N Sabhahit et al, Seismic analysis of nailed soil slopes - a pseudo dynamic approach

20) F Viel and A Jenne, Repairing the landslide Heroldsberg/Germany

21) M Vucetic et al, Dynamic failure of soil nailed excavation in centrifuge Vucetic...

22) R Yatabe et al, Stability analysis of slopes reinforced by root network

23) J M Zhou et al, In-situ test of steel nailing reinforcement for cut slope with alternating sandstone and mudstone

24) T F Zimmie and M B Mahmud, Instrumentation for centrifuge modelling of geotextile reinforced slopes

To give structure to the discussion the above papers have been divided into the two broad categories of those dealing with conventional reinforced soil, in which reinforcement is incorporated into successive layers of fill during construction, and those dealing with soil nailing.

From the point of view of internal stability this is a very important demarcation since with conventional reinforced soil the engineer has control over the quality of the fill which can be specified to have the required properties. In contrast, soil nailing has to be designed to suit the existing natural ground, or previously placed fill, which is to be reinforced. This of course requires appropriate site investigation to determine what soil or fill properties are available to the engineer.

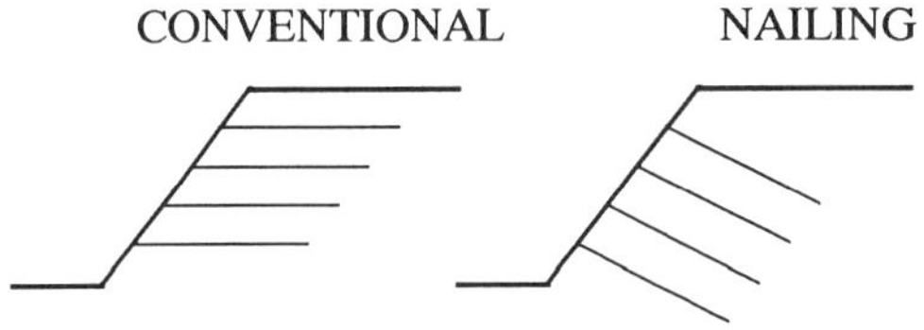

	NUMERICAL ANALYSIS	1, 6, 10
4, 8, 13, 17	EQUILIBRIUM ANALYSIS	3
2, 7, 20	CASE HISTORIES	5, 9, 12, 23
16, 24	EXPERIMENTS TRIALS	11, 18, 21
14, 15	SEISMIC	19
22	NATURAL	22

The above table categorises the 24 papers as dealing mainly with conventional reinforced soil or with soil nailing. The papers are numbered as in the program and the above summary of paper titles. Under these two categories the papers are further classified as dealing mainly with numerical analysis, equilibrium analysis, case histories, experiments or trials, seismic considerations or natural reinforcement in the form of plant roots. It is interesting to note that of the 24 papers 50% are authored by academic organisations, 33% by joint academic and commercial organisations and 17% by practitioners.

Under the classification of numerical analysis, paper 1 presents a 3-D finite element analysis of reinforced soil in which the reinforcing mechanism is modelled by employing a no length change in energy functions. Unlike 2-D analysis, where reinforcement is modelled as equivalent continuous plates, this 3-D approach allows reinforcement to be modelled as linear elements.

Paper 6 presents a parametric study of the influences of construction processes including soil removal, nail and facing installation. The paper addresses important practical considerations. Paper 10 presents the results of finite element analyses and field trials of the influence of nail bending stiffness on wall deformation and concludes that bending stiffness has an insignificant effect. The trial used 23 mm diameter nails in a 45 mm diameter hole which is uncommon outside Japan. See papers 5 and 6.

Under the classification of equilibrium analysis, paper 3 considers deep seated stability of nailed slopes and particularly the effect of face slope angle where the dominant failure mode changes from deep seated failure to toe failure. This paper gives interesting pointers for future directions. Paper 4 presents a rational procedure for assessing raking piles which helps to quantify the benefits and mechanisms of a complex system.

Paper 8 presents useful design charts for geosynthetic reinforced slopes based on the Federal Highways Administration method for extensible reinforcement while paper 13 presents an analytical method in which the reinforcement is deemed to endow the soil with a pseudo cohesion. This approach may help to introduce the concept of reinforced soil into other slope stability programs where the modelling of discrete reinforcing elements difficult. Paper 17 presents a kinematical limit analysis approach based on log-spiral failure surfaces. Although this type of approach has been current for some time it has not gained wide acceptance.

Case histories were the broad topic of 29% of the papers with paper 2 which records reinforced soil used in remedial works to a huge landslide involving 8 Mm^3 of schist rock. Paper 5 recounts the use of 25 mm diameter nails in 200 mm diameter grouted holes to stabilise a mine waste slope built on a mountain slope. The design and construction of conventional geogrid reinforced slopes to retain hazardous waste are reported in paper 7 while paper 9 presents four projects completed using ballistic soil nailing in some cases complemented by no-fines concrete buttresses at the toe of the slope. Versatility of the technique is exemplified by the facility to use the nail launcher suspended from a crane, rather than mounted on a hydraulic excavator, and speed of construction by the installation of nails at the rate of one per two minutes.

Case histories continue with paper 12 which presents a large scale in-situ test on volcanic ash reinforced with steel bars and facing units. A comprehensive monitoring system was used, for more than ten months, to record the axial loads in the bars as well as climatic conditions. Paper 20 presents a remedial works case history involving the use of geogrid reinforcement in stabilising a 10 m deep cut slope which failed shortly after construction due to infiltration of heavy rainfall. Paper 23 gives an interesting account of nailing of a slope comprising alternate layers of sandstone and mudstone.

Experiments were the topic of five papers with paper 11 comparing results obtained from a 2 m high experimental nailed wall with results from finite difference analyses. Paper 16 reports the results of an interesting tilting box experiment in which the surface stability of a soil slope of progressively increasing inclination was studied using the Schneebeli plane strain technique in which soil is modelled by a mass of aluminium rods. Paper 18 recounts model tests, using a 1.5 m high wall backfilled with high density iron ore, and investigates reinforcement intensity related to induced reinforcement tensile strain.

Paper 21 presents results from centrifuge testing used to investigate the failure mechanism and seismic stability of a 7.6 m deep nailed cut slope while paper 24, also based on centrifuge testing, investigated the influence of installing strips of nonwoven geotextile into soft soil slopes to serve the dual function of reinforcement and drainage.

Paper 14 reports a finite element analysis of the seismic stability of a 15 m high geogrid reinforced slope and considers maximum relative slope displacements. Seismic design is the topic of paper 15 which, although based on conventional pseudo static analysis, introduces the concept of designing to a serviceability limit defined by the permanent displacement resulting from direct sliding instability. Paper 19 compares the results of conventional pseudo-static analysis, in which the shear wave velocity is considered to be infinite, analysis with pseudo-dynamic analysis in which the shear wave velocity is considered to be finite.

The effects of natural reinforcement, in the form of tree and plant roots, are considered in paper 22 which concludes that root systems induce an apparent cohesion which should be taken into account in assessing slope stability.

Having briefly introduced the papers comprising this session I now invite questions from the floor

Mr Stevenson of Tensar Corporation to Dr Matsuoka - Can you relate your experiment to the US practice of geogrid reinforced veneers on steep slopes. Your experiment is based on cohesionless soil whereas a veneer on steep slopes is based on cohesive soil. Nonetheless it is a very interesting concept to stabilise these veneers. Can you give design guidelines ?

Dr Matsuoka answers by thanking Mr Stevenson for his question. I am not too familiar with US practice or design but am mainly interested in the mechanisms of surface sliding. Once the mechanisms are defined then these give a basis for design.

Professor Klosek of the Technical University of Silenic asked Professor Vuceti - What is the influence of sand friction on the wall of the box and what are the differences in 2-D and 3-D models ?

Professor Klosek responded by agreeing friction is a problem. However, side friction can be eliminated by using an isolating composite, comprising two sheets of rubber with a water based gel lubricant between the two layers, installed between the walls of the box and the soil. That this eliminated friction is verified by the fact that transverse tension cracks which developed in the soil ran the full width of the box and were perfectly perpendicular to the side walls. Had there been friction between the soil and the walls then the cracks would have curved as they neared the walls of the box. I have to give credit to Professor Tatsuoka who suggested this idea.

Professor Madhav of IIT Kanpur addressed Professor Bang - You seem to have used nails of equal length. If variable length is used it seems to be optimal in terms of overall economy. Can you give your views on this type of optimisation ?

In answer, Professor Bang agreed that variable length is a parameter to be investigated because in many cases equal lengths, although convenient, do not give optimal economic performance; its just a matter of convenience and workability.

However, to optimise performance, length has to vary from top to bottom with constant variations in length and, as far as I know, workability then becomes a critical problem. Nail length might be varied in steps. For example, a block variation can be used in which a given block comprises nails of a constant length. Nail length can be varied over the height of the slope by using three or four blocks of different block nail lengths, but in terms of static and dynamic performance the length variation could be different depending on how you consider it and which one is more critical.

Professor Madhav recounted that in optimisation work he found nails around mid height longer than those at the top or bottom, do you have a similar experience ?

Professor Bang replied that longer nails in the middle makes the problem difficult to optimise and the case for longer nails in the middle would need to be studied.

Professor Chang of Oxford University asked Professor Ehrlich were the nails connected rigidly to the facing or were they unconnected ? My opinion is that nail tension would be different depending on the connection to the facing.

Professor Ehrlich confirmed that the facing was free to move and there was no connection between nail and facing. Professor Chang then stated that if there is a strong connection between facing and nail then the facing acts as though it were anchored. However, with no connection there is a difference. Professor Ehrlich responded that in the case of conventional reinforced soil there may be rotation of the facing, however, with soil nailing construction is step-wise and any movement occurs before the facing is connected. I think my results correctly model behaviour.

Dr Gurung of Hiroshima University then asked Professor Vucetic to elaborate on how the facing panel in the soil model was set to support the excavation. If I understand correctly, as the depth of the soil model increases so stress also increases. Did this produce any lateral or vertical displacements while accelerating the centrifuge from the 1G level to the 50G level before shaking the model. If yes, how did you incorporate this in your test results ?

Professor Vucetic replied that this is a two part question. First how did we install the facing and, second, what was the rotation of the facing and movement in spinning the centrifuge up to 50G. We built the soil in the box in layers of different colours and at predetermined locations we placed the nails and filled the box completely with sand, including the part which was the excavation. While doing this we compacted the soil together with the nails. Then we put weights on top of the box, covering the surface of the sand, put it in the centrifuge, spun it and shook it so as to densify the model without excavation. Then we stopped the centrifuge, removed the weights and excavated the excavation part so that the tips of the nails that were buried there were left sticking out. Then we custom made the facing, glued the same sand on the facing and then lightly pressed the facing against the nails and then glued the nails to the facing. Then the assembly was returned to the centrifuge for testing. Rotation of the facing occurred during the spin-up and the magnitude of rotation, which was very small, corresponded to field values reported in the literature. We did measure the rotation, which I cannot recall, but it was small. We then shook the model and it failed as we described.

Professor Gassler of EH Munich then asked Professor Ehrlich if Dr Jewell was right, or not, to say that the bending stiffness of nails is a second order effect. In my opinion Dr Jewell is right, at least for near vertical walls. This is something I discovered from field tests in 1976. The fact should be considered that bending moments are mobilised too late and that you can only put in your calculation things that are mobilised simultaneously. No doubt, after a larger movement, in the post failure moment, some resistance from bending moments might prevent the full collapse of a wall but before this failure has already occurred and you might try to simulate this in your FEM analysis.

Professor Ehrlich responded that what he was modelling was the working stress condition, not the failure condition, and so you do not consider the large strains necessary to mobilise the bending moment influence. The stiffness is quite important as the results have shown.

Professor Gassler responded that stiffness helps a little, even to reduce deformation in the serviceability state, but it is very very little. Using FEM analysis is still today problematic in modelling moment failure, so, your results should be compared with field tests. At the moment it is better to err on the conservative side in giving advice to practical engineers otherwise stability could be overestimated. Thank you.

Professor Bang, in adding to the view of Professor Gassler, commented that centrifuge work was conducted in 1994 and reported in the Singapore conference in a paper by Sam

Freedman. This experimental work used two types of nail; soft lead and steel. No difference was seen in the bending moment until failure. Once failure was approached the moment effect took place. I think this coincides with what Dr Kim was reporting here.

Professor Ehrlich agreed that field studies are quite important, but lets see what numerical analyses say. Numerical work is important in giving us direction in future studies, if we do not know what we are looking for we cannot instrument properly. The yielding regions will be different depending on bending stiffness. What we want to point out in our paper is that bending stiffness should be taken into account in research.

Professor Gassler then asked Professor Leshchinsky if he had investigated the two part wedge mechanism in rotation for reinforced walls near to the vertical. This could be an unsafe mechanism, particulary for seismic loading.

Professor Leshchinsky replied that the difference between mechanisms is a question of semantics, we are talking about the same mechanism. I looked at all possible failure mechanisms and failure is basically sliding along planes. In geosynthetics it is sliding along the weak plane namely the interface between the soil and geosynthetic. Its an equilibrium approach.

Professor Gassler commented that maybe there is a misunderstanding. The sense of rotation is different in the simple circular slip line and the double wedge mechanism in rotation. We have observed from the slides of Professor Tatsuoka that the overturning moment, or toppling, seems to have happened very often during the last heavy earthquake in Japan. At the end, I agree with you that you can replace the two wedge failure mechanism in rotation by the two wedge mechanism in translation, for the practical engineer, since the latter one is much easier to handle. However, theoretically, you should also investigate rotation and maybe you will come to the same result that I have that about 15% depends on the ratio of acceleration to gravitational acceleration and on some other boundary conditions. The rotational mechanism seems to be more critical, however, from a practical point of view, 15% is not very much in soil mechanics, for example there can be that much scatter in results for soil friction. At the end perhaps we should apply a model factor of 1.1 or something like that. We will agree, no doubt, but before giving things to practical engineers, we should be absolutely sure that things are theoretically correct. Thank you.

Professor Leshchinsky responded that he investigated two types of log spiral; one very deep, which is more critical than the two wedge in rotation, so I looked at that mechanism. Indeed, it has a significant effect on vertical walls or slopes up to 70° and it is that which controls the failure, no doubt. When you increase seismicity above 0.2 then the direct sliding mechanism, the two part wedge, starts taking place. So, it might be dangerous to generalise completely. We looked at the rotational mechanism, first using the simple log spiral to find the reaction force in each reinforcement layer and therefore what is the required strength. The other one, I call a compound failure, is a rather deep log spiral which extends back into the un-reinforced soil and you can very easily fit it with a bi-wedge and get a difference of perhaps 5%. So, in this case the spiral is more critical that the bi-wedge.

Mr Zhao of the Tenax Corporation commented that upper bound analyses of heavily loaded slopes indicate that the translational failure mechanism governs while for lightly loaded slopes the rotational mechanism is more critical.

Dr Koseki of the Institute of Industrial Science, University of Tokyo, then asked Dr Pokharel to explain the assumptions behind his "no length change" analyses. Is it the same as assuming the reinforcement to be infinitely rigid ?

Dr Pokharel confirmed that this was so. This concept came from undrained analysis in which there is no volumetric change. The volumetric stiffness of the soil and water is different. We had no volume change under undrained conditions. There is some elastic volume change in the water which we neglected. Reinforcement stiffness is quite big so that is why we had "no length change" based on plastic flow and undrained analysis.

Dr Koseki then asked if it is possible to formulate the effect of rigidity in your approach. Dr Pokharel responded that the rigidity is infinite. The EI value is not considered and the length between two reinforcement nodes is constant.

Mr Lawson of Royal Ten Cate then commented, in reference to the presentation by Professor Gassler, that in his own experience Eurocode 7 was that it is a poor document for reinforced soil design. Before I give two example of this I just want to say that its important that any new code developed for reinforced soil should provide results which are consistent with existing good practice in terms of the quantities and strengths of reinforcement employed. One good example from

Eurocode 7 is the partial factor of 1.25 used on shear strength of soils. When such a partial factor is applied, one practical implication, particularly for reinforced soil walls, is that it penalises the contractor for using good quality, highly frictional, fill. Factoring down design shear strength automatically increases reinforcement design load and so there is no real incentive for the contractor to use high friction fills. When it comes to basal reinforcement of embankments on soft foundations, again, use of a partial factor of 1.25 on the frictional characteristics of soil is prescribed in Eurocode 7. In this case, when it comes to calibrating against existing good practice, using this value of 1.25, as the ratio of the disturbing moment to the resisting moment changes then it is necessary to use unique partial factors on the loads, the driving forces, to result in consistent designs. It is much better to use a partial factor of 1.0, of unity, on the soil shear strength and a higher partial factor on the driving forces, eg the loads. Then designs results become consistent with past good practice.

Professor Gassler commented that Eurocode 7 is a general code for normal applications of geotechnical work. It cannot work for reinforced soil systems. It is better to make use of the British standard BS 8006. For example, you will learn a lot more about reinforced soil reading this. It could not have been the objective of Eurocode 7 to cover reinforced soil systems in every aspect. Its only the aspect of using partial safety factors in general. All the other details have to be looked up in other codes such as BS 8006.

Finally, Professor Vuceti of UCLA asked Mr Myles, the discussion leader, has any real soil nailed structure in the field, not a model, failed to the extent that one can claim that the geometry of the failure surface is known. If yes, what type of failure mechanism was involved; was it parabolic, circular or bi-linear ?

Mr Myles responded that the reality is that in a lot of failures in soil nailed structures there is not a well defined, simple failure surface; nothing as clear as seen in centrifuge tests. There are a lot of reasons for this. One of the major problems is failure of the toe and in many cases this is not particularly appreciated in design. It is possible to get long term bearing failures, tilting and cracking, because the toe is over stressed, although I have seen, in one case, a large circular global failure which passed outside the nails. I cannot recall failures of the nails themselves;it tends to be pull out or facing failures. I do not believe there is a well defined failure mechanism, it tends to be a whole series of things which go wrong which perhaps masks the situation.

Professor Leshchinsky commented that with geosynthetic reinforced soil walls failures sometimes occur and the failure mechanisms are not too far from those used in analysis. However, it is extremely dependent on boundary conditions, and type of reinforcement, eg the spacing, stiffness etc. In fact I have just completed some research work on a full scale wall, 6 metres high with very short reinforcement, and we could not generate failure within the reinforced soil. So, if such failure does not generate, perhaps we should develop a different design methodology within the parameters of the problem we are considering. At the moment, the approach of failure mechanisms is very common in soil mechanics and quite safe.

Mr Myles, the discussion leader, finished by thanking the contributors and asking them to submit their questions in writing.

Summary discussion 'Testing methods'

Earth Reinforcement, Ochiai, Yasufuku & Omine (eds) © 1997 Balkema, Rotterdam, ISBN 90 5410 833 9

Chairman's report: Testing methods

R.K. Rowe
Geotechnical Research Centre, University of Western Ontario, London, Ont., Canada

ABSTRACT: The excellent long term performance of reinforced soil structures is highlighted. However, it is observed that current techniques for testing and their use for establishing design parameters may result in overly conservative design. In this context, nine questions relating to testing are discussed. It is concluded that there is still considerable research required to identify the reasons for the overdesign evident in many monitored reinforced soil structures. The use of low quality fills in reinforced soil structures was identified as the area where there is the greatest potential benefit for expanding the use of reinforced soil but also noted that much more needs to be known before reinforced soil structures with low quality (eg. cohesive) fills could be used with confidence in regular applications.

The Summary Discussion Sessions at IS Kyushu '96 were initiated with the intention of evaluating where we are after 30 years of experience with respect to modern reinforced soil structures and after three truly international symposia on the subject (Kyushu '88, '92 and '96). The proceedings of IS Kyushu '96 alone contain 151 papers with 32 of these papers dealing directly with testing and materials and a number of papers being related to testing and materials. This alone highlights the level of continued interest in the subject. The technical literature abounds with examples of reinforced soil structures that have performed well and indeed Kyushu '96 contributes many more papers to this collection. Of particular significance is the excellent performance of reinforced soil structures during recent earthquakes (eg. see White & Holtz, 1996; Tatsuoka et al., 1996; Frankenberger et al., 1996; Kobayashi et al., 1996; Nishimura et al., 1996; Otani et al., 1996). Recognizing that field performance (particularly under extreme conditions) represents the ultimate test, it is useful to begin any discussion on testing and materials by reflecting on how reinforced structures have behaved in the field. From that starting point we can then examine what we know and identify what we need to know and what research needs to be done in the future.

Observation of the performance of reinforced soil walls has demonstrated that under normal design conditions (and sometimes under even extreme conditions) the strains mobilized in the reinforcement are quite small (less than 1%: eg. see Benigni et al., 1996; Tajiri et al., 1996) and that the deformations at the face of the wall are also small (typically less than 1.6% of wall height: eg. see Boyle & Holtz, 1996). Even walls constructed on highly compressible deposits and with cohesive backfill (eg. Kojima et al., 1996) perform well provided they are appropriately designed and constructed with reinforcement that also provides for drainage and provided that large deflections (primarily settlement) can be tolerated.

Observation of the performance of reinforced embankments provides similar reassurance. Under normal design conditions, the strain in the reinforcement is usually smaller than anticipated and it has been observed that in many cases the strain decreases with time as pore pressures dissipate (eg. Barsvary et al., 1982; Rowe et al., 1984; Rowe & Mylleville, 1996). However, the opposite trend can be observed for "unconventional" soils that exhibit a significant visco-plastic response (eg. see Rowe & Gnanendran, 1994; Rowe & Hinchberger, 1995) since for these soils the strain in the reinforcement can increase significantly following construction as the soils shed load to the reinforcement. Design procedures for use of soil reinforcement, in conjunction with staged construction, has also proven to be very effective in the design and construction of embankments on peat (eg. Rowe & Mylleville, 1996; Oikawa et al., 1996) although it must be recognized that there will be significant settlement and care is required to control the rate of pore pressure build-up in peat during construction (see Rowe & Mylleville, 1996).

The one area where, despite successes, there are also potential problems is in the use of cohesive backfill in walls and steep reinforced slopes. Here, the potential problems appear to be developing adequate interface resistance if some geogrids are used and basic soil mechanics issues related to creep of the material and the pressure that can arise when tension cracks develop in the cohesive fill.

The lower than designed strains in the reinforcement of reinforced soil walls, slopes and embankments appears to be the result of several factors: (1) Conservatism in the selection of reinforcement parameters and the fact that the stiffness (and strength) of the reinforcement is often far greater than actually assumed in the design (this will be discussed in more detail shortly). (2) Conservatism in the selection of soil parameters where (a) the designer assumed strength of backfill is usually well below that actually mobilized in the field; (b) there is often matric suction due to the presence of moisture in the fill that gives rise to an apparent cohesion and a substantially higher than assumed strength in the fill; (c) the designer usually adopts a conservative assessment of the undrained strength of the foundation - this can involve adopting a conservative estimate of strength based on the scatter of data as well as an underestimate (or total neglect) of strength increase with depth (which can be masked by the effects of sample disturbance unless appropriate care is taken). (3) Redundancy in the reinforced soil system which is neglected in design (eg. toe resistance for walls and the presence of a root mat or crust for embankments on soft soil); and (4) conservatism inherent in the simplified analysis.

Of these issues, the first (conservatism in the selection of reinforcement) was a major discussion topic in this session (see Voskamp, 1996; Greenwood, 1996) with it being argued that the current approach to limit states design is resulting in unrealistically low strength (and stiffness) being used in design; we will return to this. With respect to conservatism regarding soil parameters there is some inevitability to engineers being conservative in neglecting apparent cohesion in design (since it is readily lost due to either drying or an increased degree of saturation). More debatable is the choice of the design friction angle - but this debate will be left for the session on Design Methods (Bolton, 1996). Less justifiable is the conservative interpretation of undrained shear strength profiles with depth for embankments on soft clay; here there is sound fundamental argument for considering the real strength increase with depth and the reason for ignoring it is usually either (a) failure to recognize that solutions exist for considering the bearing capacity on soil whose strength increases with depth (eg. see Davis & Booker, 1973; Rowe & Soderman, 1987) or (b) failure to recognize the increase in strength with depth due to sample disturbance which tends to increase with increasing depth and often masks the actual strength increase (this can be dealt with by the use of better sampling techniques). The issues of redundancy and the conservatism inherent in simplified analyses while warranting discussion are beyond the scope of this report on "Materials and Testing". Hopefully, they will be addressed in the report on "Design Methods" (Bolton, 1996).

The Summary Discussion Session on "Testing and Materials" involved panel presentations by Prof. Makiuchi (Geosynthetics Testing in Japan), Mr. P. Segrestin (Monitoring of Reinforcement Durability and Thoughts About Experience and Needs), Mr. W. Voskamp (Proposed Method to Determine the Safety Capacity of Reinforced Soil Structures During Its Lifetime) and Dr. J. Greenwood (Residual Strength: An Alternative to Stress-Rupture for Earth Reinforcement Design). In addition, there were five oral paper presentations by Profs. Palmeira (Palmeira et al., 1996), Chang (Chang et al., 1996), Mitachi (Nakamura et al., 1996), Gourc (Morel & Gourc, 1996) and Adanur (Adanur et al., 1996). The session was organized to address a number of questions; the remainder of this article will itemize these questions and the corresponding presentations/discussions.

1. Are current standard tests (ISO/CEN/ASTM) adequate?

Mr. P. Stevenson indicated that the standard ASTM tensile test (D4595) gave tensile stiffness results that were too low for high strength materials and that the sample length affects the results. He showed that for one geotextile tested, the tensile stiffness for 100 mm gauge length was less than half the value obtained for a 450 mm gauge length. He indicated that the gauge length should exceed 450 mm if meaningful results are to be obtained.

Mr. B. Myles agreed that the problem is real but a much greater problem is the poor definition of the initial part of the extension curve and he was critical of both the ASTM and ISO wide width tensile test methods. He stated they use much too high a preload and this masks quite a lot of the "softness" of the geotextiles. Consequently, people believe that the geosynthetic will take up load much quicker than they do since many geotextiles need 1% extension (or more) before they come onto what is normally called the "tensile curve".

Mr. Stevenson accepted Mr. Myles' comment and said that they looked at the initial part of the curve (with no preload). Dr. Greenwood indicated that he was not aware that ISO 10319 was inadequate for strong materials.

It is clear from the discussion that while current tensile index tests serve a useful purpose in identifying changes in the tensile characteristics of a given product, there is still some question regarding the validity of comparing tensile characteristics of different products on the basis of the existing ISO and ASTM tests both for nonwoven geotextiles (whose in-soil stiffness may be greater than the in-isolation stiffness) and for some high strength geosynthetics where gauge length may result in the tensile stiffness being underestimated.

2. How different are the stress-strain and creep characteristics of geosynthetics in unconfined tests from those in-soil, and what are the problems of doing confined tests?

Three papers in the proceedings address this question (see Adanur et al., 1996; Chang et al., 1996; Palmeira et al., 1996) and numerous questions arose from the floor. Dr. Yogarajah commented that the stress-strain behaviour of geosynthetics in both the unconfined and confined state tests are different from the stress-strain-behaviour of geosynthetics in walls and slopes. Dr. Palmeira agreed and indicated that even the confined tests are "index tests" in that they do not fully duplicate field conditions but that they do give a better indication of likely behaviour than unconfined tests (at least for some types of geosynthetics).

Questions then arose regarding the difficulty of isolating the effects of friction and boundary conditions when performing confined tests. Dr. Palmeira indicated that lubrication of the membrane had a significant effect on the results (as compared to those obtained using an unlubricated membrane). This clearly shows that there are potential problems of boundary effects (especially if an unlubricated membrane is used). What is not yet clear is what effect the lubricated membrane has on the results. Dr. Palmeira stated that much of the improvement in the mechanical behaviour of nonwoven geotextiles was due to soil becoming impregnated into the geotextile and thereby reducing fibre mobility.

Dr. Alexiew suggested that we need to carefully consider how the results of laboratory tests will be used before embarking on extensive testing. All the tensile tests being conducted confined in soil are similar to pullout tests and the tensile force is applied externally to the reinforcement, then from the reinforcement to the soil and then to the test box. The results may be applicable to equivalent situations in the structure - however, this is only in the anchoring zones; in all other zones of the reinforced structure there is the reverse situation where the soil mass is loading the reinforcement. Thus, in reality, in 70-90% of zones in the structure the soil mass is loading the reinforcement (this is why we need the reinforcement) and hence the so called confined tests are not really relevant to these regions of the soil and hence one must be very careful in applying the results from confined tests to real structures. Dr. Chang responded that they were in an early stage of their research and he wanted to understand if frictional resistance reduces creep and the results were not intended to be used in design. He agreed that his test does not simulate in-situ conditions.

In support of Dr. Alexiew's comment, Mr. Voskamp noted that in confined tests, movement between a geosynthetic and the confinement soil will take place during loading, due to the extensibility of the geosynthetic. This will give a lower load in the confined area and gives rise to the following questions: How representative is this test? What is the actual load level and what is the actual strain in the geosynthetic in the confined area? How is the different strain in the unconfined area considered? He stated that, in fact, the confining stress results in increased friction between the fibers which explains some of the short term improvement in stress-strain behaviour relative to an unconfined test. However, there is no long term data to confirm the short term results.

Prof. D. Leshchinsky indicated that except for nonwoven geotextiles, there is a problem with confined tests since people are confusing the intrinsic material properties with the effect of the interfaces (which increases the apparent stiffness). He suggested that the confined test is unnecessary in these cases.

Prof. Wu stated that
"(i) the creep/potential of geosynthetics as evaluated by the in-isolation creep tests presented in this discussion session (i.e., tests performed on a geosynthetic specimen) is misleading;
(ii) subjected to the same load, a geosynthetic can exhibit very different creep behaviour when it is embedded in a granular soil versus a cohesive soil--because of soil-geosynthetic interaction;
(iii) a soil-geosynthetic long-term interactive test method (Wu & Helwany, 1996) has been developed (and used) at the Colorado Transportation Institute to account for the effects of soil."

Dr. Greenwood responded to the effect that: "Experimenters have been struggling for over 20 years to devise a reliable and acceptable method for measuring the mechanical properties of geosynthetics in soil and I am pleased to learn of Professor Wu's success. Other laboratories should start intercomparison tests as soon as possible to confirm the reproducibility of his results, prior to establishing the method as a standard and building it into the design codes." Interest was also expressed in the cost of this test.

Mr. Voskamp did not agree with Prof. Wu's comment that creep tests performed in-isolation are misleading. He stated that "in-isolation tests give insight regarding the polymer characteristic: creep. When tests are executed in-confinement this creep may be restricted or slowed down, but at the end the in-isolation obtained results are the ultimate failure level. Because of soil-geosynthetic interaction the results of the in-confinement tests may give more optimistic results. However, there is no proof in practice that the in-confinement tests give representative values. The actual load level in the test specimen is not the same as the applied load due to the same soil geosynthetic interaction. Further, the load, soil and vertical stress vary in every application. The statements of Jonathan Wu are based on limited results and as such should not be considered valid in general terms."

In response to Prof. Wu's comment about creep in confined tests, Mr. Segrestin commented that the debate about the way in which creep tests and (even more useful) creep rupture tests should be performed in order to take the confining effect of the soil into account is certainly an important one. However, in order to avoid hasty or optimistic conclusions, it seems useful to note that this confining effect is highly dependent on the configuration of the reinforcement. For example, it is presumably notable for a bidirectional isotropic mesh, but negligible for a linear strap.

3. Do pullout and shear box tests really tell us what we need to know and what are the problems with interpretation? (See Adanur et al., 1996; Ghosh & Bhasin, 1996; Hayashi et al., 1996; Konami et al., 1996; Lin et al., 1996; Lopes & Ladeira, 1996; Segrestin & Bastick, 1996; Chang & Milligan, 1996.)

Prof. Makiuchi had presented details of the standard pullout tests in Japan and indicated the pullout box was 30 cm square and 20 cm high. The specimen width was the same as the box. Dr. Chang indicated that in his experience one can get misleading results from this test because even if you do reduce friction the distribution of vertical stress is not even and this may have a significant effect on the results of the test.

Dr. Palmeira raised questions regarding what would be the appropriate length of sample for assessing the equivalent friction angle (interaction factor) and also the boundary effects that can occur in these tests. Mr. Voskamp stated that the shear box should be large (approximately 2 x 1 m with 2 x 0.5 m height is common these days). Prof. C. Jones stated that the boundary conditions at the face of the pullout box are very important and that this is currently being examined in Europe. Prof. Wu pointed out that there is a technical paper by Sohbi and Wu in the next issue of the Geosynthetics International Journal which addresses how to obtain a consistent interpretation of pullout test results.

Dr. Lo stated that a rigid sleeve can be a problem in pullout tests and that one needs a flexible sleeve. However, he stated that a more fundamental problem is the progressive failure that occurs along the reinforcement. He has found that in the pullout box the geosynthetic moves a lot relative to the soil and hence gives a lot of progressive failure however in a field case with which he has been involved the reinforcement is approaching the limit state at much less relative movement between the soil and geosynthetic and hence there is less progressive failure and a higher apparent friction angle in the field. The pullout test gives lower friction angles as the length of the box (and sample) is increased.

Dr. Pokharel observed that the pullout test is usually performed under passive conditions whereas in the field the anchorage length is located behind a zone where active conditions prevail and hence he questioned the validity of results obtained from a pullout test for assessing properties required for determining the anchorage length in walls and slopes.

4. Is it practical to develop a meaningful test to establish the effect of construction damage to reinforcement?

Dr. Greenwood responded to this question by stating that he does not believe that one can assess the resistance of all types of geosynthetic to mechanical damage (or robustness) solely on the basis of the traditional tests of tensile strength, static and dynamic puncture and mass per unit area. There are of course large scale performance tests which measure the reduction in tensile strength under simulated site conditions, but the results are specific to the backfill used and the tests are expensive to carry out. There has been considerable experimental work in Europe directed at the development of a simpler laboratory test for resistance to mechanical damage, in particular the dynamic compression of a sample between two layers of angular fill, followed by visual examination for holes and by tensile testing. This method has recently been put forward as a European prestandard, to be supplemented by a further method related to a geotextile over soft ground. There will be a need for international experimental validation to examine whether these tests are representative and practical, and for guidance to the designer as to how he is to apply the results in practice with reference to his own site conditions and the function of the geosynthetic.

5. How does reinforcement influence the dynamic properties of the reinforced mass?

Although many papers addressed the issue of the performance of reinforced walls during earthquake, only one paper (Chen et al., 1996) investigated dynamic behaviour in the laboratory and made a number of conclusions. Dr. Shadu noted that the tests reported in this paper were performed on an "ideal" Ottawa sand and cautioned that similar tests should be performed on other, more typical, sands before extrapolating the results to field conditions.

6. How to interpret triaxial and biaxial test results?

As noted in the paper on page 111 of the proceedings, triaxial tests have been performed to illustrate the benefits of reinforcement. Dr. Pokharel noted these tests are very difficult to interpret since they begin as an axi-symmetric test but the conditions of the test change due to the presence of imperfections and strain localization which remove the symmetry. The biaxial test proposed by Morel and Gourc (1996) has the potential to provide more consistent (tractable) testing conditions although the effects of sample dimensions and boundary effects still require careful examination when interpreting the test results.

7. Are the techniques used to reduce side wall friction in controlled system wall tests (eg. RMC, Denver, Tokyo) effective at reducing side wall friction to negligible values throughout the test?

Large-scale laboratory tests on prototype walls have been performed by a number of investigators (eg. Jarrett & McGown, 1987; Wu, 1992; Huang & Huang, 1996; Tajiri et al., 1996; Tsukamoto et al., 1996). These tests are often used either to extrapolate to field conditions or to test numerical models that are then used to extrapolate to field conditions. It is noted however that these tests may be difficult to interpret and the question arises as to why. For example, on page 533 of the proceedings, the test reported by Tsukamoto et al. (1996) is reported to give a coefficient of active earth pressure of $K_a \simeq 0.1$ without any reinforcing for a wide range of applied surcharge pressures (49-294 kPa). This implies a consistently very high friction angle for a relatively uniform (C_u = 1.7) sand at 90% relative density.

In this, as in many other similar tests, lubricant and plastic membranes are put on the walls to reduce friction between the backfill and the walls. Prof. Bathurst asked the question both of this and other tests - are plane strain conditions really achieved? Since plane strain conditions are usually assumed in interpreting the results this is a key question. He noted that even when using lubricant and plastic or latex membranes between the sand and walls of the test box, the actual friction angle is still potentially significant, particularly for low ratios of test box width (L_w) to height (H). He indicated that even though people report friction angles as low as 1° for lubricated sides, his experience is that while the friction angle may start off that low, chemical interactions between the membrane and lubricant as well as penetration of soil particles into the membrane increase side wall friction with time. Dr. Gassler indicated that he fully agreed. He said that qualitatively these tests give good results but quantitatively you need to be careful when backanalyzing the results due to the side friction.

Drs. Wu and Huang agreed that this can be a problem but stated that in their tests the problem is minimal. Dr. Wu indicated that in his tests, the lubricant used was a silicon grease (manufactured by Shin-Etsu Chemical Company in Japan). The particular grease is compound KS 63G, comprising a thin silicone base oil and fine angular silica fillers. The membrane used was a thin latex membrane. The thickness of the membrane varied according to the soil type employed in the test. Fumio Tatsuoka, University of Tokyo, recommended a membrane thickness of 200 micrometer for a clean sand. For long term tests he agreed that there is a problem. Dr. Bathurst indicated that in his experience "long term" may be as little as one day. Dr. Huang agreed with Dr. Bathurst that it is important to have a wide specimen (ie. $L_w/H > 1$) in order to minimize the side wall effects and to approach plane strain conditions.

It was clear from the discussion that researchers reporting results from these types of test should provide data to support the friction angle of the wall interface over test periods similar to that of the actual test being reported (see Bathurst & Benjamin, 1988). Another factor that should be reported is the stress-strain characteristics of the soil under the same conditions of moisture content that were used in the test wall; if the sand is moist this may significantly influence the interpretation of the test results.

8. How does one assess environmental damage to reinforcing materials?

A number of papers in the proceedings (see Jailloux & Anderson, 1996; Smith et al., 1996; Greenwood & Yeo, 1996; Mak & Lo, 1996) as well as the Panel Presentation by Mr. Segrestrin, Mr. Voskamp and Dr. Greenwood address, at least in part, the issue of environmental damage. Mr. Kassner questioned the paper by Jailloux and Anderson (p. 45) and the presentation of Mr. Segrestin which discussed the performance of polyester yarn at a pH of 1 (in HCl), 10 (in Lime) and 12 (in NaOH). He asked how would steel perform under the same conditions and under what real soil conditions would one expect to find a pH of 1 or 12? Mr. Segrestin indicated that this work was for "research purposes" and that it confirmed what was already known: that polyester is degraded at very high pH but low pH is not aggressive. In contrast, steel is attacked at low pH but performs well at high pH. Dr. Alexiew cautioned that one must distinguish between external and internal hydrolysis. External hydrolysis corresponds to chemical attack but internal hydrolysis can be examined in water. He indicated that a great deal of work has been done on this issue in Germany and suggested that attendees read the paper by Wilmers (1996) which concludes that there is no hydrolysis occurring in real, normal soils including typical alkaline soils (but excluding lime stabilized soils). Dr. Alexiew also stated that they have recently exhumed a polyester geotextile that was submerged for 14 years and that tests indicate a rate of degradation of about 1% per decade.

In discussing the paper by Smith et al. (1996, page 151) and the panel presentation by Mr. Segrestin, Prof. Jones commented that in his opinion corrosion was a function of surface area and the best shape for steel reinforcement would be a circular section rather than a flat strip of the same area. He said that if you are concerned about pitting corrosion, a pit of a given size is likely to be more critical for the flat strip than a circular bar. He suggested using a circular reinforcing element with a plate anchor (as used in Japan) or a triangular anchor (as developed in the U.K.) and stated that this would minimize the corrosion problem for steel.

Mr. Segrestin replied that "Professor Jones is right when he notes that a round bar is at first more resistant to corrosion than a flat strip. It is clear that, if we compare two reinforcements having the same cross-section, for example a 40 mm x 5 mm strip and a ϕ16 mm bar both with a 200 mm^2 cross-section, an average loss of thickness of say 1 mm would have a smaller impact on the bar. The residual cross-section would be indeed 160 mm^2 for the strip and 177 mm^2 for the bar.

However, the study we made does not relate to reinforcements with the same cross-sections, but to reinforcements which are actually used by the industry, for example 50 mm x 4 mm strips compared to wire-meshes made of ϕ8 mm or ϕ10 mm bars. Our study also deals with the effects of local pitting, rather than with average uniform loss of thickness. As shown in the paper by Smith et al., such pitting proves to be more detrimental to the long-term resistance of small round bars and would justify larger sacrificial thicknesses for design."

Professor Fukuoka stated that it is important to look at case records. He first used steel bars as reinforcement in 1963 and the structure is still standing without any problems. He constructed a tall reinforced earth wall in Kyushu in 1967; and it is still standing without any problems. He used PVC in 1965 and still no problems. He recommends that we collect more case records to contribute to the discussion on durability. Dr. Greenwood responded by stating he agreed entirely that when samples are exhumed, field records should be compiled with as much information as possible on the soil environment, temperature, rainfall etc. One must then ask the question: "Given the environment to which they were exposed, would we have expected the geosynthetics to have degraded?". Exhumations which end up with the conclusion "nothing was expected to happen, and nothing happened" are of limited value.

Mr. Segrestin responded that he agreed with Prof. Fukuoka's statement that actual field records are indispensable for a complete understanding of the question of durability as emphasized in his presentation. He stated that it is not a surprise that structures using steel perform very well after 30 or 35 years; they will undoubtedly remain in service for many more decades. However, he understood that the structures mentioned by Prof. Fukuoka are made up of relatively big galvanized steel anchors or multi-anchors. Again, the discussion in his presentation and in Smith et al.'s paper is about the sensitivity of small bars, only a few millimeters in diameter, to the effects of superficial pitting and this may not be relevant to the structures described by Prof. Fukuoka.

9. How do we evaluate the design strength of reinforcement based on test data?

The panel discussion by Mr. Voskamp and Dr. Greenwood (as well as Greenwood & Yeo, 1996; Mak & Lo, 1996) addressed the question of how to assess the design strength of geosynthetic reinforcement. There is a clear implication that current limit state design procedures (eg. U.K. BS8006) are overly conservative.

In his presentation, Mr. Segrestin implied that the time-temperature shift commonly used to predict long term performance might be questioned under extreme conditions of high pH (Jailloux & Anderson, 1996). Mr. Segrestin cautioned that these are very preliminary results and they are very aggressive environments and here the Arrhenius coefficient is non-linear but they need more results.

Dr. Greenwood stated that the temperature transposition involves the use of short term tests at higher temperatures to predict longer term effects at lower temperatures. For it to be valid, the physical and chemical mechanisms operating at both higher and lower temperatures must be identical, and there should not be any change in the physical state of the polymer between them. In polyesters there is a change in physical state at about 65 degrees celsius, above which the molecules in the amorphous regions (as opposed to the crystalline regions) of the fibres become more mobile. This will, among other things, expose the molecules more to chemical attack, including hydrolysis. Extrapolation of data on the rate of hydrolysis from tests performed above 65 to temperatures below is therefore not strictly valid. Since the phase transition reduces the sensitivity to hydrolytic attack at lower temperatures, however, the extrapolation is likely to be conservative and will predict too low rather than too high a lifetime. In the absence of other data, therefore, a prediction based on a temperature transposition of this kind can be regarded as a minimum value.

Mr. Voskamp added that there can be a difference between the results obtained above and below the glass transition temperature. Long term creep testing has been executed for PET up to 60°C and time temperature shifting gave a perfect fit within that range (see Greenwood & Yeo, 1996).

Dr. Lo asked how one could experimentally determine the residual strength curves for a given material. Dr. Greenwood replied that the residual strength curve is determined in the same manner as for stress-rupture, but the test is interrupted after a fraction of the anticipated time-to-rupture and the tensile strength of the specimen determined.

A key issue in the discussion of limit state design is the identification of what one is trying to do. As noted by Dr. Greenwood, if the "factor of safety" is intended to guard against miscalculation of the overall load on the product, in other words, if the sustained load in practice could be higher than the design load, then the design process must be based on stress-rupture as before. The residual strength approach is valid where the safety factor is intended to maintain a certain reserve of tensile strength throughout the design life to guard against sudden, instantaneous, increases in load - such as seismic events.

In response to Dr. Greenwood's presentation on residual strength, Mr. Segrestin opined that "the "residual strength" approach, i.e. considering the way in which the resistance of the geosynthetic reinforcement varies before it goes to creep rupture, does not finally modify the design procedure. One must bear in mind that the "overall factor of safety", usually of the order of 1.50, essentially accounts for the uncertainties in the applied loads, potential for local overstresses and imperfections of the computation methods. If the calculations are carried out as "working stress" calculations, this means that it cannot be excluded that the actual tensile load might be up to 1.5 times larger (T_o) than calculated (T_m). If the calculations are carried out as "limit state" calculations, the "load factor" is already included in the design load (T_d). In both cases one must check that, in the extreme case, the reinforcement will not break. In other words, one must make sure that T_d is just smaller than the creep rupture strength corresponding to the required service life (all other effects such as construction damage or environmental ageing being put aside here)."

Mr. Lawson noted that there were many important points made in Mr. Voskamp's panel presentation and that he wanted to emphasize some of these points. In particular, he noted that the issues of testing, partial factors and design need to be considered together and not in isolation. One needs to keep in mind the magnitude of the load factor and the material reinforcement factor when one comes to assess the components of the material reinforcement factor otherwise one can start looking for worst case values for the various components. He expressed concern that some people (and some papers in the conference) were going to an extreme in arguing for certain materials factors; he pleaded for logic and consistency in the selection of the materials component of the partial factors used in design. Mr. Voskamp agreed and stated that one has to remember that design life curves are logarithmic and that if one is not careful in designing for a 100 year design life but using excessive partial factors one may be designing for a 1000 year design life.

Mr. Myles commented that part of the reason for conservative factors in the British Standard (BS8006) is that there was not enough evidence available at the time to adopt a less conservative approach. These factors can be revised when more evidence becomes available. He pleaded for the manufacturers to provide more information.

In response to a question from Mr. Yokota regarding reinforcement characteristics relevant to seismic loading, Mr. Voskamp noted that seismic loading of the structure results in high frequency loading during a short period. When the load remains below the stress-rupture line level for the actual conditions, no problem is to be expected especially as the residual strength of the material will be higher than the stress-rupture value at the time of loading. Tests on polyester geogrids have shown that high frequency loading leads to increased stiffness of the material without influencing the strength.

Recognizing that most of the discussion had focused on retaining walls which have a long design life and whose long term performance depends on the reinforcement performing this role for the entire life of the structure, Dr. Sharma emphasized that we need to distinguish between "short term" and "long term" applications of soil reinforcement. He stated that creep is usually not a problem when reinforcing an embankment on soft clay since it is a "short term" situation. Mr. Voskamp agreed and stated that when the stress-rupture line is used for design, the corresponding rupture load can be determined for every design service life. This means for short term applications that the allowable load is much higher than for long term application.

In response to Dr. Sharma's comment, Mr. Segrestin opined that: "Imagining creep is not an issue for short-term applications could be deceptive. It should be borne in mind that the creep rupture strength practically decreases uniformly as a function of time, in a log-log scale. A product expected to have a long-term (100 years) creep rupture strength equal for example to 30% of its short-term breaking load, might break after about 10^5 hours (say 10 years) under 35% of this load, or after about 10^4 hours (say 1 year) under 45%. Its creep rupture strength could be even already reduced to about 70% or so after only one week!".

Concluding Comments

The papers on testing presented at the conference as well as the discussion in the Summary Discussion Session highlight the fact that although we now have 30 years of experience with modern reinforced soil structures and three international reinforcement symposia in Kyushu, there are still some unanswered questions and work for researchers and manufacturers to do. Much of this work is related to identifying the reasons for "overdesign" evident in many monitored reinforced soil structures and improving the parameters (based on testing) used in design. This relates particularly to issues such as the short term stress-strain properties of reinforcement, the time dependent behaviour of reinforcement, the effects of construction damage, and environmental impact on reinforcement. Areas where there is still considerable controversy include the use of standard index tests (eg. ISO/ASTM wide width tensile tests) for assessing design parameters for reinforcement, in-soil (confined) tests, the interpretation of large scale laboratory tests, and the selection of appropriate partial factors (for use in design) that account for stress-rupture, damage and environmental effects.

While there is still much useful work to be done, it is also evident that reinforced soil structures, whether they be embankments, slopes or walls, are performing remarkably well both under static and earthquake conditions. Where they have been monitored, the strains are generally well below what would be expected. Thus much of the discussion today has focused on us being too conservative however it is very comforting to know that current test methods combined with current design methodologies do generally result in structures which may have been a little more expensive than they needed to be, but have performed their design function well and by using soil reinforcement were usually constructed at a cost well below the cost of alternative (more conventional) designs. We are doing well. Reinforced structures may not be understood to the level we would like for research, we may be able to make them more efficient and reduce cost - but they are working remarkably well and we should be pleased that over the last 30 years of research and application we have achieved a great deal.

What about the future? There are many areas for future research arising out of the discussion. However, in terms of advancing the discipline, the use of low quality fills in reinforced soil structures would appear to require the most attention. This will involve dealing with some difficult (and still not fully resolved) issues of soil mechanics as well as issues of soil-reinforcement interaction. A number of papers at IS Kyushu '96 address this issue and I expect to see even more focus on this issue when next we meet again.

References

Adanur, S., Mallick, S. and Zhai, H. (1996). Analysis of geotextile-soil interaction in pull-out tests. Proc. of the International Symposium on Earth Reinforcement, Edited by H. Ochiai, N. Yasufuku and K. Omine, A.A. Balkema, Rotterdam, 1996, Vol. 1, pp. 3-8.

Barsvary, A.K., MacLean, M.D., Cragg, C.B. (1982). Instrumented case histories of fabric reinforced embankments over peat deposits. Proc., 2nd International Conference on Geotextiles, Las Vegas, Vol. 2, pp. 647-652.

Bathurst, R.J. and Benjamin, D.J. (1987). Preliminary assessment of sidewall friction on large-scale wall models in the RMC test facility. The Application of Polymeric

Reinforcement in Soil Retaining Structures. NATO Advanced Study Institutes Series, Kluwer Academic Publishers, pp. 181-192.

Benigni, C., Bosco, G., Cazzuffi, D. and De Col, R. (1996). Construction and performance of an experimental large scale wall reinforced with geosynthetics. Proc. of the International Symposium on Earth Reinforcement, Edited by H. Ochiai, N. Yasufuku and K. Omine, A.A. Balkema, Rotterdam, 1996, Vol. 1, pp. 315-320.

Bolton, Summary Discussion Session, Panel Discussion, Proc. of the International Symposium on Earth Reinforcement, Edited by H. Ochiai, N. Yasufuku and K. Omine, A.A. Balkema, Rotterdam, 1996, Vol. 2.

Boyle, S.R. and Holtz, R.D. (1996). Prediction of reinforced soil retaining wall deformations: A review of two procedures. Proc. of the International Symposium on Earth Reinforcement, Edited by H. Ochiai, N. Yasufuku and K. Omine, A.A. Balkema, Rotterdam, 1996, Vol. 1, pp. 327-332.

Chang, K.T. and Milligan, G.W.E. (1996). Effects of the transition zone in a nailed wall model test. Proc. of the International Symposium on Earth Reinforcement, Edited by H. Ochiai, N. Yasufuku and K. Omine, A.A. Balkema, Rotterdam, 1996, Vol. 1, pp. 333-338.

Chang, D.T.T., Chen, C.A. and Fu, Y.C. (1996). The creep behavior of geotextiles under confined and unconfined conditions. Proc. of the International Symposium on Earth Reinforcement, Edited by H. Ochiai, N. Yasufuku and K. Omine, A.A. Balkema, Rotterdam, 1996, Vol. 1, pp. 19-24.

Chen, T.C., Chen, R.H., Lee, Y.S. and Pan, J.C. (1996). Dynamic reinforcing effect of reinforced sands. Proc. of the International Symposium on Earth Reinforcement, Edited by H. Ochiai, N. Yasufuku and K. Omine, A.A. Balkema, Rotterdam, 1996, Vol. 1, pp. 25-28.

Davis, E.H. and Booker, J.R. (1973). The effect of increasing strength with depth on the bearing capacity of clays. Geotechnique, 23(4), pp. 551-63.

Frankenberger, P.C., Bloomfield, R.A. and Anderson, P.L. (1996). Reinforcee earth walls withstand Northridge Earthquake. Proc. of the International Symposium on Earth Reinforcement, Edited by H. Ochiai, N. Yasufuku and K. Omine, A.A. Balkema, Rotterdam, 1996, Vol. 1, pp. 345-350.

Ghosh, C. and Bhasin, A. (1996). Displacement controlled pullout test of geotextile in granular soil. Proc. of the International Symposium on Earth Reinforcement, Edited by H. Ochiai, N. Yasufuku and K. Omine, A.A. Balkema, Rotterdam, 1996, Vol. 1, pp. 35-38.

Greenwood, Summary Discussion Session, Panel Discussion, Proc. of the International Symposium on Earth Reinforcement, Edited by H. Ochiai, N. Yasufuku and K. Omine, A.A. Balkema, Rotterdam, 1996, Vol. 2.

Greenwood, J.H. and Yeo, K.C. (1996). Assessment of geogrids for soil reinforcement in Hong Kong. Proc. of the International Symposium on Earth Reinforcement, Edited by H. Ochiai, N. Yasufuku and K. Omine, A.A. Balkema, Rotterdam, 1996, Vol. 1, pp. 363-368.

Hayashi, S., Alfaro, M.C. and Watanabe, K. (1996). Displacement controlled pullout test of geotextile in granular soil. Proc. of the International Symposium on Earth Reinforcement, Edited by H. Ochiai, N. Yasufuku and K. Omine, A.A. Balkema, Rotterdam, 1996, Vol. 1, pp. 39-44.

Huang, C.C. and Huang, B.N. (1996). Infiltration tests on reinforced clay wall and test embankment. Proc. of the International Symposium on Earth Reinforcement, Edited by H. Ochiai, N. Yasufuku and K. Omine, A.A. Balkema, Rotterdam, 1996, Vol. 1, pp. 373-378.

Jailloux, J.M. and Anderson, P.L. (1996). Long term testing of polyester yarn and product at 50°C and 23°C in different environments. Proc. of the International Symposium on Earth Reinforcement, Edited by H. Ochiai, N. Yasufuku and K. Omine, A.A. Balkema, Rotterdam, 1996, Vol. 1, pp. 45-50.

Jarrett, P.M. and McGown, A. (1988). Editors: The Application of Polymeric Reinforcement in Soil Retaining Structures. Kluwer Academic Publishers, Dordrecht, Netherlands.

Kobayashi, K., Tabata, H. and Boyd, M. (1996). The performance of reinforced earth structures in the vicinity of Kobe during the Great Hanshin Earthquake. Proc. of the International Symposium on Earth Reinforcement, Edited by H. Ochiai, N. Yasufuku and K. Omine, A.A. Balkema, Rotterdam, 1996, Vol. 1, pp. 395-400.

Kojima, K., Sakamoto, N., Tateyama, M. and Maruyama, O. (1996). Geosynthetic-reinforced soil retaining wall using clay on a very soft ground for Hokuriku bullet train yard in Nagano. Proc. of the International Symposium on Earth Reinforcement, Edited by H. Ochiai, N. Yasufuku and K. Omine, A.A. Balkema, Rotterdam, 1996, Vol. 1, pp. 227-232.

Konami, T., Imaizumi, S. and Takahashi, S. (1996). Elastic considerations of field pull-out tests of polymer strip reinforcement. Proc. of the International Symposium on Earth Reinforcement, Edited by H. Ochiai, N. Yasufuku and K. Omine, A.A. Balkema, Rotterdam, 1996, Vol. 1, pp. 57-62.

Lin, S.S., Chen, M.C., Cheng, M.Y. and Kuo, S.H. (1996). Pullout tests on geogrids buried in lateritic soils. Proc. of the International Symposium on Earth Reinforcement, Edited by H. Ochiai, N. Yasufuku and K. Omine, A.A. Balkema, Rotterdam, 1996, Vol. 1, pp. 83-88.

Lopes, M.L. and Ladeira, M. (1996). Pull-out tests for the assessment of soil-geogrids interaction - Influence of some mechanical and physical parameters. Proc. of the International Symposium on Earth Reinforcement, Edited by H. Ochiai, N. Yasufuku and K. Omine, A.A. Balkema, Rotterdam, 1996, Vol. 1, pp. 89-94.

Mak, J.C-H. and Lo, S-C.R. (1996). Towards a limit state design specification for reinforced soil walls. Proc. of the International Symposium on Earth Reinforcement, Edited by H. Ochiai, N. Yasufuku and K. Omine, A.A. Balkema, Rotterdam, 1996, Vol. 1, pp. 415-420.

Morel, J.C. and Gourc, J.P. (1996). Reinforced soil failure: Analysis at the biaxial compression test. Proc. of the International Symposium on Earth Reinforcement, Edited by H. Ochiai, N. Yasufuku and K. Omine, A.A. Balkema, Rotterdam, 1996, Vol. 1, pp. 111-116.

Nakamura, T., Mitachi, T. and Ikeura, I. (1996). Some factors affecting the results of soil-geogrid direct shear test. Proc. of the International Symposium on Earth Reinforcement, Edited by H. Ochiai, N. Yasufuku and K. Omine, A.A. Balkema, Rotterdam, 1996, Vol. 1, pp. 123-128.

Nishimura, J., Hirai, T., Iwasaki, K., Saitoh, Y., Morishima, M., Shintani, H., Yoshikawa, S. and Yamamoto, H. (1996). Earthquake resistance of geogrid-reinforced soil walls based on a study conducted following the southern Hyogo earthquake. Proc. of the International Symposium on Earth Reinforcement, Edited by H. Ochiai, N. Yasufuku and K. Omine, A.A. Balkema, Rotterdam, 1996, Vol. 1, pp. 439-444.

Oikawa, H, Sasaki, S. and Fujii, N. (1996). A case history of the construction of a reinforced high embankment on an extra soft ground. Proc. of the International Symposium on Earth Reinforcement, Edited by H. Ochiai, N. Yasufuku and K. Omine, A.A. Balkema, Rotterdam, 1996, Vol. 1, pp. 261-266.

Otani, Y., Mega, M. and Matsui, T. (1996). Damage performance of steel-reinforced earth structures in the 1995 Hyogoken-Nambu Earthquake. Proc. of the International Symposium on Earth Reinforcement, Edited by H. Ochiai, N. Yasufuku and K. Omine, A.A. Balkema, Rotterdam, 1996, Vol. 1, pp. 451-456.

Palmeira, E.M., Tupa, N. and Gomes, R.C. (1996). In-soil tensile behaviour of geotextiles confined by fine soils. Proc. of the International Symposium on Earth Reinforcement, Edited by H. Ochiai, N. Yasufuku and K. Omine, A.A. Balkema, Rotterdam, 1996, Vol. 1, pp. 129-132.

Rowe, R.K. and Gnanendran, C.T. (1994). Geotextile strain in a full scale reinforced test embankment. International Journal of Geotextiles and Geomembranes, Vol. 13, No. 12, pp. 781-806.

Rowe, R.K. and Hinchberger, S. (1995). The significance of rate effects in modelling the Sackville test embankment. GEOT-16-95, Faculty of Engineering Science, University of Western Ontario.

Rowe, R.K. and Mylleville, B.L.J. (1996). A geogrid reinforced embankment on peat over organic silt: a case history. Canadian Geotechnical Journal, Vol. 33, No. 1, pp. 106-122.

Rowe, R.K. and Soderman, K.L. (1987). Very soft soil stabilization using high strength geotextiles: The role of finite element analysis. International Journal of Geotextiles and Geomembranes, Vol. 6, pp. 53-81.

Rowe, R.K., MacLean, M.D. and Barsvary, A.K. (1984). The observed behaviour of a geotextile reinforced embankment constructed on peat. Canadian Geotechnical Journal, Vol. 21, No. 2, pp. 289-304.

Segrestin, P. and Bastick, M. (1996). Comparative study and measurement of the pull-out capacity of extensible and inextensible reinforcements. Proc. of the International Symposium on Earth Reinforcement, Edited by H. Ochiai, N. Yasufuku and K. Omine, A.A. Balkema, Rotterdam, 1996, Vol. 1, pp. 139-144.

Smith, A., Jailloux, J.M. and Segrestin, P. (1996). Durability of galvanized steel reinforcements as a function of their shape. Proc. of the International Symposium on Earth Reinforcement, Edited by H. Ochiai, N. Yasufuku and K. Omine, A.A. Balkema, Rotterdam, 1996, Vol. 1, pp. 151-156.

Tajiri, N., Sasaki, H., Nishimura, J., Ochiai, Y. and Dobashi, K. (1996). Full-scale failure experiments of geotextile-reinforced soil walls with different facings. Proc. of the International Symposium on Earth Reinforcement, Edited by

H. Ochiai, N. Yasufuku and K. Omine, A.A. Balkema, Rotterdam, 1996, Vol. 1, pp. 525-530.

Tatsuoka, F., Koseki, J. and Tateyama, M. (1996). Performance of reinforced soil structures during the 1995 Hyogo-ken Nanbu Earthquake. Proc. of the International Symposium on Earth Reinforcement, Edited by H. Ochiai, N. Yasufuku and K. Omine, A.A. Balkema, Rotterdam, 1996, Vol. 2.

Tsukamoto, Y., Ishihara, K., Higuchi, T. and Aoki, H. (1996). Active earth pressures on walls retaining geogrid-reinforced soil. Proc. of the International Symposium on Earth Reinforcement, Edited by H. Ochiai, N. Yasufuku and K. Omine, A.A. Balkema, Rotterdam, 1996, Vol. 1, pp. 531-536.

Voskamp, Summary Discussion Session, Panel Discussion, Proc. of the International Symposium on Earth Reinforcement, Edited by H. Ochiai, N. Yasufuku and K. Omine, A.A. Balkema, Rotterdam, 1996, Vol. 2.

White, D.M. and Holtz, R.D. (1996). Performance of geosynthetic-reinforced slopes and walls during the Northridge, California Earthquake of January 17, 1994. Proc. of the International Symposium on Earth Reinforcement, Edited by H. Ochiai, N. Yasufuku and K. Omine, A.A. Balkema, Rotterdam, 1996, Vol. 2.

Wilmers, W. (1996). Installation of geosynthetics in road construction - Possibilities, problems and solutions. In Geosynthetics: Applications, Design and Conservation. M.B. de Groot, G. den Hoedt, R.J. Termaat, Editors, A.A. Balkema, Rotterdam, pp. 151-152.

Wu, J. (1992). Editor: Geosynthetic Reinforced Soil Retaining Walls. A.A. Balkema, Rotterdam.

Wu, J.T.H. and Helwany, S.M.B. (1996). A performance test for assessment of long-term creep behavior of soil-geosynthetic composites. Geosynthetics International, Vol. 3, No. 1, pp. 107-124.

Earth Reinforcement, Ochiai, Yasufuku & Omine (eds) © 1997 Balkema, Rotterdam, ISBN 90 5410 833 9

The monitoring of soil reinforcements' durability: Thoughts about experience and needs

Pierre Segrestin
Terre Armée Internationale, France

ABSTRACT: Galvanized steel strips have been monitored in many real structures for 20 years, positively confirming the safety of the design procedures. This paper explains why the same requirements should also apply to steel bar mats and geosynthetics. The other topics discussed relate to the carefulness which is required when using some extrapolation methods, the relevance of creep rupture strength with regard to the allowable tensile load and why this sends back to the definition of factors of safety.

1. SUMMARY

This panelist's report is essentially about the monitoring and the prediction of the long-term performance of soil reinforcements. It is primarily based on the experience and the practice of the Terre Armee Internationale (TAI) Group of Companies and on its ongoing research.

In the first section, we will summarize how the TAI Group validates laboratory studies in the field, that is, we will present the current practice concerning the field monitoring of steel strip reinforcements. This will lead us to discuss the need for similar product-specific field testing, especially for steel grids and geosynthetic reinforcements.

In the second section, we will discuss a couple of issues regarding the validation of extrapolation theories, as well as the need for testing the characteristics of the reinforcing products which are actually relevant for design.

2. FIELD MONITORING OF GALVANIZED STEEL STRIPS

We will first describe the procedure which is used to monitor the long-term behaviour of galvanized steel strips in actual structures, and then summarize the main findings after more than 20 years of extensive experimentation.

2.1 *Monitoring procedure*

2.1.1 Purpose

The monitoring procedure consists of installing strip coupons in walls and extracting them periodically. This is at least advised for major structures and is now specified in several countries. The extraction of the coupons is usually scheduled to take place through the facing after 10, 30, 50 and 70 years.

The main purpose is to give the owner the option of making sure that everything is evolving more slowly than what was assumed for design. Should anything abnormal be revealed, enough time would be available to carry out further investigations and to decide about appropriate corrective measures. Moreover, the installation and retrieval of such strip coupons gives an opportunity to augment the data base and to further validate the theory.

2.1.2 Installation

The strip coupons are of the same type as the reinforcing strips used in the structure. Five coupons, each one metre long, are cut from the same strip, with four installed in the wall. The fifth one is used for future reference, and its breaking load and the thickness of the zinc coating are measured at the time of construction.

All strips coupons are identified, weighed and labeled before they are placed in the backfill of the

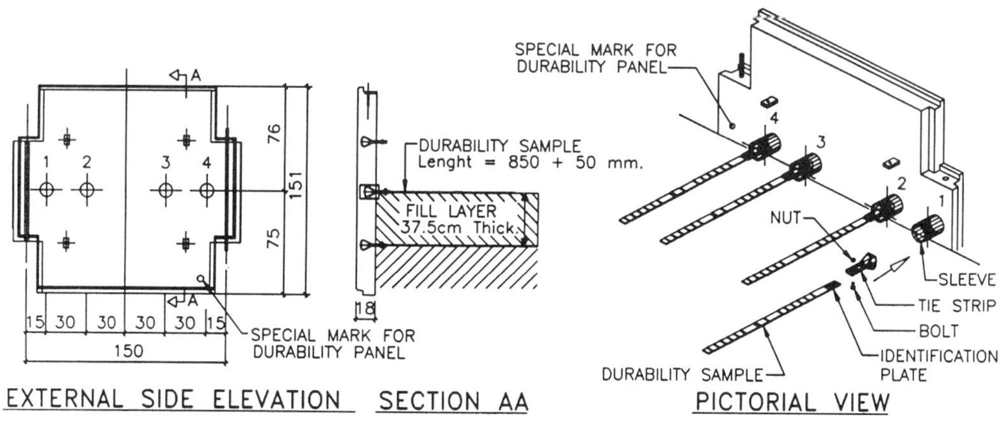

Figure 1: Installation of durability samples

Reinforced Earth structure, at the same time as the normal strips. They are equipped with tie-strips meant for future extraction and are placed perpendicular to the facing, behind sleeves made in special panels (Figure 1). These panels, located in different zones of the wall, have a discreet mark in one of their corners.

Sixteen durability samples, behind four special panels, are usually placed in a same structure, and their locations in the wall are recorded.

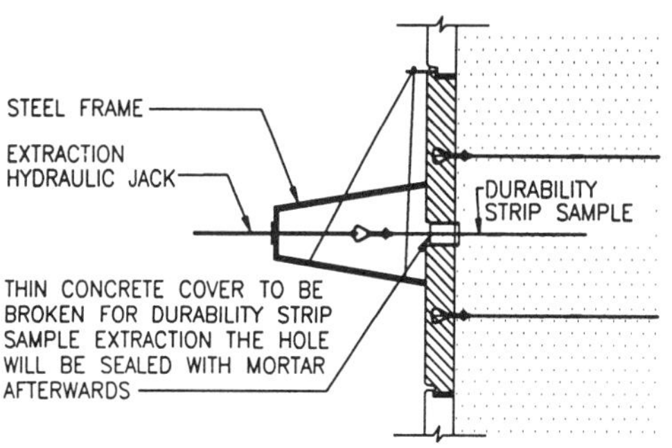

Figure 2: Extraction device

2.1.3 Retrieval

At the time of retrieval, the strip coupons are pulled-out through the facing (after breaking the concrete plug) with the help of a light extraction device placed against the panel (Figure 2). Usually four coupons are retrieved from a particular structure on the same date, one from each zone where test samples are located.

A sample of backfill is also taken in the vicinity of each test sample extracted. Care is taken to ensure that it is representative of the core of the backfill. A determination of its electro-chemical characteristics (pH, resistivity, chloride and sulfate content) is made, using appropriate testing procedure.

2.1.4 Analysis

Once extracted, the durability test samples are the subject, in the laboratory, of:

- a visual examination before and after the removal of the corrosion by-products
- the measurement of the loss of weight, equivalent to an average loss of metal thickness (zinc and steel)
- the measurement of the tensile strength. Since the initial resistance is known, the loss of resistance can be determined and correlated with the loss of steel thickness.

All manipulations, including the way the samples must be brushed, cleaned of corrosion by-products, rinsed etc... and the ways the measurements must be made and the results recorded, are described in a detailed manual.

2.2 *Findings*

Such samples have now been extracted from more than 100 walls, built with moderately aggressive backfills complying with the standards. Two thirds of these walls are 10 to 22 years old.

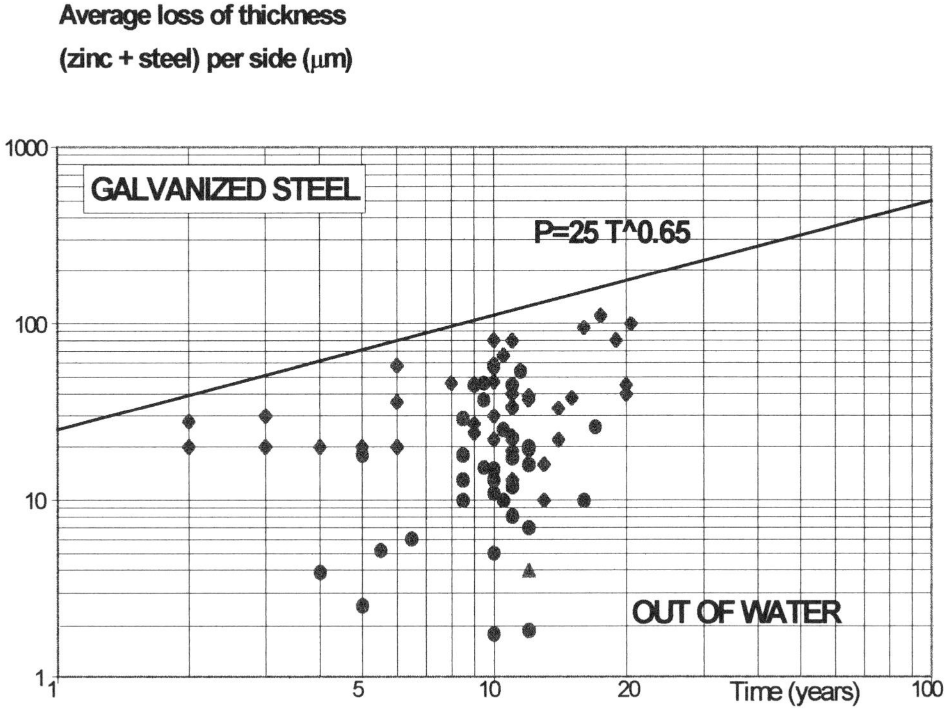

Figure 3: Results of monitoring of galvanized steel strips in actual structures

2.2.1 Loss of thickness

As the graph shows (Figure 3), in all cases the loss of thickness, represented by the dots, is well below the theoretical loss. The actual rate of corrosion is often 5 to 10 times slower.

The straight line envelope (plotted on a log-log scale) was drawn to enclose all results obtained in the laboratory, over 20 years, both from test boxes and from over 200 corrosion cells (these based on electro-chemistry techniques). The envelope, and the results shown here, represent only data obtained in soils for which the electro-chemical characteristics meet the current specifications for fill materials used in dry-land structures.

2.2.2 Loss of strength

For the coupons where steel is most corroded, the relative loss of tensile strength is found to be less than 1.7 times the relative loss of steel weight, again below the current assumptions for design.

3. FIELD MONITORING OF STEEL BAR MATS

3.1 *A product-specific issue*

We think that there is a need for collecting the same kind of information for other types of reinforcements, first for galvanized steel grids or welded bar mats made of small round bars.

It must be acknowledged that the correlation between the relative loss of strength and the relative loss of weight and, subsequently, the sacrificial thicknesses used for design, were specifically developed for flat thin strips, based on the research conducted by Terre Armee Internationale. This correlation does not apply in like manner to the round bars which are used in the mats and grids later introduced by other wall suppliers.

3.2 *The effect of local pitting*

An analysis specific to bar mats still needs to be developed. It should primarily address the question of superficial pitting. The fact is that the loss of strength is affected by local pitting, although pitting is much smaller with galvanized steel than it is with black steel.

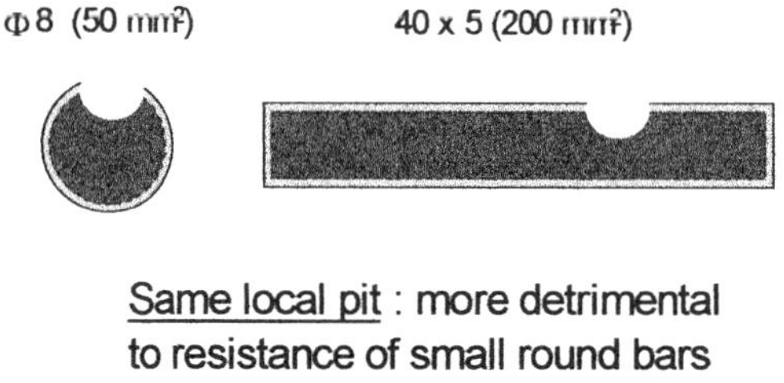

Figure 4: Effect of superficial pitting on the resistance of small round bars

Figure 4 clearly shows why the tensile strength of a small round bar such as those currently used in bar mat reinforcements, possibly 4 times smaller than a current flat strip, can be more significantly affected by a given local pit. Even if the loss of average thickness is likely to be about the same, the tensile strength of bar mats and grids made of small bars, about 8 or 10 mm in diameter, is expected to decrease more quickly than the resistance of strips about 5mm thick. This should logically result in larger sacrificial thicknesses for design for these grids.

The issue is already roughed-out, although only from a theoretical point of view, in a paper titled "Durability of galvanized steel reinforcements as a function of their shape," prepared for the Kyushu'96 Symposium by A. Smith, J.-M. Jailloux and P. Segrestin (page 151).

3.3 *The need for experimental data*

There is a clear need for more studies, which should be backed up by actual data, obtained from the monitoring of samples of grids and mats retrieved from a large enough number of existing structures.

4. FIELD MONITORING OF GEOSYNTHETIC REINFORCEMENTS

4.1 *Reasons for the monitoring*

We believe that the same policy of installing test samples in most structures should also apply to geosynthetic reinforcements, at least for structures designed for a service life in excess of 25 years. The main reason would again be to give the owner a possibility of controlling the condition of his wall.

There are important additional reasons to install test samples.

- For one, samples would permit the progressive obtaining of confirmation that the various reduction factors used in the design are still valid and conservative in the actual environment. In addition, sampling could give some insight into the question of the potential synergy among the main causes of loss of strength: construction damage, creep and environmental aging.

- Contrary to what is done with the steel coupons, however, the first retrieval of geosynthetic samples should take place right after the construction, so that the effects of the construction damages can be immediately assessed.

4.2 *Recommended procedure*

The European Standards Organization (CEN, Technical Committee 189, WG 5) is presently working on a series of standards dealing with the durability of geotextiles and geotextile-related products. It will include a standard about the "installation, on-site retrieval and laboratoy testing of samples," which will specify how such a procedure should be implemented and how the monitoring and the analysis should be conducted.

5. CAREFULNESS REQUIRED WITH EXTRAPOLATIONS

This new section will be about the limitations of the knowledge concerning the extrapolation theories dealing with chemical degradation. For this discussion, we will refer to the TAI Group's research regarding polyester-based reinforcements.

5.1 *Summary of TAI's research programme*

Our research programme about the hydrolysis of polyester is a long-term one, divided into three phases. The first phase, conducted at 95°C in 27 different media, aimed at identifying the most aggressive environments. The second phase, at 80°C, aimed at differentiating the performances of various fibers.

The third phase is now underway and it will last until about year 2010! It is much slower, since it is conducted at 50° and 23°C. This time, the main purpose is to analyze the kinetics of hydrolysis in neutral and aggressive environments by correlating the results obtained at the 4 different temperatures.

5.2 *Third phase: early findings*

No definite conclusions can be drawn yet, after only two years of the third phase, but very alkaline environments (pH = 13) confirm to be extremely aggressive, with 40% loss at 23°C.

More important with regard to our discussion, however, it appears that, at least in a very alkaline environment, the Arrhenius equation which is usually applied in order to extrapolate the results obtained at elevated temperatures does not work very well. For the results to fit, the activation energy would have to be reduced from about 25,000 cal/mole to 15,000 cal/mole below the glass transition temperature (65° C). This is explained in more details by J.M. Jailloux and P. Anderson in a paper titled "Long term testing of polyester yarn and product at 50° C and 23° C in different environments," Kyushu'96, page 45.

This might be an example of the carefulness which is required before conducting such extrapolations over either large ranges of temperatures or from only one year (or 10^4 hours) to 100 years!

6. SELECTION OF RELEVANT PARAMETERS

6.1 *Example of creep*

This last section is about the need for identifying the parameters which are actually relevant to the design, the ones for which testing methods are actually necessary. The effect of creep is one of the most critical examples.

6.2 *Irrelevance of creep strain*

It should be noted that some guidelines still define the allowable tensile strength of extensible soil reinforcements which are subject to creep, by referring to a given limit strain, for example 10% or 5% (Figure 5). This could be relevant to a serviceability criteria, but it is not relevant to an ultimate limit state criteria.

6.3 *The real issue: creep rupture strength*

With regard to ultimate limit state criteria, the question is, by definition: "Is the reinforcement susceptible to breaking before the end of the service life?" and the answer should be no. Then, rupture is the issue, and "creep rupture strength" is the only relevant creep criterion.

6.4 *The meaning of the "overall factor of safety"*

The remainder of this section is primarily directed at the people who are not too familiar with the use of "load factors" as partial factors of safety, but are used to "working stress" types of calculations (as it is generally the case in the USA). They should first recognize that there is no fundamental difference between these "load factors" and what they use to call the "overall factor of safety". It is important to understand, however, the real meaning of these factors, when considering the possibility of a rupture before the end of the service life.

The overall factor of safety is meant to cover all the uncertainties in the calculated tensile loads, such as: uncertainties in weights, surcharges, fill properties, potential for local overstresses, and inaccuracy of the calculation model. Its value is usually around 1.5 in all major Codes, independent of the type of reinforcing material. This factor means that there is a minor possibility, or a tiny probability which is accepted in advance, that the nearly constant tensile load could be actually 1.5 times larger than what is normally calculated. Even in this case, which is not totally excluded, the reinforcement must not break prematurely (Figure 6).

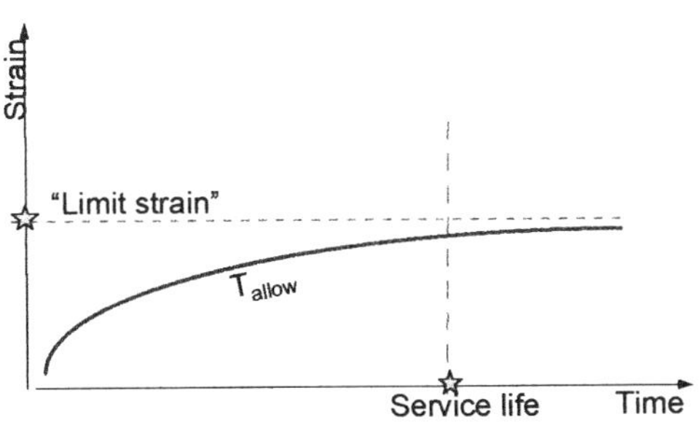

Figure 5: Allowable tensile load defined as a function of a limit strain

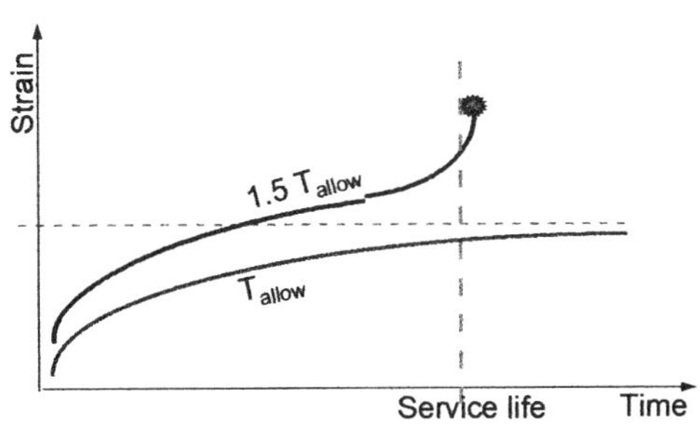

Figure 6: Creep rupture under factored load.

This is why one must assess the creep rupture strength of the reinforcement at the end of the service life and use it for the definition of the allowable tensile load.

No matter how the strength of the reinforcement evolves in the course of its life, even if this reinforcement still retains a significant residual strength a little before it finally breaks, what counts is that it does not break before the end of the service life, even if its load is 1.5 times larger than calculated.

Time to creep rupture is clearly what needs to be measured, over the longest possible period of time, in other words under the smallest possible loads, so that extrapolations remain reasonable. Of course it is desirable that creep rupture strength be measured in conditions which are representative of the confined three dimensional environment of a reinforcement in a real structure.

The question of creep rupture and more generally of the fundamental meanings of safety factors and reduction factors is thoroughly discussed in a paper titled "The need for standard factors of safety in the determination of allowable tensile loads," by P.L. Anderson, M.S. Boyd, P. Segrestin and K. Worrall (Kyushu'96, page 297). The whole issue can be summarized (or at least illustrated) as shown on figure 7. The graph explains how the reduction factors corresponding to installation damage, creep and environmental aging should be first applied. It shows how the factor of safety FS_l covering the uncertainties in loads, and the factor of safety FS_m accounting for the uncertainities in long-term tensile strength and extrapolations, should be then ideally introduced.

7. CONCLUSIONS

Monitoring the long-term behavior of soil reinforcements in actual structures and actual environments is a necessity. This is currently done for galvanized steel strips and it confirms the validity of the theory and of the design procedures. The monitoring needs to be implemented on an ordered basis for the other types of reinforcements: steel grids and geosynthetics. This should progressively help to improve our knowledge regarding issues such as the durability of small round bars, the synergy among construction damage, environmental aging and creep, and the validity of some extrapolations.

It is also necessary to state which mechanical characteristics, such as creep rupture strength, are actually relevant to the design and, therefore, need to be measured. Such clarification will eventually eliminate the confusion about the issue of safety factors.

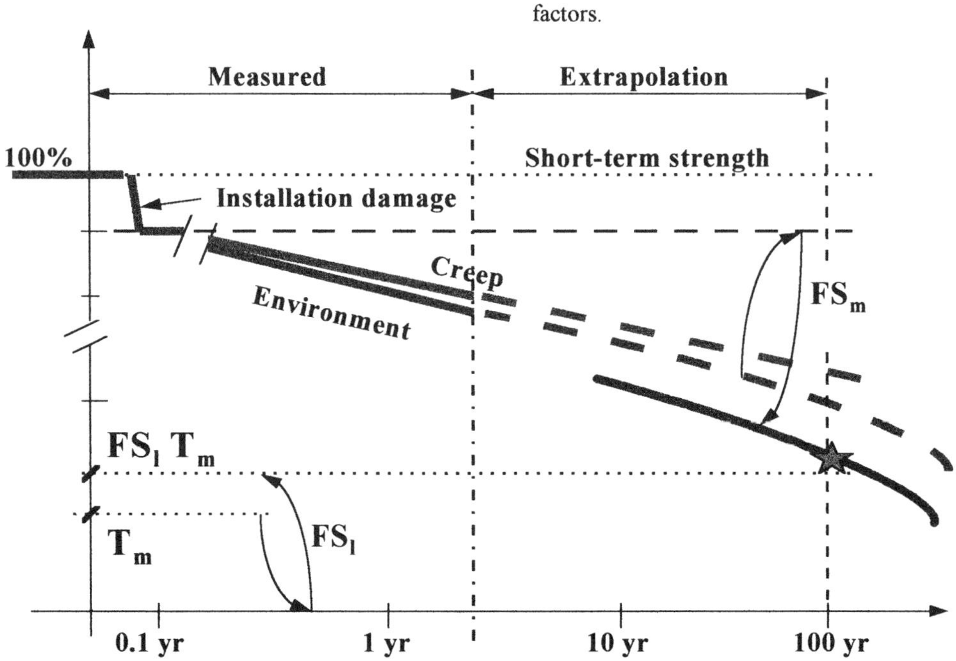

Figure 7: Allowable tensile load; applying reduction factors and safety factors

REFERENCES

Smith, A.; Jailloux, J.-M.; Segrestin, P.: Durability of galvanized steel reinforcements as a function of their shape. *Earth Reinforcement practice*. Fukuoka 1996.

Jailloux, J.-M.; Anderson, P.: Long-term testing of polyester yarn and product at 50°C and 23°C in different environments. *Earth Reinforcement practice*. Fukuoka 1996.

Anderson, P.; Boyd, M.; Segrestin, P.; Worrall, K.: The need for standard safety factors in the determination of allowable tensile loads. *Earth Reinforcement practice*. Fukuoka 1996.

Earth Reinforcement, Ochiai, Yasufuku & Omine (eds) © 1997 Balkema, Rotterdam, ISBN 90 5410 833 9

Proposed method to determine the safety capacity of reinforced soil structures during the lifetime

W.Voskamp
Akzo Nobel Geosynthetics, Netherlands

ABSTRACT: The method to calculate the design strength of geogrids as outlined in BS 8006 is described. Partial factors are used to reach the factored design strength. As the circumstances which lead to reductions in the strength can be sometimes time-dependent, a considerable additional safety margin is available in the design strength; this safety margin is not earlier quantified. In this paper a method is proposed to determine the extra safety in an earlier built structure (safety capacity). This method, based on the analysis of the creep-strain history of the structure, would allow to determine the expected additional service life of the reinforcement when the load in the reinforcement is not changed or to determine the additional load that could be allowed in the structure in relation to the lifetime.

1 INTRODUCTION

Calculation of the design strength according to BS 8006. The design strength of soil reinforcement material should be higher than the load in that material (multiplied with load factors).

This design strength is based on the long term characteristics of the material and further reductions of strength which could take place during the service life are introduced as partial reduction factors.

1.1 *Unfactored design strength*

According to the British standard BS 8006, the design strength of soil reinforcement should be based on the strength to prevail at the end of the design life. The unfactored strength of reinforcement = T_B. This design strength may be governed by considerations of serviceability or tensile creep rupture.

T_{CR} is the extrapolated creep rupture strength at the end of the selected design life and at maximum operational temperature.

T_{CS} is the extrapolated tensile load which gives rise to a maximum creep strain.

The unfactored design strength T_B is reduced by the reinforcement material factor f_m, resulting in the reinforcement design strength $T_D = T_B / f_m$.

At any instant of time during the life time of the structure, the factored design strength T_D should equal or exceed the design load.

1.2 *Partial factors*

f_m is the partial material factor for reinforcement and has two components.

$$f_m = f_{m1} * f_{m2} \tag{1}$$

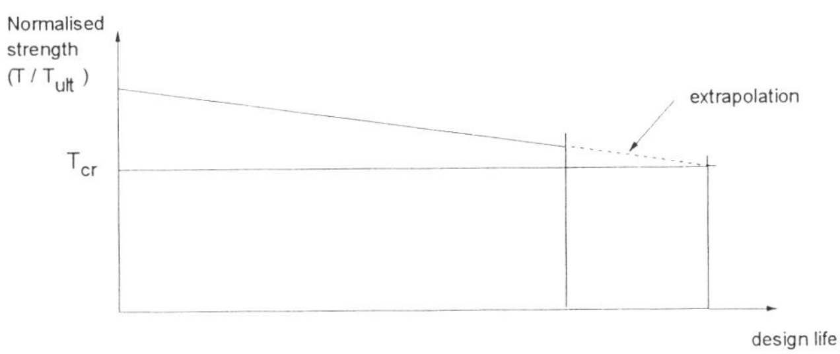

Figure 1 Stress-rupture line.

where f_{m1} = a partial material factor related to intrinsic properties of the material; f_{m2} = a partial material factor concerned with construction and environmental effects.

f_{m1} has two components:

$$f_{m1} = f_{m11} * f_{m12} \tag{2}$$

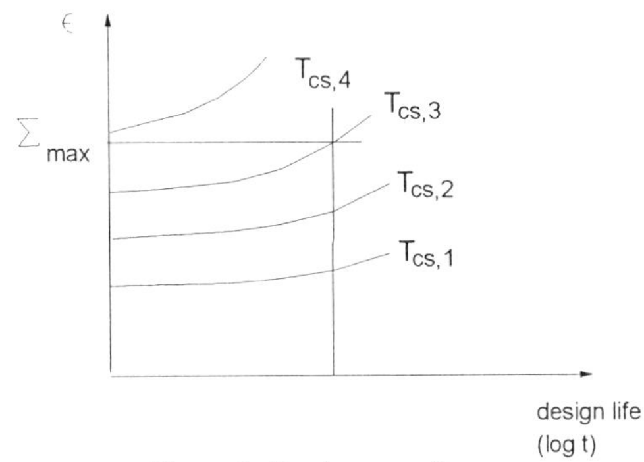

Figure 2 Isochronous lines.

where f_{m11} = a partial material factor related to the consistency of manufacture of the reinforcement and how strength may be affected by this and possible inaccuracy in assessment; f_{m12} = a partial factor related to the extrapolation of test data dealing with the base strength.

The value of f_{m1} depends on the availability of test data and accuracy of test data and quality control standards. It is a fixed value which reduces the design strength and is of the overall safety factor type. This factor results in a lower loading condition of the material during lifetime. In the following analysis we call this a type A behavior.

$$f_{m2} = f_{m21} * f_{m22} \tag{3}$$

where f_{m21} = a partial material factor related to the susceptibility of reinforcement to damage during installation in the ground; f_{m22} = a partial material factor related to the environment in which the reinforcement is installed.

The value of f_{m21} is determined based on the results of tests under actual construction conditions. The

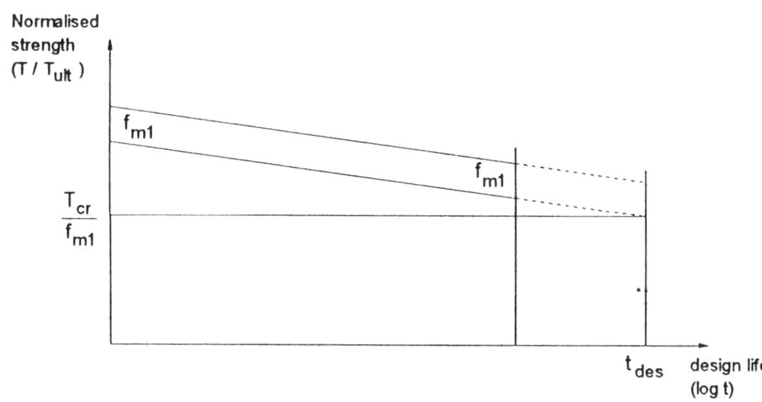

Figure 3 Stress-rupture line with calculated reduction due to f_{m1}.

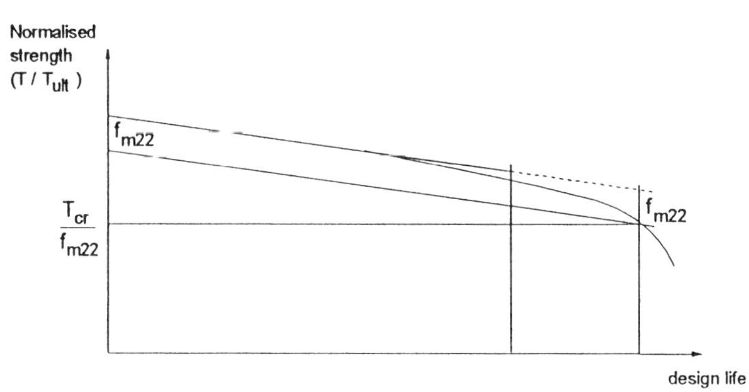

Figure 4 Stress-rupture line with calculated reduction due to f_{m2}.

conditions and test results are categorized and values fixed. These fixed values reduce the design strength and are of the overall safety factor type, (type A). f_{m22} is based on testing and extrapolation under environmental conditions. The estimated reduction in strength during the service life of the structure is determined and applied as a fixed value as factor. The strength reduction as a result of the environmental conditions is not linear in the time. We call this non linear behavior type B.

Typical examples of these circumstances are hydrolysis of polyester , oxidation of polyolefins, UV attack.

1.3 *Factored design strength*

The factored design strength T_D is:

$$T_D = T_B \;/\; f_m \;=\; T_B \;/\; (f_{m1} * f_{m2}) \tag{4}$$

and is the calculated strength to prevail at the end of the design life.

2 UNDER ESTIMATION OF THE DESIGN STRENGTH

2.1 *Description of model of analysis*

There are two aspects in this calculation which can result in considerable under-estimation of the design strength.

This is the easiest to describe when we use the expression:

$$T_D \;=\; \sigma_S \;*\; A \tag{5}$$

where T_D = the design strength (N); σ_S = the specific strength of the material (N / mm^2); A = the cross-sectional area (mm^2)

When we assume that the specific strength of the material does not change in time, the design strength is directly proportional to the cross-sectional area A of the load bearing elements of the reinforcement.

$$T_D \;=\; f(\,A\,) \tag{6}$$

2.2 *Strength reduction at the begin of the design life*

The partial factors which are fixed values and which are based on the maximum reduction of strength during installation, or accuracy of data etc. (type A), result directly in a larger cross-sectional area of the load bearing elements of the reinforcement. This results in a situation that either the cross-section (A) is over-dimensioned or in the case that all negative factors apply, in an exactly correct value of A. This means that in most cases the cross-sectional area is over-dimensioned, which automatically leads to a lower normalised ($T_D \;/\; T_{Ult}$) stress condition.

2.3 *Strength reduction developing during the service life*

When the environmental effects on the strength increase in time (type B behavior), the cross-sectional area (A) is calculated, based on the available strength at the end of the service life (acc. BS 8006). As the speed of the process increases in time, it is clear that the factored design strength T_D leads to an over-estimated cross-sectional area (A) at the begin of the life time. This over-estimate reduces in time leading to the correct cross-sectional area at the end of the service life of the structure.

The design strength T_D is based on the extrapolated creep-rupture strength at the end of the design life time (or limited by strain requirements). The creep speed and therefore the rupture strength is depending on the normalised strength. So if we over-estimate the cross-sectional area of the load bearing elements and if we over-estimate the loadings in the soil mechanical design by means of load factors, we reduce the normalised strength of the reinforcement considerably. This lower

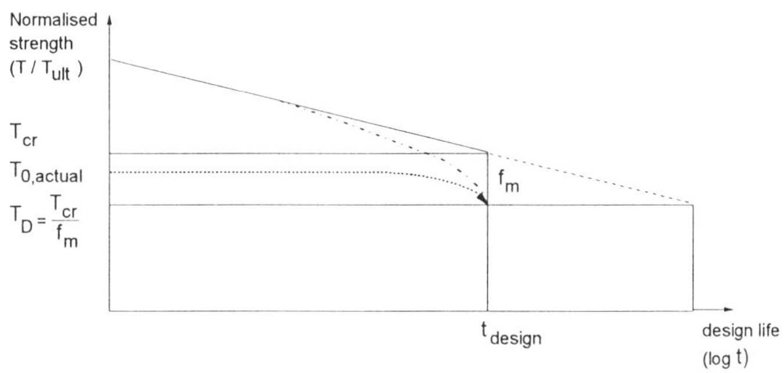

Figure 5 Stress-rupture line with T_D, actual and T_D, theoretical.

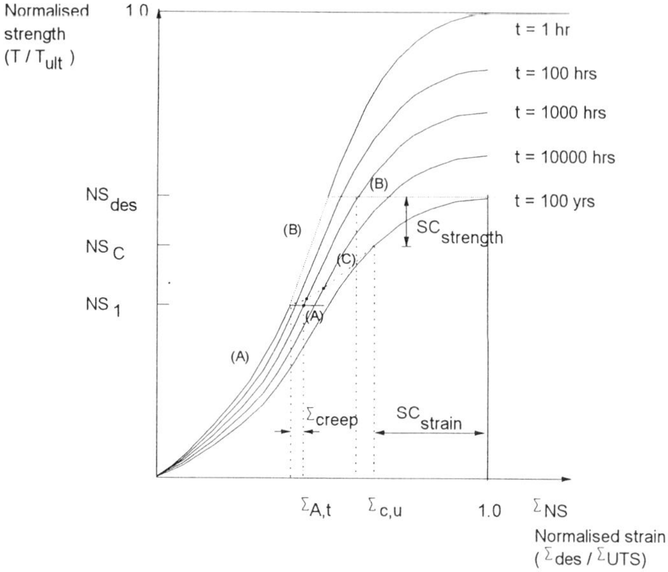

NS_{des} = factored normalised design strength (theoretical); NS_1 = factored normalised design strength (actual, due to over estimation of cross section A); (A) = stress - strain path (actual); (B) = stress - strain path (theoretical, design); (C) = stress - strain path (due to time dependent strength reduction processes)

Figure 6 Isochnous curve with actual stress-strain path.

normalised strength results in lower creep speeds, which result in longer time before rupture takes place. Or it results in under-estimating the rupture strength. This is the main reason why relatively low strains are measured in prototypes under long term loading.

3 PROPOSED METHOD TO DETERMINE THE LONG TERM SAFETY CAPACITY OF A REINFORCED SOIL STRUCTURE

3.1 *Description of method*

The creep strain development in time is shown on isochronous curves. If we use these curves we can determine together with the elapsed time, what the actual stress level has been during the service life of the structure.

The factored design strength results in a stress-strain path indicated with (B). The actual stress-strain path, as result of over-dimensioning, is indicated with (A). This over-dimensioning is the result of the fact that the reduction in strength of the virgin material has not occurred in practice (f_m factor is a "safety factor"). However, we do not know exactly at what normalised strength the material has actually been loaded.

When we remove parts of the reinforcement from a built structure or from loaded test samples, we can execute a tensile strength test. The strength and the strain at failure are found. The strain which is found ($\Sigma_{A,u}$) is lower than the strain at failure (Σ_{UTS}) when virgin material is tested. This is due to the fact that the material has been creeping. The reduction in strain (Σ_{UTS} - $\Sigma_{A,u}$) as indicated in the figure below is the creep (Σ_{creep}) that has taken place. This creep value is determined.

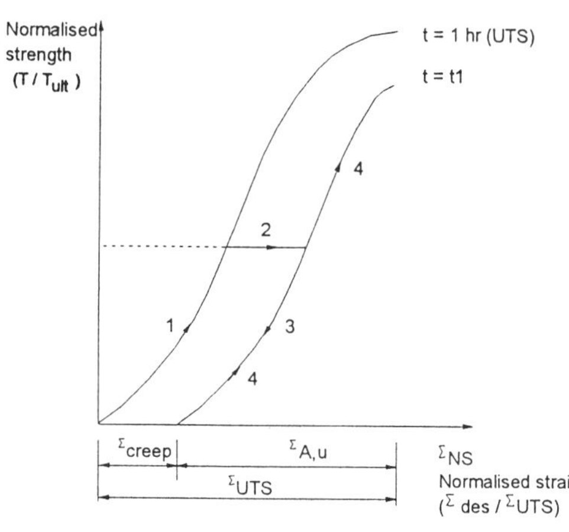

Figure 7 Creep history during loaded situation.

In the normalised stress-strain isochronous curves

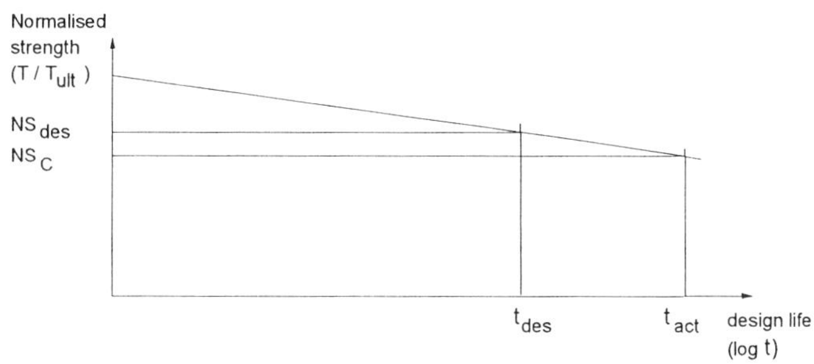

Figure 8 Stress-rupture line, safety capacity in years.

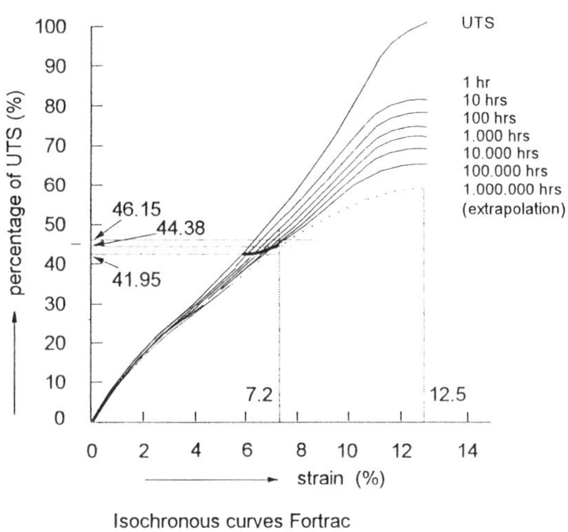

Figure 9 Isochronous curve.

we can indicate this strain. Curve (A) is the stress-strain path for the material loaded at actual load level (including f_m). Curve (B) is the stress-strain path for the material loaded at theoretical factored design level (including f_m). The creep value is measured from the t = 1 line (UTS) to the $t = t_l$ line with t_l = time over which loading has taken place. In this way the actual stress-strain condition point is found. The normalised strength for this point can be found (NS).

When a time dependent process takes place, the strain increases in time. This is indicated with the stress-strain path (C). It will now be harder to determine the strain at the end of the service life. However when more than 3 points are determined and when the theoretical description of the chemical process is known, the curve (C) can be calculated. Where (C) crosses the isochronous curve for the end of the service life, the strain at that point will be the actual strain at the end of the service life of the structure.

The actual safety capacity in strength in the reinforcement at the end of the service life of the structure is:

(Safety capacity is defined as the actual Factor of Safety available in the material as result of the multiplication of the various partial factors, while actually this factor should be 1.0, as the calculations are made in a limit equilibrium condition and load factors are applied on the design)

$$SC_{strength} = NS_{des} \;/\; NS_C \tag{7}$$

The safety capacity in time can be determined with the stress-rupture curve. The lifetime t_{act} is determined for NS_C as normalised strength value.

$$SC_{time} = t_{act} \;/\; t_{des} \tag{8}$$

The safety capacity in strain can be determined with

$$SC_{strain} = \epsilon_{des} \;/\; \epsilon_C \tag{9}$$

In case the loading of a reinforced structure will change in the future, the safety capacity of the structure can be calculated, by calculating the creep that has taken place during the service life using isochronous curves. Based on the isochronous graphs it can then be calculated what increase in loading is allowed.

With this method the use of reinforcement can be optimized. Considerable reductions in materials can be reached when sufficient data is collected in the coming years as the combined effects of the various conditions that influence the reinforcement strength can be determined.

3.2 *Example*

When we assume that the material is over-dimensioned with a factor of 1.3 and when we assume that a factor of 1.1 for a time dependent process is applied, it can be calculated that for a typical PET material the available safety capacity for strength is:

$$SC_{strength} = 60 \;/\; 44.38 = 1.35. \tag{10}$$

The safety capacity for strain is:

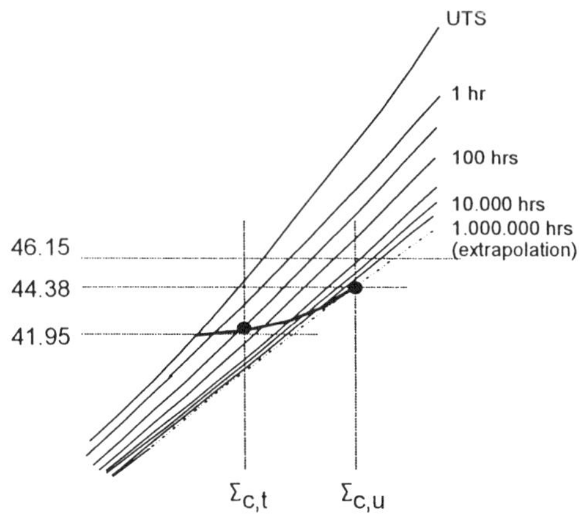

Figure 10 Detail isochronous curve of Figure 9.

$$SC_{strain} = 12.5 / 7.2 = 1.73. \quad (11)$$

The safety capacity in service life is:

$$SC_{service\ life} = 10^{16} / 10^{8} = 10^{8} \text{ hrs.} \quad (12)$$

When the loading factors which apply in the geotechnical design are taken into account the above mentioned safety capacity values will be even much more resulting in a further decrease in creep and consequently resulting in an additional safety capacity.

4 CONCLUSION AND FURTHER RESEARCH WORK

As shown above this method gives the opportunity to analyze the actual loads in the reinforcement based on the creep history of the material. With this back calculation method it is possible to get a reasonable accurate value of the load in the reinforcement.

The effects of the extra safety due to the over estimation in cross section of the reinforcement can be analyzed in the same way. At Akzo Nobel Geosynthetics research laboratories, a number of long term tests are being executed, some test points already more than 6 years. Some of these test samples will be used for validation of this method.

The author intends to verify this theory further by additional testing and detailed analysis of already available (long term) test data. When the initial findings and the method are confirmed it will form an excellent base to determine the safety capacity of already constructed reinforced soil structures. It can eliminate the uncertainty, leading to unnecessary high safety factors in the design of (semi) permanent structures.

REFERENCES

BS 8006 1996, *Code of Practice for strengthened / reinforced soils and oter fills*, BSI, U.K.

Jewell R.A., J.H. Greenwood 1988, Long term strength and safety in steep soil slopes, reinforced by polymer materials, *Geotextiles and Geomembranes* 7: 81-85.

Earth Reinforcement, Ochiai, Yasufuku & Omine (eds) © *1997 Balkema, Rotterdam, ISBN 90 5410 833 9*

Residual strength: An alternative to stress-rupture for earth reinforcement design

J.H.Greenwood
ERA Technology Ltd, Leatherhead, UK

ABSTRACT: Current procedures for applying partial safety factors to a design strength based on the extrapolated stress-rupture curve lead to an unnecessarily high factor of safety. The use of residual strength curves would lead to the definition of a more realistic design strength. Strength reductions due to mechanical damage and environmental degradation can be incorporated into the procedure.

1 DEFINITION OF UNFACTORED STRENGTH

According to the British Standard BS 8006 the design strength of soil reinforcement should be based on the strength remaining at the end of the design life. The unfactored strength of reinforcement T_B which is the lower of the two strengths T_{CR}, which is governed by limit strain, and T_{CS}, governed by serviceability.

T_{CR} is derived from materials properties as follows. The stress-rupture curve defines the time to failure t_f if a load T is applied continuously to the geosynthetic. The curve is conventionally plotted as applied load (or its logarithm) against the logarithm of time to failure (Fig. 1). If a design life t_d is specified, it is then possible to read off an unfactored strength T_{CR} which would lead to fracture at the end of the design life at the maximum operating temperature (Jewell and Greenwood 1988).

T_{CS} is the extrapolated tensile load which gives rise to a maximum acceptable creep strain ε_{max} at time t_d.

The unfactored strength T_B is reduced by the material factor of safety f_m, resulting in the reinforcement design strength $T_D = T_B / f_m$. At any instant of time during the lifetime of the structure the design load should be less than or equal to T_D. f_m is the product of further partial factors including f_{m21}, which reflects the susceptibility of the reinforcement to damage during installation in the ground in the process of construction, and f_{m22}, related to the environment in which the reinforcement is installed, including the effects of ultraviolet light, the hydrolysis of polyester and the oxidation of polyolefins. f_m is a single factor which is applied to the unfactored design strength independently of time.

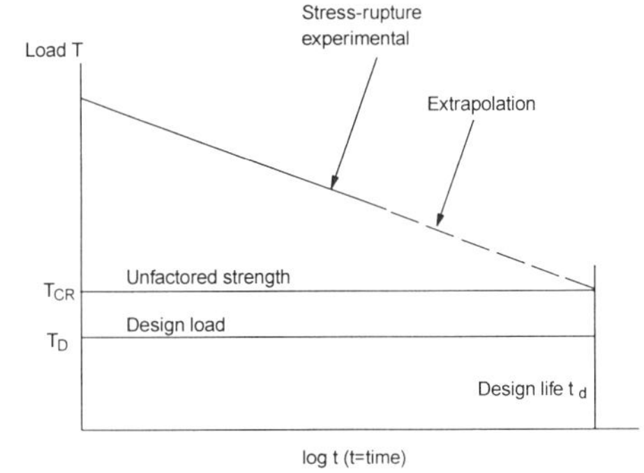

Fig. 1 Stress-rupture diagram

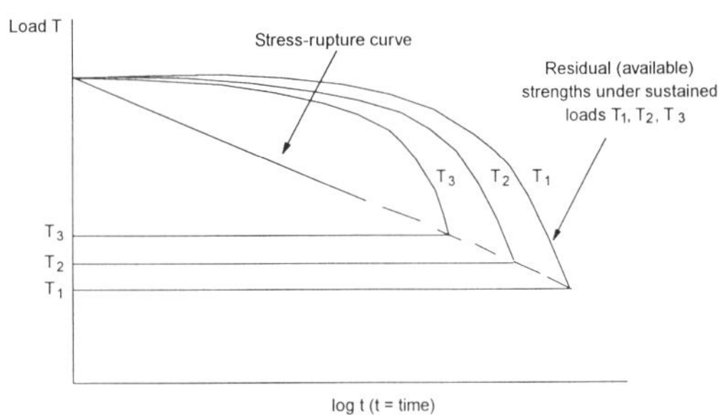

Fig. 2 Residual strengths as a function of time

2 UNDERESTIMATE OF T_D

It is the purpose of this contribution to point out that certain aspects of this procedure lead to overdimensioning due to an underestimate of T_D.

The design load should reflect the ratio between the applied load and the available strength. In spite of its appearance, the stress-rupture curve does not depict the reduction of strength with time in the same manner as, for example, degradation due to exposure to weathering. For each load level T and time t there is an available or residual strength T_R which remains close to the tensile strength of the material for most of the design life, falling to meet the stress-rupture curve at t_f at which point $T = T_B$ (Fig. 2).

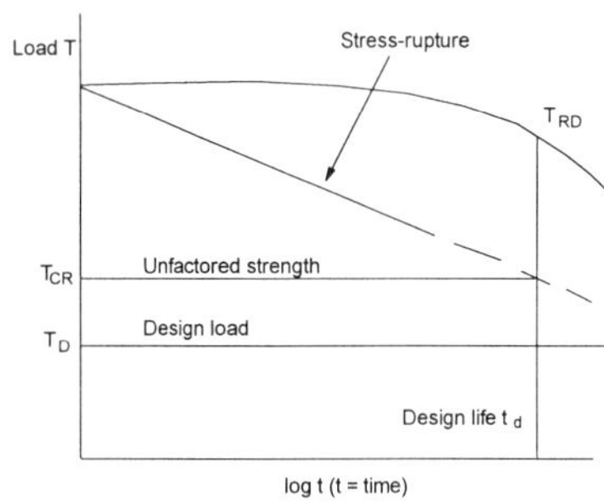

Fig. 3 Definition of T_{RD}, the residual strength at the end of design life t_d under sustained load T_D. $T_D / T_{CR} = f_m$.

If the applied load T is reduced from T_B to T_D, the residual strength T_{RD} at the design life t_d will remain close to the original tensile strength and the effective factor of safety will be increased from T_D / T_B to T_D/T_{RD} (Fig. 3).

This was pointed out by Schardin-Liedtke (1990). As she remarked, however, there is a lack of data on residual strength.

It is possible to derive an analytical expression for the residual strength based on brittle fracture crack growth principles, where it can be shown that if the stress-rupture of geosynthetics is expressed as a power law

$$T = (t/B)^{-n}$$

then for a polyester n is approximately 40 and that this can be associated with a high available or residual strength over most of the design life. Such high indices have little physical meaning, however. Horrocks (1996), working with polypropylene, has noted that stress-rupture should be regarded as a chemical process which may be associated with a long initiation time to set up the conditions for local rupture. This would leave the residual strength unchanged until shortly before failure.

3 REVISED DESIGN PROCEDURE

3.1 *Residual strength due to sustained load alone*

It is proposed that instead of using the stress-rupture curve directly, the design load could be based on the residual strength curve T_R corresponding to the anticipated design load T_D. In the absence of an analytical function to describe the residual strength curves, an iterative process will be required to derive the load T_D which, at the end of the design life, will maintain a factor of safety f_m between the applied load and the residual or available strength T_{RD} such that $T_D = f_m . T_{RD}$ (Fig. 4).

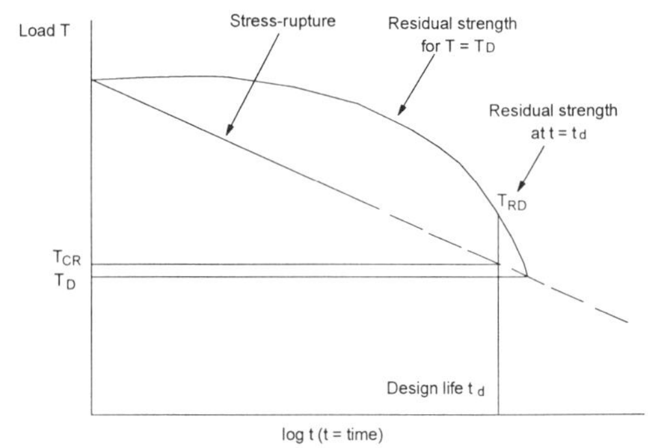

Fig. 4 Redefinition of T_D such that $T_D / T_{RD} = f_m$

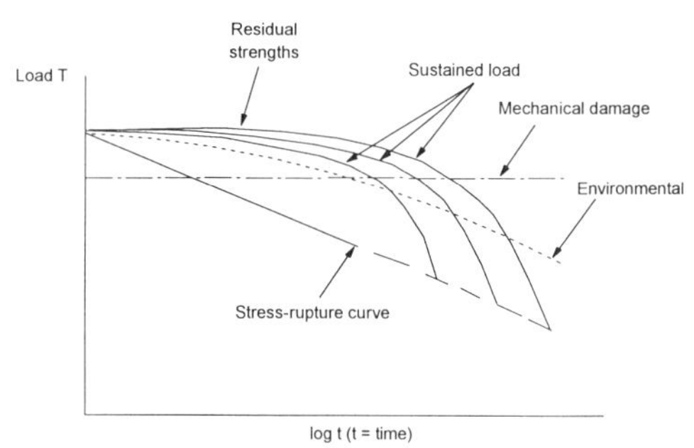

Fig. 5 Residual strengths due to sustained load, mechanical damage and the environment

3.2 *Residual strength due to sustained load, mechanical damage and the environment*

A procedure based on residual strength has the further advantage that it can easily incorporate degradations due to other effects. Fig. 5 shows the residual strength due to sustained load, mechanical damage and environmental effects individually. If all the effects - load, mechanical damage and

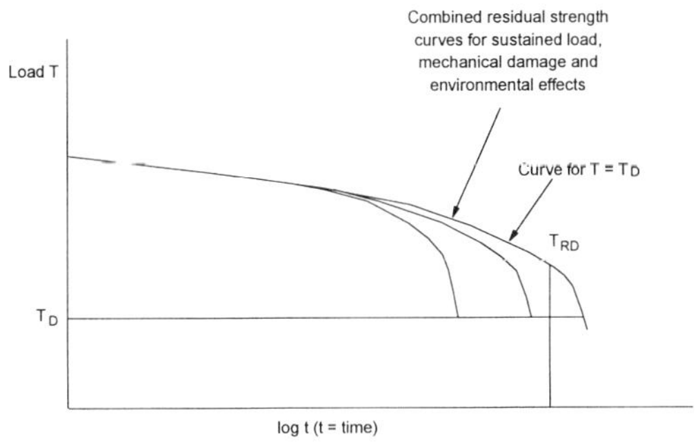

Fig. 6 Combined residual strength curves for sustained load, mechanical damage and the environment; redefinition of T_D such that $T_D / T_{RD} = f_m$

environmental - apply simultaneously but without any synergistic effects such as environmental stress cracking, the degradation curves expressed as fractions of the initial tensile strength may be multiplied to provide a set of reduced residual strength curves as shown in Fig. 6. The design load T_D may then be calculated such that, at time t_d, $T_D/T_{RD} = f_m$ in the same manner as in Fig. 4 but using the reduced curves.

ACKNOWLEDGEMENT

J H Greenwood thanks the Directors of ERA Technology for permission to publish.

REFERENCES

BS 8006: 1995. Code of practice for strengthened/reinforced soils and other fills. London: British Standards Institution.

Horrocks, A. R. 1996. The effect of stress on geosynthetic durability. In de Groot, M. B. et al. (eds), *Geosynthetics: applications, design and construction*: 629-636. Rotterdam: Balkema.

Jewell, R. A. & J. H. Greenwood 1988. Long-term strength and safety in steep soil slopes reinforced by polymer materials. *Geotextiles and Geomembranes* 7: 81-118.

Schardin-Liedtke, H. 1990. Geotextiles for the support of steep slopes: approval procedure. In G. den Hoedt (ed), *4th International Conference on Geotextiles and Related Products:* 79-85. Rotterdam: Balkema.

Summary discussion 'Design methods'

Earth Reinforcement, Ochiai, Yasufuku & Omine (eds) © 1997 Balkema, Rotterdam, ISBN 90 5410 833 9

Chairman's report: Design methods

Malcolm D. Bolton
Cambridge University Engineering Department, UK

1 THE SCOPE OF THE DESIGNER

A discussion on design methods must inevitably begin with a reconsideration of what we mean by design; there are three contrasting approaches.

Practitioners might like to reduce the scope of design to the basic decisions which must be taken on the materials to select, the manner in which they are to placed and connected, and the form and dimensions of the finished structure. So practitioners would wish to hear of new materials and fresh opportunities.

Research workers might like to focus more on the evidence which is available on the real performance of structures in comparison with the simplified mechanisms which have been developed to assist designers with their calculations. So research workers would wish to hear of new theoretical mechanisms which more closely reflect the truth.

Regulators might like to retain simplified analyses and approach real behaviour through the alternative device of probability theory and factors of safety. So regulators would like to hear of standard measures of acceptability in the face of uncertainty.

This session was organised in an attempt to satisfy all three desires - the practical, the scientific, and the philosophical. The broad range of the word "design" is reflected in the broad range of activities which may seem relevant. Consider the following list of ten things a designer might do:

- refine the project brief in relation to life-costs, completion date, robustness, and visual appeal;
- select the materials and methods of construction;
- dimension the structure using accepted calculation procedures;
- conduct a site investigation;
- requisition appropriate tests on materials;
- judge loads, earthquakes, floods and other site-specific hazards
- evaluate the software used in standard calculation procedures;
- commission expert FE analyses where conditions extend beyond current experience;
- monitor construction, and compare observations with expectations;
- develop a database of the long-term performance of structures in service.

The legal situation will generally dictate that the first five are essential, while the commercial situation may suggest that only well-established companies would entertain the last five (and demand higher fees accordingly). Public clients, and regulators, will be aware that they are taking a risk in permitting non-specialists in small companies to design reinforced soil structures, especially in relation to site-specific hazards. They may nevertheless wish to do this in order to keep design costs low.

The methodology generally chosen to reduce risk to an acceptable level is that of the Code of Practice. In the UK (with which the Chairman is familiar) a Code of Practice, such as the new British Standard BS 8006 for Reinforced Soils, has only implicit legal force unless it is mentioned in contract documents. In the UK system, Codes are quite likely to be incompatible since they are produced by independent committees, so the conflict in approach between BS 8002 Earth Retaining Structures and BS 8006 Reinforced Soils, for example, is not surprising. It may even be seen as valuable, since a monolithic and comprehensive set of interlinking Codes would undermine our perception of uncertainty and risk, and might lead to the collective loss of the professional judgements represented by the last five bullet-points for action which are listed above.

Our understanding is not perfect, and it is both honest and useful to display differences of opinion where they occur. That is what the succeeding discussion is for. It is the responsibility of Government ministries and regulators (or insurers) to assist the engineering community in maintaining a healthy proportion of independent experts and successful companies, who continually test their understanding and extend technology beyond current limits. It might be argued that a free market in design services will keep costs down while an intelligent regulatory mechanism, which recognises the different capacities of individuals and companies, is essential to keep quality up.

The tension between quality, cost and risk is reflected obliquely in discussions on Code requirements. Simple authoritative statements may lead to cheap designs today, but may block progress to cheaper or more effective designs tomorrow, or may even lead to rare but catastrophic failures if they introduce false concepts which conflict with scientific evidence or logic. More honest expressions of alternative possibilities in complex circumstances may lead to gross errors due to confusion on the part of the inexperienced. There is no monopoly of wisdom, either way.

However, research workers must recall that the final objective is the creation of new works. Wonderful analyses may be pregnant with conceptual insight but they will lead nowhere unless that insight is distilled into some simple rules for the guidance of decision-makers.

2 SOIL CHARACTERISATION

Geosynthetics are novel and exciting materials which have contributed most to the development of these continuing symposia in Kyushu. But soil is a complex material too, and it often seems to be neglected. There has been very little discussion about the level of strain at which the strength of soil is required, whereas there has been considerable anxiety about the admissible strain of reinforcement. Figure 1 points up the incompatibility between recording the ultimate tensile strength T_{ult} of steel reinforcement, which may occur at 20% strain, and the peak strength ϕ_{max} of the surrounding soil which may occur at only 1% strain. The ultimate strength of granular soils, which might logically be used with T_{ult} for safety evaluations at large strains, is the critical state friction angle ϕ_{crit}. The larger ϕ_{max} could be relevant to serviceability checks. The exceedence of 1% strains in service, for example, could be prevented by the designer checking that the 1% yield strength T_{yield} would not be exceeded due to earth pressures calculated from ϕ_{max} .

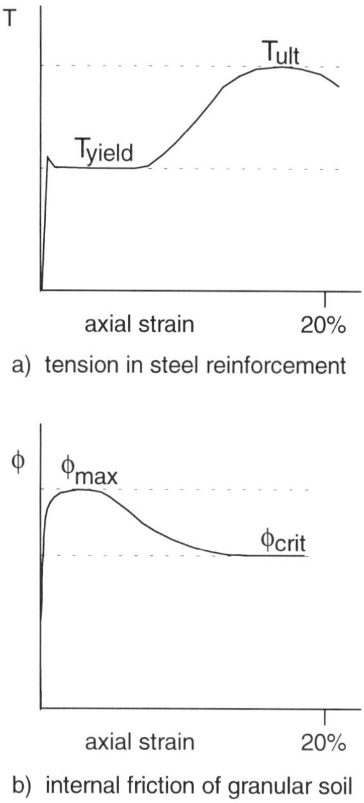

Fig 1 Strength and compatibility

There needs to be a much wider recognition that historic safety factors have had the effect of reducing mobilised angles of friction below the safety net of ϕ_{crit} and, indeed, below the level at which soil strains could be considered excessive. The new BS 8002 took the view that a mobilisation factor M = 1.2 was generally necessary to reduce the representative peak friction angles of moderate to good soils to a value which should not require any linear strain to exceed 0.5%. This limit was placed on the basis that traditional concrete or steel retaining walls would begin to show signs of distress at corresponding face rotations of about 1%. Reinforced soil structures should be more tolerant of deformation, and simple designs might best be calculated on the following basis:

- ${}_{design}\phi = \phi_{crit}$ to ensure safety;
- fill is to be well-compacted to limit strains;
- fill is to be free of fines to ensure drainage
- no further safety factor on earth pressures
- appropriate design strength for reinforcement.

At present, designers arc oftcn astonishcd to find that $\phi_{max} \approx 45°$ in low-stress tests on dense fill. The suggested approach leads them to use ${}_{design}\phi \approx 35°$, corresponding to the internal friction of loose angular fill, but then requires that it be properly compacted. The advantage chiefly lies in the justification of 35°, which shows that no other safety factors are needed.

A similar approach to design can be used with the undrained strength parameter c_u, except that a larger reduction factor of M = 1.4 was recommended in BS 8002 if strains were to be limited to 0.5% as before. The ${}_{design}c$ value of c_u/M can then be used, as before, without further factoring. For example, a well-reinforced block of width B placed on top of a deep clay layer should not settle immediately by much more than B /200 if the bearing pressure were limited to $5.14_{design}c$. Logically, larger M values should be used if tighter control of construction-induced strains is required. However, clayey soils can not escape consolidation or heave due to long-term changes of effective stress.

3 EARTH PRESSURE IS NOT A LOAD

The suggestion, made above, that earth pressure is not simply to be factored up for the purposes of selecting the required reinforcement, is difficult at first for structural engineers to accept. Engineers have become used to treating design live load values as though they were selected from probability-density functions. Stochastic live loads can logically be factored up, so that "collapse loads" in Ultimate Limit States (ULS) will be larger than "working loads" in Serviceability Limit States (SLS). Obviously, engineers would wish to assume more severe loading for ULS checks of very low probability events which have more severe consequences. Earth pressures are not stochastic live loads, however.

As a soil-structure deforms, and strains increase, the mobilised soil strength rises towards its peak and then falls to a critical state. Designs should be based on ${}_{design}\phi = \phi_{crit}$ mobilising critical state strengths, but on the way up to peak strength in Fig 1, not on the way down. In compacted granular fill, this guarantees acceptably small strain. If there were an exceptional live load on the fill, or an earthquake, or a sudden elevation of the water table, or some other loading event which causes the structure some distress, the mobilised soil strength would then increase towards ϕ_{max}.

The following logic applies:

- a safe structure in a working state mobilises an earth pressure coefficient K_{mob} based on ${}_{design}\phi < \phi_{max}$;
- if it is challenged by real live loads (such as a heavy vehicle) its earth pressure coefficient will fall as ${}_{design}\phi \rightarrow \phi_{max}$;
- the actual earth pressure may rise if the live load increases faster than the earth pressure coefficient falls, or if water pressure increases, for example;
- foreseeable live loads must be included in the analyses, and traffic loads may be factored if that is considered appropriate;
- earth pressures will fall from SLS to ULS conditions if the only cause is extension in the reinforcement, due to creep for example.

Estimation of initial earth pressures after construction may require further thought. The earth pressure coefficient behind a rigid face is often said to be $K_o \approx 1 - \sin\phi_{max}$, corresponding to conditions "at rest". It will be found that this is equivalent to mobilising an internal angle of friction $\phi_o = \phi_{max} - 11.5°$. This expression makes it easier to understand the consequences of subsequent lateral strains which erode the 11.5° margin between ϕ_o and ϕ_{max}. However, the K_o condition only applies to gentle pluviation. If the fill has been compacted to density γ by a roller of effective weight Q per unit width, extra lateral pressures up to $\sqrt{(Q\gamma)}$ may be locked in place. These larger pressures are relaxed, as explained above, if the wall face begins to move out.

The consequences of large strains also require some thought. The proposal made here is that the design point lies below ϕ_{crit}. The sudden imposition and removal of a live load, such as might be caused by an accident or explosion, should then leave the structure in equilibrium so long as the reinforcements have not also ruptured, or become disconnected from the facing panels, in the process. If design had been permitted with ${}_{design}\phi > \phi_{crit}$, however, an accidental load cycle could leave the structure generating mass-accelerations due to the imbalance between load and resistance; this would lead to catastrophic and sudden collapse in which lives could be lost.

DESIGN USING MOBILISABLE STRENGTHS

Limit Equilibrium (LE) and Finite Element (FE) analyses may seem to be in opposition, but they can come together most helpfully. LE analyses are based on equilibrium and ignore compatibility, whereas FE analyses take both into account. Failure in an FE analysis is usually recognised when the solver discovers equilibrium errors which it can not relax. At this point, there will be some failure mechanism developing in which strain is accumulating in the body without further resistance being developed. In a reinforced soil structure, some reinforcement will be extending plastically, some may be slipping relative to the neighbouring soil, and some soil zones will also be in some limiting state of stress which induces plastic shear strains, so that peak strength begins to fall towards critical states. In order for the FE analysis of failure to be reasonable, it will therefore be necessary to have a good non-linear solver, interface elements which permit sliding of extensible reinforcement, and a soil constitutive model which features peak strength dropping to critical states.

However, structures are also required to deform acceptably, and this criterion will be more demanding than the final collapse situation in a ductile body. If strains of the order of 1% are considered limiting, it might be assumed that granular fills will have reached peak strength but not yet deteriorated towards critical states, whereas reinforcing materials within the fill, and soft clays beneath, will have mobilised only a fraction of their peak strength. The ratio of the strain mobilised in these compliant materials will be some proportion of that observed in the structure as a whole, expressed perhaps as a displacement / height ratio. These strain ratios can be established for various classes of construction, and proportions of reinforcement, by FE analysis. These results can then be used in practice through LE analyses which account for both the mechanism and the strain ratios.

For example, if it happens that synthetic reinforcement in retaining walls usually extends, at thc critical positions, by 2% in FE analyses when the wall displaces by 1%, the LE analysis can be based on the tension mobilised at 2% strain. Afterwards, the foundation can be checked for equilibrium under soil strengths mobilised at 1% strain. Here, the limit which is being addressed in the LE analyses is the 1% deformation limit (SLS) rather than the collapse limit (ULS). Instead of a "factor of safety", the designer would independently check that fully softened soil would not collapse, and would provide for ductility.

CONCLUSIONS

"Symposium" means coming together. It is time to try to weave together some of the separate strands of thinking about the design of reinforced soils.

Experts can contribute to practical design by transferring ideas from FE to LE analyses in the manner described above, and then by testing this strategy in practice. There is hope that deformations can be accounted for by LE analyses based on mobilisable strengths.

Much more needs to be spoken about ductility. We have participated in a development of a material concept which is widely regarded as capable of absorbing a great deal of coincidental deformation, of behaving plastically in other words. But reinforced soils techniques often seem to based on potentially brittle details, and to neglect the effects of local corrosion on the reduction of overall extensibility, for example. Are we prepared to declare some ductility allowance, such as a minimum 5% extension to rupture following the acquisition of the design working strength? Or do we prefer to admit that some designs are brittle?

Composite materials concepts such anisotropic cohesion, and structural concepts such as anchored bulkheads, also need to be related. Resources seem to be being wasted in walls or steep slopes with compliant facings, in the sense that large earth pressures are not found to act on the facing. The frictional interaction between reinforcements and soil in the "active" zone can certainly lead to the strong reduction of lateral pressure on the facing, as would be well understood by a composite approach. This understanding needs to be transmitted in a simple way to structural engineers who have only learned the bulkhead approach.

Finally, we must draw together the concerns of technology with those of the environment. In particular, we must not set rules which make it unnecessarily hard for designers to achieve "green" faces. Vegetation, and the avoidance of harsh external lines through sculptured faces, are to be welcomed. The use of reinforcement techniques to achieve efficiency in landfill design is also to be welcomed.

Earth Reinforcement, Ochiai, Yasufuku & Omine (eds) © 1997 Balkema, Rotterdam, ISBN 90 5410 833 9

Summary of discussions

Malcolm D. Bolton
Cambridge University Engineering Department, UK

Dr Gassler asked, for the purposes of limit analyses of reinforced soils, which direction of T_{max} is the correct one, the initial reinforcement direction or the direction of the inclined slip surface which later develops. Prof Gourc remarked that the ultimate tension in biaxial tests was generally closer to the shear band direction, but depending on the relative stiffness of the soil. Dr Leschinsky reported that he had published a study in granular soil which showed that the inclination effect was not worth more than about 10% of extra strength when associated changes in the geometry of the slip surface were considered. Dr Gassler expressed some surprise, since he perceived that inclination along the shear band would offer a much bigger lever arm against rotation.

Prof Akagi raised the question of whether there was an optimum amount of deformation of reinforcement materials, or a practical maximum extensibility? Dr Bolton remarked that strength had to be mobilised, but that clients would expect some limit to be placed on service displacements, perhaps 1% of the height of the structure. Prof Jones expected that deformation should be taken into account in determining earth pressure coefficients, K_a for synthetic materials or K_o for metallic reinforcement, but he also noted that any material can be used to create a stiff structure depending on the proportion of reinforcement. Prof Ehrlich spoke of the link between compaction and the reduction of future deformation; compaction acts like a pre-stress. Dr Bolton agreed, and reminded listeners about the generally undeformed state of Reinforced Earth walls after earthquakes, as shown by Prof Tatsuoka earlier; he expected that they remained relatively undamaged because the lateral earth pressure increment of the earthquake was no worse than the original compaction machine. Prof Leschinsky underscored the need for compaction of fill to reduce the vertical shearing at reinforcement connections with wall. Dr Miles pointed out the distinction between reinforcement systems which extend initially during construction, and the long term extension of polymers which creep; he felt designers ought to monitor construction.

Prof Fukuoka drew attention to the strong reduction in soil stiffness and strength which could occur in fine grained soils by water infiltration after construction, leading to unusual displacements if that effect had not been allowed for. He felt that drainage could not always avoid the accumulation of high pore water pressure, so that a better plan was to place a geomembrane to avoid water ingress.

Prof Fukuoka also responded to the Chairman's discussion topic "Why are field tensions so small?" In the case of soft ground and embankments he showed that large tensions were induced in steel bars with steel plates. Prof Gourc said that it was necessary to remember that we have a passive reinforcement technique; it will be very difficult to correlate design values with measurements due to differences between construction methods. He also pointed out that current designs with factors of safety do not solve for the stress distribution in the various components, just the global equilibrium. Dr Bolton remarked that field structures are not necessarily being challenged as the designer assumed in the design scenario, so some component will be under-mobilised and this is likely to be the reinforcement, since the soil must move first. Prof Bathurst noted that boundary conditions are often not allowed for, eg the constraint of footings below the facing of a wall. Dr Lo had used non-linear FE analysis which showed that friction at the base cut tensions in the bottom third of a reinforced soil retaining wall, and noted that suction in compacted soils creates apparent cohesion and reduces also tensions.

Regarding soil nails, Dr Chang showed that there was considerable uncertainty on nail-soil bond, due to unknown radial effective contact stresses after construction, as modified by dilation during shearing as tension develops. There was a need to measure contact stress in some field trials, perhaps using Cambridge contact stress transducers. Dr Gassler doubted whether dilation was practically important since all movement is on the shear band, not around the nails.

Dr Sharma commented on FE analysis for design purposes. He said that there had been much development of PC-based FE packages, which include the ability to model collapse and construction sequence. Although such programs need an expert "driver", universities are producing young engineers who can do the job. Prof Leschinsky noted that design has to be safe and economical, and that reinforcement presently costs 5% to 10% of the overall project cost. If FE analysis is used he felt that overall costs would rise above 10%, so this was not practically applicable, even if it was good in research. Prof Gourc recognised that it was possible to model very complicated geometries, but wondered how to model compaction. Dr Bolton wondered whether we know enough to insert compaction stresses during layer building in FE analyses. Prof Ehrlich said that his paper with Prof Mitchell in April 1994 covered this point. Dr Pokharel warned that axisymmetric and plane-symmetric analyses can be flouted by shear bands; there was a need to model failure without assumed symmetry. Dr Bolton agreed that certain problems appeared to be symmetrical before instability set in, and recalled the Tower of Pisa.

There was a discussion about soil parameters and partial factors used in the UK design Code for reinforced soils. Prof Jones responded to the criticism of the partial factors in BS 8006 which had been made by Dr Bolton in his panel presentation. Any Code, such as BS 8006, had to be used with other Codes of Practice, eg BS 5400 Bridges, so the partial factors have to be compatible between Codes. The partial factors in BS 8006 were chosen to reproduce current designs; they can be changed at some future time within the same framework of ideas laid out in the Code. Dr Bolton had shown a worked example based on a maximum angle of soil internal friction of 45°, but this is not practical since the designer would select a lower value such as 35° since he will not know what material the contractor will obtain. Dr Bolton responded to these three points. If other Codes of Practice are not compatible with common sense, he hoped that they would also be changed. He felt that the factor on density in BS 8006 should certainly be removed, and replaced by a mobilisable strength approach to selecting earth pressures and specifying reinforcement strengths. Dr Bolton pointed out that the designer can specify densification by compaction, and that the UK Retaining Wall Code BS8002 offers advice which will lead to the recognition of the existence of high peak friction angles such as 45°. Prof Jones' advice to designers to select 35° in any event, was exactly Dr Bolton's point: this lower value corresponds to a typical critical state angle of residual strength, which is the design parameter recommended in BS 8002. If a designer makes the mistake of measuring ϕ_{max} and then using it, he could get into a lot of trouble following BS8006. Mr Lawson explained that the logic of BS 8006 was for the partial factors and ϕ_{max} values to go together; he felt that the package was OK as a calculation tool. He was concerned that ϕ_{crit} as basis of design wipes out the incentive for the contractor of using good granular fill. Dr Bolton asked whether that meant that can we utilise cheaper fills? Mr Lawson felt that poor drainage would then become a problem. Dr Bolton recommended that permeation properties then be addressed directly, rather than by hoping that the use of ϕ_{max} would somehow protect you.

Finally, Prof Leschinsky had a comment on creep reduction factors in the design of reinforcement. $T_{allowable}$ is T_{ult} x creep reduction factor x damage factor, and the creep reduction factor is significant. In order to assess this factor we keep sustained load on a strand of reinforcement in isolation in the laboratory. But in real applications, load redistribution will occur, and overstressed layers will relax, as he had demonstrated in a simple experiment. He felt that the creep reduction factor could safely be reduced in its effect.

Earth Reinforcement, Ochiai, Yasufuku & Omine (eds) © 1997 Balkema, Rotterdam, ISBN 90 5410 833 9

Geosynthetic reinforced soil: Evolution of design methods in the USA

Dov Leshchinsky
Civil Engineering Department, University of Delaware, Newark, Del., USA

ABSTRACT: An overview of past and present design methods for geosynthetic reinforced soil structures is discussed. The focus is not on details but rather on difference in assumptions and methodologies. It is concluded that current designs are not more complicated than those used in the past. They are by far less conservative with regard to reinforcement strength. Current designs also use partial safety factors in a rational way. Unlike past designs, current ones are based on a standardized testing techniques to characterize materials properties. Current designs outcome produce safe, and in many cases economical, reinforced soil structures. Simplicity of design methods promotes the application of soil reinforcing and ensures safe structures.

1 INTRODUCTION

It should be stated up-front that the scope of this paper is limited to a few, but typical, design methods currently used in the USA. The details of these methods can be found in the cited references. The writer has chosen to provide just an overview of these methods, including an historical perspective, so that the reader may focus on the current state of the state-of-practice. Hopefully, a judgement about the current state of design methods could be rendered independently by the reader. It should also be stated that the writer tries to be an active researcher; i.e., attempts to come up with solutions to problems and thus possesses a critical view of existing knowledge. However, in writing this paper, the critical view towards existing knowledge was somewhat suppressed. Instead, the perspective of the "real world," where economics plays a major role, was adopted (i.e., the quest for the absolute, but perhaps elusive and sometimes inconsequential, "truth" is intentionally being abandoned).

Reinforced soil structures constructed in the 70's caused excitement among many geotechnical practitioners. These structures were designed based on simple methods, directly extrapolating sound designs of conventional geotechnical structures. Happily, these methods produced conservative, yet economical, structures (e.g., the Glenwood Canyon walls as reported by Bell, Barrett and Ruckman, 1983). One can just wonder what would have happened if these design methods were unconservative (i.e., rendering an occasional failure) or complicated (i.e., difficult to comprehend by practicing engineers). That is, would we have today proliferating reinforced soil structures?

2 OVERVIEW OF PAST METHODS

2.1 Design of Walls

Formal design guidelines were issued by the US Forest (Steward et al., 1977), following recommendations by Bell et al. (1975). The basic approach was a direct adaptation from design guidelines developed for steel reinforced walls as presented by Lee et al. (1973, 1975). Analysis was divided into internal and external stability. The internal stability was used to determine the required tensile strength of the reinforcement, as well as its initial length. At-rest lateral earth pressures were assumed to exist and thus K_0 was used to calculate the reactive tensile force in the reinforcement needed to resist these pressures. Essentially, tieback analysis was carried out by assuming a wedge, inclined at $(45+\phi/2)$, to act as the active (i.e., unstable) soil mass. The soil behind this wedge was assumed to be stable. Therefore, the reinforcement was anchored (tied-back) into this stable soil. This anchorage ensured the capacity of the reinforcement to develop its required tensile force. The overall length of each layer was evaluated by combining the wedge width and the calculated anchorage length at each layer's elevation. The maximum combined length value, considering the lengths of all layers, was used to determine the required *uniform* length for the wall.

External stability was conducted similar to that of conventional gravity walls. The reinforced soil mass (its initial dimensions determined by internal stability analysis) had to resist direct sliding and overturning. It also had to have a sufficient margin of safety against bearing capacity failure.

The length of the final uniform layout was based on the maximum length obtained from internal and all external stability analyses. Typically, a minimum length value of $L/H = 0.8$ and a maximum spacing of $S_v = 0.45$ *m* were used in case the analyses indicated less stringent values.

2.2 Design of Slopes and Embankments

In the context of this paper, the term slopes refers to internally reinforced steep slopes. The term embankments refers to embankments over soft soil that are reinforced only along their interface with the foundation.

Limit equilibrium slope stability analyses were modified to include reinforcement (e.g., Christie and El-Hadi, 1977; Fowler, 1982; Christopher and Holtz, 1984). The modified analyses were, to start with, conservative in terms of unreinforced slope stability evaluation (e.g., Fellenius method, Two-Part Wedge method with a horizontal interwedge force resultant). These simplified analyses are easier to modify and apply; however, they do not fully satisfy the basic premise of the analysis, that is, equilibrium. Reinforcement force inclination was assumed to be either horizontal, tangential or some arbitrary value in between. Since limit equilibrium deals with global stability, the total reinforcement force, needed to ensure a prescribed margin of safety, was determined. It was assumed that this force is evenly distributed among all reinforcing layers.

Numerical complexity delayed the use of rigorous stability analyses. Lack of computer programs made the design more of an art. Consequently, this situation motivated the development of simplified design charts (e.g., Leshchinsky and Reinschmidt, 1985; Schmertmann et al. 1987).

2.3 Safety Factors and Tests

An integral part of any design is the prescribed factors of safety. Furthermore, material properties determined based on standard tests are also essential.

A bulk factor of safety applied to the geosynthetics was used. It multiplied the calculated required tensile resistance so that a geosynthetic possessing adequate tensile strength could be selected. Another factor of safety was applied in the calculations for the required anchorage length.

Standard laboratory tests were adopted from the garment industry (e.g., grab tensile strength, seam strength). These tests were far from reflecting the actual application of geosynthetic as a reinforcing material. However, they usually produced conservative results. Additional data needed for design (e.g., friction along soil-geosynthetic interface) were developed on an *ad-hoc* basis and in many cases, a conservative value was selected in lieu of conducting a test.

3 OVERVIEW OF CURRENT METHODS

In the past 15 years, extensive advanced numerical analysis, mainly finite element (FE), has been conducted on reinforced soil structures. Simultaneously, experimental research, using physical models and full scale structures, has also been conducted. Especially useful were the FE parametric investigations using a code that has been validated against experimental tests.

Numerous studies concentrated on walls, a few on slopes, and very few on embankments. These studies resulted in:

1. Refined design methods for walls;
2. Evolution of walls with new types of facings that satisfy both aesthetics and economics (e.g., segmental blocks); and
3. Little change in design-oriented analysis for slopes and embankments though generic software packages are available, as tools, for conducting rigorous and comprehensive limit equilibrium designs.

3.1 Design of Walls

Currently, two popular design manuals are used for reinforced walls in the USA. One manual is titled *DEMO 82: Mechanically Stabilized Earth Walls and Reinforced Soil Slopes, Design and Construction Guidelines* (Elias and Christopher, 1996). It was developed under the sponsorship of the Federal Highway Administration (FHWA). This manual is compatible with, but more detailed than, AASHTO guidelines to appear in 1997. The second manual is titled *National Concrete Masonry Association Segmental Retaining Wall Design Manual* (Simac et al. 1993). It was developed under the sponsorship of the National Concrete and Masonry Association (NCMA). Both manuals present comprehensive designs. The manuals address the selection of material properties, consider various aspects of construction, and assess every aspect of structural stability. Furthermore, design details related to connection strength of the geosynthetic to the facia, as well as earthquake loading and reinforcement spacing and length, are also given.

NCMA manual is limited to segmental walls whereas DEMO 82 addresses all types of reinforced walls. NCMA account for some effects of soil-facing block friction while DEMO 82 ignores these effects. Consequently, DEMO 82 uses Rankine's wedge whereas NCMA uses Coulomb's wedge. In both cases the lateral earth pressures used are much smaller than K_0 (either Rankine's or Coulomb's). For typical granular soil this implies the tensile force at-work conditions is nearly half the value used in the 70's. Both NCMA and DEMO 82 allow for a minimum reinforcement length (i.e., minimum L/H is restricted to either 0.6 or 0.7) that is shorter than the minimum values used in the 70's.

It is interesting to note that DEMO 82 is actually simpler, though more comprehensive, than the previous FHWA manual (Holtz and Christopher, 1984). For example, the previous manual required to determine the vertical stress at the facing using Meyerhof's vertical stress distribution at each reinforcement level and then calculate the horizontal stress using Rankine's K_a. DEMO 82 does not require the use of Meyerhof stress for calculating the tensile reaction but rather uses a simple overburden stress calculation. The end result is somewhat lower tensile force in the reinforcement. Furthermore, the previous FHWA manual required overturning stability evaluation; DEMO 82 requires only that eccentricity be within a certain range. Most importantly, the current FHWA manual allows the use of geosynthetic reinforcement in properly designed permanent structures (e.g., bridge abutments) and complex geometries (e.g., cascading walls).

3.2 Designs of Slopes and Embankments

Design of reinforced slopes is addressed in DEMO 82 (Elias and Christopher, 1996) and the Corps of Engineers (WES; Leshchinsky, 1995, 1996). Both designs use comprehensive stability analyses approach, examining a variety of potential failure mechanisms so as to ensure sufficient length and strength of reinforcement. Both are based on limit equilibrium analyses.

Limit equilibrium analysis can be used to assess the global stability of a reinforced slope. It does not deal with local equilibrium. If this equilibrium is being ignored, some layers might be overstressed, while others are under-stressed, potentially resulting in rupture or pullout of reinforcement layers. Such a situation may lead to a progressive failure. Both modified designs address this aspect. DEMO 82 assumes that the total required force to restore stability is uniformly distributed among all reinforcing layers if $H < 6.0\ m$. However, for $H > 6.0\ m$ an empirical distribution is used: layers embedded in bottom H/3 carry $T_{total}/2$, in middle H/3 carry $T_{total}/3$, and in top H/3 carry $T_{total}/6$. The layers within each H/3 are uniformly stressed. WES ensures analytically both local and global equilibria at a limit state stability. Local equilibrium is assessed in a similar fashion to walls; however, log spiral slip surfaces are used. This analytical approach leads to a probable range of tensile force in each layer (Leshchinsky et al., 1995). Consequently, the reinforcement strength could be selected.

There has been very little change in the design of reinforced embankments over the past 15 years. This problem is complicated; it possesses high uncertainty in defining the soft soil properties; the potential for deep-seated failures that may lead to extremely high-strength and expensive reinforcement; significant dependence of design on consolidation rate (difficult to predict) and on complex construction (which may result, for example, in mud waves). Indeed, this design is to a large extent an *art*. Many large-scale projects using geosynthetic reinforcement were successfully completed in recent years. The writer is aware of some projects where the economics of the reinforced embankment by far outweighed alternative engineering solutions. Since some of these projects are well instrumented and monitored, it is likely that the experience gained will lead to some refinements in design (e.g., what should be the inclination of reinforcement force in analysis? in embankment over soft soil using tangential instead of horizontal inclination may lead to half the required strength). It is also likely that economics will dictate the use of strip drains and staged construction.

3.3 Safety Factors and Tests

Partial factors of safety (or reduction factors) for geosynthetics are currently being used. Their value depends mainly on the application, environment of the installation site, construction, and polymer type. These reduction factors address creep, installation damage, and chemical and biological degradation. They are selected rationally so as to ensure that the required strength will be available at the end of the design life of the structure. To account for uncertainties associated with material parameters and with design assumptions, a global factor of safety is also used (typically its value is between 1.3 and 1.5). Unlike the bulk safety factor used in the 70's, the break down to partial factors allows the designer to conduct a rational design in which each "weakness" of the material is explicitly accounted for.

Whether designing a wall, a slope or an embankment, standard laboratory tests are available to determine the reinforcement properties in a realistic manner. These standards were developed by ASTM (American Society for Testing and Materials) and GRI (Geosynthetic Research Institute). They are constantly being revised. These standard tests ensure credible input for design and allow for uniform specifications of design output.

4 CONCLUSION AND DISCUSSION

The following main points can be stated when comparing past and present designs:

1. Past design of walls was based on at-rest working stress for calculating the reinforcement tensile reaction. Stability of walls was evaluated using various limit equilibrium analyses. Current design is based on active state working stress for calculating the reinforcement reactive force. There has been very little change in methods assessing the stability of walls.
2. Past design of slopes was based on limit equilibrium analysis, disregarding local equilibrium. Current designs are also based on limit equilibrium, however, local equilibrium (or stability) is considered either implicitly (DEMO 82) or explicitly (WES).
3. Past and present designs of embankments over soft soil are based on limit equilibrium and contain a considerable amount of engineering judgement.
4. In the past, a bulk factor of safety was applied on the required tensile resistance. Currently, partial factors of safety are being used to account for conditions such as the environment at the site, construction and the actual reinforcing product.
5. In the past, materials properties were loosely defined. Currently, there are standard laboratory tests to define necessary materials properties in design and specifications.
6. Past structures reinforced with geosynthetics were built in secondary and temporary applications. Confidence in current designs and material characterization allows for use of geosynthetics in important and permanent structures.

Engineers and owners desire safe and economical designs. Evolution over the past 15 years resulted in refinement and rationalization of design procedures. The refinement of designs was made possible by field tests, model tests, and advanced numerical analyses. The end result is indeed safe and economical structures.

The analytical portion of current designs is *not* more complicated than the one used in past designs. This agrees well with the philosophy that design requiring simple analysis is likely to be used. Furthermore, it is likely to be used successfully. Considering current specified backfill soil and type of structures, the writer think that further refinement of designs for walls and slopes may yield marginal results. In fact, it may result in adversely affecting constructability (if construction is ignored in design) or simply be impractical (i.e., require a "paper thin" geosynthetic that will fail during installation or a design rendering a layout where each layer possesses a different length). One has to bear in mind that the typical cost of geosynthetics is only 5 to 10% of the overall cost of a project. However, to alleviate the concerns some designers have with creep, it is desirable to address this issue in design by allowing stress relaxation to be coupled with creep. Under such coupled conditions, one may realize in a rational way that creep does not necessarily means intolerable deformations in the future.

The writer foresees a major application of geosynthetic reinforcement in conjunction with poor backfill soils. The economics of such an application seems to be very appealing. Designs for such backfills will have to be completely revamped.

REFERENCES

Bell, J.R., R.K. Barrett & A.C. Ruckman 1983. Geotextile earth-reinforced retaining wall tests: Glenwood Canyon, Colorado. *Transportation Research Record*, 916: 59-69.

Bell, J.R., A.N. Stilley & B. Vandre 1975. Fabric retained earth walls. *Proceeding of the 13th Annual Engineering Geology and Soil Engineering Symposium*, University of Idaho: 271-287.

Christie, I.F. & K.M. El-Hadi 1977. Some aspects of the design of earth dams reinforced with fabric. *Proceeding of the International Conference on the Use of Fabrics in Geotechnics*, Paris, 1: 99-103.

Christopher, B.R. & R.D. Holtz 1984. *Geotextile Engineering Manual*. Draft course notes prepared for FHWA, Contract No. DTFH61-80-C-00094.

Elias, V. & B.R. Christopher 1996. *Mechanically Stabilized Earth Walls and Reinforced Soil Slopes, Design and Construction Guidelines*. Demonstration Project No. 82, Publication No. FHWA-SA-96-071 (Final draft: October 1996).

Fowler, J. 1982. Theoretical design considerations for fabric-reinforced embankments. *Proceeding of the 2nd International Conference on Geotextiles, Las Vegas*, 3: 665-670.

Lee, K.L., B.D. Adams & J-M.J. Vageneron 1973. Reinforce earth retaining walls. *J. of the Soil Mechanics and Foundation Engineering Division*, ASCE: 99(SM10), 745-764.

Lee, K.L., B.D. Adams & J-M.J. Vageneron 1975. Reinforce earth retaining walls. Closure to discussion, *J. of the Geotechnical Engineering Division*, ASCE: 101(GT3), 745-764.

Leshchinsky, D. 1995. *Design Procedure for Geosynthetic Reinforced Steep Slopes*. Technical Report REMR-GT-120 (Temporary Number), Waterways Experiment Station, US Army Corps of Engineers, Vicksburg, MS.

Leshchinsky, D. 1996. *ReSlope (V 1.2): Computer Program for Design of Reinforced Steep Slopes and Walls.* Distributed by Industrial Fabrics Association International, (IFAI) Bookstore, 345 Ceder St., Suite 800, St. Paul, MN 55101-1088, USA.

Leshchinsky, D., H.I. Ling & G. Hanks 1995. Unified design approach to geosynthetic reinforced slopes and segmental walls. *Geosynthetic International*, 2(5): 845-881.

Leshchinsky, D. & A.J. Reinschmidt 1985. Stability of membrane reinforced slopes. *J. of Geotechnical Engineering*, ASCE: 111(11), 1285-1300.

Schmertmann, G., V.E. Chouery-Curtis, R.D. Johnson & R. Bonaparte 1987. Design charts for geogrid-reinforced soil slopes. *Proceedings of Geosynthetics '87*, New Orleans: 1, 108-120.

Simac, M.R., R.J. Bathurst, R.R. Berg & S.E. Lothspeich 1993. *National Concrete Masonry Association Segmental retaining Wall Design Manual*, National Concrete and Masonry Assoc., Hendron, VA: 250 p.

Steward, J., R. Williamson & J. Mohney 1977. *Guidelines for Use of Fabrics in Construction and Maintenance of Low-Volume Roads*. Chapter 5, USDA, Forest Service, Portland, Oregon (also revised version June, 1983).

Earth Reinforcement, Ochiai, Yasufuku & Omine (eds) © 1997 Balkema, Rotterdam, ISBN 90 5410 833 9

A search for reliability of the mechanism of geotextile-reinforced soil wall

Y.Tsukada
Public Works Research Institute, Ministry of Construction, Tsukuba, Japan

ABSTRACT: Recently, Geotextile Reinforced Soil Wall is widely constructed instead of conventional retaining walls such as gravity type concrete walls due to the establishment of design manuals and the development of geotextile materials and methods. In Japan, Public Works Research Institute established the design manual in 1992. However, there are still unknown factors on the design which should be solved. This paper points out the comparison between design assumption and actual behavior which include earth pressure on the facing, tension generated by the reinforced materials etc.

1 INTRODUCTION

The reinforcing function of reinforced soil walls is achieved by a combination of the tension effect of the reinforcing material, the resistance effect of the facing, and the virtual retaining wall effect of the monolithic reinforced zone. In other words, [1] embankment with a steep slope is supported by a facing in which the reinforcing material links the resistance forces required to maintain embankment in stable condition, and [2] the reinforcing material provides tensile resistance force based on the friction resistance in the stable embankment to guarantee balanced conditions. And [3] the soil reinforced by the reinforcing material is integrated as the virtual retaining wall, and it acts as the retaining soil for the backfill soil behind it. In line with this concept, Jones C.J.F.P. (1984)[1] and Jewell R.A. (1984) [2] proposed a design method for reinforced soil walls or steep slope reinforced embankments reinforced with a geogrid, and defined [1] and [2] as the internal stability, and [3] as the external stability. Assuming that it was necessary to evaluate the overall sliding stability including that of the foundation ground to provide external stability, the design method was completed from a rigid plastic body. However, their design methods do not necessarily fully apply in Japanese design criteria, for examples, their design methods have not considered stability during earthquakes, the application of the safety factor in various stability evaluations differs from that in Japan. Therefore some experiences have been conducted on these issues. Fig.1 shows the evaluation modes for design of reinforced soil wall. This figure also shows evaluation of the stability of reinforcing material near facing as internal stability (partial stability), and an overall stability evaluation of a case where the foundation ground is weak sandy ground. Many structures are constructed based on design work done in accordance with the evaluation modes explained above, but measurements of various experiments have confirmed that their design assumptions do not necessarily fully explain actual behavior. The subjects such as the comparison between design assumption and reality and effects of facing in geotextile reinforced wall are described as following based on by some experimental researches.

2 COMPARISON BETWEEN DESIGN ASSUMPTION AND REAL BEHAVIOR

In fact, reinforced soil structures require a certain degree of deformation in order that resistance forces act between the soil and the reinforcement material (in the case of a soil wall, the facing is included). In other words, reinforced soil walls are flexible structures which can maintain their stability while undergoing deformation because they can follow the deformation. Consequently, their characteristics differ from those of conventional concrete retaining walls; namely on the condition that the foundation ground remains stable in a case where the bearing ground under a reinforced soil wall is soft, it is

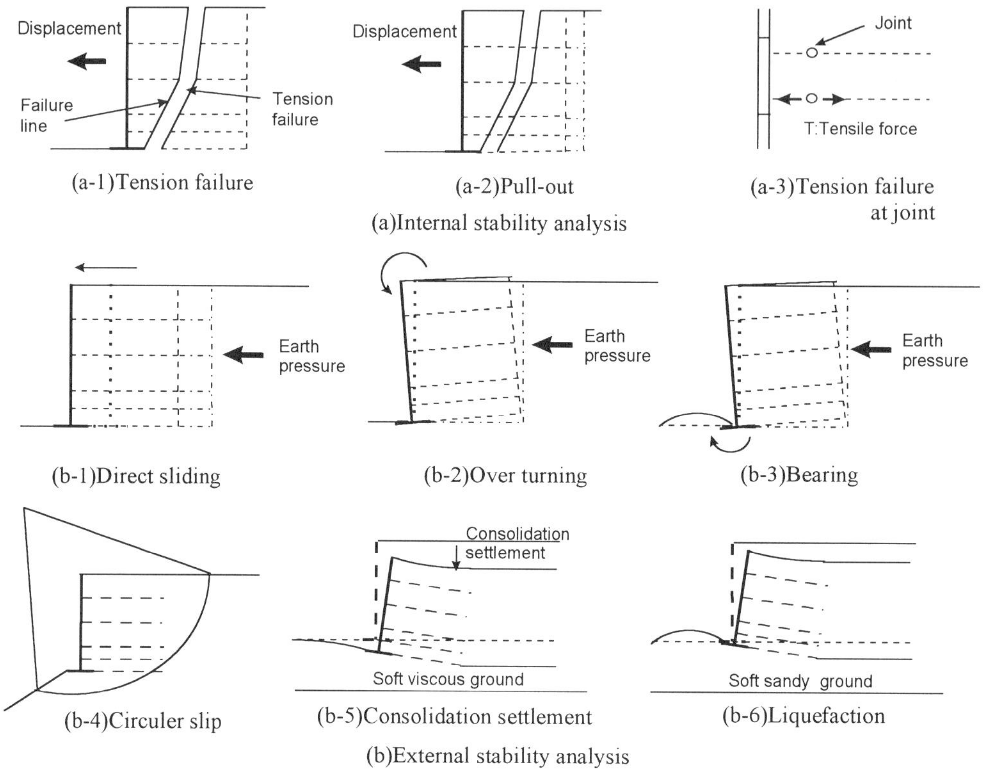

Fig.1 Failure patterns and analysis points on the design of reinforced soil wall

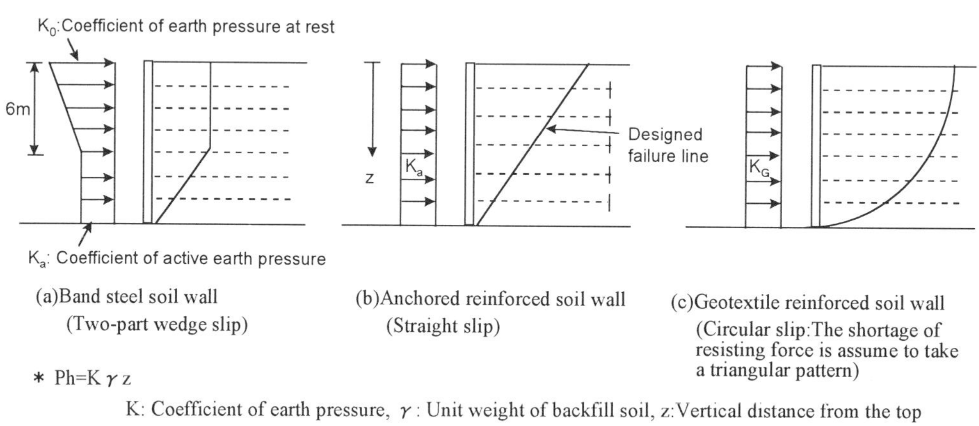

Fig.2 Analysis concepts of required tensile strength

possible to construct a structure which remains stable while it follows deformation. This characteristic as a flexible structure causes actual behavior which differs from that assumed for design purposes explained in the above. The following is a summary of comparison between design assumption and its actual phenomena.

2.1 *Earth Pressure Acting on the Facing*

It is assumed that the earth pressure acting on the rear of the facing is distributed as shown in Fig.2 along with the slip surface. In the case of flexible facing on the other hand, the backfill soil reinforced close to the facing work is deformed and resistance

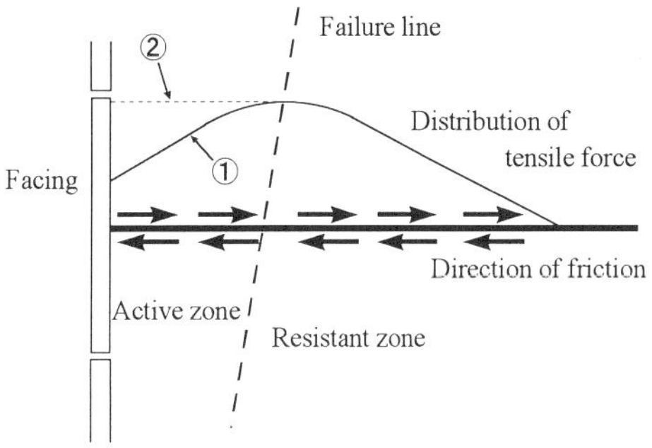

Fig.3 Tension distribution on the reinforcement

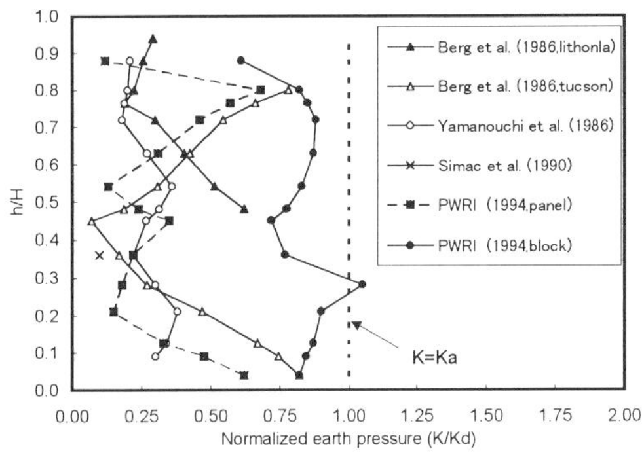

Fig.4 Results of horizontal earth pressure of wall

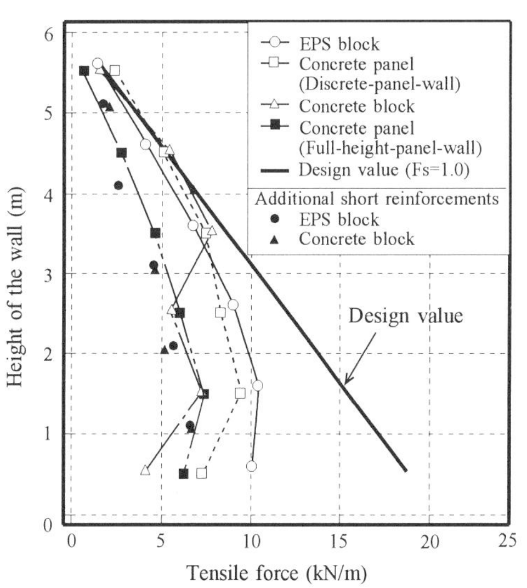

Fig.5 Distribution of maximum tensions experiments of geotextile reinforced soil walls(H=6m) at PWRI

force is generated by the reinforcing material. The tension distribution, which is presented schematically in Fig.3, results in a reduction of the earth pressure acting on the facing.

In this fig.3, when the friction stress in the active pressure area is distributed as in [1], the earth pressure is reduced, and when it is distributed as in [2], there is no friction stress between the soil and the reinforcing material, so the earth pressure is not reduced. Fig.4 represents the results of measurement results obtained in reinforced soil walls in demensionless earth pressure K/Ka vs h/H [3]

For design purposes, this earth pressure reduction mechanism is not sufficiently understood and is not used for safe side judgments right now.

2.2 *Tension Generated by the Reinforcing Material*

Fig.5 shows the distribution of tension of reinforcing material during a full-size experiment using a specimen with a height of 6 m at the Public Works Research Institute (hereinafter PWRI) [4]. In this experiment, a comparison was made of various facing categories: EPS, integrated concrete panels, divided concrete panels, and block type. In the design, the earth pressure acting on the facing was distributed in triangular form and the tension generated by the reinforcing material did, similarly, increase towards the bottom, but actual measurements have shown that the values of tension were either identical or smaller. The values of tension were particularly low towards the bottom.

2.3 *Subgrade Reaction and the Stability of the Bearing Ground*

The reinforced and integrated zone is assumed to have the same earth retaining effects as those provided by a gravity retaining wall, and ground reaction with the trapezoidal distribution shown in Fig.6 acted against the external force acting on it. Measurements of existing reinforced soil walls[5] and the experiment performed at PWRI explained in the above[4] have revealed that the distribution of subgrade reaction at the bottom surface of the reinforced zone is approximately the overburden pressure, and clearly indicate that large reaction force is generated at the bottom of the facing by the dead weight of the facing (Fig.7).

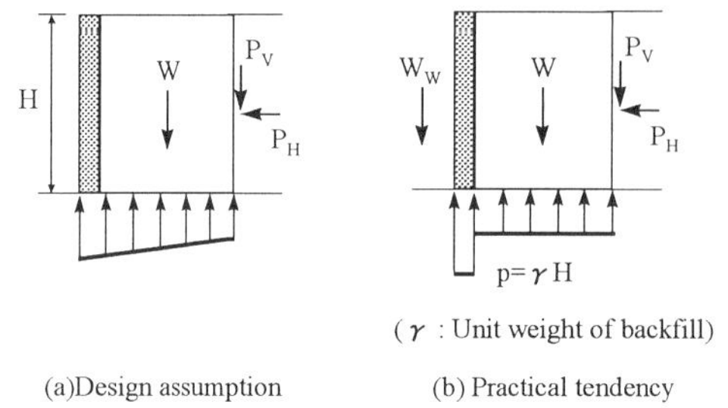

Fig.6 Assumption and practice of ground reaction

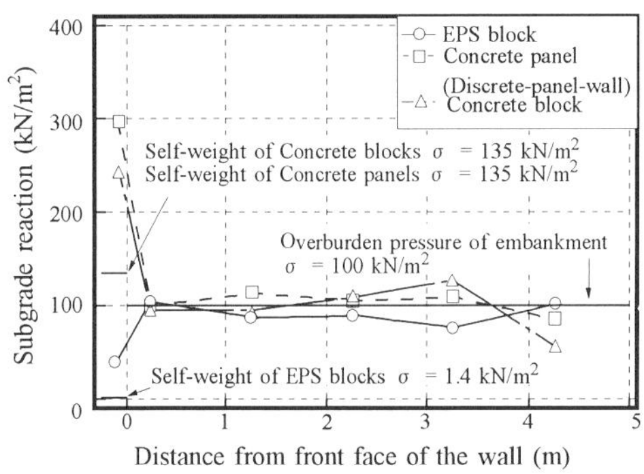

Fig.7 Distribution of subgrade reaction force (experiments of geotextile reinforced soil walls (H=6m) at PWRI)

In the design, the allowed bearing capacity was satisfied for this trapezoidal distribution ground reaction, but according to the results of a full scale reinforced soil wall on-site experiment using block type with a height of 8 m, the foundation ground consisted of relatively soft sandy soil at the surface and a soft clay layer deposited beneath the sand, conditions which do not satisfy bearing capacity conditions, but nevertheless, stability was clearly maintained. The value of the reaction force at the bottom edge of the facing was also greater than the dead weight of the facing as it was in the laboratory experiments (Fig.8).

Fig.9, Fig.10 and Fig.11 show the results of measurements of a full scale reinforced soil retaining wall conducted over a period of about 550 days. The results indicate that unlike an extremely stiff structure such as a concrete retaining wall, a reinforced soil wall will maintain stability while it deforms. It also suggests that as far as the overall stability including that of the foundation ground is concerned, because the main body of a reinforced soil wall is an embankment, it may be concluded that, as an embankment, it satisfies safety conditions based on its rotational slip.

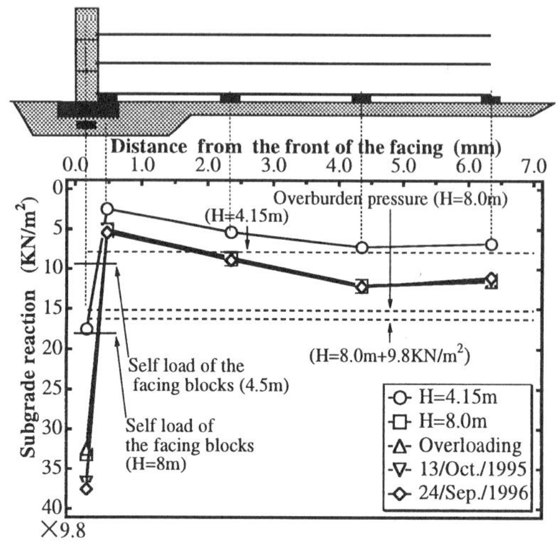

Fig.8 Distribution of ground reaction force (An experiment of geotextile reinforced soil walls (H=8m) at PWRI)

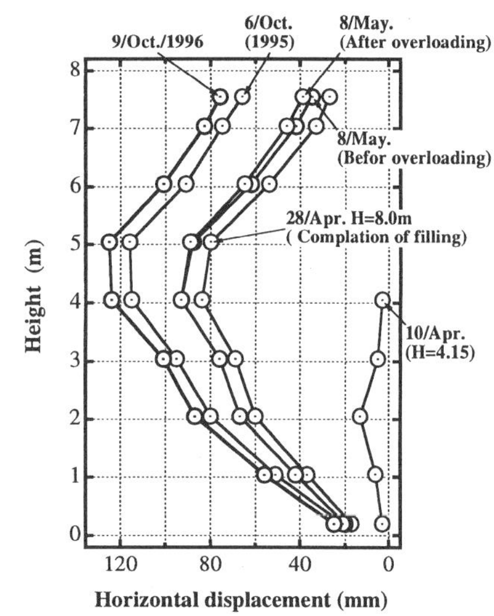

Fig.9 Distribution of horizontal displacement (An experiment of geotextile reinforced soil walls (H=8m) at PWRI)

3 EFFECTS OF THE FACING

In the current design method, it was assumed that the effects of the facing work of a reinforced soil wall included the prevention of leakage of the soil and support for the earth pressure acting on the facing. In contrast, Tatsuoka proposed that according to the

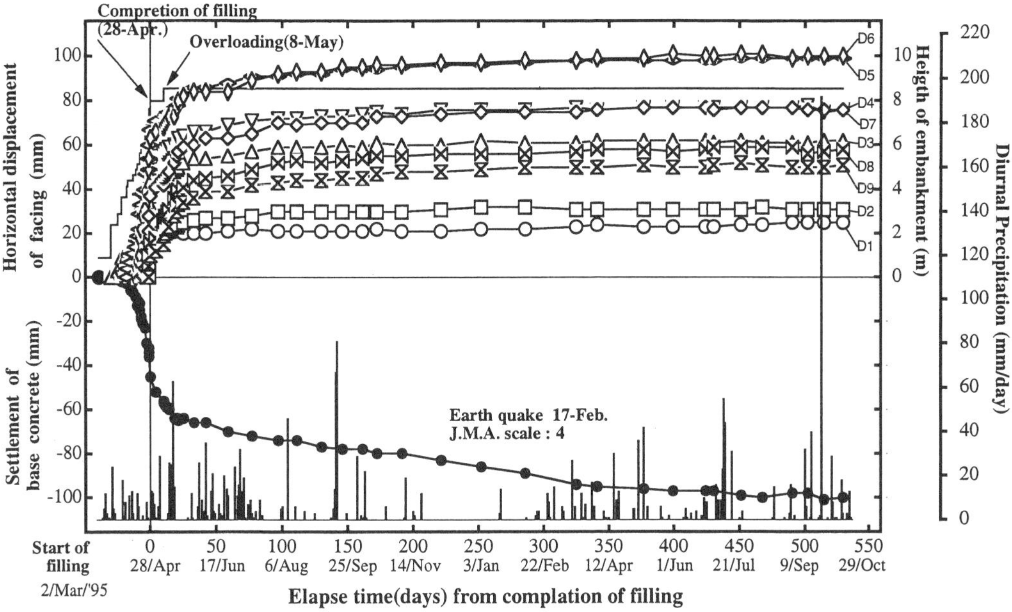

Fig.10 Horizontal displacement with Elapsed time
(An experiment of geotextile reinforced soil walls(H=8m) at PWRI)

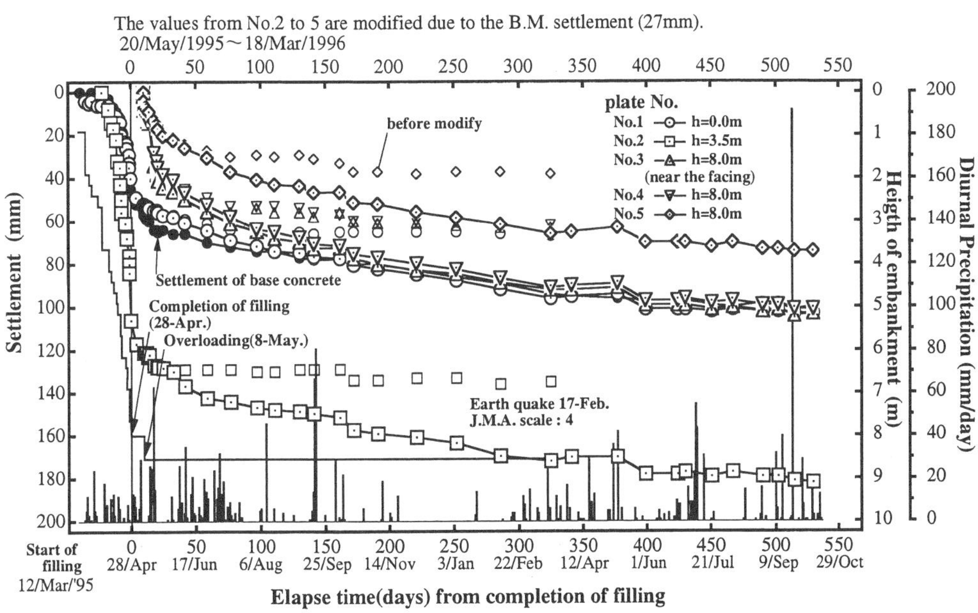

Fig.11 Settlement of embankment with elapsed time
(An experiment of geotextile reinforced soil walls(H=8m) at PWRI)

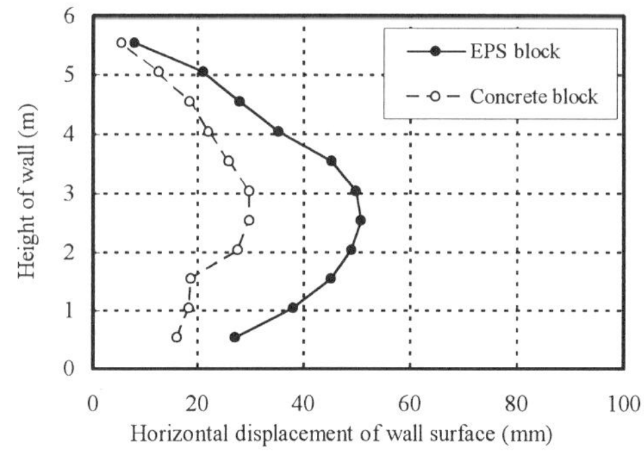

Fig.12 Horizontal displacements of an EPS block and concrete block(experiments of geotextile reinforced soil walls (H=6m) at PWRI)

type of facing, the effects can be classified local rigidity, overall longitudinal rigidity, overall shearing rigidity, overall bending rigidity, and gravity comparison of the horizontal displacements of an EPS block and concrete block based on an experiment conducted by PWRI.

Although deformation was restricted by the facing, in case of EPS block the weight effect of the facing work was small regardless of the presence of longitudinal rigidity[4]. And the tensile stress acting on the reinforcing material is shown to reach its maximum value near the facing. This trend has also been confirmed by a full scale experiments of reinforced soil wall using block facing.. And when designing a reinforced soil wall such as a block type wall, it is possible to reduce the tension acting on the reinforcing material by accounting for the weight resistance of the blocks and the friction resistance between the blocks. Similar proposals have been offered by Gotteland et al.[7] (1993) and Leshchinsky[8].

4 CONCLUSION

This paper is an outline focused on the reinforcement effect mechanism and design concepts of reinforced soil walls or reinforced embankments as geotextile reinforced soil structures. Based on numerous experiments, analyses, and on-site measurements, a design method has been created based on a concept of ultimate balance. But it has also been shown that this design method can not necessarily fully explain actual behavior.

While on one hand, real reinforced soil structures are flexible structures, one important question is how to predict the deformation, then control it through the design and execution stages. As the point of contact between the limit equilibrium method and the deformation analysis method, it is important to conduct model experiments which are able to analyzed by all researchers and designers who are concerned.

REFERENCES

1)Jones, C.J.F.P. 1984. *Design and construction methods. Proc. Symp. Polymer Grid Reinforcement*, ICE,London, pp. No.6.1,8p.

2)Jewell,R.A.,Paine,N. and Woods R.I. 1984. *Design Methods for Steep Reinforced Embankments, Symp. Polymer Grid Reinforcement*, ICE, London, pp.No.3.1,12p.

3)Yamanouchi,T., Fukuda,N. 1994.*12 State of the art of geosynthetic reinforced soil structures*, OXford & IBH Publ.Co.PVT.LTD. ed. Saxena, pp.305-335.

4)Tajiri,N.,Tsukada,Y.,Yamamoto,J.,Ochiai,Y.,Dobashi,K. 1995. *Comparison among full-scale failure experiments of geotextile reinforced soil walls with different types, 10th Geosynthetics symposium*, pp.37-46.

5)Nakajima,T.,Ooki,Y.,Oda,K.,Miyatake,H., Dobashi,K. 1995. *The fileld performance of a geotextile reinforced soil wall with concrete facings, 10th Geosynthetics symposium*, pp.47-56.

6)Tatsuoka,F. 1993. *Role of facing rigidity in soil reinforcing, Proc.Int.Symp. Earth Reinforcement Practice, Keynote Lecture, IS-KYUSHU'92, Fukuoka, Vol.2*, pp.831-870.

7)Gotteland,Ph., Grouc,J.P.,Jones, H.W.,Baraize, J. 1992. *Cellular walls associated with geosynthetics: A laboratory model study, IS-KYUSHU'92, Fukuoka, Vol.1*.pp.337-334.

8)Leshchinsky,D. 1993. *Geosynthetic reinforced steep slopes and walls: Effects of facing blocks, Proc. Geotech. Seminar Tokushima u.*, pp.95-133.

Earth Reinforcement, Ochiai, Yasufuku & Omine (eds) © 1997 Balkema, Rotterdam, ISBN 90 5410 833 9

Design assumptions and actual behavior in soil nailing

N. Kotake
Toyo Construction Co., Ltd, Tokyo, Japan

S. Tayama
Japan Highway Public Corporation, Tokyo, Japan

ABSTRACT: Design assumptions and actual behavior in soil nailing are reviewed with respect to failure and deformation modes of nailed slopes. First, failure modes assumed in the present design methods and those actually observed during construction are presented. Secondly, deformation modes and mobilized nail force are described schematically, and discussed in relation to the findings obtained in current research works.

1 GENERAL REMARKS

The soil nailing practice seems to go ahead of the theory on the reinforcing mechanism as is often the case with other practical techniques in geotechnical engineering.

The soil nailing is one of versatile techniques to reinforce in-situ soil, thanks to its simplicity and flexibility in execution works. This technique has been used to stabilize both natural and steep cut slopes on small to medium scales. In urban areas the technique has also been used to construct temporary retaining structures to support the ground close to neighboring structures that are sensitive to deformation with appropriate measures to reduce ground movement.

As for the design some well organized recommendations and technical documents have been published based on the findings from research works and practical experiences gained for the past decades. However, in the present design methods, which are all based on the limit equilibrium stability analysis, the working principal of the nailing technique as passive reinforcement is not considered in an explicit way. Important subjects that cannot be solved by the limit equilibrium based analysis include evaluation of 1) deformation of nailed slope, and 2) soil strength and nail force mobilized in service in relation to those at failure. Durability of reinforcement required for long term stability of nailed slopes and preservation of environments are also important issues.

2 FAILURE MODES

2.1 *Failure modes assumed in design*

In routine design, the stability of nailed slope is analyzed by the limit equilibrium method as for reinforced backfills. The Working Group for Design in "Research Committee on Reinforced Technology for Natural Slopes" (1996) summarized the design methods presently used in Japan. Table 1 shows possible failure modes of nailed slope assumed in design procedures. They are roughly classified into three groups with respect to external or internal failures, and global or local failures.

Group-A assumes a rigid reinforced zone to examine external failure modes with respect to sliding, overturning, bearing and circular slip failures likewise for a conventional retaining wall. For Group-B, the global stability is examined on internal sliding of either circular slip, single-wedge, or two-part-wedge failure (B-1). The local stability of reinforcement is examined along the most critical potential failure surface assumed in B-1. The local failure modes are pullout and rupture of reinforcement, and sliding out of the soil mass in the active zone (B-2). Group-C includes local failures of reinforcement and facing, which are examined based on the local equilibrium to fulfill the rigid body assumption in Group-A.

It is difficult to evaluate appropriately the density and layout of nails when the whole reinforced zone is assumed to behave as a rigid body (Group-A). In the majority of the design methods, therefore, Group-B is basically adopted to examine both global and local failures though Group-B covers part of Group-A. Among the external failures in Group-A, the circular slip failure is mostly examined as

Table 1 Failure modes assumed in the present design for nailing (Revised RCRTNS, 1996)

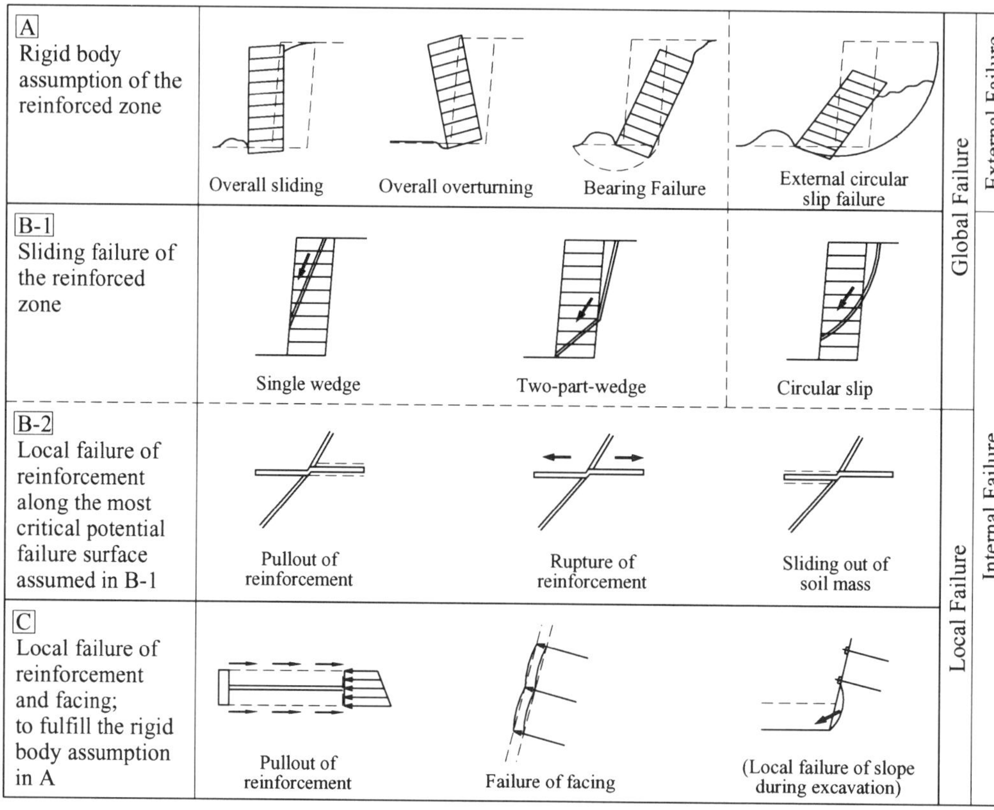

required depending on geological and loading conditions.

2.2 *Adopted design methods*

According to the fact-finding on application of soil nailing conducted by Japan Highway Public Corporation (Mouri et al., 1993), different types of design methods had been adopted as shown in Fig.1. It is seen that the single-wedge failure was adopted in a half of the total cases while the circular slip failure was adopted in 14 %. The empirical design method, which specifies a standard layout of nails to prevent relatively shallow failures without any design calculation, was adopted in approximately 20 %. This tendency may somewhat reflect the fact that the majority of collected cases were the application to rock slopes.

2.3 *Patterns of collapse actually observed*

The construction procedures of nailed slope involve a critical phase with respect to local or global stability. The Working Group for Execution in

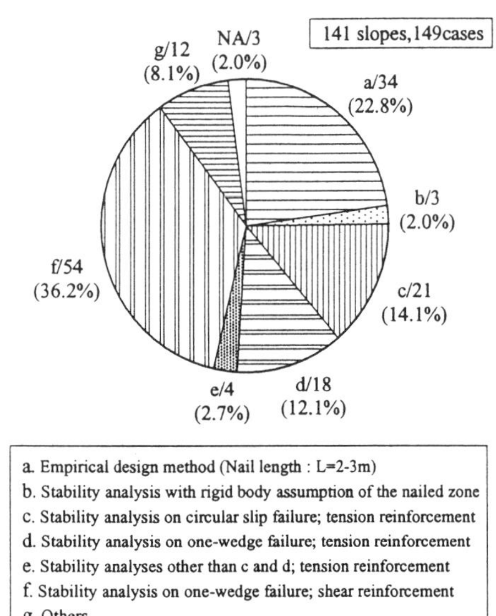

Fig.1 Adopted design methods (Mouri et al.,1993)

"Research Committee on Reinforced Technology for Natural Slopes" (1996) analyzed 1) typical failure patterns which have been observed so far during construction as illustrated in Fig.2, and 2) the probable causes for these failures as listed in Table 2.

It is considered that most of collapses occurred due to unexpected geological conditions and misjudgment on execution procedures. The overall sliding of a slope seems to have occurred either along failure surfaces that were deeper than the most probable potential failure surface predicted in design (Failure pattern-4), or those located at depths exceeding the limits of application of soil nailing.

However, the number of such external failure as above is limited. Most of the failure occurred either along shallow failure surfaces within nailed zones (Pattern-2,3), or in shallow zones which had been exposed without nailing or shotcreting after excavation below the nailed zone (Pattern-1,2,3).

Table 2 Causes of failure (RCRTNS, 1996)

Cause of failure \ Failure Pattern *			(1)	(2)	(3)	(4)
Geological Conditions	local lack of cohesion of soil		○	○		
	ground water		○	○		
	misjudgments	in-situ soil strength		○	○	○
		in-situ pullout force			○	○
	unexpected conditions	deeper failure surfaces			○	○
		sliding along interfaces			○	○
Design Analysis	insufficient nail density			○	○	
	insufficient nail length				○	○
	insufficient pullout force				○	○
Execution Procedures	over excavation in height		○	○	△	△
	insufficient grouting	in drilled hole		○	○	△
		at the head	○	○	△	
	delay of fixing nail head			○	△	
	delay of shortcreting			○	○	○

*) Failure patterns (1) to (4) in the table correspond to those shown in Fig.2.

3 DEFORMATION MODE AND MOBILIZED NAIL FORCE

The actual behavior of nailed slope strongly depends on each specific condition at a site such as geological characteristics, gradient and height of slope, layout and density of nails, rigidity of nails and facing, external loading conditions and so on. We cannot precisely evaluate deformation of nailed slope either by empirical methods or numerical analyses considering various factors as mentioned above. Under such situations the practical approach to keep deformation below allowable levels would be monitoring field behavior with feed back to design procedures and using supplementary methods to reduce shear stress of soil and to increase rigidity of the reinforced ground.

3.1 *Schematical behavior of nailed slope during construction*

Fig.3 shows schematically the deformation modes of nailed slope observed during construction and in laboratory model tests. It is supposed here for simplicity that the ground is homogeneous and nails are relatively short compared to the slope height.

1) Global behavior of nailed slope

The stress release due to excavation causes deformation of the nailed slope. The loosened zone is characterized as induced higher strains. As excavation proceeds the loosened zone grows

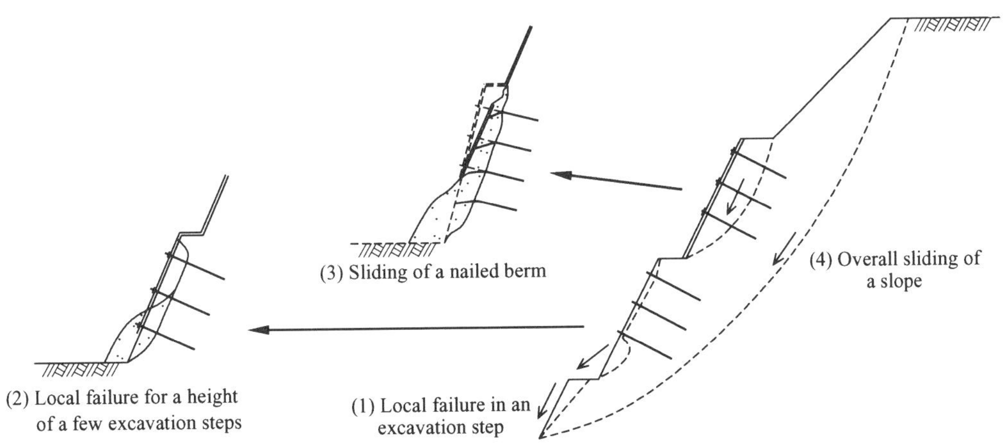

Fig.2 Typical failure patterns (RCRTNS, 1996)

backward to the backface of nailed zone. Due to the effect of reinforcement, soil strength that is larger than when the slope is excavated without reinforcing, is mobilized in the reinforced zone.

2) Displacement of slope face

Tilting is the most predominant displacement mode of nailed slope in the field. In this mode, outward horizontal and vertical displacements are maximum at the top of the slope, with the outward displacement occurring below the transient bottom of excavation. The slope tilting increases with excavation, resulting from deformation in both the nailed zone and the zone behind it.

3) Mobilized nail force

The tensile force in nail is mobilized by deformation of the nailed zone. The nail force at a certain level reaches its maximum after several excavation stages at depths below the level of the nail, whereas the slope deformation would continue by further excavation of the slope. Therefore, when the excavation is completed, the nail force at the lowest level is usually not largest, but may even be smallest. Then, the nail force would be redistributed by creep deformation in both soil and nails with an increase in the nail force at the lowest level.

Such a progressive nature of mobilization of soil strength and nail force as described above is not considered in usual design procedures. Rather, it is assumed that the design values of soil strength and nail force are mobilized simultaneously.

3.2 *Nail force distribution*

Along a given nail, the maximum tensile force T_{max} appears at a certain depth from the slope face (Fig.3). It is usually assumed that the location of T_{max} coincides with that of a potential failure plane (SCFNP, 1993). This assumption has been supported by a number of model tests in which slopes were forced to fail either by loading at the crest with a rigid platen, or by decreasing the soil

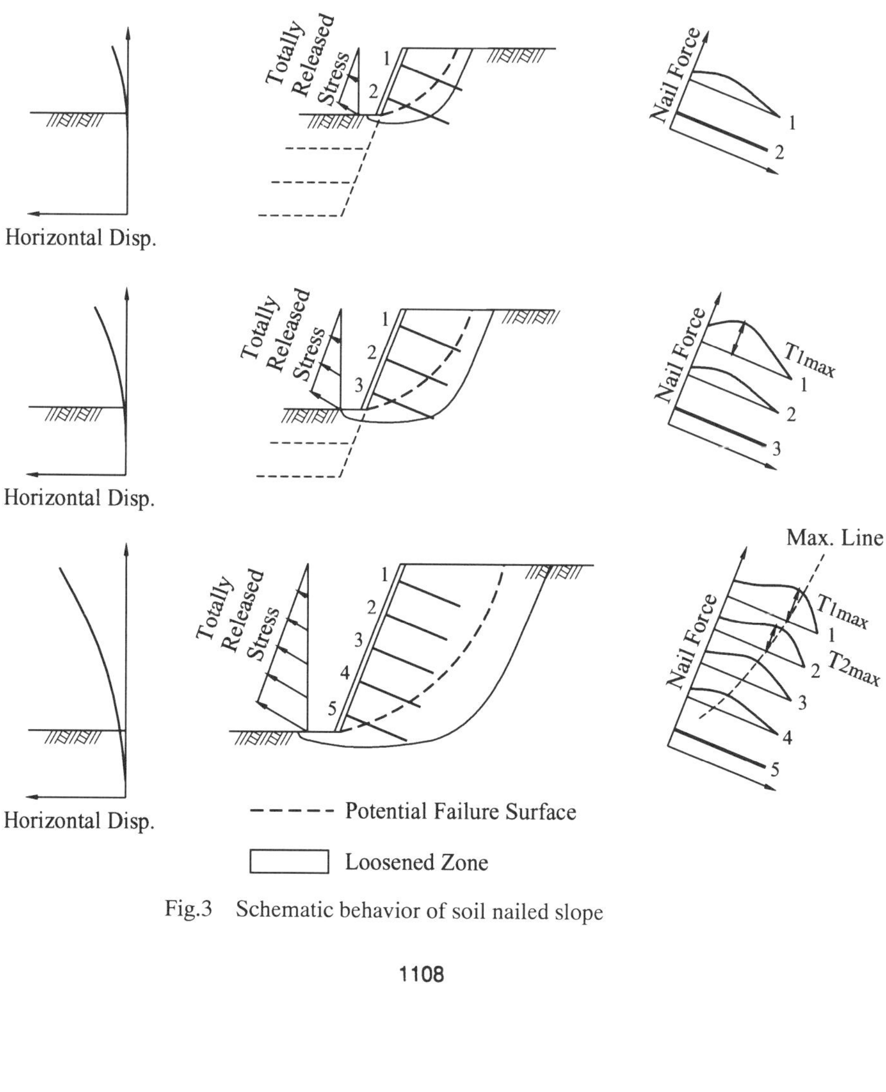

Fig.3 Schematic behavior of soil nailed slope

strength by water-saturating in 1 g model tests, or by increasing the soil weight in centrifuge tests. In some recent laboratory model tests in which the model slopes were brought to failure by decreasing the pressure on the slope face simulating the actual excavation process (Muramatsu et al., 1995), however, the location of T_{max} was found shallower than the failure plane. It was also shown that the nail force distribution be largely affected by the facing type; as the overall facing rigidity increases, the location of the maximum force T_{max} becomes closer to the slope face and the T_{max} value increases. According to this result and a number of field observations of full-scale behavior, in the design recommendation (JHPC,1995), different distributions of design nail force and T_{max} value are selected depending on the facing type.

When the direction of the potential failure plane rotates largely along a given failure plane, the effects of nail direction relative to the tangential direction of the failure plane where the nail is crossing the failure plane should be considered (Tateyama et al., 1993). The effects of the shape and dimensions of nails are also important, particularly for large-diameter nails (i.e., dowels). These factors affecting the mobilized nail force are not sufficiently considered in usual design procedures of soil nailing.

3.3 *Deformation modes far before and near or at the failure state*

As mentioned earlier, when a slope under excavation with nailing is far before the failure state, the most predominant deformation mode is tilting (Fig.4a). In many construction projects, the displacements at the top of slope are monitored during excavation, and judgments as to the current stability of the slope is made based on those observations. Based on the past experiences, it is usual that the procedure of excavation and nailing are controlled so that the outward lateral displacement at the top of slope be lower than 0.1 to 0.3 % of the current slope height.

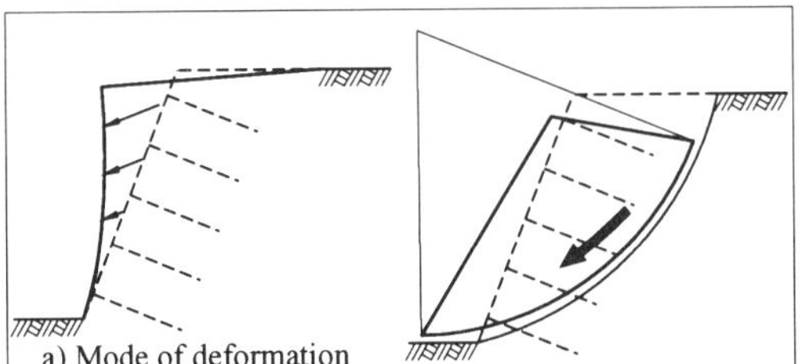

Fig.4 Modes of deformation and failure

On the other hand, the deformation mode when a nailed slope becomes near or at the failure state is often the rotational mode along a circular failure plane (Fig. 4b). In this case, as actually observed in a prototype slope (e.g., Aoki et al, 1996), the slope face near the slope bottom is pushed out largely around the slope toe.

The failure of both unreinforced and reinforced soil masses results from the progressive development of a shear band (or shear bands), which finally lead to an overall failure zone (or failure plane), or from the failure of reinforcement for a reinforced soil mass (e.g., Huang et al., 1990). Therefore, to simulate and analyze in a consistent manner both the deformation of a nailed slope during excavation and in service, at stages far from the failure state, and that near and at the failure state, a sophisticated numerical method such as the FEM analysis considering strain localization into shear bands is one of the promising methods (e.g., Kotake et al., 1996). In such analyses, a shear band width that is specific to the soil type concerned should be introduced. Further, more model tests bringing the slope to failure with careful observation will be required. It is expected that based on such researches as described above, the safety factor of a given slope can be estimated by monitoring displacements at the top of the slope during excavation.

4 PERMANENT NAILED SLOPES

Durability of reinforcement required for long term stability and preservation of environments are important issues for permanent nailed slopes.

1) Durability

There are few available information on long term durability of nails under actual working conditions, partly because of its short history.

Tayama et al. (1996) investigated the conditions of nails by sampling them by means of over-core boring from nailed slopes ten years after their construction. Though some nails exhibited partial corrosion due to insufficient grouting, most of the nails were in very good conditions. Considering the conditions of slope face that had been observed in inspection since their construction, they concluded that the nailed slope kept stability as had been designed.

Since the number of sampled nails is insufficient, more investigation will be required so as to specify design life of nails or nailed slopes based on actual performance. It should be stressed here that long term performance of nailed slope is closely related with execution control, especially in drilling and grouting. On the other hand, nailed slopes are relatively easy to maintain, compared with, for example, those stabilized by ground anchors. Though we have scarcely experienced rehabilitation of deteriorated nailed slopes, one of the proper methods may be restoring by additional nailing. Such a matter as mentioned above may be evaluated in terms of life time cost of stabilized slopes.

2) Landscape preservation

Besides the effects of facing rigidity on nail force mobilization as mentioned earlier, type of facing is also an issue on environments. Although a full-height rigid RC facing or lattice RC structure covering the whole slope face is better in stabilizing the slope, they would not be preferable for natural landscape preservation. One of the attractive methods would be the use of a facing having some degree of rigidity while allowing planting or vegetation (Horie et al., 1996).

CONCLUDING REMARKS

The nailing technique seems to be well-established in engineering practice. Currently the technique is often used in urban areas with appropriate measures to reduce deformation of the ground, being competitive to other types of earth retaining methods. Although deformation of nailed slope during cutting is the most important index to evaluate the slope stability, no reliable methods are available at present.

The seismic stability of cut slopes, being beyond the scope of this paper, is not usually examined in ordinal design procedures even in Japan. Seismic stability of reinforced soil structures was eventually highlighted after the disastrous Hyogo-ken Nanbu Earthquake. During the earthquake a number of nailed slopes were reported to have performed well (Tatsuoka et al., 1996). Reflecting this evidence, the nailing technique has been adopted for remedial and preventive works in many slopes. It is noted that aseismic design methods of nailed slopes should be developed taking into account its ductility as observed during recent large earthquakes.

ACKNOWLEDGEMENTS

Most of the topics presented in this paper are based on the activities in "Research Committee on Reinforced Technology for Natural Slopes" organized by Japanese Geotechnical Society in the period from April 1993 to March 1996. The authors greatly wish to thank Prof. Tatsuoka,F., Univ. of Tokyo, who chaired the committee, and other members for their valuable suggestions and supports in the preparation of this paper.

REFERENCES

(J) denotes in Japanese.

Aoki,H., Maruyama,O., Yonezawa,T., Taneda,N. 1996. Design Method of Reinforced Slope for Excavation (part2) -Measurement and FEM Analysis-, Symposium on Reinforcement of Natural Slopes, JGS, pp269-274. (J)

Horie,N., Kikuchi,H., Yamada,H. 1996. An application of composite long-fiber reinforced earth work method for stability and vegetation, Tuchi-To-Kiso, JGS, Ser.No.465, Vol.44, No.10, pp.33-34. (J)

Huang, C., Tatsuoka,F., Sato,Y. 1990. Bearing capacity of footing in reinforced sandy slope, Tenth South Asian Geotechnical Conference, Taipei.

Japan Highway Public Corporation, 1995. Guidance for Design and Execution of Cut-Slope Reinforcing Construction Methods (Draft). (J)

Kotake,N., Tanaka,T., Tatsuoka,F. 1996. Numerical simulation of plane strain compression failure of reinforced sand, Annual Conference, JGS, pp.2431-2432. (J)

Mouri,S, Tayama,S. 1993. Current application and subjects on reinforced slopes with steel bars, THE DOBOKU-SEKOU, Vol.34, No.7, pp39-46. (J)

Muramatsu,M, Sueoka,T, Tatsuoka,F. 1995. Reinforcing mechanism of soil nailing and effects of slope facing, JSCE, No.517/Ⅲ-31, pp93-104. (J)

Research Committee on Reinforced Technoogy for Natural Slopes, 1996. Activity Report, Symposium on Reinforcement of Natural Slopes, JGS, pp1-191. (J)

Scientific Committee of the French National Project. 1993. CLOUTERRE: RECOMMENDATIONS CLOUTERRE, 1991 (Soil Nailing Recommendations 1991).

Tateyama,M., Tatsuoka,F., Kishida,H., Urakawa,T., Tamura,T. 1993. Consideration on reinforcing effect in bar like reinforcing members, Annual Conference, JGS, pp.2787-2290. (J)

Tatsuoka,F, Koseki,J, Tateyama,M. 1996. Performance of reinforced soil structures during the 1995 Hyogo-ken Nanbu Earthquake, IS-Kyushu '96 Special Report

Tayama,S., Maeno,H., Matsuyama,H. 1996. An investigation on durability of ground reinforcement technique, Tuchi-To-Kiso, JGS, Ser.No.465, Vol.44, No.10, pp.35-36. (J)

Earth Reinforcement, Ochiai, Yasufuku & Omine (eds) © *1997 Balkema, Rotterdam, ISBN 90 5410 833 9*

Finite element analysis on earth reinforcement – Current and future

Atsushi Yashima
Department of Civil Engineering, Gifu University, Japan

ABSTRACT: The Japanese Local Task Force Committee of Asian Technical Committee for Earth Reinforcement carried out extensive survey of technical papers related to the numerical analysis on earth reinforcement which were appeared in Journals and Proceedings for last 12 years. In this paper, current state and future trend of the Finite Element Analysis on earth reinforcement will be discussed based on this survey.

1 STATISTIC ON NUMERICAL ANALYSIS FOR EARTH REINFORCEMENT

Statistics on current state of the numerical analysis is summarized based on the survey of technical papers related to the numerical analysis on earth reinforcement which were published in Journals and Proceedings for last 12 years (1985 - 1996). The details of this survey are introduced in another report written by the Japanese Local Task Force Committee of Asian Technical Committee for Earth Reinforcement in this proceeding volume.

More than 160 technical papers in which there is a keyword "analysis" were selected and reviewed from the following journals and proceedings;

- ASCE Journal of Geotechnical Engineering
- Géotechnique
- Canadian Geotechnical Journal
- Soils and Foundations
- Geotextiles and Geomembranes
- Computers and Geotechnics
- Int. J. Num. Anal. Meth. in Geomechnics.
- Proc. JSCE
- Proc. IS-Kyushu ('88, '92 and '96)

Half of the selected papers treated the deforma-

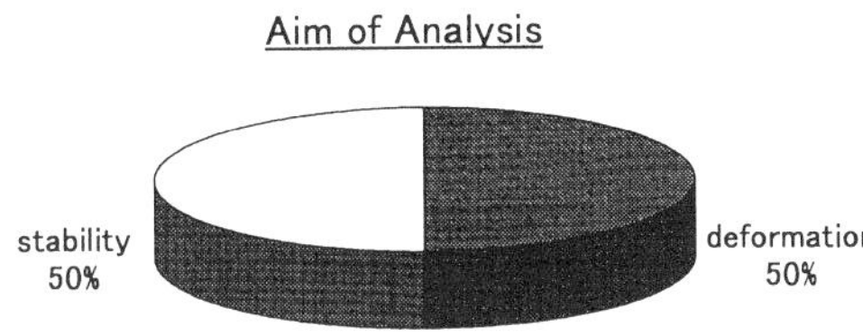

Fig.1(a)

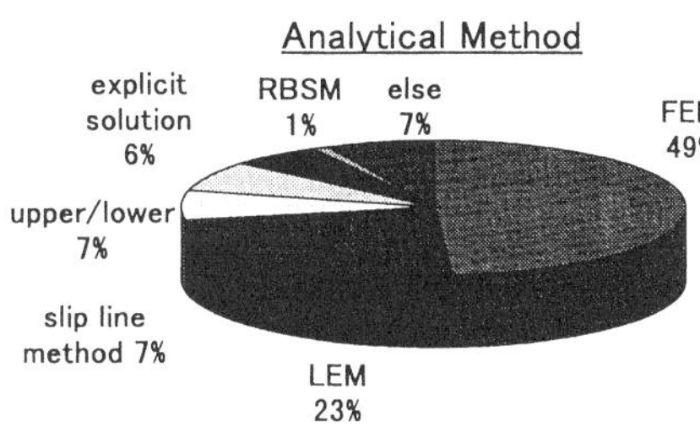

Fig.1(b)

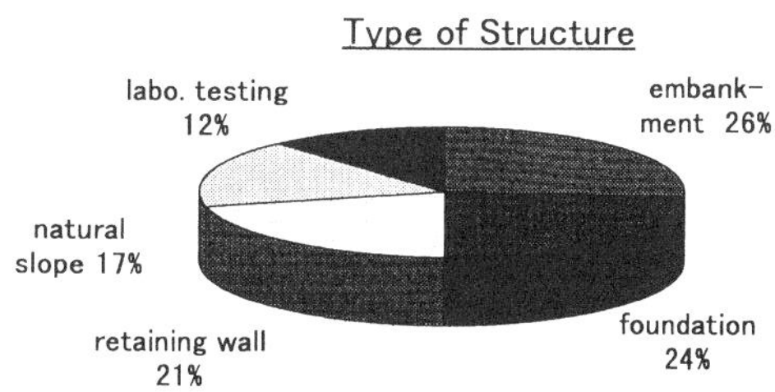

Fig.1(c)

Fig.1 Statistics on numerical analysis for earth reinforcement

tion problem and another half treated the stability problem (Fig.1(a)). It is found from Fig.1(b) that the finite element method (FEM) was used for the prediction of the deformation of the ground with earth reinforcement. On the other hand, the limit equilibrium method (LEM), slip line method, upper/lower bound theorem, RBSM, etc. were used for the assessment of the stability of the ground with earth reinforcement. Fig.1(c) summarizes the type of structure treated in the technical papers. Deformation/stability problems on embankment and foundation subsoil were mainly investigated by the numerical analysis. From this statistic, the finite element analysis on embankment and foundation subsoil with earth reinforcement is found to have been studied intensively for last 12 years. In this report, therefore, the finite element analysis will be discussed as a main subject.

2 CONTRAST BETWEEN LEM AND FEM

For the analysis by the limit equilibrium method (LEM) which is commonly used for the assessment of the stability of the ground with earth reinforcement, the design strength parameters are only needed. The construction process should be reflected in determining design strength parameters. We have to choose undrained, partially drained or fully drained strength of subsoil materials based on the construction process and permeabilities of subsoils. The compressive, extensive or simple shear strength should be carefully used based on the stress condition of the subsoil near the failure state.

For the analysis by the finite element method (FEM) which is used for the prediction of the deformation of the ground with earth reinforcement, on the other hand, we have to predict the elastic behavior, plastic (viscoplastic) behavior and failure (ultimate limit state) based on a correct initial condition (Fig.2). For this purpose, the correct initial condition, constitutive model and stable numerical algorithm must be prepared before the analysis.

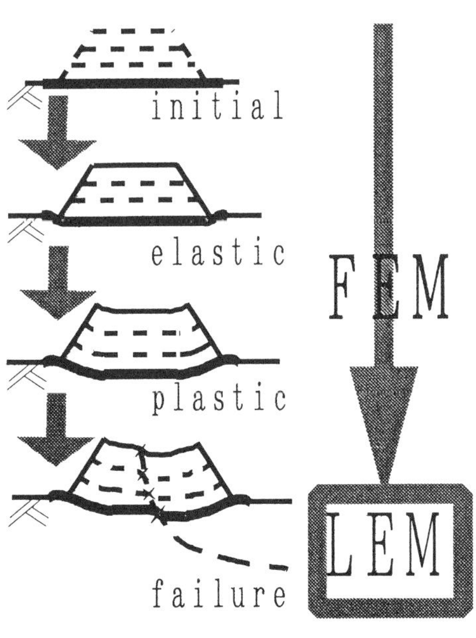

Fig.2 Contrast between LEM and FEM

3 CURRENT STATE OF FE ANALYSIS

3.1 *Merit of FEM in earth reinforcement design*

The finite element analysis is a more powerful analytical tool for solving boundary value problems than the limit equilibrium method.

The finite element analysis

- offers deformation, stress-strain distribution which informations are highly required in designing some of the important civil engineering structures,
- helps engineers to understand real mechanism in earth reinforcement,
- provides additional information to fully understand the complex interaction behavior which will be reflected in easy-to-use design method (refines the current design chart),
- validates a simplified design method,
- takes account of construction process which is one of the most dominant factors in determining reinforced soil behavior,
- identifies potential failure planes,
- is easily applied in observational method,
- etc.

3.2 *Difficulty in FEM based earth reinforcement design*

Why is the finite element method still far from being a practical tool for designing the earth reinforcement? The answers for this questions are;

- The correct initial condition is difficult to be postulated.

- The finite element analysis is poor at predicting failure (ultimate limit state).
- The finite element analysis is rather expensive and is usually limited to the mainframe computer.
- The mechanism of complex interaction is needed before analysis (It is impossible to obtain new mechanism which was not implemented a pri-o-ri in the modeling).
- How the complex system is going to fail is needed to know before modeling that failure.
- The finite element analysis is often thought to be a black box (refuses engineers to participate in the whole design process).
- The soil, reinforcement, and interaction properties under operational conditions are difficult to determine from the results of standard laboratory tests on component materials.
- Sophisticated constitutive model proposed has been not implemented for earth reinforcement analysis.

3.3 *Recent progress in FE analysis on earth reinforcement*

In 1980, Ohta et al. (1980) published the paper entitled "Investigation of Soft Foundations with Surface Reinforcement". In their paper, they discussed many important subjects on the prediction of the deformation of the embankment constructed on the clay foundation with surface reinforcement by the finite element analysis. In this chapter, we will understand the current state of the finite element analysis by answering the question "what is the progress from the paper by Ohta et al. (1980)?".

Soil modeling

In the paper by Ohta et al., the Cam-clay type elasto-plastic model with stress induced anisotropy (Sekiguchi and Ohta 1977) was used for the clayey foundation. For the analysis of the earth reinforcement, Mohr-Coulomb yield surface, Drucker-Prager yield surface, Modified Cam-clay model (Hird et al. 1990), elasto-viscoplastic model, nonlinear elastic model, etc. have been proposed and used for the foundation material. From now on, we should use a sophisticated constitutive model for foundation and embankment material.

Reinforcement modeling

In the paper by Ohta et al., an elastic thin layer element was used to model the behavior of reinforcement material. Based on the laboratory/in-situ tests for reinforcement materials, many constitutive models considering elastic, visco-elastic or plastic properties have been proposed. To model the reinforcement itself, the bar element, truss element or thin layer element has been used.

To obtain the reasonable prediction of the behavior of reinforced ground, the interaction behavior between soil and reinforcement material should be modeled appropriately. The special elements which simulate the interaction behavior have been proposed. A simple thin-layer element to model soil-geotextile interaction was proposed by Handel et al. (1990). The anisotropic element was designed by the homogenization technique. The joint (Kutara et al. 1986; Ochiai et al. 1988; Matsui and San 1989) or linkage elements has been used to reproduce the slippage between soil and reinforcement material. The nodal compatibility-slip model was proposed by Rowe (1984). Rowe has succeeded in predicting the deformation of the reinforced embankment with this special treatment in many case studies. Otani and Ochiai (1993) introduced a special rectangular element with high cohesion to model the interaction between soil and reinforcement.

Soil-water coupling

In the paper by Ohta et al., the soil-water coupling behavior was treated by introducing the variable of pore water pressure in the finite element formulation. In most papers appeared in journals and proceedings, fully undrained behavior has been considered for clayey foundation and fully drained behavior has been considered for sandy foundation. During the construction process, the generation and dissipation of the pore water pressure takes place at the same time at different places in the analyzed region. The soil-water coupled analysis is therefore recommended in the application to the reinforcement soft foundation.

Hardware

In the paper by Ohta et al., the parametric study on imaginary reinforced embankment and application to trial embankment were carried out using the mainframe computer. In most technical papers, mainframe computer has also been used for the numerical analysis. For the last decade, we have experienced a great progress in hardware of the personal computer (PC) and rapid development of pre- and post-processor. Now, the two-dimensional soil-water coupled analysis on earth reinforcement can be carried out in PC.

Static or dynamic

In the paper by Ohta et al., the static problem was only treated. Many reinforced earth structures have been reported to confirm their high seismic stability during recent large earthquake disaster. The reinforced soil is found to have high ductility during earthquakes. For the assessment of the stability of the reinforced soil during earthquake, the limit equilibrium method is commonly used. Only few papers dealt with deformation problems of the reinforced soil during earthquake by the finite element method. Uzuoka et al. (1996) tried to analyze the dynamic response of the embankment with/without reinforcement on liquefiable ground (Fig.3). They used effective stress based dynamic finite element method with beam element for reinforcement material.

Initial condition

To predict the deformation of the reinforced soil, the correct initial condition should be postulated. In the paper by Ohta et al., normally consolidated as well as overconsolidated stress state were assumed as the initial stress condition for the clayey foundation. The effect of the anisotropic consolidation was also taken into account. In most papers, on the other hand, much attention has not been paid to postulate the initial stress condition.

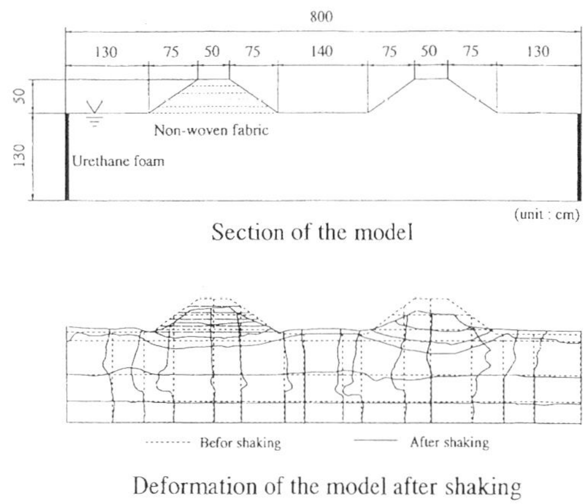

Fig.3 Dynamic response of the embankment with/without reinforcement on liquefiable ground (Uzuoka et al. 1996)

In the finite element analysis, the correct initial stress condition must be established before the analysis.

Infinitesimal or finite strain

It is well known that the restriction of the lateral movements of the ground surface beneath the embankment load can be effective in increasing the bearing capacity of the foundation and decreasing

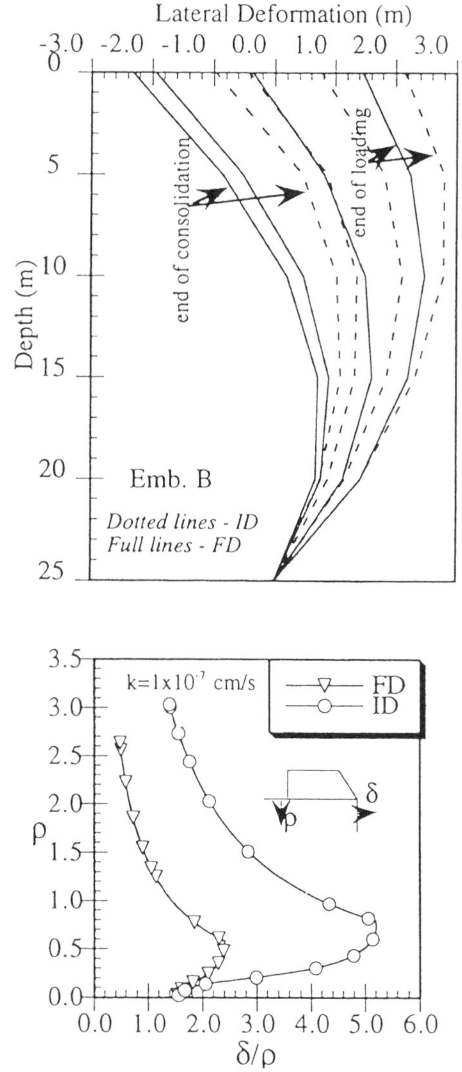

Fig.4 Infinitesimal strain theory vs. finite strain theory in predicting lateral deformation of embankment (Fernando 1996)

the undrained settlement. In the finite element analysis of the reinforced embankment, therefore, the prediction of the lateral displacement of the foundation is a very important subject. In the paper by Ohta et al., the infinitesimal strain was assumed. Fernando (1996) carried out a comparative study on infinitesimal strain theory vs. finite strain theory in predicting lateral deformation of embankment. It is found from Fig.4 that the infinitesimal strain theory fails to predict acceptable lateral deformation when significant geometry changes occur. To assess the stability of the reinforced soil which is close to the failure state by the finite element method, we have to study

- finite strain theory,
- strain localization,
- progressive failure,
- stable numerical algorithm,
- etc.

4 EXPECTED DIRECTION FOR FE ANALYSIS

4.1 *Purpose of FE analysis*

The purposes of the finite element analysis on earth reinforcement are summarized as;

- Qualitatively predict all features of composite structures (complex earth reinforcement) by using **only** the results of standard laboratory testing carried out on **component** materials.
- Provide a back analysis tool in observational method.
- Validate a simplified design method and refine the current design chart.
- Help engineers to understand real mechanism in earth reinforcement.
- Offer detailed informations to engineers in designing some of the important civil engineering structures.

4.2 *Acceptable Modeling (at least)*

- Composite modeling (e.g. by homogenization) is acceptable if there are numerous reinforcement elements. On the other hand, the interface elements are recommended to model a few reinforcement elements in the structure.
- Soil-water coupled analysis is recommended in the application to the reinforced soft ground.
- A linear or nonlinear elastic constitutive model is still acceptable for predicting a small strain distribution (e.g. less than 0.5~1%). To predict larger deformation up to failure, on the other hand, elasto-(visco)plastic constitutive model should be used.
- Two-dimensional idealization is acceptable in the form of a sheet or grid. Where strip or anchor reinforcement is used, the equivalent stiffness and equivalent interaction model in 2-D should be proposed based on a real 3-D behavior.

4.3 *Future Trend for FE Analysis*

Based on the previous discussions, the future trend for the finite element analysis on earth reinforcement is summarized as;

- PC-based finite element code as well as PC-based pre- and post-processor should be developed **(practical design tool, Computer Aided Design)**.
- Dynamic analysis is necessary.
- Three-dimensional analysis is necessary (e.g. for strip or anchor).
- The finite element analysis qualitatively evaluates new applications (new materials, new structures).
- To predict failure, finite strain theory, strain localization (shear banding), progressive failure, post peak behavior, stable numerical algorithm, etc. should be studied and implemented into the analytical tool.
- Sophisticated constitutive model should be used. At the same time, the optimum procedure for obtaining correct material parameters in the constitutive model from laboratory tests should be proposed.
- More attention should be paid to postulate the initial condition of the analysis.
- At least, deformation analysis should be carried out prior to a practical design in the case of the reinforced soft ground.

ACKNOWLEDGMENTS

The author wishes to express sincere thanks to all the members of the Japanese Local Task Force Committee of Asian Technical Committee for Earth Reinforcement for their devotion in surveying technical papers related to the numerical analysis on earth reinforcement.

REFERENCES

Fernando,G.S.K. 1996. Non-linear consolidation deformation analysis of soils under embankment loading. Dissertation for Degree of Doctor of Engineering, Nagoya University.

Handel,E., Schweiger,H.F. and Yeo,K.C. 1990. A simple thin-layer element to model soil- geotextile interaction. Performance of reinforced soil structures, British Geotechnical Society, 317-321.

Hird,C.C., Pyrah,I.C. and Russell,D. 1990. Finite element analysis of the collapse of reinforced embankments on soft ground. Géotechnique, 40:4, 633-640.

Kutara,K., Gomadou,M., Takeuchi,T. and Maeda,S. 1986. Deformation analysis of geotextiles in soils using the finite element method. Geotextiles and Geomembranes, 4, 191-205.

Matsui,T. San,K.-C. 1989. An elastoplastic joint element with its application to reinforced slope cutting. Soils and Foundations, 29:3, 95-104.

Ochiai,H. et al. 1988. Pull-out behavior of polymer grid in soils and its analytical method. Memoirs of the Faculty of Eng., Kyushu Univ. 48:2, 125-139.

Ohta,H., Mochinaga,R. and Kurihara,N. 1980. Investigation of soft foundations with surface reinforcement. Proc. 3rd Australia - New Zealand Conf. on Geomechanics, Wellington, 360-365.

Otani,J. and Ochiai,H. 1993. Bearing capacity of the reinforced soils by rigid plastic finite element analysis. Proc. 2nd Asian-Pacific Conf. on Computational Mechanics, Sydney, 455-460.

Rowe,R.K. 1984. Reinforced embankment: Analysis and design. Journal of Geotechnical Engineering Division, ASCE, 110:GT2, 231-246.

Sekiguchi,H. and Ohta,H. 1977. Induced anisotropy and time dependency in clays. Constitutive Equations of Soils, Proc. Specialty Session 9, 9th Int. Conf. SMFE., Tokyo, 229-238.

Uzuoka,R., Mihara,M., Oka,F. and Yashima,A. 1996. Behavior of reinforced earth embankments on liquefiable sandy ground. Proc. Int. Sympo. on Earth Reinforcement. IS-Kyushu, Fukuoka, 273-278.

Earth Reinforcement, Ochiai, Yasufuku & Omine (eds) © 1997 Balkema, Rotterdam, ISBN 90 5410 833 9

Limit state design of reinforced soils

Malcolm D. Bolton
Cambridge University Engineering Department, UK

INTRODUCTION

Limit state design codes should aim to assist the designer in preventing failures by:
(i) defining the envelope of hazards which the structure must survive,
(ii) clarifying the various modes of failure which should be checked,
(iii) adopting a robust and holistic approach to the engineering, based on sound principles such as equilibrium but which also includes good connection details, construction practices, and quality control procedures.

The new British Code of Practice for Reinforced Soils, BS 8006, claims to have achieved this. It has made a valuable contribution, but it has also introduced some features which are inconsistent and inadvisable. This is discussed below.

DESIGN SCENARIOS

A design code should generate design scenarios which are both onerous and believable, e.g. severe loads, high water tables, poor quality fill, aggressive groundwater, corroded or degraded reinforcement. This requires judgement, both from the code-drafter and the code-user. High standards demand more than a check-list of actions. Guidance must be given on the precise design conditions. This is sometimes presented as a matter of selecting "load combination factors", but it is more helpful to create "design scenarios".

For example, consider the question of whether a reinforced soil structure should be designed on the basis of the worst believable degradation of the reinforcement, or the worst credible surcharge, or the 1000 year earthquake, or all three simultaneously. It is expensive in materials (but cheap in design fees) to design every structure for the worst corrosion ever seen anywhere, *and* the biggest earthquake ever experienced anywhere. It saves in construction costs (but is more expensive in design time) to take a more scientific view based on data relevant to the particular site.

The aim should be to ensure firstly that material deterioration does not cause failure until the design life of the structure has been exceeded, and secondly that when failure does occur it will not be catastrophic. A more accurate (less conservative) design always demands more information - on groundwater chemistry, and the actual deterioration of similar materials in the ground, for example. However, some essential aspects will remain a matter of judgement - the likelihood of road salt or spillages reaching vulnerable materials, or the future performance of a new polymer. The same qualifications apply to the information which can be obtained on seismicity or other live loading for a structure at a given location.

BS 8006 issues a check-list of actions together with some strong prescriptive rules, such as the required thickness of sacrificial metal to compensate for corrosion in various steels, the presumed efficiency of different classes of joint, and the electrochemical properties of fill in relation to the selection of reinforcement materials.

It does not deal with seismicity. It is, however, interesting to reflect on the general principles which might apply to the specification of design scenarios which could include such exceptional environmental actions. Consider the following proposal.

In terms of loading, a retaining structure faces hazards of three main types - compaction during its construction, live load surcharge in normal service, and rare environmental incidents such as a severe earthquake or erosion at the base due to flooding. In terms of material properties, a structure tends to deteriorate during its design life. The designer should derive design values of superimposed surcharge and material deterioration, each of which could reasonably be expected to represent the limit of the hazard during the design life of the structure. A design flood leading to erosion, and a design earthquake will also be specified which could reasonably be expected to occur only once in the design life of the structure.

These influences should be considered to co-exist in certain patterns as exemplified below. The structure should be able to survive construction and service life loading together with lifetime deterioration without exhibiting a serviceability failure. It should not collapse suddenly due to a design earthquake *or* a design flood even while carrying design surcharge and after design lifetime deterioration, but either of these conditions may cause permanent deformations which represent a serviceability failure. The structure, its materials and connections must, however, be designed to maintain sufficient ductility and continuity to survive all credible design scenarios without collapsing suddenly.

Although this particular wording may be contentious, it should be clear that it does set a standard. It draws attention to the avoidance of sudden collapse, which will lead the designer to consider how to maintain continuity and ductility of the structure. It specifies believable but extreme events with return periods equal to the design lifetime of the structure, rather than extraordinarily severe events which almost never happen. It then asks for the design to assure survival under a simple combination of these extreme events. If there is no relevant database from which probabilities of failure could be derived, they can play no part in the process of decision-making. However, the recommended approach demands that sudden collapse through loss of stability (ULS failure) should always be preceded by obvious deformations (SLS failure) which would give the owner time to evacuate those in danger.

MODES OF FAILURE

It has become natural in limit state design codes to discriminate between deformation (serviceability limit states, SLS) and collapse (ultimate limit states, ULS). Unfortunately, the original understanding of the distinction has been eroded. Many structural engineers actually design to limit deformations in steel and concrete, but erroneously say that they are "designing against ULS". It seems important for the designers of reinforced soil structures to be very clear about the difference.

Soil is inherently brittle, especially when it is well-compacted, stiff or dense. It will mobilise a peak strength (expressed as a secant angle of internal friction) ϕ_{max} at about 1 or 2% axial strain in a triaxial test, which will then drop to the ultimate strength ϕ_{crit} as the soil ruptures on slip bands. Only about 5 particle diameters of slip will be sufficient to "fully soften" the soil in this way. If a soil structure has been designed to mobilise $\phi > \phi_{crit}$ at collapse, then if it collapses it will collapse violently with mass-accelerations, since its resistance will be dropping as it deforms. The new British Code of Practice on Earth Retaining Structures, BS 8002, regards this as inherently dangerous: it demands that $_{design}\phi \leq \phi_{crit}$.

BS 8006, takes a different approach - it was written by a different committee, with no liaison. It permits ϕ_{max} in ULS calculations for reinforced soil zones on the grounds that the reinforcement will prevent slip bands forming, and ϕ_{crit} in unreinforced zones. But it also factors up all soil densities and live loads so that in reality much smaller angles of friction would be mobilised at failure if failure actually occurred. It is self-evident that soil can not ever weigh 30 kN/m^3, which is the typical unit weight derived for granular fill after applying a partial factor of 1.5 as BS 8006 requires for retaining walls. BS 8006 can not therefore be described strictly as a limit state design code. By the time the partial factors have been applied, the design scenario is not believable; the code has introduced arbitrary safety factors which are unrelated to the parameters to which they have been applied.

Soil is also inherently compliant; it needs to suffer large strains if it is to mobilise even its peak strength. BS 8002 took the view that a displacement / height ratio of 0.5% was as large as the owners and users of retaining walls would generally accept. It deduced that a 0.5% strain in a triaxial test, or a 1% shear strain in a plane test, represented the serviceability limit for soils. It went on to propose for moderate to good soils that the angle of internal friction at 1% shear strain could be presumed to be $\phi_{1\%} = \tan^{-1}(\tan\phi_{max}/M)$ where the mobilisation factor $M = 1.2$ would be sufficient to guard against such strains. In that case, we can apply the constraint $_{design}\phi \leq \phi_{1\%}$. What has then been derived in BS 8002 is a design soil strength which satisfies both safety (ULS: $_{design}\phi \leq \phi_{crit}$) and serviceability (SLS: $_{design}\phi \leq \phi_{1\%}$) constraints. It is a mobilisable soil strength.

Interestingly, BS 8006 adopts a similar approach to the selection of a mobilisable reinforcement strength. The design tensile strength is taken to be the lower of the ultimate (or creep rupture) strength factored down only for uncertainty, and the tension mobilised at the serviceability limit of extension in service (typically 0.5% to 1%). The BS 8002 approach to the mobilisable strength of soil, and the BS 8006 approach to the mobilisable strength of reinforcement are, most appropriately, compatible.

Finally, bond failure must be averted. If the roughness of the reinforcement exceeds the mean particle size it may be taken that the interface friction matches the internal friction of the soil, so that $_{design}\mu = {}_{design}\tan\phi$. Otherwise, shear box tests have to be performed sliding soil against a block of reinforcement fixed in the lower half of the box, to give a smaller value of interface friction. This approach is set out in BS 8006, which recommends factoring down the interface friction by 1.3, bringing the friction parameter down almost exactly to the $_{design}\tan\phi$ value which could be derived via BS 8002. BS 8006 also permits pull-out tests to determine an equivalent coefficient of interface friction μ^* which can take advantage of arching of stress onto the reinforcement due to dilation around strips or rods. It does not, however, seem to be certain that an isolated pull-out test can fairly represent the degree of arching in the field, especially in earthquake loading.

It is interesting to compare earth pressures generated by the alternative design approaches in BS 8002 and BS 8006. Consider a well-compacted fill with typical properties $\phi_{max} = 44°$ at $\gamma = 20$ kN/m^3, and $\phi_{crit} = 35°$. BS 8002 would derive $_{design}\phi = 35°$, and $_{design}K_a = 0.27$. The mobilisable active earth pressure at 5 m depth would then be 27 kPa. BS 8006 would use $\phi_{max} = 44°$ to get $K_a = 0.14$ but would apply this to soil with $\gamma = 30$ kN/m^3 to give an earth pressure in reinforced fill at 5 m depth of 21 kPa. This would be later factored up again by 1.1 if the consequences of failure would be life-threatening, so reinforcement capable of withstanding 23 kPa would be provided. This is not very different from the BS 8002 result.

A much larger discrepancy is revealed when dealing with the unreinforced backfill. BS 8006 then imposes $\phi_{crit} = 35°$ to get $K_a = 0.27$, whilst still factoring up the density as before; the design earth pressure at 5 m becomes 45 kPa which would correspond to an earth pressure coefficient of 0.45 in real soil. BS 8002 would derive the same earth pressure as before, 27 kPa. Considering the lack of clarity which already exists regarding the correct earth pressure coefficients to use with real soil, it seems a little unfortunate that BS 8006 chose to factor up soil density instead of factoring down soil strength.

BS 8006 shows a complete compendium of limit equilibrium analyses, with valuable references. It seems to demand some redundancy in the making of equilibrium checks, but perhaps this is no bad thing. Its main weaknesses seem to be:

- soil strength data may be fitted either by tangent parameters (c', ϕ') or by a secant peak angle of friction ϕ_{max} - different partial factors are proposed for these alternative approaches, and this is inconsistent;
- soil density is strongly factored up to an unbelievably high value - this undermines reliability, produces unbelievably high earth pressures, and causes difficulty in dealing with effective stress analyses under water where the unit weight of water must equally be factored up (though this seems not to have been recognised);
- there is no recognition that soil also needs to strain to mobilise strength - peak soil strength is used in design calculations for earth pressures and a small strength reduction factor of 1.35 is advocated in calculating the bearing pressure of soft clays whereas deformation considerations might have led to a mobilisation factor between 1.5 and 2.0 (this unconservatism presumably compensates for the excessive conservatism referred to immediately above);
- there is neglect of the well-known question regarding the impossibility of mobilising bond on both top and bottom surfaces of sheets of reinforcement;
- although certain ductile metals are listed as usable, and there is an admirable exposition of connection details for example, the vital importance of ductility and continuity are under-emphasised.

A very simple design approach now suggests itself, however, based partly on BS 8002 and partly on BS 8006. Let us design so that the strength of soil nowhere exceeds $_{design}\phi$ and so that the strength of reinforcement nowhere exceeds $_{design}T$, each of which satisfy criteria both of safety and serviceability. Reduce the frictional bond to $_{design}\mu$ where measurements on smooth materials can be used to define a lower bound. Satisfy rigorous but believable design scenarios, but do not apply any other factors of safety. Instead, simply demonstrate in each scenario that an equilibrium free-body diagram can be constructed for the reinforced soil zone acted upon by backfill pressures and surcharges, and held in place by mobilisable bearing pressures on its foundations. Then, demonstrate that internal stresses could be mutually in equilibrium without exceeding the design strengths discussed above. Wedges or other failure mechanisms, earth pressure diagrams and stress analyses - indeed any validated limit equilibrium analysis - could then be used to verify that the structure has internal equilibrium. If there is a stress distribution inside the structure which satisfies equilibrium under these conditions, the structure can neither collapse nor deform excessively.

ENGINEERING

It is very much to be welcomed that BS 8006 contains a good account of connection details, construction methods and quality control procedures. The reliability of the finished product is strongly dependent on the intelligent management of the design process and the skill of site personnel. Two practical issues, which are fundamental to reinforced soils, are not given sufficient significance: compaction pressures in fill, and the ductility of reinforcement and connections.

Reinforced soil structures work by trapping high earth pressures in a zone which is laterally confined by reinforcements which must extend; they thereby develop tension which maintains stability. This fundamental mechanism comes into play whether the structure is resisting gravity, surcharge or earthquake. Compaction pressures locked into retained fill during construction are not inherently a bad thing. They "proof-test" the structure against some future surcharge in service, or an earthquake, either of which could create identical earth pressures in a quite different way. Only if the earth pressures during construction far exceed those predicted for design live loading or earthquake loading will it be advisable to consider how to reduce them.

Ductility and continuity are essential to a safe structure. There is danger in permitting sacrificial steel around connections which are already weak; the ultimate tensile strength of the connection might then be smaller than the yield strength of the parent reinforcement. Such a tie would fail without much elongation. Progressive failure has been observed in centrifuge tests on reinforced soils with brittle ties. It seems highly desirable to set a ductility target in terms of the extension to failure of reinforcement / connection / panel systems. A stretch of 5% to fracture of deteriorated materials, following the achievement of the design value of reinforcement tension, would alert the authorities and avoid catastrophe in the event of an unforeseen overload.

CONCLUSION

Limit state design offers a flexible framework within which clear standards can be set. The new UK Code, BS 8006, has made some useful contributions. It is much preferable to use the concept of mobilisable strength rather than adopting arbitrary partial factors. Mobilisable strength is simpler to understand, easier to apply, and more faithful to the fundamental principles of mechanics. It encompasses both deformation and strength criteria. All that is then required are realistic but severe design scenarios, and a rigorous check for global and local equilibrium.

BIBLIOGRAPHY

1. BSI: *'Code of Practice for Earth Retaining Structures'*, BS 8002, 1994.
2. BSI: *'Code of Practice for Reinforced Soils'*, BS 8006, 1996.
2. Bolton, M.D.: 'What are partial factors for?', *DGF Bulletin* 10 / 3, 565-583, Danish Geotechnical Society, Copenhagen, May 1993.
3. Bolton, M.D.: 'Codes, standards and design guides', *Retaining Structures*, Thomas Telford, 387-402, 1993.
4. Bolton, M.D.: 'Design methods', *Prediction and Performance in Geotechnical Engineering*, 50-71, Thomas Telford, 1993.

Earth Reinforcement, Ochiai, Yasufuku & Omine (eds) © 1997 Balkema, Rotterdam, ISBN 90 5410 833 9

Soil reinforcement applications: Promising directions for the future

J.P.Gourc
IRIGM, University Joseph Fourier, Grenoble, France

ABSTRACT : Retaining structures with the association of soil and geosynthetics now can be considered as conventional earthworks. But new applications of this technique of reinforcement, related to the versatility of geosynthetics, are coming to light. Some of them are presented here, reinforcement of waste and poor geomaterials, reinforcement by geocells structures, reinforced soil raft. The development of these concepts, already experimented in the field needs an accompaning research to quantify and to adapt the efficiency of the reinforcement to the problem.

1 INTRODUCTION

The application of geosynthetics in reinforcement is not exclusively the use of horizontal sheets associated to soil body in embankments and retaining structures. Recent experimental and research works open promising directions for the future of soil reinforcement. Some examples are presented below. It is reasonable to think these applications will be able to bring a new development of the market of geosynthetics.

2 REINFORCEMENT OF WASTE AND POOR GEOMATERIALS

2.1 *Landfills and waste*

To find new sites for storage of waste is more and more difficult. Consequently the landfills become bigger and bigger. Vertical expansions of waste disposals imply new concept of dumping. It is now frequent to notice sliding of slopes in landfills because these embankments are steeper and steeper, and furthermore the geotechnical properties of waste are not well known. Geotechnical survey of waste is up to date.

Reinforcement is a new solution to stabilise dump : the first application was the use of reinforcement of soils for dikes surrounding the dump (figure 1) but reinforcement of waste itself can be proposed.

Here is the cross section figure 2 (Artières et al, 1994) where the domestic waste is reinforced by successive sheets of geosynthetics, in this case dual function geosynthetics, woven structure for reinforcement and non-woven structure for drainage of leachate. Of course, the functions of the geotextile have to be considered as temporary because durability

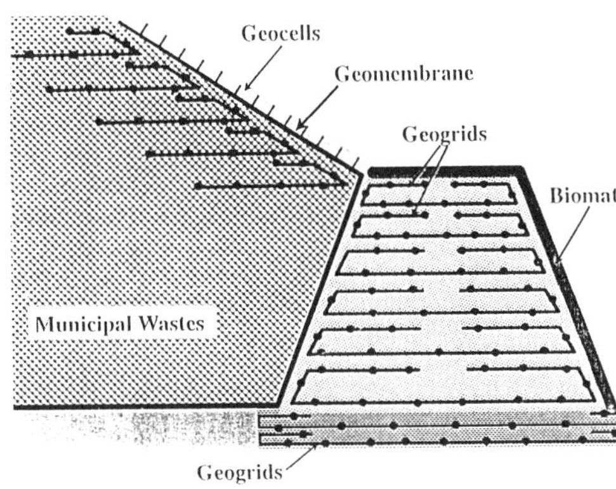

Figure 1 : Reinforcement of dikes and waste in a landfill

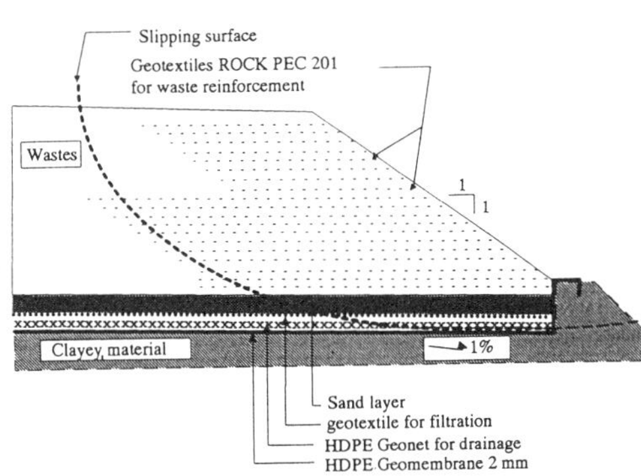

Figure 2 : Cross section of a landfill with reinforcement of waste

of polymers in these conditions are problematical. Domestic waste is generally an alkaline material (pH > 9). But it is worth noting that, for domestic waste, a significative improvement of mechanical properties is observed with elapsed time, correlated to a chemical + physical evolution of the material which is not properly consolidation observed for soils.

2.2 *Reinforcement of slags of incineration*

New requirements of the French Ministry of Environment involve from the year 2002 a treatment of all the solid waste before disposal, for instance incineration. But from four tonnes of waste, it remains 1 tonne of bottom ash. In the present time these slags resulting from incineration are dumped in landfills. According to present european regulations (January 1992), these slags can be used for public works, depending of the heavy metals content. The quality of ash is of course correlated to the type of domestic waste : the increasing use of sorting process after collection is very efficient to improve the environmental safety of slags.

However the use of slags of incineration is only acceptable after several months of "maturation". During this period, chemical-physical evolution of the material is observed, with an exothermic reaction.

In the present time, this geomaterial is free of charge for the contractor, because its use is stimulated by the environmental authorities.

The mechanical properties of the ash are quite good, the average value of the friction angle ϕ is higher than 30° and the cohesion higher than 5 kPa. Several applications were carried out in the region of Grenoble, where the backfill in ash was associated to reinforcement by geogrids or geotextiles. The main application was the remediation of roads in landslide areas. In this case, it is efficient to lighten the backfill by the way of layers of old tires, not entirely filled with ash or even separated of the ash layers by geotextiles. Here is the cross section (figure 3) of a slag backfill reinforced by geogrids in polyethylene (Bossoney et al, 1995) and on the figure 4 a photo of a backfill combining ash, old tires and woven geotextile in polypropylene.

These embankments were monitored but no significative increase of the temperature was observed.

Old tires constitute an huge problem for the environment, the fire risk of this kind of storage is important. We mention an example of application related by Prof. Wu from Denver University : old tires are cut up, and the small pieces can be used as a light granular geomaterial. To protect a landslide in Colorado, a backfill in cut tires reinforced by geogrids was built (figure 5).

Finally, the spreading of the reinforcement of these specific materials is founded on a better knowledge of the specific mechanical behaviour of these materials and of their interfaces with additive geosynthetics.

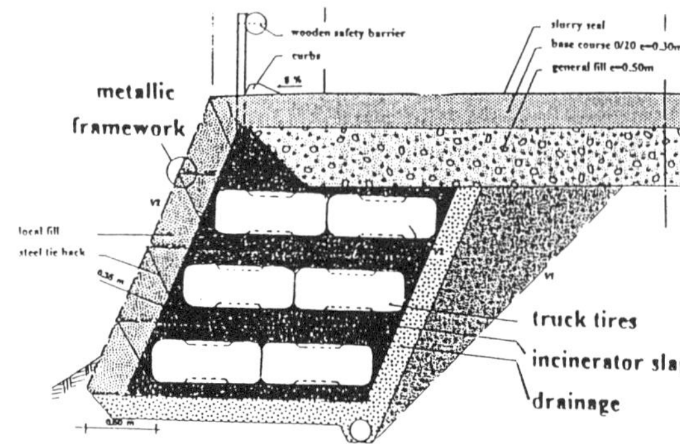

Figure 3 : Cross section of a reinforced backfill in ash of incineration

Figure 4 : Backfill combining ash, old tires and geotextile

Figure 5 : Light backfill in pieces of tire with geogrid reinforcement (Wu)

Figure 6 : Retaining structure in clay reinforced by geocell for a landfill

Figure 7 : Clay barrier reinforced by metallic gabions

3 REINFORCEMENT BY GEOCELLS STRUCTURES

Since several years, geocells, initially used for erosion control of slopes to improve vegetalisation of the facing, are experimented for retaining structures.

The layers of geocells are lied horizontally and filled with compacted soil. One interest of this technique is the confinement of poor mechanical or cohesive soils. In the past, mattress of geogrids were proposed to create a rigid raft for embankments on weak soil. An example of experimental retaining structure with geocells is presented at the conference Kyushu IS 96 (Gourvés et al, 1996) : the wall is 5.4 m high with a width corresponding to two or three cells and is anchored by means of longer sheets of geocells. The experience of reinforcement of a very wet soil demonstrated the capability of this type of reinforcement to support large strains.

Another recent application (figure 6) was the construction of a vertical dike in clay for a landfill with reinforcement by Geocells. In this case, the function of this structure is only to act as a retaining wall, but a new relevant application could be the use of this technique to produce vertical clay barriers for landfills.

In many countries, the environmental regulations require a five meters wide clay barrier for waste disposal, and to build a such artificial barrier on steep slopes is very difficult. Geocells could be one of the only ways to success to obtain economically very steep or even vertical mineral barriers. It would be however necessary to specify for geocells impermeable sheets of geosynthetics.

Here it is an example of an alternative technique for vertical clay barriers (figure 7), in Germany, reported by L. Wechter : the body of the barrier is in clay reinforced by a metallic grid and the double facing is in gabions used as protection and filter-drain.

However, the development of geocells technique needs a better knowledge of the mechanical behaviour of this three dimensional reinforcement. Interesting results on pull-out tests and simple shear tests in a large box performed for different directions of the geocells layers were published (Gourvés et al, 1996).

4 REINFORCED SOIL RAFT

4.1 *Foundation upon a heterogeneous soil (figure 3)*

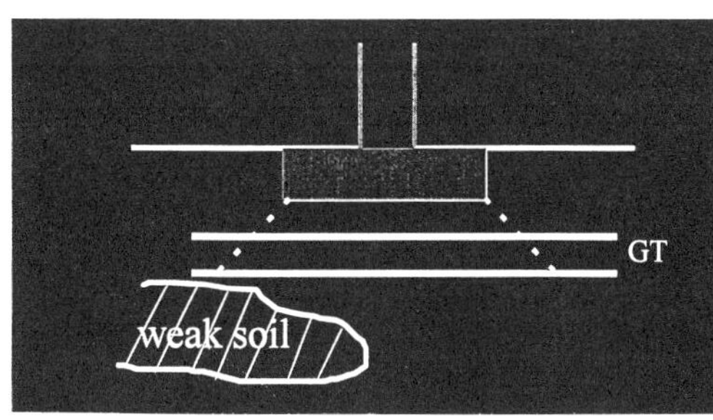

Figure 8 : Reinforced soil raft

The basic concept is to realise a mattress of soil below the foundation, rigid enough to be able to spread the localised overburden, so that differential settlements will be restricted.

It is typically an important application of reinforcement of soil. But surprisingly, reinforced soil raft is not a widespread technique, when the cases of potential use are numerous. In the literature, a lot of researches on this topic are displayed, numerical approaches or small scale models, but almost all the studies are focalised exclusively on the bearing capacity problem. And, still today, to evaluate the settlements in a serviceability state is a difficult process on account of the lack in the interaction soil-reinforcement knowledge.

The results obtained by different researchers are

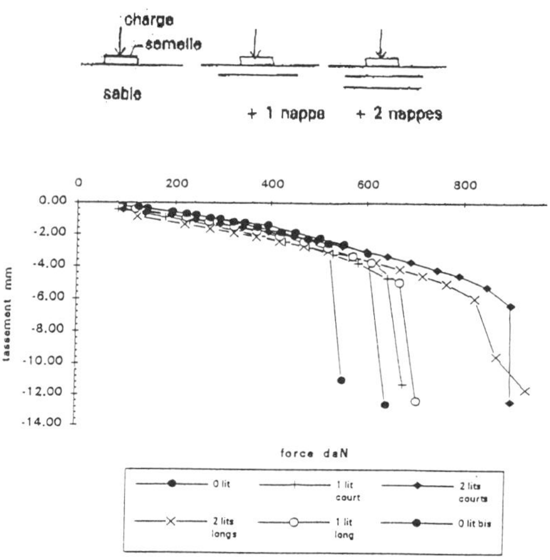

Figure 9 : Punching test in centrifuge (Dubreucq)

also often controversial. We exhibit here some recent diagrams obtained by Dubreucq by means of centrifugal tests in Lcpc Nantes on sand reinforced by geotextile. The sheets of geotextiles (figure 9) are short (the width of the foundation) or long (three times the width of the foundation).

Clearly there is no influence of the length of the reinforcement. Only the central part of the geotextile is efficient along all the test.

We believe important to encourage research (and large scale experimentation) on this topic likely to open a lot of applications.

4.2 *Foundations upon sinkholes (figure 10)*

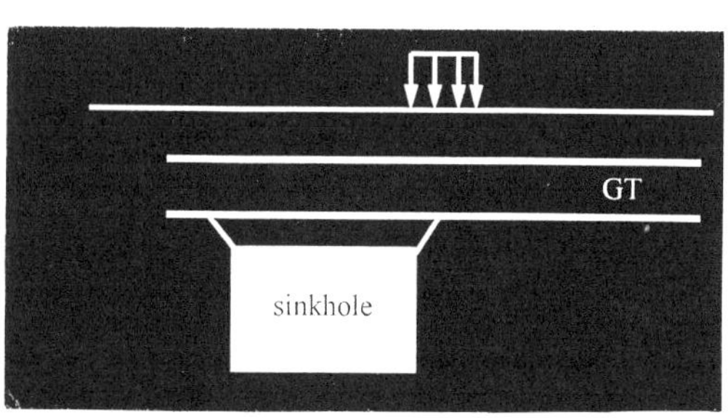

Figure 10 : Foundation on sinkhole

The development of sinkhole can result of many reasons (Kempton et al, 1996), mining work, gypsus cavities, ...

The challenge is generally, not to prevent any settlement, but only to avoid a catastrophic subsidence above the holes which weren not localised. As in the former case of heterogeneous soil foundation, the proposal is to include a reinforced

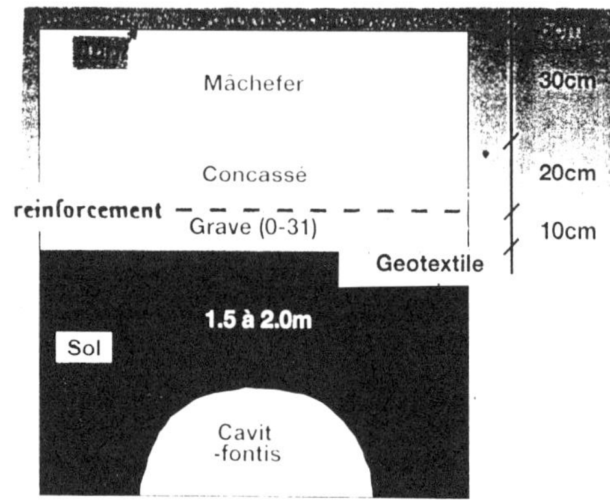

Figure 11 : St Etienne Stadium : reinforcement to prevent sinkhole

layer providing bending strength to support the load above the void.

For the design of this reinforced structure, the first idea was to propose a mechanism where the geosynthetic acts as a membrane (with exclusively tensile strength). It is analytically proven that the membrane effect needs large deformations to be effective. This being so, catastrophic subsidence is avoided, but the structure is not yet in situation to fulfil its function (for instance regular traffic for a road).

Different researches are in progress to evaluate a combined behaviour of soil and geosynthetic, with a more sophisticated approach than the membrane effect. In France, a large programme is initiated by Irigm, with Lcpc and the french companies for railways and highways with different requirements in the two different applications (very small differential settlements allowable for railways).

Tests of subsidence in large pits will be carry out to calibrate the numerical approach.

It is really a large market. For example, more than 300 km of the new track for the Tgv train are concerned.

To demonstrate the versatility of this technique, finally we present a particular application : for the world cup of football in 1998 in France, the stadium of St Etienne was remediated. Mining galleries are localized beneath the lawn, being the cause of local settlements. A reinforcement by geogrid of the subgrade was carried out (figure 11).

The technique of reinforced soil raft has still other applications like the reinforcement of caps of landfills subjected to local differential settlements. But the challenge here is different, it is to keep the water-tightness of the cap liner, to avoid tensile breaking of the geomembrane or bending cracks of the clay mineral barrier.

REFERENCES

Kempton, G.T., Lawson, C.R., Jones, C.J.F.P.,

Demerdash, M. 1996. The use of geosynthetics to prevent the structural collapse of fills over areas prone to subsidence, 1st European Conference on Geosynthetics, Maastricht, Nederland : Balkema.
Bossoney, G., Dieny, P., Gotteland, Ph. 1995. An innovative ecotechnical alternative : an embankment of incinerated municipal solid waste reinforced with geosynthetics, *2nd French Conference on Geosynthetics, Rencontres 95*. Beaunes, France.
Gourvés, R., Reiffsteck, P., Vignon, J.F. 1996. *3rd International Symposium on Earth Reinforcement*. Fukuoka, Kyushu, Japan : Balkema.
Artières, O., Delmas, Ph., Correnoz, M., Oberti, O. 1994. New concepts in waste disposal, Improved by geosynthetics. *5th International Conference on Geosynthetics*. Singapore.

Asian Technical Committee reports on earth reinforcement

Earth Reinforcement, Ochiai, Yasufuku & Omine (eds) © *1997 Balkema, Rotterdam, ISBN 90 5410 833 9*

Asian Technical Committee for earth reinforcement – Review of current state in earth reinforcement practice

H.Ochiai
Department of Civil Engineering, Kyushu University, Fukuoka, Japan

1 INTRODUCTION

Earth reinforcing techniques have become a useful and economical solution to many problems in geotechnical engineering practice. In recent years, considerable interest on this technique has been promoted both fundamental and practical studies as well as the development of various types of reinforcing materials. It follows that new techniques for earth reinforcement and their applications to geotechnical engineering practice are developing rapidly. Especially in Asia, there are many regional and difficult subsoils and many projects for earth reinforcement practice have been running around the whole area of Asia. But it is considered that the activities in each country so far have not been exchanged among Asian countries and these information should be exchanged in order to improve this technique.

The two international symposia for earth reinforcement, IS Kyushu '88 and '92, were held in Fukuoka, Kyushu, Japan, which had provided successful contributions for the development of earth reinforcement technique. The third symposium, IS Kyushu '96 was just finished last November in 1996. Not only these activities but also any other activities in Asia for earth reinforcement practice have been successfully organized.

2 OBJECTIVES

Under these circumstances, The Asian Technical Committee (Asian TC) has been organized under the Japanese Geotechnical Society. The objectives of this committee are:

1) to continue in developing and spreading the knowledge of earth reinforcement practice;
2) to facilitate the activities of earth reinforcement research in Asia; and
3) to support IS Kyushu '96.

3 ACTIVITIES

The contents of the activities are summarized as follows:

1) The establishment of Asian TC by Asian oriented members;
2) Construction of network system using electric mail(e-mail) among Asian TC members:
3) Exchanging of current situations of research and practice on earth reinforcement in each country ; and
4) Supporting IS Kyushu '96 for both technical and management points of view. organizing Asian TC members meeting at IS Kyushu '96.

In order to complete these activities, a Supporting Committee has been organized under the Japanese Geotechnical Society in Japan.

4 NEWSLETTER (ERA News)

And also, in order to exchange the information about each activity on earth reinforcement in Asia, Newsletter (**ERA News**) has been started to publish under this committee in every three or four month (three or four issues in each year) from 1996.

Followings are the possible contents of this ERA News:

1) MEMBERS NEWS:any news about earth reinforcement activities in Asia or some essays by the members;
2) DIARY OF EVENTS: information of upcoming conferences and symposia related to the topic on earth reinforcement ;
3) TECHNICAL REPORT: reports on the case histories of earth reinforcement practice including its design or construction methods in each countries; and
4) THE OTHERS

5 MEMBER OF THE COMMITTEE

Followings are the member of the committee:
Prof. Ochiai,H.: Chairman
Kyushu University, JAPAN
Prof. Otani,J.: Secretary
Kumamoto University, JAPAN
Members
Prof. Bergado, D.T.
Asian Institute of Technology, THAILAND
Dr. Cho,S-D.
Korea Institute of Technology Construction, R. KOREA
Dr. Fukuda, N.
Fukken Co.,Ltd., JAPAN
Prof.Guler, E.
Bogazici University, TURKEY
Prof. Huang C.C.
National Cheng Kung University, TAIWAN, R.O.C.
Prof. Karunaratne,G.P.
National University of Singapore, SINGAPORE
Prof. Kabir, M.H.
Bangladesh University of Engineering and Technology, BANGLADESH
Prof. Madhav, M.R.
Indian Institute of Technology at Kanpur, INDIA
Prof. Yang, Can-Wen
China Academy of Railway Sciences, P.R.CHINA
Dr. Yeo, K.C.
Civil Engineering Department, HONG KONG
Prof.Emiritus Wiseman, G.
Israel Institute of Technology, ISRAEL

6 JAPANESE SUPPORTING COMMITTEE

In order to enhance these activities, the Japanese Supporting Committee for Asian TC has been organized in the Japanese Geotechnical Society. The task of this committee is to support any activities by Asian TC. In this committee, three working groups have been set as follows:

1) Group I : TESTING METHODS
2) Group II : DESIGN METHODS
3) Group III : NUMERICAL ANALYSIS

The final reports for the results of each activity will be published as a technical paper here in the proceedings.

Each title of the papers are listed as follows:

Group I: Testing methods of reinforcements for use in reinforced soil structures:
Group II: Present state and perspective of design methods of reinforced soil structures
Group III: Current state on numerical analysis of reinforced soil structures

7 MEMBER OF THE JAPANESE SUPPORTING COMMITTEE

Ochiai,H.: Chairman
Kyushu University, Japan
Prof.Otani,J: Secretary
Kumamoto University, Japan
Dr. Fukuda, N.
Fukken Co.,Ltd., Japan
Prof. Hayashi, S.
Saga University, Japan
Mr. Hirai, T.
Mitsui Petrochemical Industrial Products Ltd., Japan
Dr. Kodaka, T.
University of Tokyo, Japan
Dr. Kumagai,K.
Maeda Corporation, Japan
Prof.Kuwano,J.
Tokyo Institute of Technology, Japan
Dr. Miki, H.
Ministry of Construction, Japan
Dr. Murata, O.
Railway Technical Research Institute, Japan
Mr. Otani,Y.
Hirose & Co.,Ltd., Japan
Mr. Sakaguchi,M.
Taisei Corporation, Japan
Dr. Ogisako, E.
Shimizu Corporation, Japan
Mr. Tayama, S.
Japan Highway Public Corporation, Japan
Prof. Yashima, A.
Gifu University, Japan
Prof. Yasufuku,N.
Kyushu University, Japan
Mr. Yamamoto, A.
Obayashi Corporation, Japan
Mr. Yokota,Y.
Maeda Kohsen Co., Ltd., Japan

ACKNOWLEDGMENT

In order to complete the activities by not only the Asian TC but also the Japanese Supporting Committee, Many researchers and practitioners in Asian countries including Japan have given to us many information including testing and design method as well as current research activities in each country. Especially, the following reports by the Japanese Supporting Committee could not been completed without those contributions. Thus, I, as a chairman of the committee, would like to acknowledge those sincere contributions.

Earth Reinforcement, Ochiai, Yasufuku & Omine (eds) © *1997 Balkema, Rotterdam, ISBN 90 5410 833 9*

Testing methods of reinforcements for use in reinforced soil structures

S. Hayashi – *Institute of Lowland Technology, Saga University, Japan*

T. Hirai – *Research Laboratory, Mitsui Petrochemical Industrial Products Ltd, Saitama, Japan*

J. Kuwano – *Department of Civil Engineering, Tokyo Institute of Technology, Japan*

Y. Yokota – *Technical Department, Maeda Kosen Co. Ltd, Fukui, Japan*

ABSTRACT : Japanese standards for Index tests on reinforcements and tests for evaluating interaction behavior of reinforcements and soil have been presented. Tests on both metallic reinforcements as well as geosynthetics have been covered. Relevant design guidelines and manuals published by other Japanese institutions have been also referred.

This report is the result of activities carried out by Japan Working Group on Testing Method, one of the working groups of Japanese Supporting Committee of the Asian Technical Committee for Earth Reinforcement (Chairman: Professor H. Ochiai).

1 INTRODUCTION

1.1 *Background*

To optimally select the reinforcements for construction of reinforced soil structures, it is important to fully understand the physical properties of reinforcements and to evaluate them as per design requirements. For this purpose, testing of reinforcement from various viewpoints is required. Such viewpoints include the functions and purposes required of the reinforcements, work environment at the time of construction of soil structures, and long-term durability against environmental factors. Also, it is essential to select the reinforcements that best suit the particular job specifications. This requires the testing of individual reinforcements. In addition, since reinforcements are embedded inside the soil, it is also important to determine the engineering properties arising out of interaction between reinforcement and soil. With these perspectives, methods of testing reinforcements as given in design guidelines and manuals published by various Japanese institutions have been presented in this report.

1.2 *Classification*

As indicated above, test items in this report have been broadly classified into tests for determining the characteristics of individual reinforcements known as Index tests (section 2) and tests for evaluating friction and other interaction behavior of soil and reinforcements (section 3). In view of the differences in engineering properties due to varied forms and shapes of reinforcements, test items in each of these broad categories have been further subdivided into tests on metallic reinforcements and tests on geosynthetics (see Table 1). Index tests on both metallic reinforcements and geosynthetics have been further subdivided into physical, mechanical and durability tests as given in Table 2. The contents of these tests will be described in this report, with reference to the representative manuals, guidelines and reports mentioned in Table 3.

Table 1 Classification of reinforced soil structure by material and shape of reinforcement

Material	Shape of reinforcement	Type of reinforced soil structures
Metal	Strip	Terre Armée, YORK-Type retaining wall
	Steel bar & anchor	Multiple anchored wall, TRRL anchors
	Grouted steel bar	Root piles, soil nailing, Micro piles, Multi-Anchorage system Reference: NATM, Ground anchor
	Iron grid	Wire wall, etc.
Fiber	Fiber	Fiber reinforced soil
Geosynthetic	Geotextile	Geotextile reinforced soil, Belt type reinforced soil
	Geogrid	Geogrid reinforced soil

Table 2 Physical, mechanical and durability tests

(a)Test items

Physical Properties A	1.Unit weight: Determination of mass per unit area and mass per unit length 2.Determination of thickness (at specified pressures) and diameter 3.Determination of size of strand
Mechanical Properties B	1.Tensile strength: Tensile test, Wide width tensile test 2.Creep test 3.Dynamic strength: Cyclic load test, Relaxation test 4.Others: Survivability (Installation damage) test, Puncture test, Tear test, Impact test, etc.
Durability C	1.Weathering test 2.Chemical degradation 3.Biological degradation and others

(b)Test implementation

Material	Shape of reinforcement	Physical properties A			Mechanical properties B				Durability C		
		1	2	3	1	2	3	4	1	2	3
Metal	Strip	Δ	Δ		□	O	O	O		O	O
	Steel bar	Δ	Δ		□	O	O	O		O	O
	Iron grid	Δ	Δ	O	□	O	O	O		O	O
Fiber	Fiber	Δ	Δ		O	O				O	O
Geosynthetic	Geotextile	Δ	Δ	Δ	□	□	O	O	O	O	O
	Geogrid	Δ		Δ	□	□	O	O	O	O	O

□ The test which must be always implemented beforehand to fined a design value
O The test which becomes necessary according to the situation (the environment condition, the building condition)
Δ The test which is implemented for the purpose of quality control of the reinforcements

2 INDEX TESTS ON REINFORCEMENTS

2.1 *Introduction*

For design and construction of reinforced earth structures, it is important to select the reinforcement according to the functions it is required to serve. For this purpose, it is necessary to conduct the index tests for determining the engineering properties of individual reinforcements. In this section, test items have been subdivided into tests on metallic reinforcements (section 2.2) and tests on geosynthetics (section 2.3). Section 2.4 deals with the determination of design strength from the index tests.

2.2 *Tests on metallic reinforcements*

Most of the test methods on metallic reinforcement materials are stipulated in Japanese Industrial Table 1 Classification of reinforced soil structures by material and shape of reinforcement.

Standards (JIS), and therefore this report only mentions the relevant number of these standards. In general, following types of metallic reinforcements are used in reinforced soil structures:

1. Metallic strip (M-1): JIS G 3101.
2. Steel bar with anchor plate (M-2): JIS G 3101.
3. Grouted steel bar (M-3): JIS G 3536.
4. Grouted steel bar and Metal others (R-M1): JIS G 3122.

(1) *Unit weight (A-1) and Diameter (A-2)*

Unit weight and Diameter are required for controlling the volume and other physical properties of reinforcements.

1. Metallic strip (M-1): JIS G 3193.
2. Steel bar with anchor plate (M-2): JIS G 3191.
3. Grouted steel bar (M-3): JIS G 3536. and JIS G 3109.
4. Grouted steel bar and Metal others (R-M1): JIS G 3122.

(2) *Tensile strength (B-1)*

Tensile properties of reinforcements such as maximum tensile strength, tensile rigidity, stress-strain relationship, etc., are necessary for design of reinforced soil structures (see JIS Z 2241.).

(3) *Creep (B-2)*

Since tension is applied to the reinforcements in reinforced earth structures for a long period of time, it is necessary to determine the deformation properties of reinforcements under sustained loading. Hence, a creep test is conducted to determine the permissible tensile

Table 3 Referred standards, manuals and report

(a) Application	Notation
Slopes & Excavations	AA
Embankments	BB
Wall Structures	CC
Foundations	DD

(b) Kind of reinforcement	Notation
Steel bar with anchor plate	MA
Metallic strip	MS
Geosynthetics (Geogrid)	GG
Geosynthetics (Geotextile)	GT
Grouted steel bar	MC
Metal (others)	MO
Geosynthetics (others)	GO

(c) Reference

Material	Code	Kind of Reinforcement	Application	Reference
Metal	(M-1)	MS	CC	PWRC (1990.05)
	(M-2)	MA	CC	PWRC (1994.10)
	(M-3)	MC	AA	JGS (1995.10)
	(R-M1)	MC, MO	AA	JGS (1995.03)
Geosynthetic	(G-1)	GG, GT, GO	BB, CC, DD	PWRC (1994.02)
	(G-2)	GG, GT	CC	PTRI (1992.11)
	(G-3)	GG, GT, GO	---	JIS L 1908
	(R-G1)	GG, GT	---	JGS (1994.01)

strength of reinforcements from load-strain-time behavior. Following are the relevant standards.
1. JIS Z 2271.
2. JIS Z 2272.

(4) *Dynamic strength (B-3)*

When reinforcements are laid in locations where load is applied repeatedly, strength of reinforcements is determined from repeated load tests. The relevant standards are given below.
1. JIS G 3536.
2. JIS G 3109.

(5) *Chemical properties (C-2)*

In view of the need to maintain the functions of reinforcements in the soil for an extended period of time, anti-acid and anti-alkali tests are conducted to determine the chemical resistance. The relevant standards are as follows:
1. JIS G 0591.
2. JIS Z 2371.

2.3 *Geosynthetic reinforcements*

Tests on both surface state and lattice-state reinforcements are described based on the design and construction manuals, guidelines and standards mentioned in Table 3.
However, geosynthetic materials will not be described here owing to the large variety of materials used as well as numerous synthetic fibers and resins from which these materials are manufactured.

(1) *Mass per unit area (A-1)*

Mass per unit area is tested for controlling the volume and physical properties of reinforcements.
1. Geotextile and Geogrid (G-1): The mass per unit area of 3 square test samples of 20 cm x 20 cm is measured, with the average to be expressed in g/m^2.
2. Geotextile and Geogrid (G-3): JIS L 1908.

(2) *Thickness (A-2)*

Thickness is tested for classifying reinforcements as well as for controlling the volume and physical properties of reinforcements.
1. Geotextile (G-1): The thickness of a 10 cm x 10 cm test sample under the pressure of 20 g/m^2 is measured. The average value of 5 test samples determines the standard thickness.
2. Geotextile and Geogrid (G-3): JIS L 1908.

(3) *Apparent opening size and size of strand (A-3)*

To evaluate the filtration characteristics of geosynthetics, apparent opening size test and test of size of strand are conducted.
1. Geotextile (G-1): Glass beads of the standard grain diameter are placed on the geotextile and screening is done. This process is continued for the beads of successively smaller diameter until percentage of glass beads passing through the geosynthetic exceeds 5 %. A grain size distribution curve is plotted and apparent opening size is read from this curve (size corresponding to 5 % passing).

2. Geogrid (G-1): Lattice shape is an index for showing the size of apertures, and it measures the distance between the centers of adjoining strands.
3. Geotextile and Geogrid (G-3): JIS L 1908.

(4) *Tensile strength (B-1)*

To evaluate the maximum tensile strength, tensile rigidity and stress strain relationship, a tensile test is conducted.
1. Geotextile and Geogrid (G-1, G-2, G-3):
Test equipment: Strain-controlled tensile tester
Test environment: Temperature → 23± 2℃
Humidity → 50± 20%
Test conditions: Refer Table 4.

(5) *Creep properties (B-2)*

Creep test is conducted to determine the elongation of the geotextile under the constant load for a long period of time.

1. Geotextile and Geogrid (G-1, G-2):
Test equipment: It should not have extraneous effects such as vibration during the test.
Test environment: Temperature → 23± 2℃
Humidity → 50± 20%
Test conditions: Refer Table 5.

(6) *Dynamic strength (B-3)*

Dynamic strength is the ability of geotextile to withstand repetitive loading before undergoing failure.
1. Geotextile and Geogrid (G-2):
(a) Momentary loads such as earthquakes:
A load of 30 % of the rupture strength is applied to 3 test samples of more than 30 cm long for 24 hours each and thereafter a load test is repeatedly conducted 3 times at a cycle below 1 Hz. The maximum load that will not result in rupture or generate strain of more than 15 % is determined.
(b) Train loads:
This test is conducted in a rigid box with geosynthetic embedded inside the soil. Loading is applied with the help of rigid plate of size 40 cm x 40 cm. Minimum loading level is maintained at 1.6 tf while cyclic load amplitude of 1.6 tf is applied at 20 Hz frequency and 1.5 million cycles are given. After the test, 10 test samples of geosynthetic are selected randomly and tested for tensile strength.

Table 4 Tensile test condition

	Test details	G-1	G-2	G-3
Geotextile	Test speed	1%/min (of the gauge length)	5%/min (of the gauge length)	20 ± 5%/min (of the gauge length)
	Width of specimen	20 cm or more	5 cm or more	(a) 5 cm or more (b) 20 cm or more
	Length of specimen between grips		30 cm or more	10 cm or more
Geogrid	Test speed	1%/min in the gauge length	5%/min in the gauge length	20± 5%/min in the gauge length
	Width of specimen	20 cm or more	Single strand	20 cm or more
	Length of specimen between grips		30 cm or more	20 cm or more
Geotextile and Geogrid	Number	5 pieces or more	20 pieces or more	5 pieces or more
	Test report	Calculate the tensile strength (kN/m)	Calculate the tensile strength (kN/m)	Calculate the tensile strength (kN/m)

Note : As per G-2, breaking strength is taken at Percent elongation of 15 (%) or less.

Table 5 Creep test condition

	Test details	G-1	G-2
Geotextile and Geogrid	Load steps	10–90% of tensile strength 5 steps or more	
	Test specimens	3 Pieces or more	3 pieces or more
	Test period	about 1000 hours	500 hours or more
	Length of specimen		30 cm or more
	Evaluation method	by Creep-Limit line chart.	by Elongation-Time line chart
Geotextile	Width of specimen	20 cm or more	5 cm or more
Geogrid	Width of specimen		One strand

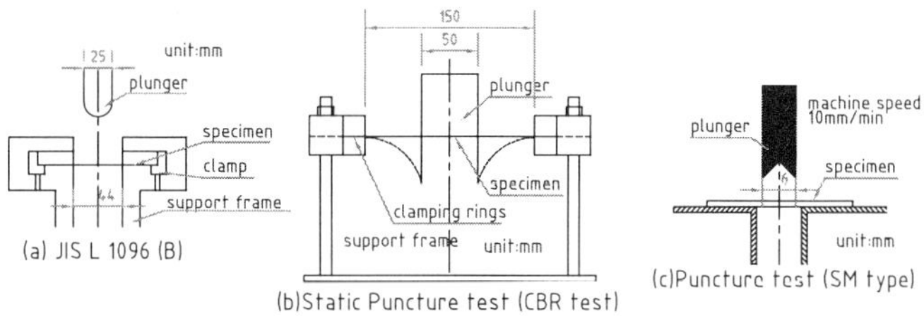

Fig.1 Examples of Test Apparatus

Table 6 Weathering test condition

(a) Outdoor exposure test

Time to begin the test	Select amongst March, April, September, and October
Exposure time	100-4000 hours
Number	5 pieces or more at each exposure interval

(b) Accelerated exposure test

Source of light kind	WS (Sunshine) type
Temperature of black panel	63±3°C
Spray cycle	18 minutes for 120 minutes
Exposure time	100-10000 hours
Number	5 pieces or more at each exposure interval

(7) *Breakage test (B-4)*

Reinforcements may get damaged by the impact of or due to the protrusion of rocks caused by the passage of heavy vehicles and/or construction machinery during the construction work. To determine the extent of damage to reinforcements due to such impacts or protrusions, a penetration test, a tear test and an impact test are conducted.

1. Geotextile and Geogrid (G-1):

(a) Puncture test: Breaking strength is tested by quasi-static penetration of rod against the reinforcement material via a compression testing machine (see Fig. 1).

(b) Tear test: The trapezoidal method recommended by JIS L 1908.

(c) Impact test: A rolling compaction test is conducted using the actual banking soil material to be used, and the resistance to damage is determined from the strength retention of the test samples after the test.

2. Geotextile and Geogrid (G-2):

(a) Impact test: Using a crusher run (C-40), 2m x 2m reinforcement material is laid covered by 30 cm depth of soil and then it is rolled over 4 times by a bulldozer, 2 times by a rammer and 6 times by a vibration roller. After digging the material out, samples are selected randomly and checked for damage and are compared with the original tensile strength.

(8) *Weather resistance (C-1)*

The strength of reinforcement material may deteriorate due to exposure to the natural conditions such as ultraviolet rays and cycles of dry and humid conditions due to climatic variations. To examine such effects, the outdoor exposure test and accelerated exposure test are conducted.

1. Geotextile and Geogrid (G-1):

(a)Outdoor exposure test: This test follows JIS L 1410 .

(b)Accelerated exposure test: This test follows JIS L 1415.

For details of these tests, refer Table 6.

(9) *Chemical properties (C-2)*

Both anti-acid and anti-alkali tests are conducted to determine the chemical resistance.

1. Geotextile and Geogrid (G-1): JIS K 7114 is followed.

Solutions used: refer Table 7.

2. Geotextile and Geogrid (G-2): Anti-alkali test.

More than 10 test samples are dipped in pH12 solution for 700 days at a curing temperature of 20℃ followed by a tensile test to evaluate the deterioration.

(10) *Biological resistance (C-3)*

To determine the extent of deterioration to the strength of reinforcements in soil caused by microorganisms, a bacteria resistance test is conducted.

1. Geotextile and Geogrid (G-1): Tests are conducted with reference to JIS Z 2911.

Table 7 Kind of soaking liquid for medicine test according to the environment

Soil condition ↓ / Medicine name →	Sodium chloride	Calcium hydroxide	Sodium hydroxide	Hydrochloric acid	Sulfuric acid
General condition A (pH = 6 - 8)	3.0%	0.01% pH = 11	0.01% pH = 11	0.01% pH = 3	0.01% pH = 3
General condition B (pH = 5 - 9)	3.0%	0.1% pH = 12	0.1% pH = 12	0.1% pH = 2	0.1% pH = 2
Special condition (pH ≤ 4, pH ≥ 10)	3.0%	Saturated solution	10%	10%	10%

2.4 *Design strength of individual reinforcements*

The design strength of the reinforcement is determined by applying partial factors of safety to the maximum tensile strength value of the reinforcement determined from the laboratory test (see Section 2.2(2) and 2.3(4)). Partial factors of safety are applied against the load conditions of actual design and according to the actual site conditions. Numerical values of these partial factors of safety are evaluated from the dynamic, creep, durability tests, etc., which are conducted as described in Section 2.2 and 2.3. The procedure to determine the design strength of reinforcements as suggested in various manuals is given below.

(1) *Metallic reinforcements*

All metallic reinforcements use materials according to the Japanese Industrial Standards with permissible stress to be taken directly as given in these standards.

1. Metallic strip (M-1) and Steel bar with anchor plate (M-2): Permissible stress for strips plated with zinc on materials confirming to JIS G 3101 is taken as 137 MN/m^2.
2. Grouted steel bar and metal others (R-M1): Materials as per JIS G 3122 are used and the permissible tensile stress given in this standard is adopted.

Example: SR 235 General strength 137 MN/m^2
In a corrosive environment 117 MN/m^2
SD 390 General strength 206 MN/m^2
In a corrosive environment 186 MN/m^2

(2) *Geosynthetic reinforcements*

Depending upon the guideline manuals, partial factors of safety for creep, durability, dynamic loads, etc., are applied on maximum tensile strength to obtain the design tensile strength.

1. Geotextile and Geogrid (G-1): Design tensile strength T_A is obtained as follows:

$$T_A = \frac{T_{max}}{F_{cr}\,F_D\,F_C\,F_B} \qquad (1)$$

Where, T_{max} = maximum tensile strength (as per G-1),
F_{cr} = safety factor for creep,
F_D = safety factor for durability,
F_C = safety factor for damage during construction work, and
F_B = safety factor for reduction of strength at joints.

2. Geotextile and Geogrid (G-2): For design of reinforced soil structures for railways, the design strength is determined for each load condition as given below:

General load $T_{AG} = \alpha_1 .\ \alpha_2 .\ \alpha_3 . T_{max}$ (2)

Temporary load $T_{AT} = \alpha_1 .\ \alpha_2 .\ \alpha_5 . T_{max}$ (3)

Earthquake load $T_{AE} = \alpha_1 .\ \alpha_2 .\ \alpha_4 . T_{max}$ (4)

Where, T_A = design tensile strength,
T_{max} = maximum tensile strength (as per G-2),
α_1 = anti-alkali reduction coefficient,
α_2 = construction time reduction coefficient,
α_3 = creep reduction coefficient,
α_4 = momentary load reduction coefficient, and
α_5 = train load reduction coefficient.

3 TESTS FOR SOIL-REINFORCEMENT FRICTION BEHAVIOR

3.1 *Introduction*

To understand the behavior of reinforcement inside the soil, it is important to evaluate the friction and other interaction properties of reinforcement and the soil. Concepts and methods for evaluating these properties differ depending upon the purpose and design of reinforced earth structures as well as the kind of reinforcement used (see Table 8). In section 3.2, the basic testing method of JGS (1995.10) has been introduced which is used for evaluating the friction

Table 8 Evaluation of soil-reinforcement friction behavior

Code	Method of Design and Construction	Usage	Evaluation of friction behavior
(M-1)	Terre Armée Reinforced Retaining Wall	Wall Structure	The friction coefficient is evaluated from the test result
(M-2)	Multiple Anchored Retaining Wall	Wall Structure	The validity of the calculation is verified from the test result
(M-3)	Ground Anchor	Slope	As per detailed testing method (see Section 3.2)
(R-M1)	Multipile Reinforced Earth	Slope, Excavation	Details not given
(G-1)	Geotextile Reinforced Earth	Embankment, Wall Structure	Details not given
(G-2)	Reinforced Railroad/road with Rigid Facing Method	Wall Structure	Details not given
(R-G1)	As per JGS (1994.01)		As per detailed testing method (see Section 3.3)

properties of metallic reinforcements and soil for reinforced earth structures. The Geotextile Standardization Committee, Japanese Geotechnical Society, prepared a new draft of the Society's norm JGS (1994.01) for evaluating the friction properties of geosynthetic reinforcements and soil by direct shear and/ or pull out tests to be used in reinforced earth walls and banking structures. This has been introduced in Section 3.3.

3.2 *Ground Anchors*

As of this date, a pull out test for determining the value of the circumferential surface frictional resistance necessary for the design of reinforcements (grout material) for use in reinforced earth structures for stabilization of natural and artificial slopes has not been established and the rock bolt test method used in ground anchoring or the NATM (New Austrian Tunneling Method) is being used as a reference. This report introduces the basic test method as given in JGS (1995.10) and can be applied for soil nails, micropiling, dowelling, etc. The basic test refers to the testing conducted to understand the behavior and to evaluate the marginal pull-out strength of the anchor in reinforced earth structures, and to determine the various constants to be used in design of such structures.

(1) *Restriction and application*

In designing ground anchors, the marginal pull-out strength of anchors is defined as follows: "The marginal pull-out strength of anchors refers to the strength necessary to destroy an anchor when the anchor's ultimate destruction mainly depends on the condition of the ground. The ultimate pull-out strength shall, as a rule, be verified with a basic test, and a test shall be definitely conducted in the case of destruction of at least an important structure in the surroundings. The marginal pull-out strength of anchors differs depending upon the condition of the ground, the method of construction and the types of anchors used and hence tests shall be conducted under conditions close to the conditions in which the anchors will actually be used. If it is impossible to do so, the results of testing shall be applied to the design with due consideration to the difference in conditions."

The circumferential surface frictional resistance of the unit area of the anchor against the ground, based on which the marginal pull-out strength will be determined, is calculated depending upon the kind of ground and its condition; however, since it differs from one construction method and anchor type to another, as a rule it is determined from the results of a basic test.

(2) *Basic test*

A basic test is conducted prior to the design. It is conducted to evaluate the frictional resistance and the marginal pull-out strength required for designing and examining the behavior of the anchor.

1. Apparatus

The equipment for the above test consists of pressurizing equipment, counterforce equipment and the measuring instruments. The test equipment should conform to the following specifications and be selected to suit the type of testing, the purpose, the maximum planned test load and the condition of the site.

(a) Pressurizing equipment

(i) It must have a nominal capacity of more than 1.2 times the maximum planned test load and be capable of maintaining a constant load for the specified duration.

(ii) It should also be capable of increasing or decreasing the load according to the planned load stages.

Note 1: Generally an oil pressure jack and an oil pressure pump are used as pressurizing equipment. In selecting an oil pressure jack, it should have sufficient capacity and stroke, and particularly for basic testing it should be

Table 9 Frictional resistance of anchor

Kind of ground			Frictional resistance (kgf/cm^2)
Bedrock	Hard rock		15 ~ 25
	Soft rock		10 ~ 15
	Weathered rock		6 ~ 10
	Mud stone		6 ~ 10
Gravel	N value	10	1.0 ~ 2.0
		20	1.7 ~ 2.5
		30	2.5 ~ 3.5
		40	3.5 ~ 4.5
		50	4.5 ~ 7.0
Sand	N value	10	1.0 ~ 1.4
		20	1.8 ~ 2.2
		30	2.3 ~ 2.7
		40	2.9 ~ 3.5
		50	3.0 ~ 4.0
Viscous soil			1.0 c (c : cohesion)

capable of loading 0.9 times of the tendon yield load. A further requirement is that the equipment should be capable of maintaining, increasing and decreasing loads smoothly at a constant speed, and it should be calibrated before use.

(b) Counterforce equipment

This should be strong enough to support the maximum planned test load and should not be any impediment to testing.

Note 2: The basic test is conducted by using a counterforce stake installed in a position outside the impact zone or by using a counterforce board and making a counterforce on the ground behind the counterforce board. In the method using a counterforce board, if the anchor body has shallow overburden, the counterforce may work as holding the load thereby breaking the ground behind the counterforce board, making it impossible to determine the true marginal pull-out strength. However, if test error can be minimized by properly choosing the overburden so as to reduce the effect of load holding, and by properly selecting the size, strength and rigidity of the counterforce board, it is possible to accurately measure the friction resistance between the anchor body and the ground, and accordingly this method is generally used.

(c) Measuring instruments

A load weight scale, a displacement meter and a timer, all of which should be verified as having the prescribed degree of precision.

Note 3: A load weight scale generally uses a Bourdon pressure gauge incorporated in pressurizing equipment or a load cell. The measuring precision of load cells may deviate depending upon the setting method. Hence, it is important to set them after making the surface smooth and to prevent the eccentric load.

Note 4: A displacement meter should have enough stroke to accommodate the expansion of the anchor, as otherwise it will require refilling which will impair the measuring precision. Therefore the maximum displacement must be accurately anticipated and a displacement meter should be selected to meet the maximum displacement. In a basic test, generally precision in the range of 0.1 mm is required. Attention should be paid so that the fixed point will not be displaced due to the load.

Note 5: As for timing devices, one conventional clock or watch for showing the standard time and one for measuring the time elapsed from the start of testing will be sufficient.

2. Conditions for testing

(a) For designing permanent anchors, a basic test should always be conducted.

(b) For designing temporary anchors, it is desirable to conduct a basic test.

Note 6: When planning and designing anchors, it is important to conduct a basic test to determine the various constants for use in design as well as the appropriateness of the anchoring method against the subject fixing ground and the structure. In view of its purpose, it is desirable for a basic test to be conducted before planning and designing, but in practice, it is difficult to do so because of the site conditions, and hence tests are usually conducted after starting the main construction work.

Nevertheless, the basic test should be conducted before the final review of the design so that the changes can be made on the basis of the basic test. In cases where temporary anchors do not heavy loads, or when anchors are fixed to a rock base, hard clay ground or dense sandy ground, a basic test may be omitted. When designing anchors without using a basic test, they must be designed according to Table 9, matching similar ground. The adequacy of the design must be verified with a suitability test as soon as possible after completing installation to verify whether or not the anchor design and installation are appropriate from the point of view of load-displacement properties by applying a load on the actual anchor to be used. For the following kinds of ground, it is desirable to conduct a basic test even for temporary anchors:

(i) Organic soil
(ii) Cohesive soil with a liquid limit > 50 %
(iii) Sandy soil of *N* value < 15.

3. Test anchors

Test anchors should be determined after considering the actual conditions under which they will be used. This includes the fixing ground for the anchor to be used in the basic test, anchor elements and the method of installation. However, the fixing length of the anchor and the sectional area of the tendon may be changed on the judgment of the head engineer.

Note 7: In many cases the frictional resistance of anchors varies depending upon anchor elements and method of installation. Hence it is desirable that the anchor used for basic testing is as close as possible to the anchor actually

Table 10 Maintenance time of load

Soil type → / Nature of load ↓	Viscous soil	Sandy soil	Bedrock	Note
When virgin load	Constant load of 15 minutes or more	Constant load of 10 minutes or more	Constant load of 5 minutes or more	However, if displacement is not steady, load is maintained until displacement stabilizes.
When load in the Stress history	Constant load of 3 minutes or more	Constant load of 2 minutes or more	Constant load of 1 minute or more	

used. If the fixing ground for the anchor, the anchor type, the diameter of the drilled hole and the method of drilling the hole vary, tests should be conducted for each different combination of factors. But, for example, if a test is conducted with an anchor element of the same fixed length as the real anchor, the maximum planned test load may become large, and this would require large test equipment and would make testing difficult. In such a situation when an exact matching test of a large anchor is conducted, it may not be possible to maintain the load under the restriction of the permissible tensile strength of the tendon until the anchor body can be pulled out, so in some cases the marginal pull-out strength can not be determined. In such a case, the test should be conducted by shortening the anchor body fixing length or increasing the sectional area of the tendon according to the judgment of the head engineer. When conducting a basic test by shortening the anchor body fixing length and applying the result thereof to a longer fixing length, there is a possibility of overestimating the per unit area friction resistance, so it is necessary to make sufficient studies in this respect. Assuming, for example, the marginal value of the circumferential friction resistance (circumferential friction strength) distributed around the body of a stretch type anchor is constant without any relation to the anchor fixing length, the marginal pull-out strength (T_{ug}) of the anchor can be calculated by the following formula with the anchor body fixing length (l_A) as a variable:

$$T_{ug} = \frac{E \cdot A \cdot \alpha}{k_s} \cdot \tau_{ug} \cdot \tanh(\alpha \cdot l_A) \qquad (5)$$

$$\alpha = 2 \cdot \sqrt{k_s / (E \cdot D)} \qquad (6)$$

$$E = (E_C \cdot A_C + E_S \cdot A_S) / A \qquad (7)$$

$$A = A_C + A_S \qquad (8)$$

Where,

τ_{ug} = marginal value of the circumferential friction resistance,
T_{ug} = marginal pull-out strength of the anchor,
k_s = counterforce coefficient of the shearing ground,
E_c = elasticity coefficient of the ground,
A_c = sectional area of the anchor body,
E_s = elasticity coefficient of the tendon,
A_s = sectional area of the tendon, and
D = diameter of bore hole.

As for the tendon, its sectional area should be determined so that a load can be maintained until the anchor body can be pulled out. It is also important that the strength and rigidity of the counterforce equipment is checked.

4. Method of loading and items of measurement

(a) The initial load should be approximately 10 % of the maximum planned test load.

(b) The load should be divided into 5 to 10 stages between the initial load and the maximum planned test load, with load control to be conducted in multiple cycles. When the marginal pull-out strength can not be confirmed even at the maximum planned load, load should be increased until pull-out is achieved, after first reducing the load down to the initial load. However, loading should not exceed 0.9 times of the yield load of the tendon.

(c) The load in each cycle should be maintained at a constant value for the times shown in Table 10.

(d) The following items should be measured at each load stage:
(i) Load
(ii)Amount of displacement
(iii) Time

Note 8: The maximum planned test load should be determined with reference to the value of the circumferential friction resistance of the anchor body against conventional similar ground. However, the difference between maximum planned load and actual load may vary up to two times and hence the tensile strength of the tendon should be appropriate. Loads should be applied as shown in Fig. 2 according to a multi-cycle method to determine the relationship between the load, residual displacement and the amount of reversion of elasticity. It is also desirable to have as many load stages as possible to upgrade the precision of testing. Where the marginal pull-out strength cannot be confirmed even at the maximum planned test load, loading should be conducted in increasing load stages until the anchor body can be pulled out, but not exceeding

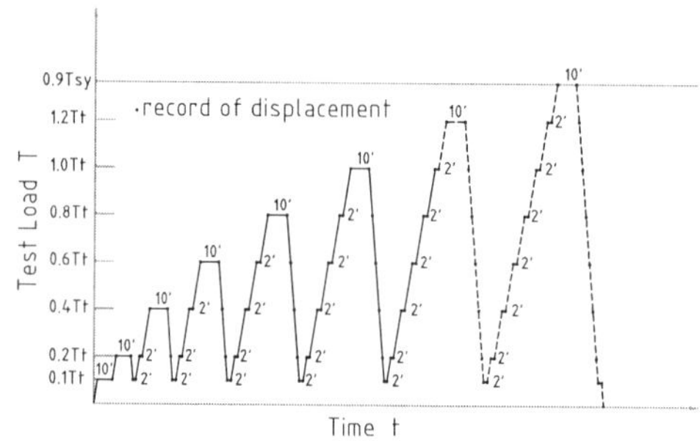

Fig. 2 Planning Chart of Loading

Table 11 Load speed (constant)

During loading	Planned maximum test load / 10~20	tf/min
During unloading	Planned maximum test load / 5~10	tf/min

0.9 times of the yield load of the tendon. Furthermore, the load should not be increased or decreased rapidly, but at a nearly constant speed. Table 11 should be used as a guide for loading speed. For judging whether or not displacement has stabilized at each load stage, it is convenient to plot the amount of displacement per minute. Guidelines such as those given in Table 12 are being used as standards in some other countries.

From the results of conventional testing, generally a change of less than 1 mm of the amount of displacement during the last 3 minutes of load retention may be used as a guideline for stable displacement.

Where there is more than 1 mm change in the amount of displacement, the load should be retained for an additional 1 minute, and load retention should be continued until the condition that change in displacement during the preceding 3 minutes is less than 1 mm/3 minutes is satisfied. Also, while under load, the amount of displacement of the anchor head and the amount of displacement of the counterforce equipment in each load stage should be measured after every 1 minute. While unloading, the approximate value of the amount of displacement of the anchor head should be measured at each of the load stages.

5. Summary of test results and judgment

(a) The load on the anchor head-displacement curve, the load-elastic displacement curve, and the load-plastic displacement curve should be plotted.

(b) The tendon free length and the tendon friction loss are obtained from the load-elastic displacement curve.

(c) The marginal pull-out strength is obtained from the load-plastic displacement curve. However, where the maximum test load does not reach the marginal pull-out strength, the maximum test load is deemed to be the marginal pull-out strength at the time of design.

Note 9: The results of the tests should be summarized, as mentioned above, in the form of load-displacement curves. The displacement should be divided into elastic displacement (δ_e) and plastic displacement (δ_p), and should be plotted in the form of a load-elastic displacement curve and a load-plastic displacement curve; Where δ_p is the amount of residual displacement at the point where unloading was reduced to the initial load, and de is the total amount of displacement δ_c at each load stage minus δ_p. The marginal pull-out strength is the load where the load-plastic displacement curve in Fig. 3 completely turns downward; however, generally the maximum value of the test load where the amount of displacement stabilizes may be deemed to be the marginal pull-out strength. When the anchor body cannot be pulled out, the shield load of the tendon is deemed to be the marginal pull-out strength. The amount of friction loss is calculated as the balance between the initial load and the cross point of the load coordinate (coordinate T) and the extension of the straight line portion of the load-elastic displacement curve (see Figs. 3 and 4). The free length of

Table 12 Different standards for steady displacement

Name of Standard	Permissible value	Observation time
DIN	0.5 mm or less	for 15 minutes from 5 minutes
PTI	1.0 mm or less	For 10 minutes from 1 minutes
SIA (Schweizer Norm)	2% or less of δ_e	For 5 minutes from 0 minute (rock and sandy soil) For 15 minutes from 0 minute (Ground with little cohesion and Over consolidatied clay)
BSI	5% or less of δ_e	For 15 minutes from 0 minute

Note 1 : displacement is considered as steady if the amount of displacement of anchor head during observation time is within the permissible value.

Note 2 : Point of time after which the load was maintained almost constantly is called 0 minutes in the observation time.

Note 3 : δ_e = Elastic displacement of anchor head

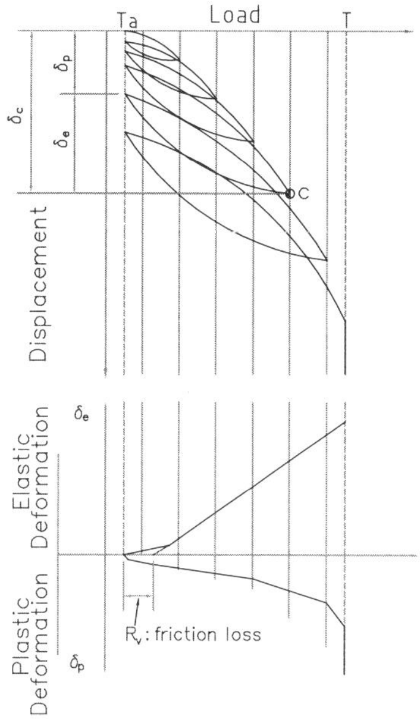

Fig. 3 Load-displacement curve

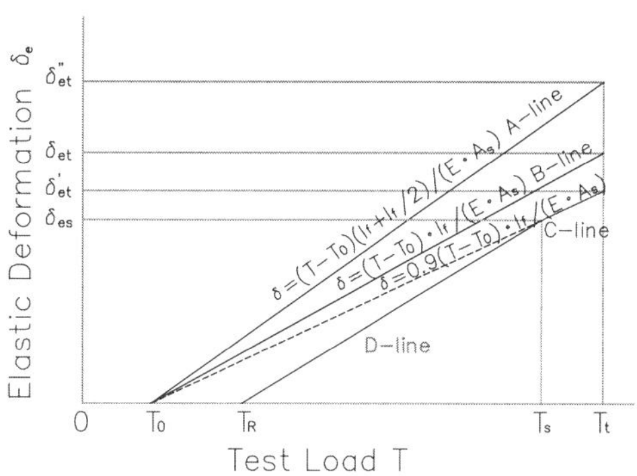

Fig.4 Evaluation Diagram

the tendon is obtained from the slope of the straight line portion of the load-elastic displacement curve, and is calculated according to the following formula:

$$l_{sf} = k \cdot E \cdot A_s = \Delta\delta_e / \Delta T \cdot E \cdot A_s \tag{9}$$

Where, l_{sf} = free length of the tendon,
k = the slope of the straight line portion of
load-elastic displacement curve,
E = Young's modulus of elasticity of the tendon, and
As = the sectional area of the tendon.

3.3 *Soil-geosynthetic friction behavior*

(1) *Introduction*

JGS (1994.01) describes the testing procedures for measuring the frictional resistance of geosynthetic to soil. The standard test methods consist of a direct shear test for determining the interface friction parameters, c_s and δ_s and a pull-out test for determining the pull-out friction parameters, c_p and δ_p, where c and δ refer to the adhesion and angle of friction, respectively. The test results obtained by direct shear test or pull-out test can be used in the design of geosynthetic reinforced soil structures (also see Hayashi et al. (1994)). The total frictional resistance of geosynthetic embedded in soil may be a combination of sliding, rolling and interlocking of soil particles and geosynthetic surfaces and shear strain within the geosynthetic specimen. Hence, the frictional properties of a geosynthetic can be expressed only with reference to the soil material and geosynthetic used in the testing. Therefore tests should be performed under site-specific soil conditions. The tests are applicable to a wide variety of geosynthetics such as geowoven, geononwoven, geoknit, geogrid and other geotextile related products. In general, the direct shear test can be used for a geosynthetic that separates the soil such as

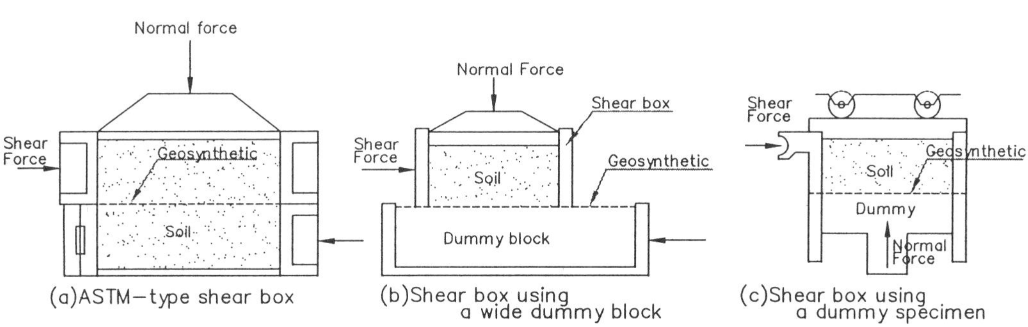

Fig. 5 Examples of Shear Box

geowoven, geononwoven and geoknit, and that develops a necking under the tensile force, such as geonet. The pull-out test is applicable to a geogrid through which super and sub-stratum soil is continuously in contact. Both tests should be conducted at a temperature of 20 ± 5℃ and a relative humidity of 50 ± 5 %. Direct shear test and pull-out test are described in Section 3.3 (2) and 3.3. (3) respectively in this report.

(2) *Direct shear friction test*

1. Apparatus

(a) Shear box: The size should be large enough in comparison with both the maximum soil particle size and the opening size of geosynthetic being tested.
Note 1: A rectangular container is recommended for geogrid and geonet. Placing more than four ribs of specimen are required in the shearing direction. The examples of the shear boxes are shown in Fig. 5.
Note 2: For a large shear box (more than 10 cm in thickness), their side walls should be coated with a silicone grease and a thin latex rubber membrane should be used to reduce the wall friction.
Note 3: A rigid dummy spacer such as wooden block may be placed either in the upper or the lower shear box.

(b) Normal stress loading device: It should be capable of applying a normal confining pressure up to 200 kPa.

(c) Shear force loading device: It must be capable of applying a horizontal shear to the specimen at a constant rate of displacement, and of moving more than 15 % of the length or the diameter of shear box.

(d) Force measuring devices: They should have an accuracy of ± 1 % of the maximum shear or normal forces.

(e) Displacement indicators: They must be capable of measuring normal and horizontal deformations of the specimen to an accuracy of ± 0.1 mm.

(f) Compaction rammer: A rammer by which the soil is compacted in the container.

(g) Multi-sieve device: For testing dry sands, this device can be used in such a manner that the height of multi-layered sieves, of which the opening sizes are 3 to 5 times larger than the maximum grain size of soil, is controlled to produce the required density of the soil specimen.

2. Preparation of test specimens

(a) The maximum soil particle size must be less than 1/10 of the height of shear box.

(b) The moisture content of the soil used in the test should be adjusted during compaction.

(c) A uniform soil specimen at the required density is prepared by multi-sieving method for dry sands or by a rammer compaction method for other soils.

(d) Prepare a geosynthetic specimen having sufficient size to facilitate clamping and testing. For an anisotropic material, a minimum of five specimens shall be taken in each direction.

(e) When geosynthetic is to be used in the wet or submerged condition and is anticipated to get changed in its strength, size, surface texture or other properties, put the specimen in the wet atmosphere or soak it in water for minimum 24 h prior to the testing.
Note 4: For specimens tested in the saturated state, leave the wetted specimen for 15 min. and then place it in a shear box within 30 min.

(f) Assemble the shear device after compacting the soil and placing the geosynthetic in the testing box.
Note 5: In the case of using a dummy block, gluing or clamping of the geosynthetic specimen to dummy substratum is an acceptable technique.
The examples of fixing methods are given in Fig. 6. To maintain the sufficient friction, coarse sand is adhered to the surface of block.

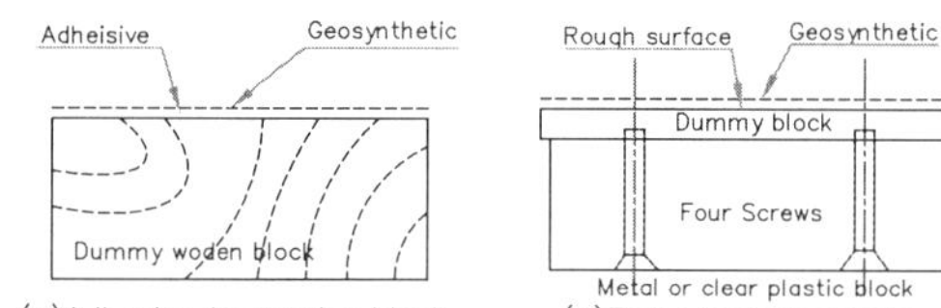

(a)Adhesion to wooden block (b)Fixing by four screws

Fig. 6 Examples of Fixing method of geosynthetic

3. Test procedure

(a) Apply the normal compressive stress, σ_n to the soil specimen.

(b) Record the normal displacement, ΔH, during the compression.
Note 6: For coarse grained soils, the compression may be finished in 10 min.
Note 7: Compression time can be estimated in the following manner: After determining the most steep tangent line of ΔH-log t curve plotted in the semi-logarithm graph, the line is slid in parallel 3 times on the time axis, and the compression time is given as the intercept of the line versus ΔH-log t curve.

(c) After the compression, the shear test is conducted using a constant rate of displacement. In the absence of any specification, a standard displacement rate of 1 mm/min. is used.

(d) Record the shear force as a function of displacement and run the test until the shear measuring device shows a peak value, or until the shear displacement reaches 15 % of the length or diameter of shear box.

(e) At the end of the test, moisture content of the soil specimen is determined.

(f) In this manner, several tests at different normal compressive stress each with new specimen are conducted.

4. Calculations

(a) Plot the test data as a graph of apparent shear stress calculated for each recorded value of shear force versus horizontal displacement.

(b) Plot applied normal stress σ_n, versus maximum shear

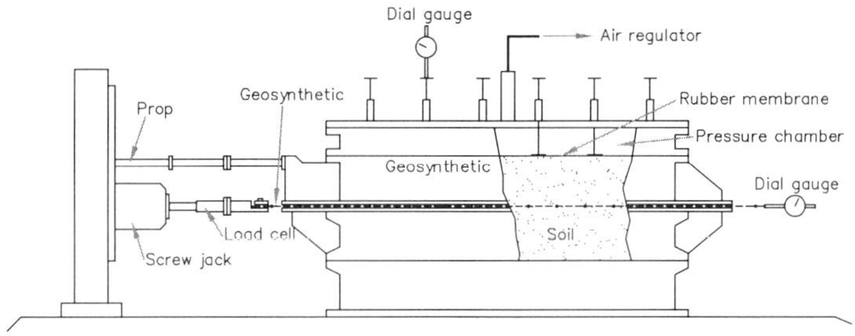

Fig. 7 An example of pull-out test apparatus

frictional strength, τ_{Smax}, for each test conducted.

(c) Connect the data points with a best fit straight line using the following equation, and determine the shear friction parameters (c_S, δ_S) of the soil-geosynthetic interface.

$$\tau_{S\max} = c_S + \sigma_n \tan\delta_S \tag{10}$$

(3) *Pull-out test*

1. Apparatus

(a) Pull-out test apparatus: The size of the upper and lower boxes must be larger than 30 cm in width (B_B), 30 cm in length (L_B), and 20 cm in height.

Note 8: A typical layout of the apparatus is shown in Fig. 7. For reducing the wall friction, the same procedure may be followed as described in 3.3 (2) 1 (a) Note 2.

(b) Normal stress loading device: as directed in 3.3 (2) 1 (b). A rubber pressure bag is used as a standard loading device.

(c) Pull-out device: It should be capable of moving about 20 % of the length of the box, at a constant rate of displacement. The standard displacement rate is 1mm/min.

(d) Clamping device: It must be wide enough to hold the whole width of specimen, and able to apply the uniform pull-out stress to a specimen without sliding or any damage.

(e) Pull-out force measuring device: It should have an accuracy of ± 1 % of the tensile breaking strength of geosynthetic used in the test.

(f) Displacement indicators: as directed in 3.3 (2) 1 (e).

(g) Miscellaneous equipment: Wires and guide-tubes for measuring elongation of geosynthetic specimen.

Note 9: Friction between a wire and soil may be reduced using a guide-tube. A stainless steel wire of about 0.2 mm in diameter and thin flex or metal guide-tube is recommended.

(h) Compaction rammer: as directed in 3.3 (2) 1 (f).

(i)Multi-sieve device: as directed in 3.3 (2) 1 (g).

2. Preparation of test specimens

(a) A soil specimen is prepared in a similar way as described in 3.3 (2) 2 (a) to (c).

(b) Prepare a geosynthetic specimen having a slightly short width than that of pull-out test box, and having length in general at least 20 % longer than that of test box. In the case of the apparatus in which the rear of a geosynthetic specimen is unable to be hung down to the outside, prepare a specimen at least 1.2 times longer than the limit length of pull-out resistance mentioned in 3.3 (3) 4(d). For an isotropic material, a minimum of five specimens shall be taken in the same direction.

(c) When the geosynthetic is to be used in the wet or submerged conditions, a specimen is prepared in a similar way as described in 3.3 (2) 2 (e).

(d) After compacting the soil in the lower box, a geosynthetic specimen is placed on the soil and connected to the clamps. A wire covered with a guide-tube is tied at each position for measuring its elongation and drawn out towards its rear side. Then compact the soil in the upper box.

Note 10: It is desirable to select five or six measuring positions at regular intervals within the box, and two positions at the front of specimen. Examples of measuring the elongation are shown in Fig. 8.

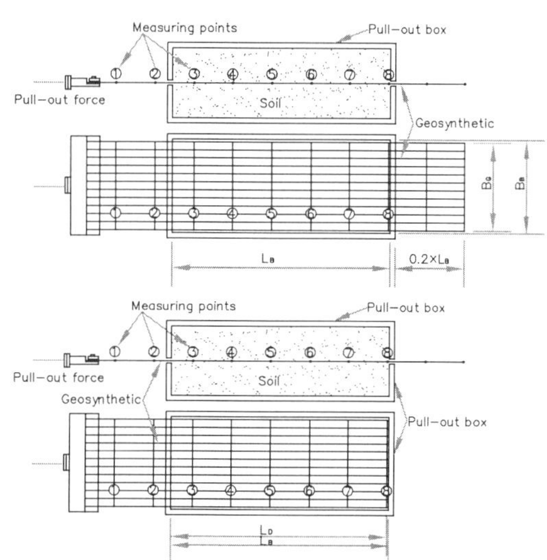

Fig. 8 Examples of measuring points of displacement

3. Test procedure

(a) The compression process of soil specimen is the same as directed in 3.3 (2) 3 (a) to (c).

(b) After the compression, a pull-out test is carried out at the constant rate of deformation.

(c) Record pull-out force, F_T, and displacement, X_i, at each measuring point under a constant normal stress, σ_n.

Note 11: A normal displacement shall be recorded during the test.

(d) The pull-out test should be conducted until the strain at the clamping reaches 20 % or until the break failure of the geosynthetic.

(e) At the end of the test, determine the moisture content of the soil specimen.

(f) In this manner, tests at different compressive stresses for more than four new specimens are conducted.

4. Calculations

(a) A pull-out force distribution in the box can be considered as shown in Fig. 9(a), where T_i' is a resistance of longitudinal rib between nodes.

Therefore a resultant force at each node, $T_i = T_i' + \tau_{ij}$ x node interval, can be determined as shown in Fig. 9(b).

Note 12: For geosynthetics without nodes, any point selected arbitrarily can be taken as a node.

(b) The distribution of elongation, X_i, and strain, ε_{ij}, can be obtained from the measured displacements at each node, as shown in Fig. 9 (c) and (d).

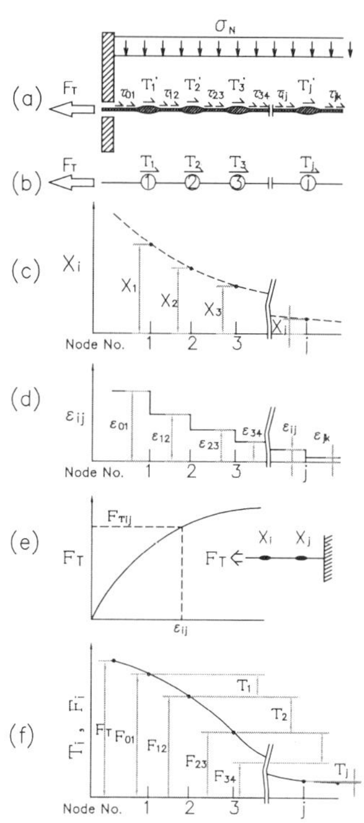

Fig. 9 Procedures of pull-out test data analysis

(c) As shown in Fig. 9 (f), the tensile force, F_{ij}, corresponding to the strain, ε_{ij}, can be estimated from an index stress-strain curve shown in Fig. 9 (e), which is obtained by testing the geosynthetic under no lateral confining pressure.

(d) A limit length of pull-out resistance, L_T is given by the following equation:

$$L_T = \frac{F_U}{2(c + \sigma_n \cdot \tan\phi)} \tag{11}$$

where, F_U = breaking strength of geosynthetic (kPa),
c = apparent cohesion of soil (kPa), and
ϕ = angle of shearing resistance of soil

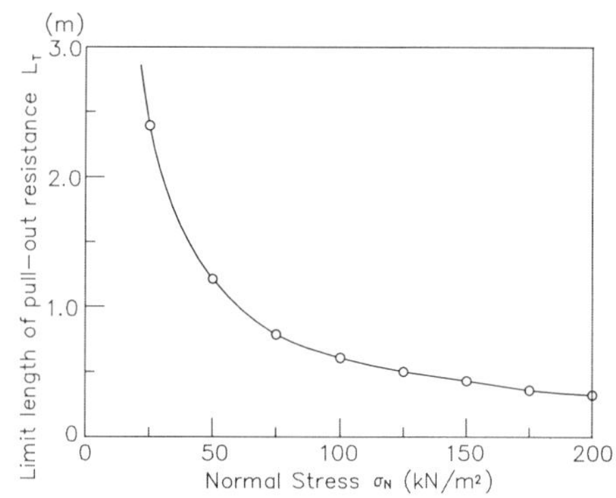

Fig. 10 An example of limit length of pull-out resistance vs. normal stress

Note 13: The relation between the limit length of pull-out resistance and the normal stress for Toyoura sand is shown in Fig. 10.

(e) The effective length of pull out resistance, L_E, and the maximum effective force of pull out resistance, F_{TE}, are defined as follows:

(i) In the case that the geosynthetic was entirely pull-out with no breaking:

$$L_E = L_B \quad \text{(the length of pull-out box)} \tag{12}$$

$$F_{TE} = F_{Tmax} \quad \text{(the maximum pull-out force)} \tag{13}$$

(ii) In the case that the geosynthetic was pull out associated with breaking or that the elongation of geosynthetic reached almost 20 % of the length of pull-out box but no movement of the rear part of geosynthetic occurred:

$$L_E = L_T \tag{14}$$

$$F_{TE} = F_{Tmax} - F_R \tag{15}$$

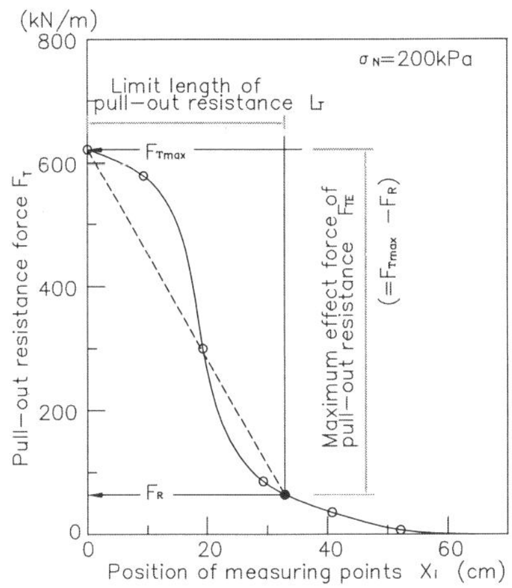

Fig. 11 An example of determining maximum effective force of pull-out resistance

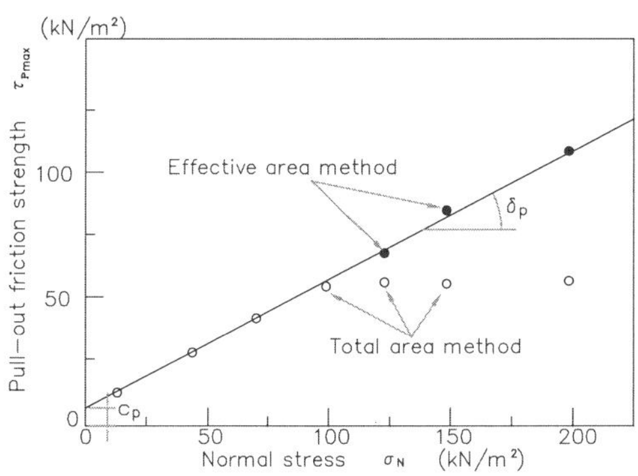

Fig. 12 An example of determining pull-out frictional parameters

where, F_R = The pull-out force generated at the rear outside of the limit length of pull-out resistance.

Note 14: The method (i) and (ii) above are called as total area method and the effective area method respectively.

Note 15: An example of calculating the maximum effective force of pull-out resistance, F_{TE}, is shown in Fig. 11.

(f) The pull-out frictional strength, τ_{pmax}, is expressed by the following equation:

$$\tau_{p\max} = \frac{F_{TE}}{2 \cdot L_E} \qquad (16)$$

(g) The pull-out friction parameters (c_p, δ_p) of geosynthetic in the soil are given as follows:

$$\tau_{p\max} = c_p + \sigma_n \cdot \tan\delta_p \qquad (17)$$

Note 16: An example for calculating the pull-out frictional parameters is shown in Fig. 12.

4 CONCLUSIONS

This report presents both index tests as well as the tests for evaluating the interaction behavior between soil and reinforcement as recommended by different Japanese institutions. Test items on both metallic as well as geosynthetic reinforcements have been discussed. Index tests are covered through relevant standards, reports and manual references. To evaluate soil-reinforcement interaction behavior, a basic test for metallic reinforcements and a direct shear and pull-out test for geosynthetic reinforcements have been described.

ACKNOWLEDGEMENTS

Authors would like to express their heart-felt thanks to Dr. J.T.Shahu for his support and to all members of the Japan Working Group of ATC on Earth Reinforcement for their valuable discussion.

REFERENCES

Public Works Research Center (PWRC). 1990.05. *Construction design method, construction manual and specifications for Reinforced Retaining Wall (Terre Armée).* (in Japanese)

Public Works Research Center (PWRC). 1994.02. *Design and construction manuals of geotextile-reinforced soil structures.* (in Japanese)

Public Works Research Center (PWRC). 1994.10. *Construction design method, construction manual and specifications for Multi-anchor-type reinforced retaining wall.* (in Japanese)

Japanese Geotechnical Society (JGS). 1994.01. Determination of soil-geosynthetics frictional behavior by Direct shear and/or Pull-out test. *A new draft of the Japanese Geotechnical Society's norm JSF T 941. Tuchi-to-kiso.* Vol.42, No.1 :92-102. (in Japanese)

Japanese Geotechnical Society (JGS). 1995.10. *Ground anchor design and construction standards and explanation.* (in Japanese)

Japanese Geotechnical Society (JGS) 1996.03. Activity report - Research committee on reinforcement technology for natural slopes. *Proc. symp. on reinforcement of natural slopes.* (in Japanese)

Railway Technical Research Institute (RTRI). 1994.03. *Reinforced railroad/road with rigid facing method (RRR).* (in Japanese)

Railway Technical Research Institute (RTRI). 1992.11. *Design standard for railway earth structure.*(in Japanese)

S. Hayashi, K. Makiuchi, H. Ochiai, M. Fukuoka and T. Hirai 1994. Testing methods for soil-geosynthetic frictional behavior - Japanese standard. *Proc. Fifth International Conference On Geotextiles, Geomembranes And Related Products.* Singapore. Vol.1. 411-414.

Japanese Industrial Standard(JIS). Investigated by Japanese Industrial Standards Committee. Published by Japanese Standards Association. (in Japanese)

JIS G 0591 *Method of 5 per cent sulfuric acid test for stainless steels*
JIS G 3101 *Rolled steel for general structure*
JIS G 3109 *Steel bar for prestressed concrete*
JIS G 3122 *Steel bars for concrete reinforcement*
JIS G 3191 *Shape, dimension, weight and tolerance for hot rolled steel bar and bar-in-coil*
JIS G 3193 *Dimensions, mass and permissible variation of hot rolled steel plates, sheets and strip*
JIS G 3536 *Uncoated stress-relieved steel wires and strands for prestressed concrete*
JIS K 7114 *Method of testing chemical resistance properties of plastics*
JIS L 1410 *Outdoor exposure test for plastic construction materials*
JIS L 1415 *Accelerated exposure test of plastic construction materials*
JIS L 1908 *Test methods for geotextiles*
JIS Z 2241 *Method of tensile test for metallic materials*
JIS Z 2271 *Method of tensile creep test for metallic materials*
JIS Z 2272 *Method of tensile creep rupture test for metallic materials*
JIS Z 2371 *Methods of neutral salt spray testing*
JIS Z 2911 *Method of test for fungus resistance*

Earth Reinforcement, Ochiai, Yasufuku & Omine (eds) © 1997 Balkema, Rotterdam, ISBN 90 5410 833 9

Present state and perspective of design methods of reinforced soil structures

H. Miki – *Public Works Research Institute, Ministry of Construction, Tsukuba, Japan*
N. Fukuda – *Fukken Co., Ltd, Consulting Engineers, Hiroshima, Japan*
O. Murata – *Railway Technical Research Institute of Japan Railway Group, Tokyo, Japan*
S. Tayama – *Research Institute, Japan Highway Public Corporation, Tokyo, Japan*
A. Yamamoto – *Technical Research Institute, Obayashi Corporation, Tokyo, Japan*
Y. Otani – *Hirose Co., Ltd, Tokyo, Japan*
Y. Yokota – *Maeda Kosen Co. Ltd, Fukui, Japan*

ABSTRACT: The soil-reinforcing construction method has a history of over 30 years. The recent 10 years or so have seen great progress in researches and in the application of the method to soil structures in the field. Besides, various design and construction manuals have been made available, generalizing the method. The reinforcing mechanism in such soil structures, including that many such structures demonstrated their capability of withstanding severe ground motions in recent great earthquakes, has not yet elucidated entirely, and still a number of problems await solutions.

Under the circumstances, the working group of Design Method , one of the working groups of Japanese Supporting Committee of the Asian Technical Committee for Earth Reinforcement, compiled this paper mainly based on cases of reinforced soil structures in Japan to report the current situation and perspective of design methods, in particular, of geosynthetic-reinforced soil structures and soil nailing.

1 INTRODUCTION

There have been proposed many reinforced-soil construction methods for the construction or reinforcement of embankments, retaining walls, foundation grounds, etc. and for slope cutting, excavation, etc. Their history is now over 30 years. Available as reinforcing materials are steel, plastic, and other new materials, and they take various forms such as sheet, grid, strip, and bar. Besides, in case of wall-facing structures, their rigidity and weight effect vary depending on their types, structures, and materials, and hence their reinforcing effect upon walls varies.

Regarding the design methods of such reinforced-soil structures, although they are commonly based on the limit equilibrium method, various design concepts are proposed for each type of structure, and for one and the same type of structure, individual design methods concerned propose their own concepts. However, they are not necessarily reflecting such reinforcing mechanisms adequately.

This report is compiled to summarize the present state and perspective of the design methods mainly for geosynthetic-reinforced soil structures and soil nailing for natural-ground.

2 GEOSYNTHETIC-REINFORCED STRUCTURES

2.1 *General remarks*

Geosynthetic-reinforced soil construction methods are applied to embankments, retaining walls, foundations of structures, soft grounds, etc. and enable us to construct soil structures as effective as the conventional concrete structures, by reinforcing soil and giving it ductility and flexibility against deformation. However, their design methods are based on the limit equilibrium method, and do not necessarily make the most of the properties of reinforced soil, though the construction records of such reinforced soil structures are abundant.

Expected of engineers in this scientific field is to develop "more rational" and "more economical" design methods under "more unified design concept and construct a "concept of combining simple design methods and detailed design methods," as well as to elucidate the reinforcing mechanisms.

Besides, new reinforced soil structures are being developed, and new applications, tried.

Discussed in this chapter are the present state and perspective of geosynthetic-reinforced soil structures, of which the points are as follows:

(i) Subjects of the current design methods based on the limit equilibrium method
(ii) Evaluation of seismic stability and its effect
(iii) Subjects in designing reinforced soil structures as permanent structures
(iv) Design methods under special conditions
(v) Utilization of low-quality surplus soil from construction sites
(vi) Reinforcing method of surface layers of soft grounds

2.2 *Subjects of current design methods based on limit equilibrium method*

The last decade has seen concentrated technical development of geosynthetic-reinforced soil structures, and data from follow-up inspections and surveys in the field have been building up. However, the reinforcing mechanisms have not yet elucidated fully, the design methods for such construction methods reflecting them inadequately.

Left outside the consideration of the current design methods are reinforcing effects not yet elucidated adequately such as the effect of slope-facing structures, the integration effect of reinforcements and soil into one body, and the effect of exclusion of seepage water. From this point of view, it can be said that the current design methods stand on the safe side, leaving a margin in the factor of safety. This margin was demonstrated by reinforced-soil structures which withstood the severe ground motion during the Hyogo-ken southern earthquake far exceeding their design seismic forces. Although such practical design methods have been made available as mentioned above, the design method is, by its very nature, a realm of symplasm. Calculation is made possible by reducing an actual phenomenon into a simple model. Accordingly, there remains many subjects such as going one step farther in pursuit of "veiled reinforcing mechanisms" and improving the technique for the prediction of behavior, typically of deformation (numerical analysis).

Principal technical subjects of the current design methods based on the limit equilibrium method are as follows:

(1) *Effect of Slope-facing structures*

The effects of slope-facing structures are classified by Tatsuoka (1993) as follows: (i) local rigidities, (ii) overall longitudinal rigidity, (iii) overall shearing rigidity, (iv) overall bending rigidity, and (v) gravity resistance, and slope-facing structures of concrete are highly appreciated for their contribution toward the stability of slopes. However, in case that geosynthetic reinforcements 1 m long or so are arranged densely in the area adjacent to the slope face, or in case that sandbags are piled up, each layer of sandbags wrapped in geosynthetic reinforcements and sandbags and reinforcements forming a multi-decker sandwich, such a flexible slope-facing structure can be regarded as a integrated structure, which contributes to the stability of embankments considerably. But we can cite few case examples of which the design took such effect into account. To address this subject, the following two approach would be worthwhile:

(i) Assuming that some anisotropic apparent cohesion is exerted in a integrated zone, to consider c and ϕ of the integrated zone in the circular slip analysis (PWRI 1988, Kutara et al. 1991).

(ii) In case that a integrated zone has relatively high rigidity, to reckon it to exert an retaining effect as a lean-to type retaining wall (Miki et al. 1991, Miki et al. 1992).

Though both the above methods are yet in a study stage, we expect that more rational design concepts with due regard to such effects of slope-facing structures will be proposed in near future.

Regarding, in particular, the reinforced embankment construction method, expected is the establishment of an evaluation method of apparent cohesion in a pseudo retaining wall consisting of densely arranged sandbags and geosynthetics. If such an evaluation method is made available, its evaluation results can easily be reflected in the design of embankments through the circular slip analysis. In addition, as for the reinforcing effect of low-rigidity, nonwoven-fabric-type geosynthetics laid closely which is difficult for the conventional stability analysis methods to explain, a more rational interpretation will become available.

Besides, it is necessary for a slope-facing structure to function as a pseudo rigid body that geosynthetic reinforcements be laid closely enough to make the reinforced zone integrated and make it, as if it were a mass of stabilized soil, exert apparent, anisotropic cohesion. In this regard, a large-sized triaxial compression test was conducted with different densities of reinforcement arrangement (Kutara et al. 1991). As the results of this test, it was confirmed that apparent, anisotropic cohesion was exerted in accordance with the tensile strength of, and the spacing between, reinforcements, and the propriety of the following equation was verified:

$$c_R = \beta \, T_f \sqrt{K_p}/2 \, \Delta H \tag{1}$$

where c_R: apparent cohesion;
β: coefficient of correction;
T_f: tensile strength of geosynthetic reinforcements;
K_p: coefficient of passive earth pressure; and
ΔH: spacing between reinforcements.

Though β is usually set at 1.0, the value of 0.43 was found in the above test.

In another study (Miki 1996a), sandbags and geosynthetic reinforcements were piled closely and alternately to construct a full-size slope-face structure (Fig. 1), the critical condition of the slope at the verge of failure was determined by cutting the geosynthetic reinforcements (cutting at 9) the circular slip analysis of the slope was performed to evaluate the effect of the slope-facing structure. As the results of this study, the safety factor of this embankment at failure appeared to be near 1.0 in case that the anisotropic, apparent cohesion in the zone of the virtual retaining wall, 45 cm thick, of piled sandbags was taken into account based on the result of large-sized triaxial compression tests with sandbags (c = 51 kN/m^2), whereas the safety factor was found to be 0.62 in case that the anisotropic, apparent cohesion was not taken into account (c = 2.0 kN/m^2), thus the former case being well consistent with the actual event.

Therefore, if a method is made available in future to

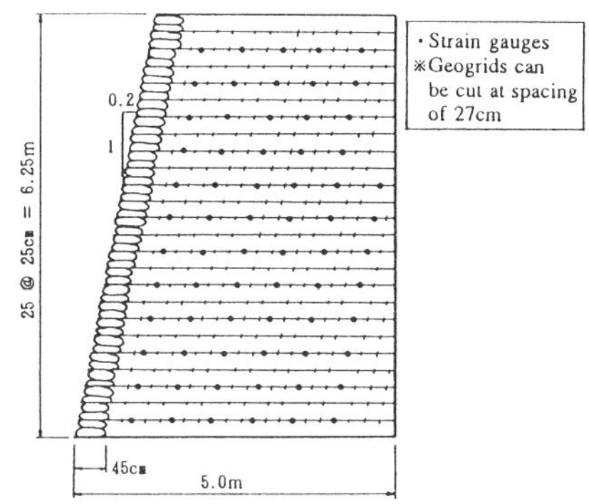

(a) Cross sectional view of steep slope embankment

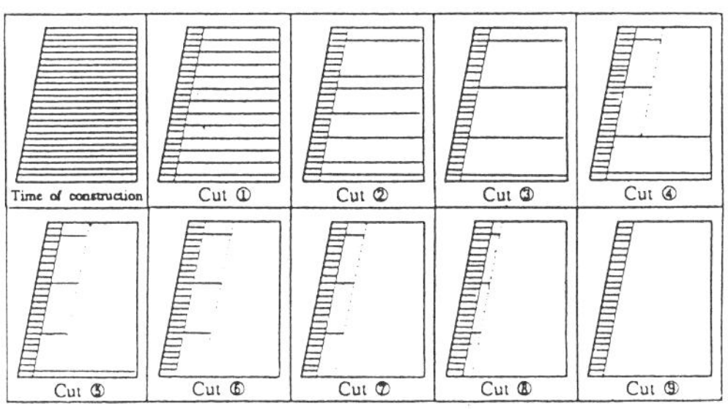

(b) Cutting stages of geogrids

Fig.1 Full-scale failure test of geogrid reinforced steep embankment by cutting the reinforcements (Miki et al. 1992).

evaluate properly the anisotropic, apparent cohesion in the zone of a virtual retaining wall consisting of sandbags and geosynthetic closely arranged, the effect of such slope-facing structures can easily be reflected in design through the circular slip analysis.

(2) *Drainage effect of geosynthetics*
If geosynthetic with a drainage function is used to reinforced embankments, the effect of excluding seepage water can be expected when raining, in addition to the tensile reinforcing effect and the above-mentioned apparent cohesion effect (Kutara et al. 1990, Tatsuoka et al. 1983, Sato et al. 1984). If seepage water can be excluded even to a certain degree, it contributes considerably to the stability of embankments because the decrease in soil strength due to seepage can be delayed and restrained. At present, however, the seepage water exclusion effect is treated merely as an allowance outside the calculation of the safety factor, and we need more studies before we will be able to give it due treatment in design. If in future our study progresses to such an extent as we can take the seepage water exclusion effect into consideration in design, the usage of geosynthetic with a drainage function in top layers of embankments may begin to gain wide acceptance in reinforced embankment construction methods, too.

(3) *Prediction of deformation*
To predict the deformation of embankments and the tensile force acting on geosynthetic by using continuum FEM analysis, etc. which make up for the shortcomings of limit equilibrium models, essential is an analytical model which well represents the nonlinear deformation property and the dilatancy property of soil and the interaction property (frictional property) between soil and geosynthetic. In case of reinforced embankments, a very complex model is needed, and hence no verifiable method of such prediction has not yet established. Under the circumstances, one of the most important subjects in future is the development of a numerical simulation method which enable us to predict at a practical level of precision not only the deformation behavior of embankments during their construction, but also their post-construction behavior during rainfall or an earthquake.

2.3 *Evaluation of seismic stability and its effect*

(1) *Present state*
In regions of frequent earthquakes, armoring structures with seismic design is an important subject. As for reinforced embankments and retaining walls, seismic design methods have been established based on the concept of limit equilibrium-seismic forces are substituted with static loads by performing various laboratory and field experiments.

The seismic forces are handled by the method of determining the distribution of earth pressure acting on wall-surfacing structures during earthquakes based on experimental results (Terre Armee), or the seismic coefficient method of substituting horizontal accelerations with seismic coefficients. In design, on the other hand, an extra amount of reinforcements is provided, or the pull-out resistance between reinforcements and soil is calculated on the safe side, or the design safety factor in the stability analysis is set high, as the measures against earthquakes.

Although the current design methods do not necessarily reflect adequately the reinforcing mechanisms ascertained by experiments and tests, the recent severe earthquakes, Loma Prieta earthquake 1989 (Collin et al. 1992), Northridge earthquake 1994 (White & Holz 1996), Kushiro Offshore earthquake 1993 (Fukuda et al. 1994), and Hyogo-ken Southern earthquake 1995 (Tatsuoka et al. 1996a), revealed the high performances of reinforced soil structures (Photo 1). Some of them had been designed under ordinary conditions, which further testifies to the high performances of reinforced soil structures. This, on the other hand, suggests that the current design methods under ordinary conditions contain some allowance.

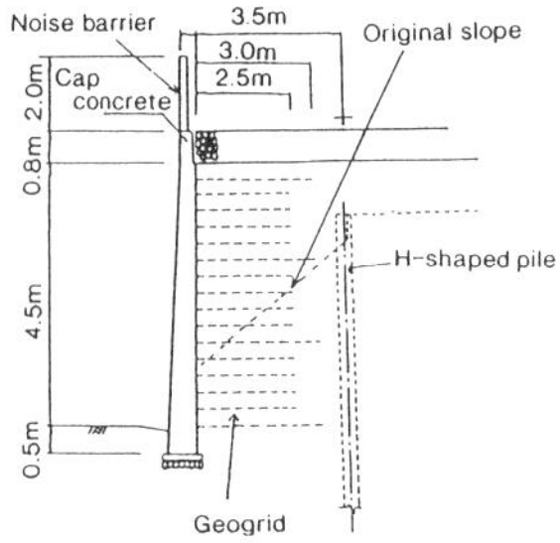

Photo 1 Scene in front of the RRR-walls in Tanata (Tatsuoka et al. 1996a).

(2) *Future problems*

In order to reflect the seismic performance of reinforced soil structures in design, the following problems have to be solved:

(i) Setting of seismic forces: The incidence of earthquakes such as the Hyogo-ken Southern earthquake, a near field earthquake, is once in a thousand years in magnitude and type, and the maximum acceleration of the Hyogo-ken Southern earthquake recorded in the area hit most severely (magnitude VII) was over 600 gals. However, the design seismic coefficient for reinforced soil structures is 0.2 or so at the highest. The interpretation of this difference is one problem, and another is how to establish the concept of equivalent acceleration which reflects the ductility of reinforced soil structures.

(ii) Handling of deformation: The limit equilibrium method gives no consideration to the deformation of reinforced soil structures. A certain degree of deformation of a structure, though far from causing its total destruction, can occasion it functional disorders and give it an unstable appearance. Thus, we need some measures in design to contain such deformation within an allowable extent in which the necessity of repair does not occur. An approach to such measures would be to review the concept of the factor of safety of the limit equilibrium method.

(iii) Numerical analysis methods: Seismic response analysis would be an effective means to interpret the behavior of structures during earthquakes and reflect the results in design. It is necessary to study how to make proper analytical models and how to reflect the analytical results in design.

(iv) Measures against liquefaction: Reinforcing an embankment on a soft sandy ground has no effect of restraining its subsidence if liquefaction occurs in the ground during an earthquake and it subsides, but such effect can be expected as prevention of the failure of, or cracks in, the embankment (PWRI 1989). Besides, with damage so restrained, we can relatively easily secure the means of transportation to cope with a state of emergency immediately after an earthquake, and can restore damaged structures with relatively small cost and in relatively short period of time (Matsuo et al. 1995). Such a design concept we need to build (Fig. 2).

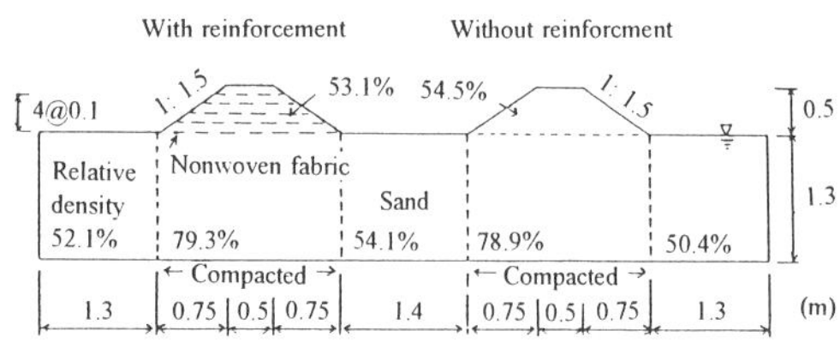

(a) Initial condition before tests

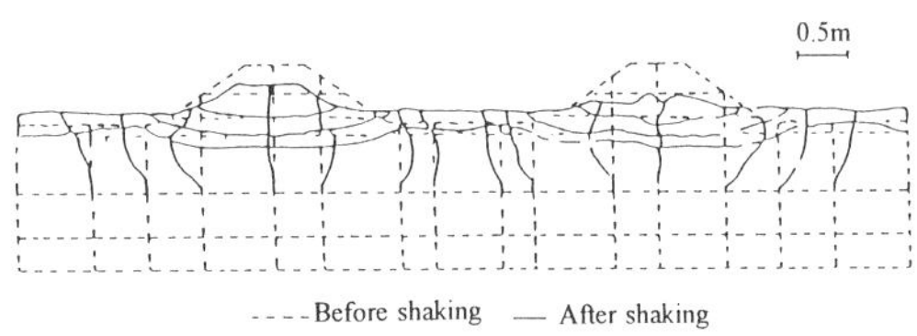

(b) Final condition after tests at 240 gal

Fig.2 Shaking table tests on geosynthetic reinforced embankment on a liqufactive ground (PWRI 1989)

2.4 *Subjects in designing reinforced soil structures as permanent structures*

The number of reinforced embankments and soil walls constructed as permanent structures for roads, railways, housing sites, etc. has recently been increasing. In these applications, required are not only the economy but also the performance, and hence the following matters are important:

(i) Durability: The anticorrosiveness in case of metal reinforcing materials and the durability of wall surfaces in case of reinforced soil walls should be secured.

(ii) High stability and rigidity: It is necessary for the deformation of structures not to become excessive under various load conditions such as traffic load, rainfall, earthquake, etc. This is especially true of important structures such as bridge abutments.

(iii) Checkup of seismic stability and systematization

of seismic design: To design reinforced soil structures as permanent structures, their seismic design methods have to be systematized as those for reinforced concrete or steel structures are done. This subject has taken up in 2.3 *Evaluation of seismic stability design method.*

(iv) Environment: Permanent reinforced soil structures have to be designed so that they can be in harmony with their environment including sceneries. Wall-facing structures, or wall facings, harmonious with various environment are also needed. Vegetative facings are friendly with environment.

Regarding the subject (ii), the deformation of a reinforced soil structure after being put in service can be minimized by providing, for example, a rigid wall-facing structure. This is applicable to bridge abutments which, as shown in Fig. 3, support girders and hence are allowed only small deformation (Tateyama et al. 1994). For the same purpose of minimizing the deformation of structures after being put in service, a prestress, pre-load, reinforced soil wall construction method as shown in Fig. 4 is also devised (Tatsuoka et al. 1996b), and would be applied to abutments to support girders of longer spans.

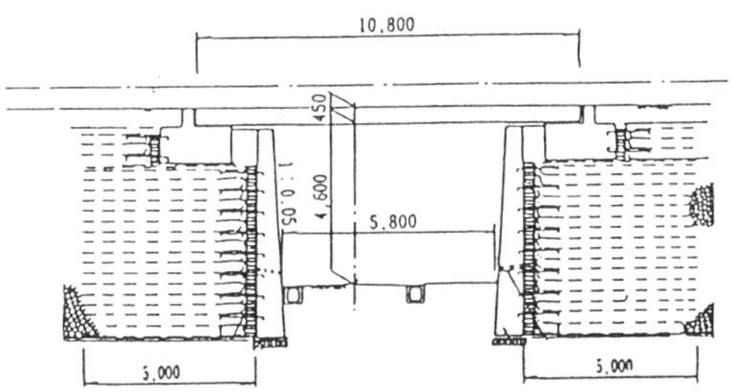

Fig.3 Cross section of geosynthetic-reinforced soil bridge abutments at Shinkansen Yard, Nagoya (Tateyama, et al., 1994).

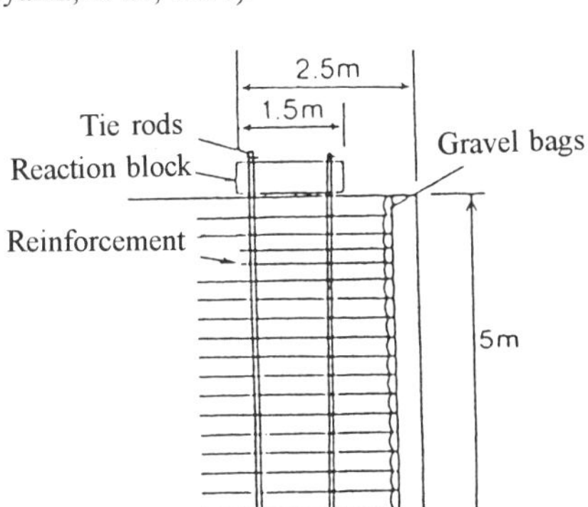

Fig.4 Typical PL/PS GRS retaining wall (Tatsuoka, et al., 1996b.)

The subject (iv) is also of considerable importance, since in order for reinforced soil structures to exist for a long period of time, they have to be compatible with their environment including sceneries. Requests for vegetation have been increasing recently, and accordingly the development of methods of vegetation and, at the same time, of providing high durability of paragraph (i) is a subject.

2.5 *Design methods under special conditions*

As far as the height of embankments, external forces, ground conditions, etc. are normal, the limit equilibrium method would continue to be practiced most commonly in future. Under special conditions, however, such as very large scale, existence of an important structure nearby, use as bridge abutments to support girders, soft ground, etc., it may become necessary to perform deformation prediction by a numerical analysis method, etc., in addition to the checkup of stability by the limit equilibrium analysis method.

Fig. 3 show an example of the application of such design method to an abutment supporting girders. Against the large local load acting from a girder-supporting foundation onto the top of the abutment, or embankment, the closely arranged geosynthetic reinforcements and the rigid wall-facing structure minimize the deformation of the abutment. The specifications of such a wall-facing structure such as required strength of, vertical spacing between, and length of, reinforcements, and configuration of the wall-facing structure (thickness, arrangement of reinforcements, etc.), necessary to secure a given factor of safety, can be determined by the limit equilibrium analysis. However, because this was the first application of the geosynthetic-reinforced soil structure to such abutments, and also because it was a railway bridge which requires strict limitation on allowable deformation, the deformation was predicted by numerical analysis, and the calculated results were compared with the measured data obtained by conducting static loading tests. Besides, the measurement was continued after the bridge had been put in service to confirm the safety of the bridge (Tateyama & Murata 1994). It was confirmed through these analysis, test, and measurement that in case of the bridge structure, or configuration, shown in Fig. 3, the deformation of abutments supporting girders of spans of 10 m or so fell within an allowable extent if the safety factors of such abutments had been secured by the limit equilibrium analysis. Thereafter to date, more than 10 bridges have been constructed by using the above technique.

In the case of a large-scale reinforced soil wall constructed on a very soft ground (Kojima et al. 1994), numerical analysis was performed in advance to predict the consolidation settlement of the ground and the deformation of the embankment, and the height of the pre-loading embankment, the leave-it-alone period, the time of construction of the wall-facing structure, and so on were determined by using the predicted values as reference data.

Thus, under the special conditions, numerical analysis is needed in our practical affairs. To predict the deformation of reinforced embankments and tensile forces acting on geosynthetic by using continuum FEM analysis, etc. which make up for the shortcomings of the limit equilibrium analysis method, essential is an analytical model which well represents the nonlinear deformation property and the dilatancy property of soil and the frictional property between soil and geosynthetic. In case of reinforced embankments, a very complex model is needed, and hence no verifiable method of such prediction has not yet been established. Under the circumstances, one of the most important subjects in future is the development of a numerical analysis method which enable us to predict the deformation behavior of embankments at a practical level of precision, and the precision can be improved by practicing comparative study with data of laboratory tests and field measurement. Besides, the improvement of precision of the numerical method leads to the rationalization of the limit equilibrium method.

2.6 *Utilization of low-quality surplus soil from construction sites*

The volume of surplus soil from construction sites has recently been increasing as urban development, utilization of underground, and so on have vigorously been carried forward. Particularly in urban areas, it has become difficult to secure disposal lands of such surplus soil. Under the circumstances, utilization of such soil is hoped for so that construction of buildings and structures can smoothly be carried forward. In the Metropolitan area, of the annual total volume of surplus soil of about 120 million m^3, approx. 70% is silt, Kanto loam, cohesive soil, etc. of high water content and low quality. Accordingly, how to use these low-quality soils is the point of the promotion of the surplus-soil utilization.

As a technique to utilize such soils of high water content and low quality, the embankment construction method using geosynthetics to improve drainage and strength is very effective. An embankment being constructed with cohesive soil of high water content is apt to lose its stability due to an excessive pore water pressure, if occurred, during its construction. In this case, by using geosynthetics as horizontal drainage materials, excessive pore water pressure is dissipated and the soil strength improves through its consolidation, as the construction progresses. Thus, such use of geosynthetics make the otherwise difficult construction works simple and economical. Stabilization with cement, lime, etc. becomes unnecessary, surplus soil without any treatment can be carried from the site of its generation to the site of its utilization, and while piling up the soil, soil strength develops by the selfweight effect of piled soil and the drainage effect of the geosynthetics. In this method, it is not purported to add strength to soil by the stabilization of soil or the tensile reinforcing effect of geosynthetics, but the original strength of soil is derived. In this meaning, it is very rational and economical.

However, because the design method of the current embankment construction method using cohesive soil of high water content and geosynthetic to improve drainage and strength is proposed based on a theory applicable to saturated soils, it is applicable to only such soils. It is necessary in future to establish a theoretical method applicable to a little bit unsaturated soils, too. Besides, hoped for are not only methods applicable to embankments on land, but also methods applicable to underwater embankments as underwater lands being reclaimed from the sea.

Moreover, it is difficult with the conventional method to construct stable, steep slopes with low-quality surplus soil from construction sites, but if such construction becomes practically possible by using geosynthetics, the prospects of such construction method will be bright in the high social mood for recycling.

Regarding the geotextile-tube dehydration method (Miki et al. 1996) which is under development as one of the researches for high-grade soils, basic

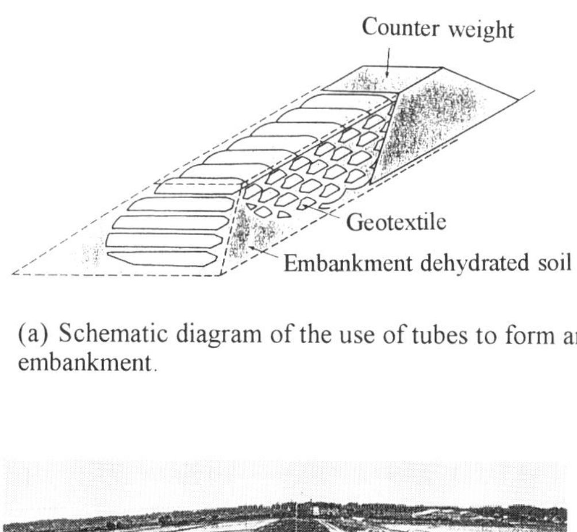

(a) Schematic diagram of the use of tubes to form an embankment.

(b) Piled up geotextile tubes.

Fig.5 Tests of geotextile tube dehydration method (Miki. et al., 1996).

experiments are being conducted to apply the method to envelop-and-pour-type dikes; i.e., pouring sludge of Kasumigaura (a vast creek in the Kanto plains) into large-sized tubes of which the periphery and length are 10 m and 20 m, respectively (Fig. 5). In addition, construction works are being carried forward, deposited silt in dams utilized by this method, and construction of multi-natural revetments by this method are under way. The evolvement of the sacked dewatering method in future is expected.

2.7 *Reinforcing method of top layers of soft grounds*

In Table 1, applications of geosynthetics to embankments on soft grounds are classified into four categories.

In the soft-ground top-layer reinforcing method for cover soil, proposed are a number of design concepts which take the bearing-capacity mechanism into consideration. However, since the influence of such construction methods is significant, the tensile strength of reinforcing materials is in reality determined based on experience in the past. Accordingly, hoped for is the establishment of more rational design methods through analyses of construction cases to date.

In the soft-ground top-layer reinforcing method for embankments, it is made clear from the feasibility study made by the Public Works Research Institute that in case of a deep soft ground or a high embankment of which the given stability factor against slip can not be secured by sand drains alone, using geogrids as an auxiliary method is effective from an overall point of view. Therefore, hoped for is the development of applications unique to geosynthetics and competitive with the conventional construction methods.

Regarding the combined method of the geosynthetic reinforcing method and the CDM, proposed is a method that, with an idea similar to the pile-net method, the whole foundation ground of an embankment is improved with piles by the CDM of a low improvement rate (10-30%), and the shortage of

Table 1. Category of geosynsethics application for countermeasures of soft ground.

Application	Construction method (reinforcements)
Surface reinforcement for reclamation	Sheets laying method (geowovens) Nets laying method (geonets) Sheets and ropes method (geowovens etc) Sheets and bamboos method (geowovens)
Surface reinforcement for embankment	Planar reinforcements (geotextiles) Single layer type (Steel rods with anchor-plates) Multi-layer type 3-D reinforcements (Geomats, Geocells)
Piles combined with reinforcement	Piles-nets method (Steel wires, Geotextiles) Deep cement mixed piles-nets method (Geonets, Geogrids)
Subgrade reinforcement	Improve bearing capacity (Geogrids) Prevent boiling of mud in railway subgrade (Geocells, Geononwovens, Geowovens, Geomembranes)

stability against circular slip are made up for with geogrids of high strength, or geosynthetic is used as a countermeasure against the settlement differential between on the improved soil bodies and on the unimproved soil (PWRC 1993, PWRI 1992, Murata et al. 1986, RTRI 1987).. Although such method combined with the CDM of a low improvement rate has not yet established as a design method, it is expected to come into practical use as a new and economical approach to the improvement of soft grounds.

Furthermore, hoped for as a countermeasure against earthquakes are studies to apply geosynthetic-reinforced soil construction methods to the restraint of subsidence of embankments, during earthquakes, constructed on soft sandy grounds which threaten liquefaction when hit by an earthquake. This subject has taken up in 2.3.

3 GROUND REINFORCEMENT

3.1 *General remarks*

The ground reinforcement has a variety of uses such as used to construct stabilizing structures of natural slopes and cut slopes, retaining walls, excavation works and reinforcing structures of foundation grounds. In this chapter, we discuss this method, focusing on stabilizing structures of natural slopes and cut slopes.

The ground reinforcement has empirically been used in construction fields since long time ago, and laboratory experiments and test construction started in 1985 or so, the method recognized as a ground reinforcement.

The reinforcing principle of this method is: when excavation of, or change of the load on, a natural ground begins to cause its deformation, tensile, compressive, shearing, bending, and other forces act on its reinforcements. At the same time, the reinforcements exert passive resisting forces, restraining the deformation and stabilizing the natural ground. In the current design method of ground reinforcement, however, the limit equilibrium method is commonly used, and therefore it is hardly good enough to reflect the above reinforcing principle adequately (Research Committee for Ground Reinforcement, JGS 1996). In order to incorporate the design in the actual behavior of reinforced natural grounds, the following have to be made properly: (i) setting of design parameters of natural grounds, (ii) evaluation of the functions of reinforcing materials, (iii) introduction of a design method which takes account of deformation, and (iv) evaluation of the effects of slope-facing structures.

On the other hand, the following are the subjects attracting attention recently: construction works with monitoring, seismic stability and its evaluation method, durability as permanent slope , construction method taking account of the landscape, application of new materials and new construction methods, and so on.

3.2 *Stability analysis*

(1) *Setting design parameters of natural grounds*
The strength parameters of the soil and the pull-out resistance of the reinforcements used in designing a ground reinforcement are usually determined by their individual tests. However, because the strain levels, at which the pull-out resistance of the reinforcements, the shearing resistance of the soil, and the passive resistance of the soil against the reinforcements are different, it is necessary to determine such design parameters, taking account of these strain conditions. For instance, in case that the peak value is used as the pull-out resistance of the reinforcements, a strength corresponding to the same strain level in stress-strain relation of soil shall be used. Proposed in the U.K. is a method of using residual strengths as strength parameters of grounds (Department of Transport, UK 1994), which takes into account decrease in the strength parameters to be caused by deformation.

(2) *Evaluation of functions of reinforcements*
When the natural ground with reinforcement is deformed, the deformation exert tensile, compressive, shearing and bending forces on reinforcement laid in the ground. These forces vary, depending on the specifications of reinforcing materials, e.g. shape, tensile rigidity, compressive rigidity, bending rigidity, shearing strength and coefficient of friction, the properties and conditions of grounds, e.g. coefficient of deformation, strength parameters and moisture condition, and the interaction between soil and reinforcements. At present, in most cases, the reinforcing effects of reinforcements are evaluated based on their tensile resistance. There are few cases taking account of tensile, compressive, shearing and bending resistance mentioned above. In France, the multicriteria method concerning the pullout failure, failure by plastification of the soil before the nail, the tensile and shearing strengths of the bar, and reinforcement plastification caused by the bending moment (Scientific Committee of French National Project CLOUTERRE 1993).

(3) *Introduction of design method taking account of deformation*
In the current design method, stability is checked by the limit equilibrium method, but deformation is not taken into account. Accordingly, at present, as the means to reduce the deformation of natural grounds and adjacent structures, the overall safety factor against the ultimate failure condition is set high, or consideration is given to the arrangement of reinforcements. Regarding the stability of a natural ground under excavation, tried is a method of estimating the displacement from the properties of the ground and the height of wall by using empirical formulas based on measured data (Scientific Committee of French National Project CLOUTERRE 1993).

Being performed as post-analysis using numerical analysis, i.e. FEM, are the stability evaluation of ground reinforcement and the evaluation of axial forces of reinforcements, based on the strains and the distribution of local safety factors. In future, when the ground reinforcement is applied to natural grounds of important structures or adjacent to structures, such evaluation method will be needed more and more for the estimation of deformation in the design stage.

The deformation analysis of ground applied of the ground reinforcement have the same subject, e.g. correct initial condition, constitutive model of ground materials, boundary condition and so forth, as deformation analysis commonly recognized in the field of the geotechnical engineering. In addition, the interaction between soil and reinforcements has attracted significant attention. Accordingly, the following specific subjects for the deformation analysis of the ground reinforcement have to be addressed:

(i) Modeling taking account of the dilatancy effect restricted by relative displacement between soil and a reinforcement or a shearing zone along the boundary caused by such relative displacement

(ii)As for the expression (composite material) of the assumptions for a pseudo-retaining wall and the unification effect of a group of piles.

(iii) Reducing the three-dimensional behavior of linear reinforcements into a two-dimensional model

(iv) Modeling of pressure-bearing grounds and facing units

In addressing the deformation problem in future, the construction works with monitoring by back analysis would be an effective means to narrow down the gap between the precision in surveys and tests to determine the parameters and the precision of analytical models on one side and the precision of actual design on the other side.

(4) *Evaluation of effectiveness of facing units*
It is ascertained by experiments, etc. that facing units have large effectiveness on restraining the deformation of the ground, provided the facing units and the reinforcements connect together tightly (Mutamatsu et al. 1995). Consequently, several design methods which take account of the effectiveness on facing units, are proposed (Japan Highway Public Corporation 1995, RRR Construction Method Association 1993).

Fig. 6 illustrates the design concept employed in the method, giving consideration to the effectiveness of the facing as well as the effectiveness of the reinforcements.

In this design method, as shown in the equation (2), the resistance force of a reinforcement, T_{pa}, is the lowest among the pull-out resistance of the reinforcement due to the sliding mass, T_{1pa}, the pull-out resistance mobilized in the resisting area, T_{2pa}, and the tensile strength of the reinforcement, T_{sa}. The earth

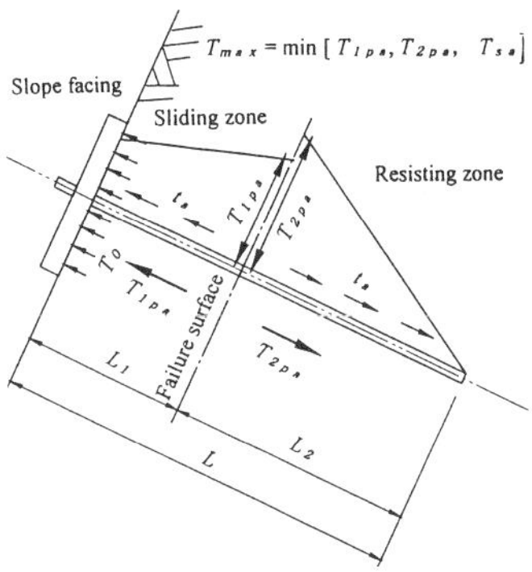

Fig.6 Schematic distributions of resisting forces on the nail (Japan Highway Public Corporation 1995).

pressures of the sliding mass acting on facing unit, To, is considered in calculating T_{1pa}.

$$T_{pa} = \min. \{T_{1pa}, T_{2pa}, T_{sa}\} \quad (2)$$

3.3 *Construction works with monitoring*

It is difficult to grasp the soil properties of a natural ground in advance and precisely, moreover it is difficult to predict the deformation of the natural ground during the phase of its reinforcing works and set managerial criterion values. Important is, therefore, construction works with monitoring; i.e., to conduct measurement of the displacements at the top of slope, the displacements in the ground, the axial forces exerted on the reinforcements, etc. and to review the managerial criteria and the design based on the measured data.

An approach using "back analysis" has recently come into use. Fig. 7 shows such a case (Tayama et al. 1996a, Japan Highway Public Corporation) and the arrangements of measuring instruments and apparatus. Fig. 8 is the flow chart of the construction works with

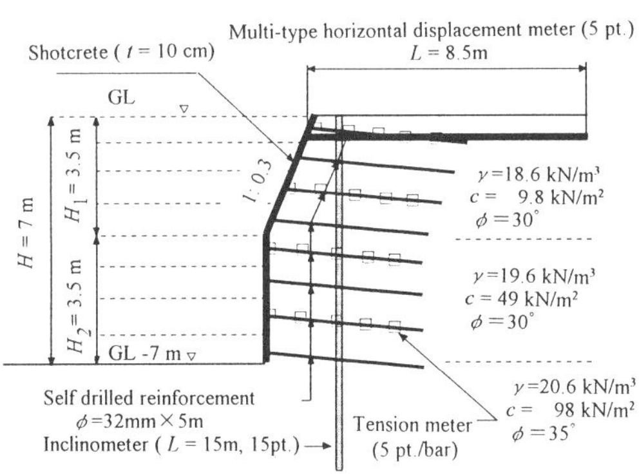

Fig.7 Cross section of test cut-slope and instrumentations (Tayama et al., 1996a).

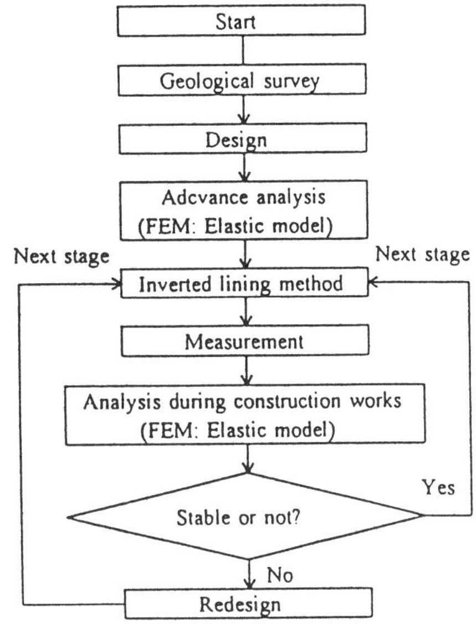

Fig.8 Flow chart of construction control system by applying backward analysis (Tayama, et al., 1996a).

monitoring regarding the case above. At the end of data each stage, FEM back analysis is conducted based on the so far obtained to review the soil parameters (mainly the modulus of elasticity), and FEM analysis is conducted again based on the soil parameters which was revised to evaluate the stability of the reinforced soil structure at the next stage or at the completion of works. This approach, however, has many problems such as the stability-evaluation method.

Although it is necessary to pursue a method of construction works with monitoring required high precision like the approach above, the measurement and analyses to be performed in such an approach are costly and the merit of economy of the ground reinforcement is reduced. Accordingly, hoped for is research and development of a simpler method; for example, based on the measurement of displacement at only the top of the slope.

3.4 *Seismic stability and its evaluation*

There were many ground reinforcement structures in the area hit by the Hyogo-ken Southern earthquake, and reports have been made to confirm their high seismic stability (Matsui et al. 1995, Tsukuda et al. 1996, Fujii et al. 1996, Otani et al. 1996). Besides, the ground reinforcement has been adopted to restore damaged structures, its advantage derived well. However, the current seismic design methods do not necessarily appreciate the earthquake-resistant performance of ground reinforcement. Accordingly, to be considered in making seismic design of ground reinforcement are the following:

(1) *Necessary conditions*
Because ordinary cut slopes are not armored with seismic design, the necessary conditions for the seismic design of reinforced cut slopes have to be set.

(2) *Earthquake-resistant performance of ground reinforcement*
When a reinforced cut slope begin to deform under a seismic load, the tensile forces of the reinforcements increase and they function to maintain the shearing strength of the soil without reducing the restricting pressure and to bear the seismic load. Besides, ground reinforcement has high toughness in the meaning that its deformation at failure is larger than that of unreinforced ground. Since ground reinforcement has excellent seismic stability in both the strength and the toughness, it is not too much to say that it has larger effect in the seismic condition than in the stationary condition. The high toughness of ground reinforcement is not directly reflected in the stability analysis by the limit equilibrium design.

These features taken into account, described below are the subjects of the current design method in order to establish, in future, a seismic design system which can precisely evaluate the earthquake-resistant performance of ground reinforcement.

(1) *Setting of seismic loads*
Levels of seismic forces, failure modes of ground reinforcement, effects of vertical seismic load are to be determined and set by occurrence probability of earthquake types such as near field earthquake and great earthquake, their incidences taken into account. Therefore, the limit state design method would be an effective means.

(2) *Modification of seismic coefficient method*
There is a tendency that the higher the cut slope is, the more the horizontal acceleration of the slope face is amplified. Accordingly, the applicable range of the seismic coefficient method, the applicability of modified seismic coefficient method, and the modification method have to be studied. Model tests and seismic response analyses assuming seismic conditions are effective means, and also seismic observation of long, large slopes is hoped for.

(3) *Optimization of seismic response analysis*
Toward more detailed seismic analysis, a method of modelling reinforcements in two dimensions which is capable of taking account of three-dimensional reinforcing effect and non-linear seismic response analysis which enable us to analyze the strain area of soil and reinforcements during earthquakes of large deformations are needed.

Given the above subjects, we have to clarify the failure mechanisms and earthquake-responding properties and conditions by surveying ground reinforcement structures damaged by earthquakes and making comparative study of results of shaking-table tests and numerical analyses.

3.5 *Durability as permanent structures*

In case that a reinforced ground with steel materials like reinforcing bars applies for a permanent slope, the durability of the reinforcements has to be given careful consideration. In Europe, conceptions such as corrosion prevention with plastic sheaths, corrosion allowance, and so on, are introduced into guidelines (Scientific Committee of French National Project CLOUTERRE 1993, Gassler 1988), etc. In Japan, although some guidelines or the likes prescribe corrosion allowances, their base data seem poor. Recently Japan Highway Public Corporation extracted some reinforcing bars using over-core boring at nine permanent slopes more than 10 years old, and examined their corrosion conditions through material tests, etc (Tayama et al. 1996b). As the result, pointed out as the causes of their corrosion were shortage of grout around the top parts of the reinforcing bars, short covering of their lower parts by grout, etc. However, because their survey was limited, data have yet to be collected and accumulated.

3.6 *Landscape*

Because the ground reinforcement enables to construct steep cut slopes, which means the reduction of topographical alteration and the reduction of cut over areas of forests, it can be said to be friendly to scenery. Because the reinforcements are driven in the ground, concerned with the landscape are the surface structure, or facing, and the shape of the cut slope. At present, available as scenery-friendly surface structures, instead of the conventional sprayed mortal or plain concrete surfaces, are wall-surface vegetation (vegetation blocks or vegetated frames arranged grid-form), vegetative facings (continuous fiber-reinforced walls), and cosmetic finishes (cosmetic blocks resembling stone-masonry retaining walls, or relief wall surfaces).

Described below are the present subjects and the perspective regarding the landscape problem:
(1) Both the water retentivity for vegetation and the drainage function for its stability have to be given to a facing structure to be vegetated. Besides, structure which will be easy and economical to maintain is required.
(2) When a slope is steep and high, it give us oppressive impression. A certain proper gradient has to be found from this aspect.
(3) Demand for facing structures constructed with precast members is expected to increase due to the necessity of laborsaving execution of works. Such members have to be developed in such colors and finishes as match the various construction sites.

Besides, design of ground reinforcement shall includes landscape design which takes account of natural environment such as climate and social environment (Miki 1996b).

3.7 *Application of new materials and new construction methods*

New construction methods, for instance, of using, as core materials, materials with drainage function, or glass or vinyl fibers of high anti-corrosiveness and high strength, of expanding the tips of reinforcements have recently been developed. Also the construction method of forming soil-cement anchors of 40-50 cm diameters and place core material in the center of each anchor have been developed. For these methods, the accumulation of construction cases is awaited.

ACKNOWLEDGEMENTS

The activities of the Asian Technical Committee of Earth Reinforcement are under the guidance of Prof. H. Ochiai of Kyushu University, the Chairman of the committee, and Prof. J. Otani of Kumamoto University, a Secretary. The authors acknowledge with thanks the support of Dr. K. Kumagai, Technical Research Institute of Maeda Corporation and Mr. T. Konami of Okasan Kogyo Co. Ltd. for the activity of the working group.

REFERENCES

Collin, J.G., Chouery-Curtis & R.R.Berg. 1992. Field observations of reinforced soil structures under seismic loading, *Proc. Int. Symp. Earth Reinforcement Practice (IS KYUSHU '92)*, Balkema, 1: 269-274.

Fujii, T. Tsuoka & M. Muramatsu: Behavior of soil nailed walls and rootpile structures caused by the 1995 Hyogo-Southern earthquake. 1996. *Proc. Symp. Ground reinforcement Construction Methods*, Research Committee for Ground reinforcement Construction Methods of JGS: 295-298. (in Japanese)

Fukuda, N., N. Tajiri, T. Yamanouchi, N. Sakai & H. Shintani. 1994. Applicability of seismic design methods to geogrid reinforced embankment, *Proc. 5th Int. Conf. Geotextiles, Geomembranes and Related Products,* Singapore, 1: 533-536.

Gassler, G. 1988. Soil-nailing theoretical basis and practical design, *Proc. Int. Geotech. Symp. Theory and Practice of Earth Reinforcement (IS KYUSHU '88)*, Fukuoka, Kyushu: 283-288.

Japan Highway Public Corporation: *Guidance for design and execution of cut-slope reinforcing construction methods (draft)*, June 1995. (in Japanese)

Kojima, K., M. Tateyama, O. Maruyama, M. Miura & T. Yonezawa. 1994. Analysis and construction of cohesive-soil, rigid-surface, reinforced retaining wall on soft ground, *Proc. 9th Symp JC-IGS*: 1-7. (in Japanese)

Kutara, K., H. Miki, Y. Hayashi, Y. Nagano, K. Yamada, M. Nakano, T. Takasago, S. Takahashi, H. Shido, K. Iwasaki & T.Sueishi. 1990. Behavior of test embankment reinforced with Spun-bonded Non-woven fabrics under artificial rain condition, *Proc. 5th Symp. JP-IGS*: 15-22. (in Japanese)

Kutara, K., H. Miki, T. Hirai & H. Kawasaki. 1991. Large-sized triaxial compression test of reinforced soil with FRP geotextile, *Proc. 26th Annual Conference on SMFE*: 2185-2188. (in Japanese)

Matsui, T., K. Kobayashi, T. Kumada & Y. Otani. 1996. Damaged of reinforced earth structures by steel reinforcements in the 1995 Hyogoken-Southern Earthquake, *Tuchi-to-Kiso, JGS,* 44- 2: 76-78. (in Japanese)

Matsuo, O., T. Shimazu, J. Koseki, N. Fukuda, T. Fujii, N. Tajiri. 1995. Analysis of behavior and effect of geosynthetic reinforced embankments at the occurrence of liquefaction, *Proc. 6th U.S.-Japan Workshop on Earthquake Disaster Prevention for Lifeline Systems*, Technical Memorandum of PWRI: 57-68.

Miki, H. 1996a. *Study on reinforcing mechanism of embankments with geotextile and its analysis method*, Dr. Thesis, Tokyo University. (in Japanese)

Miki, H. 1996b. Harmonizing road earth works with environment and Landscapes, *Tuchi-to-Kiso*, JGS, 44- 6: 5-8. (In Japanese)

Miki, H., K. Kudo, M. Taki, N. Fukuda, T. Iwasaki & A. Nishimura. 1991. Full-scale model test of geogrid-reinforced embankments, *Proc. 6th Symp., JC-IGS*: 56-62. (in Japanese)

Miki, H., K. Kudo, M. Taki, N. Fukuda, K. Iwasaki & J. Nishimura. 1992. Experimental study on full-scale geogrid reinforced embankment model, *Proc. Int. Symp. Earth Reinforcement Practice, IS KYUSHU '92*, Balkema, 1: 269-274.

Miki, H., T. Yamada, H. Kokubo, I. Takahashi, & T. Sasaki. 1996. Experimental study on geotextile tube dehydration method of dredged soil, *Proc. 1st European Geosynthetics Conference EURO-GEO 1*, Maastricht, Netherland: 933-94.

Muramatsu, M., T. Sueoka & F. Tatsuoka. 1995. Reinforcing mechanism of soil nailing and effects of slope facing, *Jour. Geotechnical Engineering*, JSCE, 517-III-31: 93-104. (in Japanese)

Murata, O., M. Tateyama, S. Moriyama & S. Nozawa. 1986. Loading test of improved soil mass by Deep Mixing Method, *Proc. 21st Annual Conf. SMFE*: 1981-1982. (in Japanese)

Otani, Y., T. Ogata & T. Matsui. 1996. Behavior of Reinforced Earth Structures in the 1995 Hyogo-Nanbu earthquake, *Proc. 31st Annual Conf. JGS*: 2417-2418. (in Japanese)

Public Works Research Center. 1988. *Design and Construction Manual of Reinforced Soil (Terre Armee) Wall Construction Method.* (in Japanese)

Public Works Research Center. 1993. Manual of a *Rational Design and Construction, Reinforced Soils with Geotextiles.* (in Japanese)

Public Works Research Institute. 1989. Research of seismic subsidence of embankments and countermeasures, PWRI report. 2769: 55-106. (in Japanese)

Public Works Research Institute. 1992. *Report on collaborative study of rational design method of geotextile-reinforced soils (Part 2)*: 385-404. (in Japanese)

Railway Technical Research Institute. 1987. *Design and Construction Manual of Agitating & Mixing Foundations, Mechanical Agitating Method.* (in Japanese)

Research Committee for Ground Reinforcement Construction Methods of JGS. 1996. *Proc. Symp. Ground reinforcement Construction Methods*: 83-152. (in Japanese)

RRR Construction Method Association. 1993. *Construction method of steepening the slopes of existing embankments.* (in Japanese)

Sato, T., F. Tatsuoka & S. Yamada. 1984. About test embankment of Kanto loam reinforced with nonwoven fabric (Continued report), *Proc. 19th Annual Conf. JSSMFE,* Matsuyama: 1171-1174. (in Japanese)

Scientific Committee of the French National Project CLOUTERRE. 1993. *Recommendations Clouterre 1991* (English translation: Soil Nailing Recommendations 1991).

Tateyama, M., O. Murata, K. Watanabe & F. Tatsuoka. 1994. Geosynthetic-reinforced retaining walls for bullet train yard in Nagoya, *Recent Case Histories of Permanent Geosynthetic-reinforced Soil Retaining Walls*, Tatsuoka & Leshchinsky (eds.), Balkema: 141-150.

Tateyama, M. & O. Murata. 1994. Permanent geosynthetic-reinforced soil retaining walls used for bridge abutments, *Proc. 13th Int. Conf. SMFE,* New Delhi: 1245-1248.

Tatsuoka, F. 1993. Roles of facing rigidity in soil reinforcing, *Proc. Int. Symp. Earth Reinforcement Practice (IS KYUSHU '92)*, Balkema. 2: 831-870.

Tatsuoka, F., T. Sato, T. Iwasaki, S. Yamada & S. Naito. 1983. Behavior of test embankment of Kanto loam reinforced with nonwoven fabric, *Tsuchi-to-Kiso*. 31-9: 21-28. (in Japanese)

Tatsuoka, F., M. Tateyama & J. Koseki. 1996a. Performance of soil retaining walls for railway embankments, *Special Issue of Soils and Foundations*: 311-324.

Tatsuoka, F., M. Tateyama, T. Uchiyama & J. Koseki. 1996b. Geosynthetic-reinforced soil retaining walls as important permanent structures (Mercer lecture), *Proc. 1st European Geosynthetics Conference EURO-GEO 1*, Maastricht, Netherland: 3-24.

Tayama, S., T. Nagayoshi & H. Maeno. 1996a. Examination on application of structures excavation with aid of soil nailing method, *Nihon Doro Kodan Researc. Institute Report.* Japan Highway Public Corporation, 33: 34-44. (in Japanese)

Tayama, S., H. Maeno & H. Matsuyama. 1996b. Survey on durability of natural-ground reinforcing construction methods, *Proc. Symp. Ground Reinforcement Construction Methods*, Research Committee for Ground Reinforcement Construction Methods of JGS: 331-334. (in Japanese)

The Department of Transport (U.K.). 1994. Design methods for the reinforcement of highway slopes by reinforced soil and soil nailing techniques, *Design manual for roads and bridges*, 4-1, Part 4, IIA 68/94.

Tsukuda, C., A. Yamamoto & N. Fukuda. 1996. Stability analysis of cut slope reinforced with steel bars at north area of Awaji Island during Hyogoken Southern Earthquake, *Proc. Symp. Ground Reinforcement Construction Methods,* Research Committee for Ground reinforcement Construction Methods of JGS: 291-294. (in Japanese)

White, D. M. & R. D. Holz. 1996. Performance of geosynthetic-reinforced slopes and walls during the Northridge, California earthquake of January 17, 1994, *Special Report of Int. Symp Earth Reinforcement. IS Kyusyu '96*: 97-104.

Earth Reinforcement, Ochiai, Yasufuku & Omine (eds) © 1997 Balkema, Rotterdam, ISBN 90 5410 833 9

Current state on numerical analysis of reinforced soil structures

J.Otani – *Department of Civil and Environmental Engineering, Kumamoto University, Japan*

A.Yamamoto – *Technical Research Institute, Obayashi Corporation, Kiyose, Japan*

T.Kodaka – *Institute of Industrial Science, University of Tokyo, Japan*

N.Yasufuku – *Department of Civil Engineering, Kyushu University, Fukuoka, Japan*

A.Yashima – *Department of Civil Engineering, Gifu University, Japan*

ABSTRACT: In recent years, considerable interest in the earth reinforcement techniques has promoted both fundamental and practical studies as well as the development of various types of reinforcing materials. It follows that new techniques for this earth reinforcement and their applications to geotechnical engineering practice have been also running around the whole area of Asia. In this paper, the purposes are : 1) to investigate what kind of numerical analysis is conducted for this decade; 2) to investigate what kinds of models including soil, reinforcing material, and the interaction have been used; to clarify what is indispensable modeling at least; and 4) to clarify the task of numerical analysis to the design method. It is noted that this paper was prepared by the working group of Japanese Supporting Committee for Asian TC.

1 INTRODUCTION

Earth reinforcement technique has been used around the world in this three decades starting from the middle of 1960's. It is no doubt that many research and practice have been conducted under this development. Especially, the quick development of computer machine made the complicated numerical analysis possible. For example, finite element analysis was capable of solving the boundary value problems with complicated boundary conditions and material properties. Meanwhile, many conferences and symposia on the theme of earth reinforcement also have been held, and for instance, three consecutive IS Kyushu(International Symposium on Earth Reinforcement in Kyushu) in every four years('88, '92 and '96) have accepted many technical papers concerning the numerical analysis on earth reinforcement. Under these circumstances, it is expected to make clear what we know and what we do not know about numerical analysis for earth reinforcement at current stage.

In this paper, the objectives are

1) to investigate what kind of numerical analyses have been conducted;
2) to investigate what kinds of models including soil, reinforcing material, and its interaction have been used;
3) to clarify what is indispensable modeling; and
4) to clarify the task of numerical analysis to the design method.

In order to complete these purposes, First of all, the technical papers, which have been published so far, concerning earth reinforcement with numerical analysis are reviewed and based on this paper review, the statistical discussion about current state of numerical analysis is conducted. Then, the various modelings for reinforced soil including soil, reinforcing material, and its interaction are summarized by introducing some of the papers. And finally, the current state of the numerical analysis on earth reinforcement is summarized.

2 STATISTICS OF CURRENT STATE

Statistics on current state of numerical analysis is summarized based on the survey of technical papers for last 12 years(from 1985 to 1996). The papers with the keywords of “Earth Reinforcement” and “Numerical Analysis” were selected from following Journals and Proceedings:

1) Journal of Geotechnical Engineering, ASCE
2) Geotechnique
3) Canadian Geotechnical Journal
4) Soils and Foundations
5) Geotextiles and Geomembranes
6) Computers and Geotechnics
7) Int. J. Nume. Anal. Meth. in Geo.
8) Jour. of Geotechnical Engineering, JSCE
9) Proc. IS-Kyushu ('88, '92, and '96)

The total number of papers which was reviewed in this paper was 192. The purpose of these researches are listed as follows:

1) development of design chart;
2) proposal of new design method;
3) development of new numerical modeling and method;
4) analysis of case study;

5) search for the reinforcing mechanism;
6) application of new reinforcing materials; and
7) new feasibility of earth reinforcing technique.

These papers are classified into 13 items as shown in Table 1 and as a results of paper review, the papers in each item are classified. Here in this chapter, each item is discussed separately.

2.1 *Analysis Type*

There are two types of analysis which are deformation analysis and stability analysis. As shown in Fig.1, it is easily realized that the number of papers for each analysis is exactly the same between these two analyses. About 90% of the deformation analysis are finite element(FE) analysis and the rest of them are shared with Rigid Body Spring Model(RBSM) and explicit solutions such as elastic solution. Most of studies used FE analysis are the applications of constitutive equation or interaction model in order to simulate real behavior of reinforced soil structures. Recently, the use of limit state design is about to discuss for geotechnical engineering practice and in this design method, two different limit states which are serviceability limit and ultimate limit. As FE analysis can offer the result of deformation, it is possible for this results to be a useful information for design purposes, especially for serviceability limit on the limit state design. As far as the papers reviewed in this paper, not many papers has not been discussed on this point of view. When the reinforced soil structures are applied to more important soil structures, its deformation property will be able to be controlled strictly. In future works, the residual deformation after earthquake is also expected to be verified. Therefore, the purpose of FE analysis should be covered these design purposes.

Table 1 Results of paper review

Item				Number	
Object	Analysis Type		Deformation	90	
			Stability	91	
	Problem		Static	151	
			Dynamic	10	
	Structure		Embankment	51	
			Retaining Wall	41	
			Natural Slope	32	
			Foundation	46	
			Laboratory Testing	23	
Numerical Analysis Method	Analysis method		Explicit Solutions	10	
			FEM	84	
			BEM	0	
			RBSM	2	
			DEM	0	
			Slip Line Method	12	
			Upper / Lower Bound Method	12	
			Limit Equilibrium	40	
			Else	12	
	Dimension		2-dimentional	156	
			3-dimentional	6	
MaterialProperties	Soils		Sand	79	
			Clay	35	
			c and ϕ Material	61	
			Rock	8	
	Reinforcing Material	Geosynthetics	Geogrids	71	137
			Geonet	1	
			Non-woven	23	
			Woven	20	
			Geocell	3	
			Membrane	10	
			Else	9	
		Steel	Steel Bar	39	50
			Steel Strip	11	
			Else	0	
	Else			3	
Modeling of Interaction between Soil and Reinfocing Material			Individually tereated	81	
			Composite Material	21	
			Whitout Interaction	31	
			Pull-Out Strength is used	37	
			Else	4	
Compared with Experimental Results			Yes	73	
			No	88	
Year					
Area			Asia	89	
			EU	33	
			America	41	

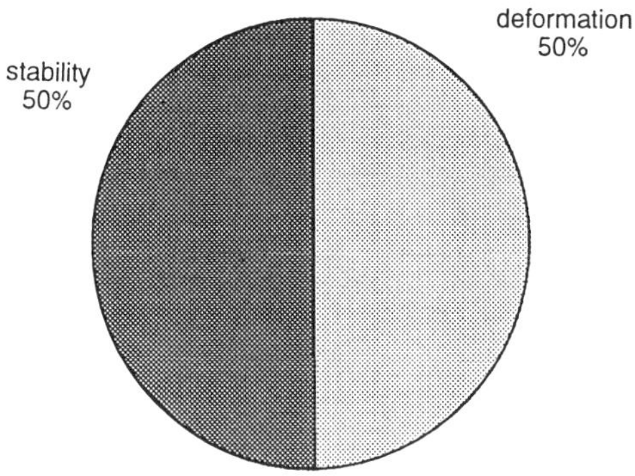

Fig.1 Analysis type

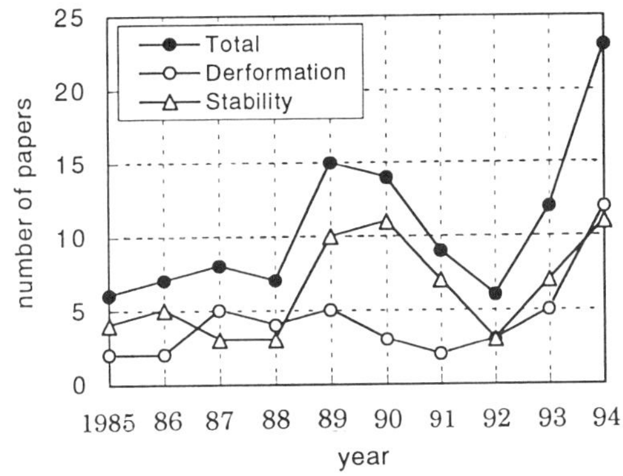

Fig.2 Change of number of papers

Meanwhile, for the stability analysis, about 36% of the papers concerning the stability analysis are limit equilibrium method following FE analysis(32%) and the others(27%) including slip line method and limit analysis with upper and lower bound methods. It is noted that most of limit equilibrium method are the purpose of design calculation, so that it seems somehow different from other numerical analysis.

Figure 2 shows the change of number of papers during the period of 1985 to 1994. It is easily realized that the number of papers concerning numerical analysis is increased after 1992.

2.2 *Problem Type*

Figure 3 shows the percentage of whether the problem conducted is static or dynamic. More than 90% of the papers have solved static problem. Most of papers are from Japan and U.S.A. for dynamic problem and the method of analysis are mostly FE analysis.

2.3 *Applications*

Figure 4 shows the applications of structures. In the earth reinforcement method, there are four typical applications which are embankment, wall structures, foundations, and natural slopes. Although there are a little difference in its percentage, those are almost the same percentage except the papers solving laboratory testing.

2.4 *Analysis Methods*

Figure 5 shows the variation of the methods on numerical analysis. About half number of total papers conducts FE analysis. The limit equilibrium method is the second which percentage is 23% following slip line method(7%), upper/lower method(7%), explicit solution(6%) and RBSM (1%). In the FE analysis, the ground (soil) is modeled with linear elastic, nonlinear elastic, elasto-plastic, or elasto-viscoplastic materials, and reinforcing material is modeled by truss, beam, or thin layer element. The friction property is mostly considered in the interaction model.

2.5 *Dimension*

As shown in Fig.6, 96% of numerical analyses are 2-dimensional analysis. 3-dimensional analysis is a rare case(4%). The structures of 3-dimensional analysis is the case of natural slopes. Here, the method of analysis used is FE analysis and the reinforcing material is steel bars. Although the case of using geosynthetics can be simulated in plane strain condition, for the case of natural slopes with steel bars, its dimension should be 3-dimension and its behavior deeply depends on the deformation and the axial stress acting steel bars in 3-dimension.

2.6 *Soils*

Figure 7 shows the soils which have been solved in the numerical analysis. The sandy ground is major object because the major applications such as embankment and retaining wall are used sandy materials, It is noted that the soil has been modeled with c-ϕ material(33%) because of its generality.

2.7 *Reinforcing Materials*
(1) Geosynthetics

As shown in Fig.8, the major geosynthetics solved in the numerical analysis is geogrid(51%) following nonwoven(17%) and woven(15%) geotextiles, respectively. It is noted that some of the papers deal with not special materials but any types of material. Thus the percentage of this category might be unreliable.

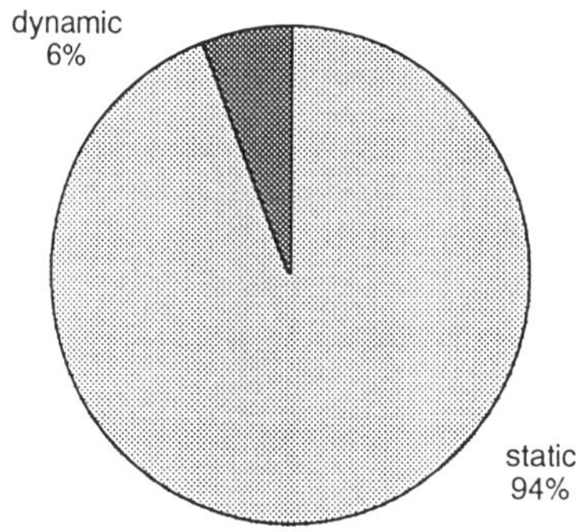

Fig.3 Problem type

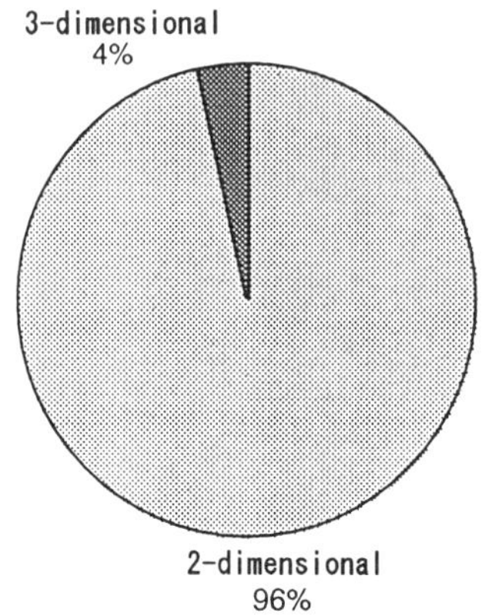

Fig.6 Dimension of analysis

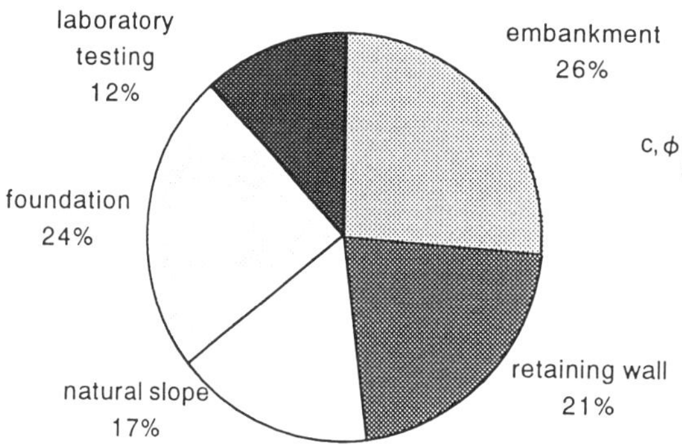

Fig.4 Applications

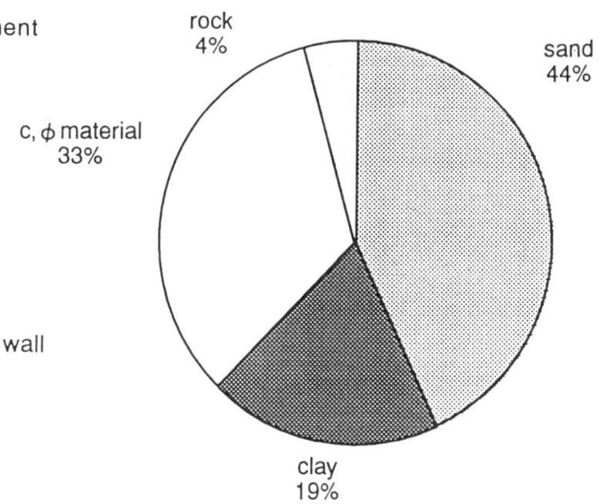

Fig.7 Soil type

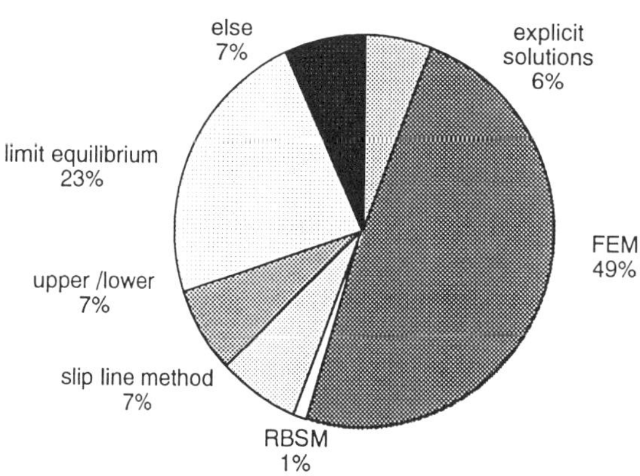

Fig.5 Analysis methods

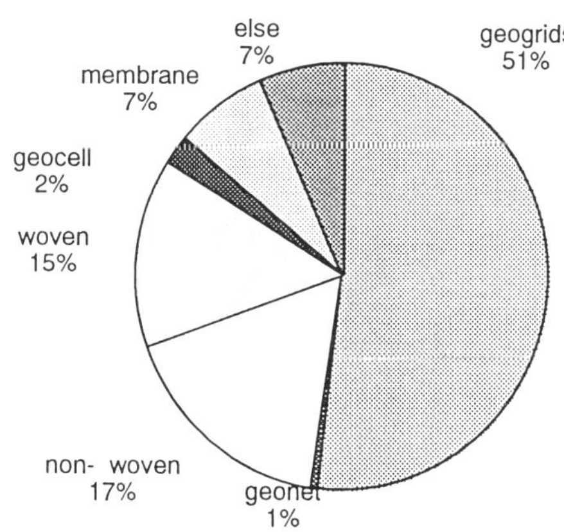

Fig.8 Reinforcing material (geosynthetics)

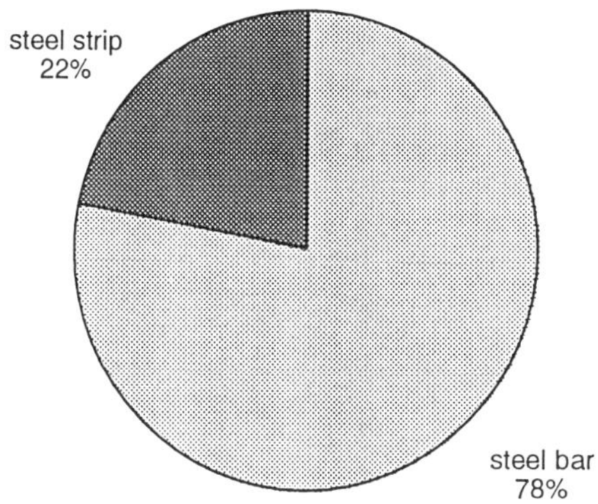

Fig.9 Reinforcing material (steel)

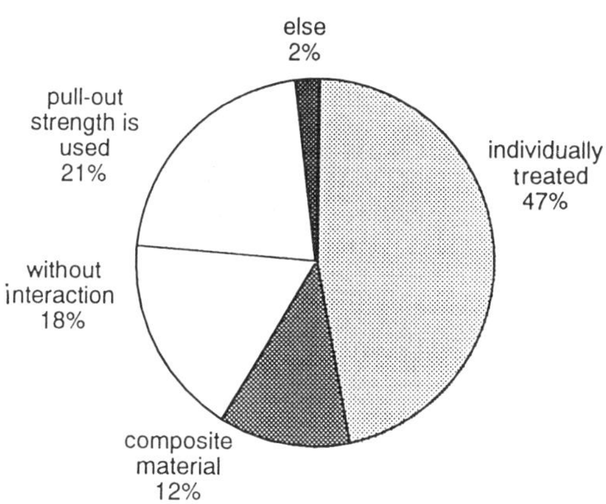

Fig.10 Model of interaction behavior

(2) Steel

As shown in Fig.9, 78% of the analyses deal with steel bars for soil nailing. In most cases, the structures used the steel bar are natural slopes and in some cases, it is applied to embankment, retaining wall and foundation. The structure used steel strip is retaining wall only (e.g. Terre Armee method).

2.8 *Interaction*

As shown in Fig.10, 47% of numerical analyses for reinforced soil structures consider interaction between soil and reinforcing material. The importance of these interaction models in numerical analysis of reinforced soil structure will be understood strongly and widely. 12% of interaction modeling in numerical analysis have been used composite material to the zone reinforced with reinforcement materials. Especially, the number of published paper related to composite material is increased after 1991. 21% of interaction is pull out testing. The results of laboratory pull-out test have been used for numerical analysis with fairy large number of percentage (21%), especially in Japan.

3. HIGHLIGHTS OF SOME OF THE RESEARCHES

As realized from Fig.5, most of the numerical analyses have been conducted with finite element method, so that some of papers concerning finite element analysis are summarized with their prominent points of view. Here, the analyses are divided into two categories which are deformation analysis and stability analysis. It is noted that slip line method is also discussed on the stability analysis.

3.1 *Deformation Analysis*

In order to show its modeling and purpose of the analysis, some of the technical papers are introduced.

(1) One of the pioneer works (Ohta et al., 1980)

Transverse surface reinforcements at the bottom of embankments placed on very soft foundations were found to reduce the amount of deformation of the foundations and to improve the bearing capacities through the elasto-plastic finite element analysis on an idealized model of soft foundation as well as on a field trial embankments. The constitutive equation employed is an infinitesimal elasto-plastic stress-strain relation which is reduced to the Original Cam Clay model under conditions of isotropic initial stress state and which is able to describe the anisotropic behavior of clay including the complicated responses to the rotation of principal stress direction, so called Sekiguchi-Ohta Model. The computed program used is the one originally written by Akai and Tamura (1976) which employs the backward finite difference scheme so as to ensure better stability in computations.

A trial embankment was placed on a very soft layer of peat and clay. The trial embankment, 386.3m long was divided into 4 sections; natural ground section, surface reinforcement section with sand drain treatment, sand compaction pile section and chemical pile section, each of which was about 50m. The cross sectional views of natural ground section and surface reinforced section are shown in Fig.11. All experimental parameters needed to analyze are easily determined from a few empirical and experimental considerations. The sand layer were treated as if they were overconsolidated with an overconsolidation ratio of 10. The surface steel strips reinforcement was replaced by a 1.2m thick elastic band. The performance of a field trial embankment demonstrating the effectiveness of surface reinforcement was back-analyzed, with

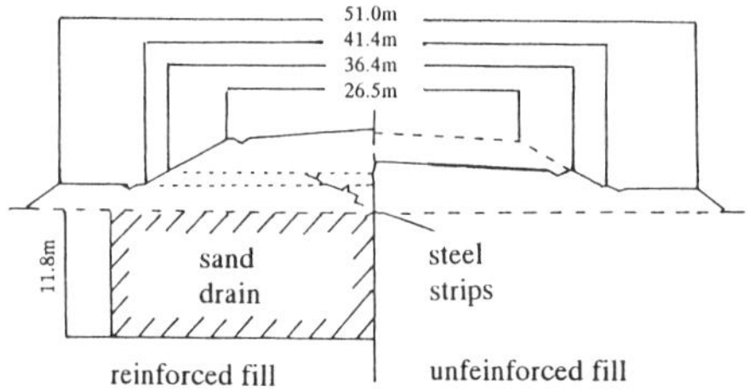

Fig.11 Cross sectional views of trial embankment (Ohta et al, 1980)

reasonable agreement, by means of the presented finite element technique. As the results of the investigations, it is concluded that the transverse surface reinforcement is one possibility technique to improve the undesirable characteristics of soft foundations.

(2) Deformation Analysis with Design Purposes (Rowe, 1984)

A numerical technique for the analysis of geotextile reinforced embankments has been outlined. This technique permits consideration of soil-reinforcement interaction slip at the soil-fabric interface, plastic failure within the soil and long deformation. The application of the approach was illustrated by reference to an embankment constructed on a soft peat deposit.

The results were obtained using a plane strain, nonlinear elasto-plastic soil structure interaction analysis to allow for large deformation. An incremental finite element technique was used in the analysis, in which the soil is assumed to be nonlinear elastic-plastic material. In the plastic state, the soil was assumed to have a non-associated flow rule of the form proposed by Davis with a dilatancy angle ψ. The model used will not permit tensile stresses for a purely friction material. Large deformation were taken into account by updating the nodal coordinates. This approach is only approximated since it neglects rotational effects. The geotextile was treated as a structural membrane with axial stiffness but negligible flexural rigidity. The soil and fabric were examined separately but were related by conditions of compatibility and equilibrium and the soil-fabric interface, the displacement of the soil and fabric were assumed to be compatible until the shear stress reached a limiting shear stress defined by a Mohr-Coulomb failure criterion. Once this shear stress was attained, slip occurred at this point. As the results, the suggested approach does provide a reasonable means of modeling the construction and performance of geotextile reinforced embankments on soft organic deposits.

(3) Deformation Analysis with interaction model (Kutara et al., 1986)

This paper introduces a no-tension FEM analysis of the deformation of geotextile-reinforced embankments. The influence of the discontinuous plane between the geotextile and the soil must be taken into consideration in such an analysis. To do so, the joint elements with a thickness of 't' were introduced between the geogrid and the soil. The geotextile was converted to plane truss elements which would resist tensile stress but not flexural stress for purposes of analyzing the deformation behavior of the composite body uniting the geotextile and the soil. Fig.12 shows the geotextile and the soil replaced with joint elements and plane truss elements. In this analysis, joint elements are provided for the upper and lower surfaces of the discontinuous plane between geotextile and the soil (see Fig.12). No-tension analysis was conducted with a model prepared by regarding joint elements with a thickness of 't' as the domain where the friction resistance between the geotextile and the soil was mobilized. The plane truss elements were connected by pins, because the geotextile is resistant

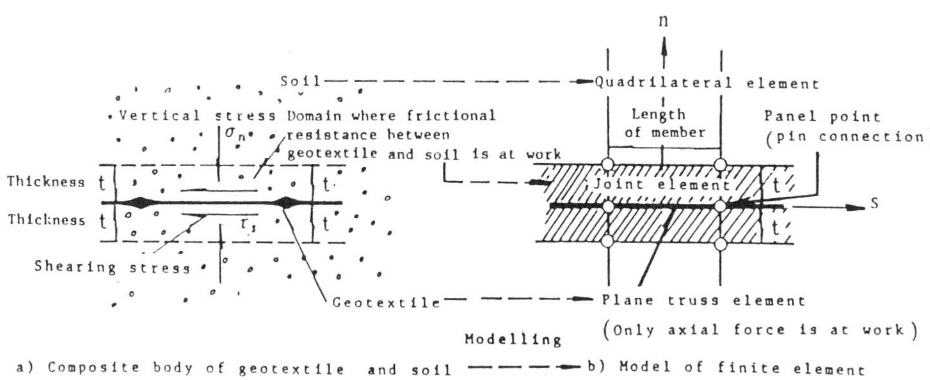

Fig.12 Modeling of geotextile and soil (Kutara et al., 1986)

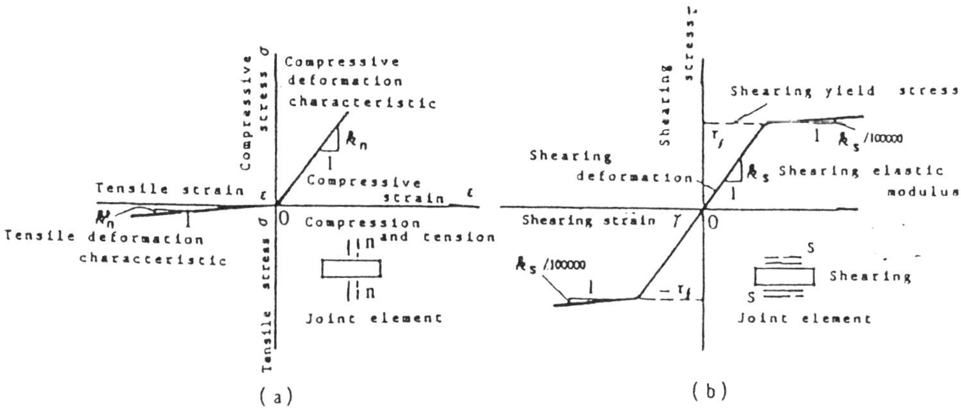

Fig.13 Deformation characteristics of joint element (Kutara et al., 1986)

to tensile stress in the axial direction, but has no resistance to bending stress. It was assumed that joint elements have no resistance to tensile stress of geotextile in the vertical direction, but transmit the compressive strength to the soil elements. Further, the joint element is assumed to behave as a nonlinear deformation against the shearing stress in the horizontal direction for the geotextile. Deformation characteristics of joint elements are shown in Fig.13. The results of the analysis were compared with not only those from pull-out tests for polymer grids but also with large model tests. The analytical results for the strain distribution of the geotextile in the soil and the deformation of the fill were in good agreement not only with those of the pull-out tests for geotextiles in soils but also with those of large model tests.

(4) Development of Design Charts Based on Deformation Analysis (Ogisako & Ochiai, 1990)

Finite element analysis for geogrid reinforced-soil retaining walls were conducted using the method which is capable of taking into account the displacement dependence property of the pull-out resistance of the geogrid in soils. A modeling of the geogrid reinforced soil presented here was a combination of the joint element expressing the property of discontinuous plane with the truss element transmitting the axial force only. This truss element whose ends were connected by the pin joint was used for modeling of geogrid. This interaction property was evaluated based on laboratory pull-out test in which the mobilizing process of the interaction behavior was modeled. The wall was modeled by beam element which has a stiffness for bending moment. In the analysis, Duncan-Chang model was used as a soil model. Based on the analysis results with changing height of the wall, and the spacing and length of the reinforcement, the design chart was developed.

3.2 *Stability Analysis*

There are several methods for stability analysis such as upper bound method, lower bound method, slip line method, finite element method(displacement method), and limit equilibrium method. The purpose of stability analysis is to obtain the limit load. Table 2(Otani et al.(1994)) shows the necessary conditions for obtaining the true limit load and the meaning of the solution in each method. As shown in this table , there are five conditions which should be satisfied in order to obtain true limit load. Although the limit equilibrium analysis is conducted quite a few cases, it should be noticed that the results from this analysis is unreliable without any theoretical considerations.

(1) Limit Equilibrium Method

As shown in Table 2, all necessary conditions are unknown for limit equilibrium method. But when the real failure mechanism is taken into account, the

Table 2 Necessary conditions for limit load (Otani et al., 1994)

	upper bound	lower bound	slip line	FEM (disp.)	limit equilibrium
equilibrium	unknown	satisfy	satisfy	partially satisfy	unknown
compatibility	satisfy	unknown	unknown	satisfy	unknown
constitutive equation	satisfy	satisfy	satisfy	satisfy	satisfy
force boundary	unknown	satisfy	satisfy	partially satisfy	unknown
displacement boundary	satisfy	unknown	unknown	satisfy	unknown
relation with limit load	upper bound	lower bound	lower bound	upper bound	unknown

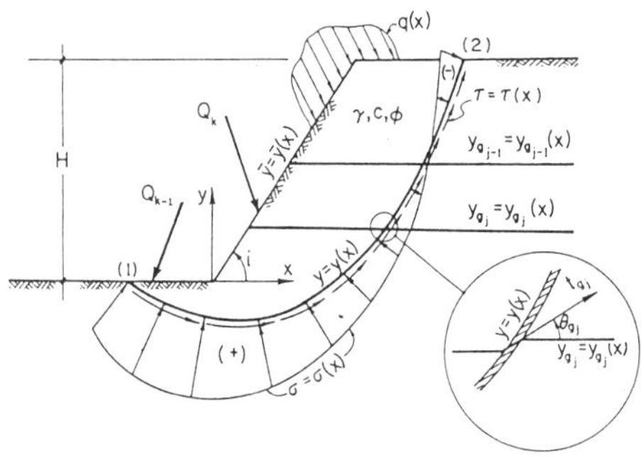

Fig.14 Basic definitions and conventions (Leshchinsky and Reinschmidt, 1985)

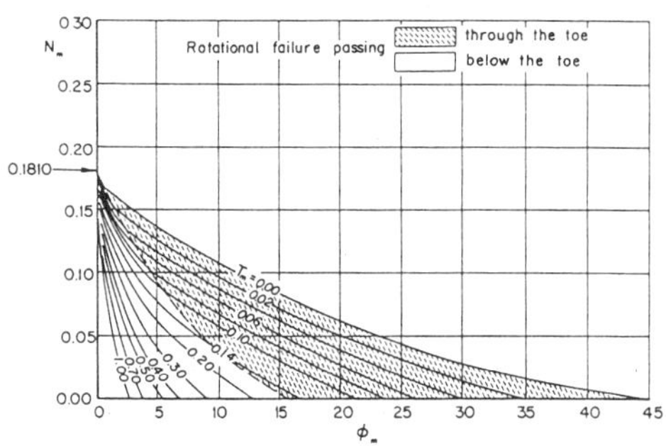

Fig.15 Stability chart for Case i = 45° and γ_g=0.50 (Leshchinsky and Reinschmidt, 1985)

obtained result should be upper bound or true. The limit equilibrium method (LEM) has been widely used in stability analysis because of its simplicity and also is used for design calculation directly.

Leshchinsky and Reinschmidt (1985) proposed a method of stability analysis based on LEM for geomembrane(geotextile) reinforced soil structures. The main purpose of their study was to prepare design charts. Figure 14 shows the analytical domain and boundary conditions for the analysis. Theoretical framework for the study was based on a kind of LEM developed for unreinforced plain slope(Baker and Barber, 1978). The slip surface and the normal stress along the slip surface were expressed in terms of unknown functions y = y(x) and τ = τ(x), respectively. The failure mechanism y(x) and the internal stress τ(x) were solved simultaneously using a variational limiting equilibrium approach. When the conventional LEM are concerned (e.g. Jambu, Bishop, Fellenius methods, etc), they are implied by the assumption regarding direction or line of action of the interslice force, and hence the obtained solutions are lacking a complete physical interpretation. However, the solution by this method is verified as an upper bound solution in plasticity. They incorporated the reinforcing forces as external forces into the above LEM.

Figure 15 illustrates the typical stability charts based on their proposed methodology. Nm, ϕm and Tm in the figure denote the cohesion of soil, the internal friction and assumed tensile strength of reinforcement, respectively. These values are normalized by the unit weight and the slope height and are further mobilized by the factor of safety Fs (e.g. Fs=1.5 for the chart in Fig.15). This figure shows that stronger the membrane the deeper the failure. The required reinforcing force in a design can be determined under arbitrary c and ϕ and the assumed Fs. Since several assumptions are introduced at the stage of incorporating the reinforcing force, it can not be avoided that the solution always deviate from the absolute solution.

(2) Finite Element Method(FEM)

There are two main methodologies in stability analysis using FEM. One is the elasto-plastic deformation analysis following the stress history from initial condition to the limit state. The other is the limit analysis based on the upper or lower bound theorems on plasticity. The former is an extension of the deformation analysis mentioned in 3.1. However, the large deformation and strain localization are required to be considered in the case of failure problem, this type of analysis should be more complicated.

Elasto-Plastic Finite Element Method

Kotake et al.(1997) simulated the results from small plane strain compression tests on sand reinforced by planar material. The soil and reinforcement were modeled by Mohr-Coulomb and the ordinary truss element, respectively. The softening type of constitutive model was applied to the shear band area to explain not only the post-peak behavior but also the occurrence of shear band formation passing through the reinforcement.

It should be regretted for the very few applications of the promising elasto-plastic deformation analysis method based on the large deformation theory in the analysis of the reinforced soil structures.

Rigid Plastic Finite Element Method(RPFEM)

In this method, the stress history until arriving at limit state can not be presented, however, the arbitrary failure mechanism, the loading system, and the stress distribution can be simultaneously obtained as the upper bound solution without any additional assumptions which are essential for LEM. Furthermore, it has been proved that the solution obtained by the minimization process is the

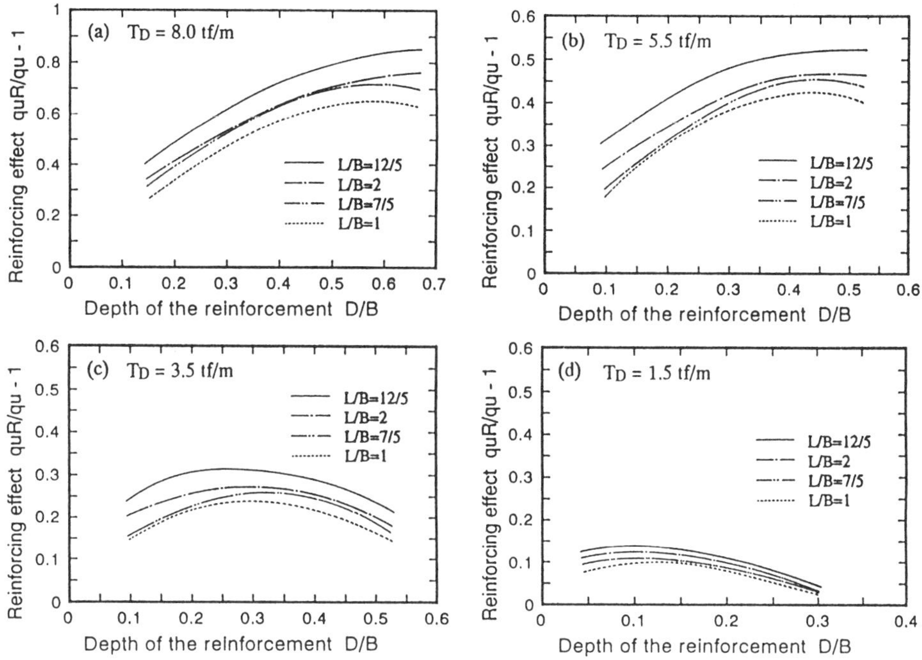

Fig.16 Reinforcing effects of geogrid foundation ground (Otani and Ochiai, 1993)

absolute solution at the limit state. When the RPFEM is applied to the stability problem of reinforced soil structure, modeling of reinforced soil is delicate, since it is quite difficult to know the behavior of reinforcement and its interaction with surrounding soil at the limit state.

Otani and Ochiai(1993) firstly used rigid plastic finite element method for the problem of earth reinforcement. In their paper, the new reinforced soil model was developed using composite theory with reinforcing material and the surrounding soils. Not only the bearing capacity but also the failure mechanism for the reinforced foundation ground have been discussed based on the analysis results. One of the results are shown in Fig. 16. In this figure, L: length of reinforcing material; B:width of foundation; D:depth of placing reinforcing material, and the improved bearing capacities were evaluated quantitatively. The conclusions of this study are listed as follows: (1) The bearing capacity of geogrid reinforced foundation ground is increased as the depth and the length of the reinforcement increases, but there are an optimum depth in order to mobilized the maximum reinforcing effect; (2) There is also an optimum number of geogrids for multiple type of earth reinforcement structures.

Asaoka et al.(1994) assumed that the length between arbitrary soil elements touching the reinforcement did not change at the limit state of soil mass. Under this 'no length change' constraint condition, the reinforcement suppresses the plastic flow of soil along reinforcement by keeping the nodal distance constant. This assumption is relied upon the fact that the rigidity of reinforcement is considerably large compared to the soil, especially at limit state of soil. The 'no length change' condition was mathematically formulated as a linear constraint condition on velocity field, and was incorporated in to the RPFEM by Lagrange multiplier method. As the real reinforcing materials never appear in this computational work, artificial model, e.g. truss element, joint element, etc., are not directly used to explain the behavior of reinforcement. Such elements are generally used in most of the computational works by FEM for the reinforced soil. Therefore, the real material constants for reinforcements are not required except the soil constants to solve the reinforced soil system. This methodology is very simple and proved to be acceptable, yet essential. It should be noted that the more material constants in the analysis, the more indefinite factors affecting the result, although the modeling seems to be close to the reality. In stability analysis by LEM, the effect of reinforcement is introduced as a resisting external force. The tensile force acting on reinforcement is a typical internal force which restricts the soil deformation. Therefore, the value of this internal force can never be decided beforehand. One of the most significant advantage incorporating the constraint condition is that the tensile force distribution along a reinforcement can be computed very well along with the well defined failure mechanism, the loading system (or factor of safety)

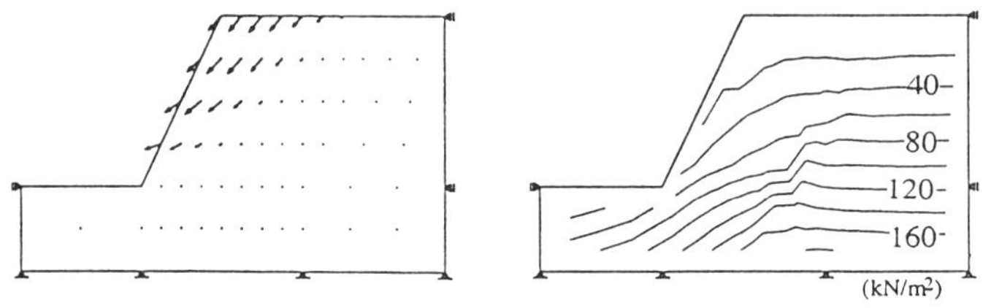

(a) Velocity Vectors (b) Mean Stress Contours

Fig.17 Velocity vectors and mean stress distributions in the case of c-ϕ soil without reinforcing material (Asaoka et al., 1994)

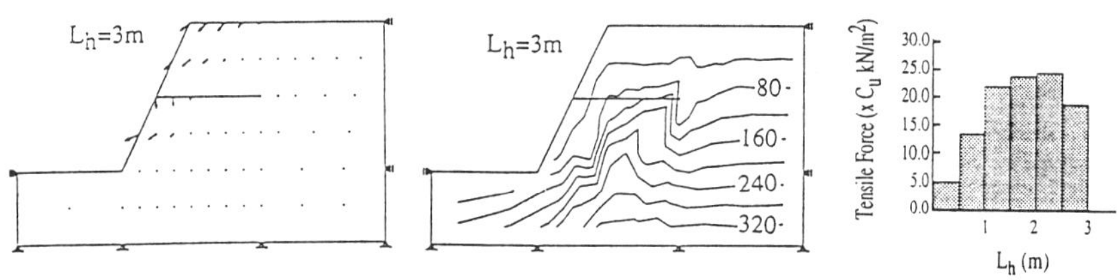

Fig.18 Velocity vectors, mean stress distributions and tensile force along reinforcements of different lengths in case of c-ϕ soil(Asaoka et al., 1994)

and the mean stress distribution, simultaneously. The major findings by Asaoka et al.(1994) were as follows: 1) The reinforcement is more effective in sand material, i.e. c-ϕ material compared to the reinforcement in purely cohesive soil. 2)In sand material, restraining of soil by reinforcement increases the mean confining stress distribution in soil element along the reinforcement. The increase in stability is attributed to the increase in mean confining stress. 3)In purely cohesive soil, the acting reinforcing force is not directly effective in the stability. Figures 17 and 18 show the calculation results on an unreinforced and reinforced sand slopes, respectively. Allowing the length between two consecutive nodes to keep constant, the mean stress will rise which eventually lead to an increase in the absolute strength of the frictional material. The high tensile force in the reinforcement should be attributed to this reason. Thus, the reinforcing force acts like an anchor in such deep zone with high mean stress distribution or the zones close to the slope faces if slopes have rigid panel facing providing high confinement effect to the frictional materials (Kodaka et al, 1995). Kodaka et al.(1995) introduced the 'no bending' constraint condition to the aforesaid methodology in order to explain the role of facing. The 'no bending' condition assumes that the angle formed by three nodes along the rigid facing does not change, i.e. no bending. This linear constraint condition which is similar to the 'no length change' condition mentioned above, was incorporated to the RPFEM. In their study, the propriety of the methodology was illustrated through the 1g medium scale model test on sand slope and the subsequent simulation. The ultimate load, the failure mechanism and the tensile force distributions along the reinforcement obtained by the analysis well support the result of the model tests. They also performed the deformation analysis using linear elastic finite element method (LEFEM) introducing both 'no length change' and 'no bending' constraint conditions on the deformation field in order to simulate the deformation of reinforced soil slope. Even though this simulation covered only the initial tangent of load - settlement curve, the result showed that the method was good enough in practical use. Figure 19 shows the results of both the model test and their simulation. The small circles, the horizontal line and the inclined line show the results of the model test observation, RPFEM (i.e. ultimate load) and LEFEM (i.e. initial tangent of load - settlement curve) results, respectively. The computational results well explain the model test results without any complicated modeling for reinforcement. In addition, the usefulness of the methodology as a tool for design could be seen.

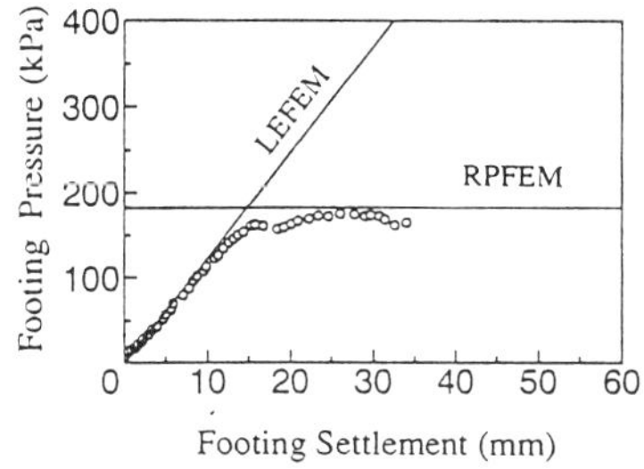

(i) **Type A** (Plain Slope Model)

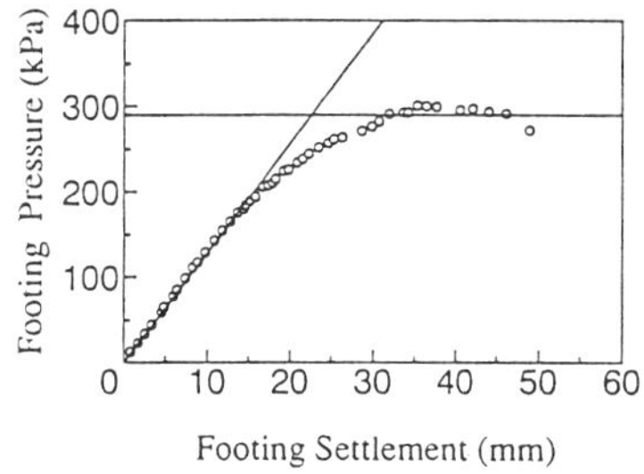

(ii) **Type B** (Reinforced Slope without Facing)

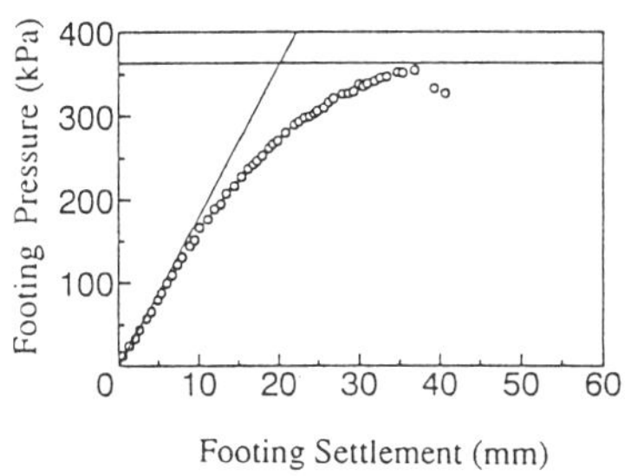

(iii) **Type C** (Reinforced Slope with Panel Facing)

a. Mild Slope (1V:0.5H) Models

Fig.19 Results of footing pressure - settlement curve and failure load with model test observations (Kodaka et al., 1995)

CONCLUSIONS

Current state on numerical analysis in earth reinforcement was reported by conducting paper review and also summarizing some of the papers. Following conclusions have been drawn:

(a) For deformation analysis

(1) Most of the analysis are FE analysis.

(2) In the FE analysis, there are two types of modelings: (i) separately modeling with soil, reinforcing materials, and its interaction; (ii) composite modeling or the model without interaction model.

(3)Although plenty of numerical analyses have been conducted, the analysis with the purpose of discussing more details of serviceability limit on limit state design has not been done so far. This should be the future works.

(b) For stability analysis

(1) Although the limit equilibrium analysis has been conducted with large number of papers, most of them are a part of design calculation. And also, the result by this analysis only should be evaluated quantitatively because that is never followed with theoretical solution.

(2) The FE analysis has been used for the purpose on the stability of reinforced soil structures, such as elasto-plastic and rigid plastic FEMs. The result of these analyses are able to count on quantitatively and should be effective.

The numerical analysis is a power tool to solve boundary value problem with difficult material property and boundary conditions as the computer machine is improved. Therefore, it is concluded that the numerical analysis should be used in order to search real deformation property and failure mechanism of reinforced soil structures.

ACKNOWLEDGMENTS

Authors would like to express their sincere thank you to all the authors who have published the technical papers which we reviewed here in this paper. And also we would like to give our gratitude to not only the member of the Japanese Supporting Committee but also all the members in Asian Technical Committee.

REFERENCES

Asaoka, A., Kodaka, T. & Pokharel, G. 1994. Stability Analysis of Reinforced Soil Structures using Rigid Plastic Finite Element Method. Soils and Foundations, 34(1):107-118.

Kotake, N, Tanaka, T., Tatsuoka, F. & Yamauchi,H. 1997. Numerical simulation of strain localization and failure in reinforced sand. Proc. of Int. Symp. on Deformation and Progressive Failure, IS Nagoya, (to appear).

Kodaka, T., Asaoka, A., and Pokharel, G. 1995. Model Tests and Theoretical Analysis of Reinforced Soil Slopes with Facing Panels. Soils and Foundations, 35(1):133-145.

Kutara,K., Gomadou,M., and Takeuchi,T. 1986. Deformation Analysis of Geotextiles in Soils using the Finite Element Method. Geotextiles and Geomembranes 4:191-205.

Leshchinsky, D. & Reinschmidt, A.J. 1985. Stability of membrane reinforced slopes. Journal of Geotechnical Engineering Division, ASCE, 111(11):1285-1300.

Ogisako,E. and Ochiai,H. 1990. Analysis and design for geogrid reinforced-soil wall. Jour. of Construction Engineering, JSCE, Vol.421,VI-13:105-114 (in Japanese).

Ohta,H., Mochinaga,R., and Kurihara,N. 1980. Investigation of Soft Foundations with Surface Reinforcement. Proc. of 3rd Australia-New Zealand Conference on Geomechanics: 360-365.

Otani,J. and Ochiai,H. 1993. Bearing capacity of the reinforced soils by rigid plastic finite element analysis. Proc. of the 2nd Asian-Pacific Conf. on Computational Mechanics:445-460.

Otani,J., Ochiai,H., and Li,X.X. 1994. Bearing capacity analysis of reinforced foundation ground by geosynthetics. Proc. 8th Int. Conf. on Computer Methods and Advances in Geomechanics, U.S.A.:1397-1402.

Rowe,R.K. 1984. Reinforced Embankments: Analysis and Design. Jour. of Geotechnical Engineering, ASCE, Vol.110,No.2:231-246.

Miscellaneous

Erratum

The figures 2,3 and 4 in 'Creep behaviour of model fabric reinforced brick faced earth retaining walls' by M. Isabel M. Pinto and T.W. Cousens on page 460 of Volume 1 are not correct. Please replace by the figures below.

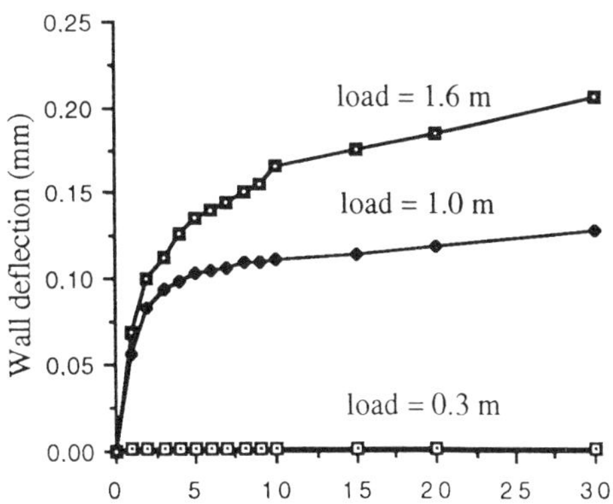

Fig. 2 - Wall deflection vs time (30 min. test)

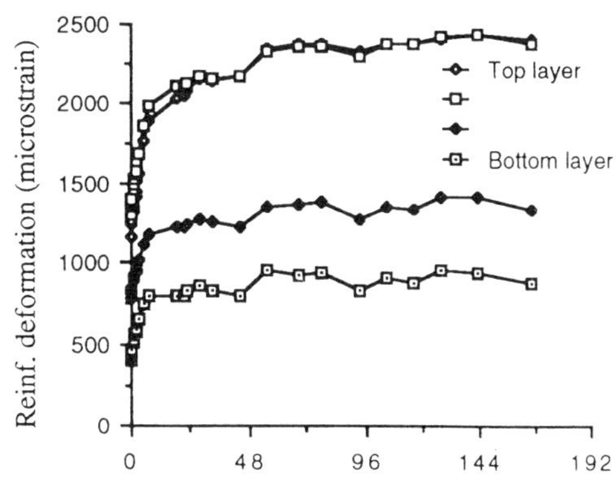

Fig. 4 - Reinforcement deformation vs time

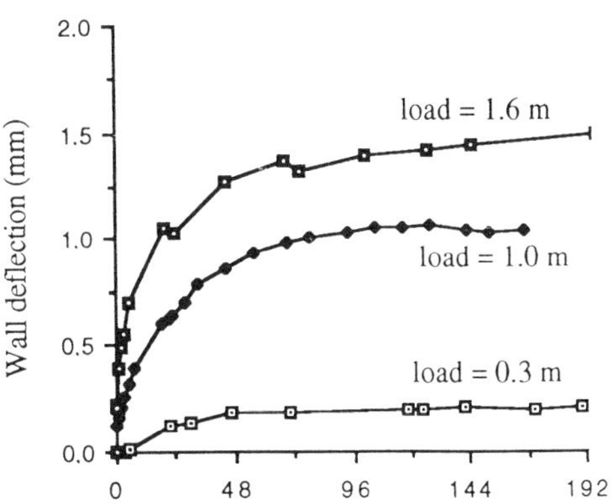

Fig. 3 - Wall deflection vs time

Earth Reinforcement, Ochiai, Yasufuku & Omine (eds) © 1997 Balkema, Rotterdam, ISBN 90 5410 833 9

List of participants

Boyd, M.
Groupe TAI, Pacific Region
The Reinforced Earth Company
PO Box 742,
Gosford NSW 2250 Australia

Karunaratne, G.P.
Geotechnical Engineering Consultant
6 Hampson Avenue
South Coogee 2034 Australia

Lo, S.C.R.
University College,
The University of New South Wales
ADFA, Canbera Australia

Alam, A.K.M.M.
Public Works Department
PWD Eden Bldg., Sub-Division I,
Bangladesh Secretariat Dhaka Bangladesh

Kabir, M.H.
Department of Civil Engineering
Bangladesh University of
Engineering and Technology
Dhaka - 1000 Bangladesh

Abramento, M.
IPT - Instituto de Pesquisas Tecnologicas
do Estado de Sao Paulo
Agrupamento de Estudos Geotecnicos
C.P. 7141 CEP 01064-970
Sao Paulo, SP Brazil

de Mello, L.G.
University of Sao Paulo
Rua Des Vicente Penteado 134 Brazil

Ehrlich, M.
Graduate School of Engineering
Federal University of Rio de Janeiro
P.O.Box 68506 21945-970,
Rio de Janeiro Brazil

Nascimento, N.A.
College of Engineering
Federal University of Parana
Rua Augusto Stellfeld, 8731202
Curitiba-89430-140 Brazil

Palmeira, E.M.
University of Brasilia
Department of Civil Engineering, FT
70910-900 Brasilia, DF Brazil

Alfaro, M.C.
Department of Civil Engineering
Royal Military College of Canada
Kingston, ON K7K5LO Canada

Bathurst, R.J.
Royal Military College of Canada
Civil Engineering Department
Kingston-Ontario K7K 5L0 Canada

Rowe, R.K.
Geotechnical Research Centre
Department of Civil Engineering
University of Western Ontario
London, Ontario Canada N6A 5B9 Canada

Guo, Z.Y.
The Department of Road and Traffic Engineering
Tongji university
1239 Sipin Road, Shanghai 200092 China

Luo, D.Z.
Shang Xi Province Road Bureau
Beiduan 29, Ling-Yuan Road
Xian, Shang Xi Provice 710068 China

Wen, Y.F.
Geotechnical Department
China Institute of Water Resources and Hydropower Research
20 West Chegongzhvang Road,
Beijing 100044 China

Ibrahim, A.M.M.
Housing and Building Research Center
8 Ghonaim Street, Hassan Mohamad,
El-Haram, Giza Egypt

Saad, M.
Military Technical College
29 Khalil Radwan, El-Zeiton,
Gharbia, Cairo Egypt

Bastick, M.
Groupe TAI
36, Avenue Raymond Poincare - 75116
Paris France

Gourc, J.P.
IRIGM-University Joseph Fourier
BP 53, 38041 Grenoble Cedex 09 France

Lerat, P.
CERMES, ENPC/LCPC; La Courtine,
93 167 Noisy le Grand Cedex France

Reiffsteck, P.
LERMES-CUST
Blaise Pascal University
BP 205 - 63 174, Aubiere Cedex France

Segrestin, P.
Groupe TAI-Terre Armee Internationale
36 Avenue Raymond Poincare
75116 Paris France

Alexiew, D.
Huesker Synthetic GmbH & Co.
Fabrikstr. 13-15
D-48712 Gescher Germany

Gassler, G.
Department of Civil Engineering(FB02)
Fachhochschule Munchen
Karlstrasse 6,
D-80333 Munchen Germany

Kassner, J.
Huesker Synthetic GmbH & Co.
Fabrikstr. 13-15
D-48712 Gescher, Germany

Wichter, L.E.
Brandenburg Technical University
Cottbus, Karl-Marx-Str.17,
03044 Cottbus Germany

Anguilar, M.J.P.
Ministry of Communications Public Works and Transport
Ba INGLES 5a Calle, Case #37
La Celba, Atlantida Honduras

Cowland, J.
GEO/ Hong Kong
8th Floor, Civil Engineering Building
101 Princess Margaret Road, Ho Man Tia
Kowloon Hong Kong

Lam, T.S.E.
Civil Engineering Department of Hong Kong Government
2/F, Civil Engineering Building,
101, Princess Margaret Road
Homantin, Kowloon Hong Kong

Malone, A.W.
Geotechnical Enginnering Office,
15/F Civil Engineering Building,
101 Princess Margaret Road,
Kowloon Hong Kong

Raybould, M.
Scott Wilson Kirkpatrick (Hong Kong) Limited
38th Floor, Metroplaza Tower 1,
223 Hing Fong Road,
Kwai Fong Hong Kong

Shen, C.K.
Dept.of Civil & Structural Engineering
Hong Kong University of Science & Technology
Clear Water Bay,
Kowloon Hong Kong

Yeo, K.C.
Geotechnical Engineering Office
Civil Engineering Department
12/F Civil Engineering Building
101 Princess Margaret Road, Homantin
Kowloon Hong Kong

Chandra, S.
Department of Civil Engineering
Indian Institute of Technology
Kunpur - 208 016 India

Madhav, M.R.
Department of Civil Engineering
Indian Institute of Technology Kanpur
P.O.-I.I.T., Kanpur- 208 016 India

Nagaraj, T.S.
Indian Institute of Science
Bangalore 560 012 India

Sreekantiah, H.R.
Department of Civil Engineering
Karnataka Regional Engineering College
Surathkal (D.K.), Shrinvivasnagar-574 157
Karnataka State India

Makmur, A.
Research Institute for Water Resources Development
J1, Ir. H. Juanda 193
Bandung 40135 Indonesia

Salehi, M.
Soil Mechanic Laboratory
Khak-Azam Company
No. 23 Moemeny Lane Maleke-Ashtar St.
Teheran Iran

Cancelli, A.
Department of Earth Sciences
University of Milan
Via Mangiagalli, 34
I-20133 Milano Italy

Rimoldi, P.
TEMAX SPA -Geosynthetics Division
Via Industria 3
22060 Vigano (LC) Italy

Simonini, P.
Instituto di Costruzioni Marittime e di Geotecnica
Universita di Padova
Via Ognissanti, 39
35129 Padova Italy

Abe, J
Civil Engineering Department, Kyushu Branch
Toda Constraction Co., Ltd.
13-12, Shiragane 2-chome, Chuo-ku,
Fukuoka 810 Japan

Adachi, E.
Osaka Branch
Tokyo Printing Ink MFG Co., Ltd.
1-28 Tamatsukuri Honmachi Tennoji-ku
Osaka 543 Japan

Adachi, S.
Fukuoka Branch Office
CTI Engineering Co., Ltd.
18th Fukuoka Building
2-1-10 Watanabe Ave, Chuo-ku,
Fukuoka 810 Japan

Akagi, T.
Department of Civil Engineering
Tokyo University
2100 Kujirai, Kawagoe
Saitama 350 Japan

Akai, T.
Evaluation Technology Department, Geotextile Group
Industrial Technology Research Institute of
Osaka Prefecture
2-7-1 Ayumino Izumi
Osaka 590-02 Japan

Amano, S.
Civil Engineering Technology Department
Obayashi Corporation
2-2-9 Hongo, Bunkyo-ku,
Tokyo 113 Japan

Amemiya, S.
Public Works Materials Group
Tokyo Printing Ink MFG Co., Ltd.
2-7 Tabatashinmachi, Kita-ku
Tokyo 114 Japan

Aoki, H.
Design Division, Head Office
Japan Railway Construction Public Corporation
Sanno Grand Bldg., 2-14-2 Nagata-cho
Chiyoda-ku, Tokyo 100 Japan

Arai, K.
Department of Architecture and Civil Engineering
Fukui University
3-9-1 Bunkyo
Fukui 910 Japan

Arioka, Y.
Public Works Materials Group
Tokyo Printing Ink MFG Co., Ltd.
8149 Syounai-cho, Miyakonojo-shi
Miyazaki 855-01 Japan

Asaoka, A.
Department of Civil Engineering
Nagoya University
Chikusa-ku
Nagoya 464-01 Japan

Chai, J.
Department of Civil Engineering
Saga University
1 Honjo-machi
Saga 840 Japan

Chijiiwa, T.
Kyushu Branch
Oriental Consultants, Co., Ltd.
Fujii Bldg. 3-10-24 Hakataekimae, Hakata-ku
Fukuoka 812 Japan

Demizu, S.
Construction Material Section
Kuraray Co., Ltd.
12-39, 1-chome, Umeda, Kita-ku,
Osaka 530 Japan

Dong, J.
Institute of technology,
Tokyu Construction Co., Ltd.
3062-1, Soneshita, Tana, Sagamihara-city,
Kanagawa 229 Japan

Fujii, T.
Geotechnical Division
Fukken Co., Ltd.
2-10-11 Hikari-machi, Higashi-ku
Hiroshima 732 Japan

Fujimoto, H.
The Association of Geological Engineering
in Miyazaki Prefecture, Soil Testing Center
2043 Hongo-kitakata,
Miyazaki 880 Japan

Fujimoto, S.
Meidaikogyo Corporation
3-43, Funakoji
Beppu 874 Japan

Fujinami, T.
Japan Highway Public Corporation
Tanjin Mitsui Bldg.
2-14-13 Tenjin, Chuo-ku
Fukuoka 810 Japan

Fujio, Y.
CTI Engineering Co, Ltd.
Highway and Bridge Division
18 Fukuoka BLDG 1-10,
Watanabe-Dori 2-Chome, Chuo-ku
Fukuoka 810 Japan

Fukahori, T.
Kyushu Branch
Chuo Kaihatsu Corporation
6-3-27 Torikai Johnan-ku
Fukuoka 814-01 Japan

Fukatsu, N.
Matsumoto Corporation
1-1-19 Maidashi, Higashi-ku
Fukuoka 812 Japan

Fukuda, M.
Osaka Soil Test Laboratory
4-3-2 Itachibori Nishi-ku
Osaka 550 Japan

Fukuda, N.
Research & Development Laboratory
Fukken Co., Ltd.
2-10-11, Hikarimachi, Higashi-ku
Hiroshima 732 Japan

Fukuda, Y.
Chuo Fukken Consultants Co., Ltd.
8-29-36 Nishimiyahara 1-Chome, Yodogawa-ku
Osaka 532 Japan

Fukuoka, M.
The Science University of Tokyo
Simo 5-15-12, Kita-ku
Tokyo 115 Japan

Fukuya, M.
Department of Civil Engineering
University of Yamaguchi
2557 Tokiwadai
Ube 755 Japan

Furuta, M.
Material and Construction Method Department
Public Works Research Institute
Ministry of Construction
1, Asahi, Tsukuba, Ibaraki 305 Japan

Furuya, H.
Obayashi Corporation
Technical Research Institute Civil 1st Division
640 Shimokiyoto 4-chome Kiyose city
Tokyo 204 Japan

Gibo, S.
University of the Ryukyus
Land Conservation
Senbaru, Nishihara-Cho
Okinawa 803-01 Japan

Gose, S.
Structure Division II
CTI Engineering Co., Ltd.
No.9 Chuo Bldg.
4-9-11 Honcho Nihonbashi Chuo-ku
Tokyo 103 Japan

Gurung, S.B.
Department of Civil and Environmental Engineering
Hiroshima University
4-1 Kagamiyama 1-chome, Higashi-Hiroshima
Hiroshima 739 Japan

Haga, H.
Engineering Division, Civil Engineering
JDC Corporation
4-9-9, Akasaka, Minato-ku
Tokyo 107 Japan

Hara, H.
University of the Ryukyus
Department of Civil Engineering
Senbaru 1 Nishihara-cho
Okinawa 903-01 Japan

Hara, T.
Traffic Structure Division II
CTI Engineering Co., Ltd.
No.9 Chuo Bldg.
4-9-11 Honcho Nihonbashi Chuo-ku
Tokyo 103 Japan

Harada, T.
Okasan Kogyo Co., Ltd.
Sumitomo Shibadaimon Bldg.
2-5-5 Shibadaimon, Minato-ku
Tokyo 105 Japan

Hasegawa, A.
Department of Civil Engineering
Hachinohe Institute of Technology
88-1 Daikai Myo, Hachinohe
Aomori 031 Japan

Hattori, S.
Osaka Road Engineer Co., Ltd.
4-50, 4-chome, Miyahara, Yodogawa-ku
Osaka 532 Japan

Hayashi, K.
Institute of Technology
Penta-Ocean Co., Ltd.
1534-1 Yonkucho, Nishinasuno
Nasugun 329-27 Japan

Hayashi, S.
Department of Civil Engineering
Saga University
1 Honjo
Saga 840 Japan

Hayashida, T.
The Kumamoto Branch
The KKM Corporation
475 Oyamamachi
Kumamoto 862 Japan

Hazarika, H.
Japan Foundation Engineering Co., Ltd.
15-17, Sakuragaoka-cho, Shibuya-ku,
Tokyo 150 Japan

Hirai, T.
Civil Engineering Materials Group ,
Mitsui Petrochemical Industrial Products, LTD.
No.9 Kawarai-cho, Kuki-City
Saitama 346 Japan

Hirama, K.
Civil Engineering Department 1
Obayashi Corporation
640 Shimokiyoto 4-chome, Kiyose-shi
Tokyo 204 Japan

Hirano, T.
Kyushu Branch
Fudo Construction Co., Ltd.
11-16, Hakataekimae 2-chome, Hakata-ku
Fukuoka 812 Japan

Hirano, Y.
Iron & Steel Div.
Sumitomo Corporation
2-1-1 Hitotsubashi, Chiyoda-ku
Tokyo 101 Japan

Hirao, K.
Department of Civil Engineering
Nishi-Nippon Institute of Technology
1633 Aratsu, Kanda, Miyako-gun
Fukuoka 800-03 Japan

Hiro-oka, A.
Department of Civil Engineering
Kyushu Institute of Technology
1-1 Sensui, Tobata-ku
Kitakyushu 804 Japan

Hirooka, T.
Hirooka Architechtural Design Office
2-3-10-420 Tenjin, Chuo-ku,
Fukuoka 810 Japan

Hisatake, M.
Department of Civil Engineering
Kinki University
3-4-1 Kowaka, Higashi-Osaka
Osaka 577 Japan

Hiyoshi, Y.
Okasan Kogyo Co., Ltd.
Tokyo Sumitomo Shibadaimon Bldg.
2-5-5 Shibadaimon, Minato-ku
Tokyo 105 Japan

Hoki, K.
Taiyo Kogyou Corporation
4-8-4 Kigawahigashi Yodogawa-ku
Osaka 532 Japan

Horiguchi, T.
Geotop Corporation
17-8 Nihonbashi-hakozaki-cho, Chuo-ku
Tokyo 103 Japan

Hotta, M.
Terre Armee Method Department
Kawasho Corporation
World Trade Center Bldg. 27F
4-1, Hamamaatsu-cho, 2-chome, Minato-ku
Tokyo 105-62 Japan

Hyodo, M.
Department of Civil Engineering
University of Yamaguchi
2557 Tokiwadai
Ube 755 Japan

Ichikawa, K.
Nittoc Construction Co., Ltd.
Technical Department, Technical Division
13-18 Akashi-cho, Chuo-ku
Tokyo 104 Japan

Iizuka, A.
Department of Civil Engineering
Kanazawa University
2-40-20 Kodatsuno
Kanazawa 920 Japan

Imabayashi, T.
Civil Engineering Research Section
Kyushu Electric Power Co., Inc
2-1-47 Shiobaru, Minami-ku
Fukuoka 815 Japan

Imanishi, H.
Kyushu Laboratory
Geo-Research Institute Osaka
Otemon-daiichihoki Bldg., 3-5-1 Otemon, Chuo-ku
Fukuoka 810 Japan

Imoto, D.
Kyushu Branch, Department of Civil
Tokyu Construction Co., Ltd.
2-1 Gionmachi, Hakata-ku
Fukuoka 812 Japan

Inoue, H.
Nissaku Co., Ltd.
4-12 Kyobashi 2-chome, Chuo-ku
Tokyo 104 Japan

Inoue, M.
Dobokubu
Tokai Kogyo Co., Ltd.
1-12-15 Akasaka, Chuo-ku
Fukuoka 810 Japan

Iryo, T.
Department of Ocean Civil Engineering
Kagoshima University
1-21-40 Kourimoto
Kagoshima 890 Japan

Ishida, M.
Civil Engineering Products Division
Taiyo Kogyo Corporation
4-8-4 Kigawahigashi Yodogawa-ku
Osaka 532 Japan

Ishihara, K.
Science University of Tokyo
Department of Civil Engineering
2641 Yamazaki, Noda City
Chiba 278 Japan

Ito, Y.
Fukuoka Doro Engineering Co., Ltd.
8-4-2 Watanabedori, Chuo-ku
Fukuoka 810 Japan

Itoh, M.
Civil Engineering Department
Maeda Corporation
10-26, Fujimi 2-chome, Chiyoda-ku
Tokyo 102 Japan

Iwabuchi, H.
Hirose & Co,. Ltd.
Toyo Central Bldg.
4-1-13 Toyo, Kohto-ku
Tokyo 135 Japan

Iwasaki, K.
Civil Engineering Materials Bussiness Department
Mitsui Petrochemical Industrial Products, Ltd.
Ueno TH Bldg. 8F
39-10 Yushima 3-chome, Bunkyo-ku
Tokyo 113 Japan

Iwata, B.
Katahira and Engineers Inc.
Wajun Bldg.
2-22-2 Koishikawa Bunkyo-ku
Tokyo 112 Japan

Iwata, H.
Technical Department
Nihon-Suiko Consultant Co., Ltd.
523-12 Goryomachi
Kumamoto 862 Japan

Jamalludin, D.
Department of Civil Engineering
University of Yamaguchi
2557 Tokiwadai
Ube 755 Japan

Jiang, Y.
Institute of Environmental Systems
Kyushu University
6-10-1 Hakozaki, Higashi-ku
Fukuoka 812 Japan

Kakita, E.
Kyushu Regional Branch
Takenaka Corporation
2-20, 4-chome Tenjin Chuo-ku,
Fukuoka 810 Japan

Kamon, M.
DPRI, Kyoto University
Gokasho, Uji
Kyoto 611 Japan

Kamuro, K.
CTI Engineering Co, Ltd.
Highway and Bridge Division
18 Fukuoka BLDG 1-10,
Watanabe-Dori 2-Chome, Chuo-ku
Fukuoka 810 Japan

Kanai, S.
Geotop Corporation
17-8 Nihonbashi-hakozaki-cho, Chuo-ku
Tokyo 103 Japan

Kanemitsu, H.
Dai-ichi Fukken Co., Ltd.
3-5-8, Hakataeki-Minami, Hakata-ku
Fukuoka 812 Japan

Kasahara, K.
Maeda Kosen Co., Ltd.
Technical Department
Okinunome, Harue-cho, Sakai-Gun
Fukui 919-04 Japan

Kato, Y.
Mitani Sekisan Co. Ltd.
Technical Devision
2-19-6 Yanagibasi Taito-ku
Tokyo 111 Japan

Kawai, Y.
Research Institute, Geotechnical Section
Japan Highway Public Corporation
1-4- Tadao, Machida City
Tokyo 194 Japan

Kawamura, M.
Department of Civil Engineering
Toyohashi University of Technology
Tempaku-cho, Toyohashi 441 Japan

Kawamura, S.
Soil Brain Corporation
1035-6, Kuriya, Ooaza
Tokuyama 745 Japan

Kazama, M.
Department of Civil Engineering
Tohoku University
Aoba-ku, Sendai 980-77 Japan

Kidera, S.
West Japan Engineering Consultants, Inc.
Geotechnical Engineering Department
Kyushu Electric Power Co.,
1-47, 2-Chome, Siobaru, Minami-ku
Fukuoka 810 Japan

Kikuchi, H.
Nittoc Construction Co., Ltd.
Development Department Technical Division
13-18 Akashi-cho, Chuo-ku
Tokyo 104 Japan

Kikuchi, N.
Okasan Kogyo Co., Ltd.
Tokyo Sumitomo Shibadaimon Bldg.
2-5-5 Shibadaimon, Minato-ku
Tokyo 105 Japan

Kimoto, M.
Hokubu-Kyushu Road Maintenance Co., Ltd.
2-5-25 Mizuki
Dazaifu 818-01 Japan

Kinoshita, K.
Department of Construction
The City of Fukuoka
2-5-31 Daimyou, Chuo-ku
Fukuoka 810 Japan

Kitamura, R.
Department of Ocean Civil Engineering
Kagoshima University
1-21-40 Kourimoto
Kagoshima 890 Japan

Kitazono, Y.
Department of Civil & Environmental Engineering
Kumamoto University
2-39-1 Kurokami
Kumamoto 860 Japan

Kitazume, M.
Geotechnical Engineering Division
Port and Harbour Research Institute
Ministry of Transport
3-1-1 Nagase, Yokosuka 239 Japan

Kobaru, A.
Uemura Corporation
2-4-30 Daimyou, Chuo-ku
Fukuoka 810 Japan

Kobayashi, K.
Hirose and Company
SA Bildg, 7th floor
17-12, 2-chome, Kiba, Kootoo-ku
Tokyo 135 Japan

Koga, T.
Taisei Kiso Sekkei Co., Ltd.
6-6, 1-Chome, Harumachi, Hakata-ku
Fukuoka 816 Japan

Kogure, K.
Department of Civil Engineering
The National Defense Academy
1-10-20 Hashirimizu, Yokosuka
Kanagawa 239 Japan

Kohata, Y.
Geotechnical Engineering Group
Railway Technical Research Institute
2-38-8 Hikari-cho, Kokubunji
Tokyo 185 Japan

Kojima, K.
Geotechnical group
Railway Technical Research Institute
2-8-38 Hikari-cho Kokubunji
Tokyo 185 Japan

Konami, T.
R&D, Okasan Kogyo Co., Ltd.
Sumitomo Shibadaimon Bild. 6F
2-5-5 Shibadaimon Minato-ku
Tokyo 105 Japan

Kondo, Y.
Koa Kaihatsu Co., Ltd.
6644-5 Okazaki, Isehara
Kanagawa 259-11 Japan

Konishi, N.
Kuraray Co., Ltd.
Shin-Nihonbashi Building
3-8-2 Nihonbashi, Chuo-ku
Tokyo 103 Japan

Kono, I.
Department of Environmental and Civil Engineering
Okayama University
2-1-1 Tsushima-naka, Okayama-shi
Okayama 700 Japan

Koreeda, K.
Kiso-Jiban Consultants Co., Ltd.
2-16-7 Hara Sawara-ku
Fukuoka 814 Japan

Koseki, J.
Institute of Industrial Science,
University of Tokyo
7-22-1, Roppongi, Minato-ku,
Tokyo 106 Japan

Kotake, N.
Design Department
Toyo Construction, Co. Ltd.
2-6-3 Hitotsubashi, Chiyoda-ku
Tokyo 101 Japan

Kudou, M.
Department of Civil Engineering
Oita National College of Technology
1666 Maki, Oita 870-01 Japan

Kumada, T.
Hirose & Co,. Ltd.
Toyo Central Bldg.
4-1-13 Toyo, Kohto-ku
Tokyo 135 Japan

Kurahara, R.
Umebayashi Construction Co., Ltd.
1-4-35 Maiduru-machi, Oita-city
Oita 870 Japan

Kurose, M.
Taiyo Gijutsu Kaihatsu Co., Ltd.
9-9 Ieno-machi
Nagasaki 852 Japan

Kusakabe, I.
Kozogijutsu center
3-5-7, Hakataekimae, Hakata-ku
Fukuoka 812 Japan

Kutami, A.
Fukuokan Branch Manager
Road Engineering
2-7-1 Watanabe-dori, Chuo-ku
Fukuoka 810 Japan

Kuwajima, K.
Tokuyama College of Technology
Department of Civil Engineering and Architecture
3538 Kume, Tokuyama
Yamaguchi 745 Japan

Kuwano, J.
Department of Civil Engineering
Tokyo Institute of Technology
2-12-1 Ohokayama, Meguro-ku
Tokyo 152 Japan

Kuwatsuka, S.
Mitsui Petrochemical Industrial Products, Ltd.
Nakanoshima-Mitsui Bldg.
4-18 Nakanoshima 3-chome, Kita-ku
Osaka 530 Japan

Li, X.X.
CTI Engineering Co, Ltd.
Highway and Bridge Division
18 Fukuoka BLDG 1-10,
Watanabe-Dori 2-Chome, Chuo-ku
Fukuoka 810 Japan

Machihara, H.
Department of Technical Office
Fukui University
3-9-1 Bunkyo
Fukui-city 910 Japan

Maeda, Y.
Department of Civil Engineering
Kyushu Kyoritsu University
1-8 Jiyugaoka, Yahatanisi-ku
Kitakyushu 807 Japan

Makiuchi, K.
Department of Transportation Engineering
Nihon University
7-24-1 Narashinodai, Funabashi
Chiba 274 Japan

Maruo, S.
Technical Research Center
Tekken Corporation
9-1 Shinsen, Narita
Chiba 286 Japan

Maruyama, K.
Okasan Kogyo Co., Ltd.
Sumitomo Shibadaimon Bldg.
2-5-5 Shibadaimon, Minato-ku
Tokyo 105 Japan

Matsuda, H.
Department of Civil Engineering
Yamaguchi University
2557 Tokiwadai
Ube 755 Japan

Matsuda, Y.
Yachiyo Engineering Co., Ltd.
Tokyo Division, Structural Department
1-10-23 Nakameguro, Meguro-ku
Tokyo 153 Japan

Matsuhisa, Y.
Kisojiban Consultants Co., Ltd.
Yamaguchi Office
795-8 Ooaza-Hirai
Yamaguchi-City 753 Japan

Matsui, K.
CTI Engineering Co, Ltd.
Highway and Bridge Division
18 Fukuoka BLDG 1-10,
Watanabe-Dori 2-Chome, Chuo-ku
Fukuoka 810 Japan

Matsui, T.
Department of Civil Engineering
Osaka University
2-1, Yamadaoka, suita
Osaka 565 Japan

Matsumoto, K.
Shinwa Techno Co., Ltd.
The development construction section
in the engineering department
1084-1 Kouramen Saza Kitamatsuuragun
Nagasaki 857-03 Japan

Matsumoto, N.
Civil Engineering Materials Business Department
Mitsui Petrochemical Industrial Products, Ltd.
Ueno TH Bldg. 8F
39-10 Yushima 3-chome, Bunkyo-ku
Tokyo 113 Japan

Matsumoto, T.
Kyushu Branch
Obayashi Corporation
Fukuoka Takeda Bldg.
9-12 Simokawabata-machi, Hakata-ku,
Fukuoka 812 Japan

Matsunaga, S.
Kyushu Yousin Construction Co., Ltd.
Hakatako-center Buld. 5F
12-1 Okihamamachi Hakata-ku
Fukuoka 812 Japan

Matsuoka, H.
Department of Civil Engineering
Nagoya Institute of Technology
Gokiso-cho, Showa-ku
Nagoya 466 Japan

Matsushima, H.
Dainichi Consultant Inc.
3-1-21 Yabuta-Minnami
Gifu 500 Japan

Mihara, T.
Technical Research Center
The Kansai Electric Power Co., Inc.
11-20 Nakoji 3-chome, Amagasaki
Hyogo 661 Japan

Miki, H.
Soil Mechanics Division
Public Works Research Institute
Ministry of Construction
1 Asahi, Tukuba-shi
Ibaraki 305 Japan

Minegishi, K.
Department of Transportation Engineering
Nihon University
7-24-1 Narashinodai, Funabashi
Chiba 274 Japan

Misawa, K.
Okasan Kogyo Co., Ltd.
Sumitomo Shibadaimon Bldg.
2-5-5 Shibadaimon, Minato-ku
Tokyo 105 Japan

Misumi, K.
Department of Ocean Civil Engineering
Kagoshima University
1-21-40 Kourimoto
Kagoshima 890 Japan

Mitachi, T.
Department of Civil Engineering
Hokkaido University
North 13, West 8, Kita-ku
Sapporo 060 Japan

Mitsue, T.
Chubu Kyushu Road Maintenance Co., Ltd.
Fukuoka Z Bldg.
3-18-28 Hakataekimae, Hakata-ku
Fukuoka 812 Japan

Miura, N.
Department of Civil Engineering
Saga University
1 Honjo-machi
Saga 840 Japan

Miyashita, T.
Sapporo Road Engineer Co., Ltd.
Sapporo Center Bldg. 5F
West 6-2, North 5, Chuo-ku
Sapporo 060 Japan

Miyata, N.
Onoda Chemico Co., Ltd
Fukuoka Office
4-2-31 Tenjin Chuo-ku
Fukuoka 810 Japan

Miyata, Y.
Department of Civil Engineering
National Defense Academy
1-10-20 Hashirimizu, Yokosuka
Kanagawa 239 Japan

Miyatake, H.
Construction Engineering Division
Public Works Research Institute,
Ministry of Construction
Asahi-1, Tsukuba 305 Japan

Miyazaki, K.
Technical Research Institute
Nishimatsu Construction Co., Ltd.
2570-4, Shimotsuruma, Yamato
Kanagawa 242 Japan

Monther, A.H.
Civil Engineering Department
Nagasaki University
1-14 Bunkyo-machi
Nagasaki 852 Japan

Moroto, N.
Department of Civil Engineering
Hachinohe Institute of Technology
88-1 Myo Obiraki, Hachinohe
Aomori 031 Japan

Motoda, K.
Kyushu Branch, Civil Engineering Department
Kajima Corporation
3-12-10 Hakataekimae, Hakata-ku
Fukuoka 812 Japan

Munaf, Y.
Institute of Industrial Science,
University of Tokyo
7-22-1, Roppongi, Minato-ku, 106 Japan

Murata, H.
Department of Civil Engineering
University of Yamaguchi
2557 Tokiwadai
Ube 755 Japan

Murata, S.
Depatment of Civil Engineering
Kumamoto Institute of Technology
4-22-1 Ikeda
Kumamoto 861-11 Japan

Nabeshima, Y.
Department of Civil Engineering
Osaka University
2-1, Yamadaoka, Suita
Osaka 565 Japan

Nagamatsu, Y.
West Japan Engineering Consultants, Inc.
Geotechnical Engineering Department
Kyushu Electric Power Co.,
1-47, 2-Chome, Siobaru, Minami-ku
Fukuoka 815 Japan

Nagami, H.
Hiroshima Office
Nippon Zeon Co., Ltd.
Seiko Hiroshima Bldg.
7-1 Mikawa-cho, Naka-ku
Hiroshima 730 Japan

Nagase, H.
Department of Civil Engineering
Kyushu Institute of Technology
1-1 Sensui, Tobata-ku
Kitakyushu 804 Japan

Nagashima, H.
Technical Division
Nihonchiken Co., Ltd.
25-25, Morooka 5-chome, Hakata-ku
Fukuoka 816 Japan

Nagatomo, F.
Japan Port Consultants Ltd.
Musunoki 7th Bldg.
1-1 Hie-machi, Hakata-ku
Fukuoka 812 Japan

Nagatomo, H.
Department of Civil Engineering
Oita National College of Technology
1666, Maki, Oita 870-01 Japan

Nakai, T.
Department of Civil Engineering
Nagoya Institute of Technology
Gokiso-cho, Showa-ku
Nagoya 466 Japan

Nakajima, T.
Department of Geotechnical Engineering,
Research Center
Sato Kogyo Co., Ltd.
12-20 Honcho 4-Chome, Nihonbashi, Chuo-ku
Tokyo 103 Japan

Nakamura, K.
Higashi Kanto Road Engineering Co.
177 Naganamibara-machi, Inage-ku,
Chiba 263 Japan

Nakamura, M.
Civil Engineering Research Department
Fujita Corporation
74 Ohdana-cho, Tsuzuki-ku
Yokohama 224 Japan

Nakamura, R.
Shinwa Techno Co., Ltd.
Geotechnical Engineering Laboratory
1856-1 Ohaza Tokuman, Kubota, Sagagun,
Saga 849-02 Japan

Nakamura, T.
Department of Civil Engineering
Tomakomai National College of Technology
433 Nishikioka, Tomakomai
Hokkaido 059-12 Japan

Nakane, A.
Institute of Technology
Tekken Corporation
9-1 Shinsen, Narita
Chiba 286 Japan

Nakano, H.
Geotechnical Institute Co., Ltd.
2-15-23-102, Haruyoshi, Chuo-ku
Fukuoka 810 Japan

Nakano, Y.
Fukuoka Branch
Teikoku Consultant
4-29-46, Chiyo, Hakata-ku
Fukuoka 812 Japan

Nakata, Y.
Department of Civil Engineering
University of Yamaguchi
2557 Tokiwadai
Ube 755 Japan

Nakatani, T.
Consultants Daichi Co., Ltd.
2-20-16 Koihonmachi, Nishi-ku
Hirosima-city 733 Japan

Nakaya, T.
Construction Material Section
Kuraray Co., Ltd.
12-39, 1-chome, Umeda, Kita-ku,
Osaka 530 Japan

Nakazono, K.
Token Geotech Co., Ltd.
5-16-13 Watanabedori, Cyuo-ku
Fukuoka 810 Japan

Nanbu, Y.
Civil Engineering Construction Planning Department
Obayashi Corporation
4-33 Kitahamahigashi Chuo-ku
Osaka 540 Japan

Nemoto, T.
Shikoku Branch Office
Sumiko Consultants Co., Ltd.
1-6, Sobiraki-cho
Niihama 792 Japan

Nishi, S.
Fukuoka Division
Tanaka Co., Ltd.
No.6 Green Bldg.
2-12-9 Hakataekimae Hakata-ku
Fukuoka 812 Japan

Nishida, K.
Research and Development Laboratory
Matsuo Construction Co., Ltd.
4-27, 1-chome, Tafuse
Saga 840 Japan

Nishigata, T.
Department of Civil Engineering
Kansai University
3-3-35 Yamate-cho, Suita,
Osaka 564 Japan

Nishikawa, K.
Kisojiban Consultants Co., Ltd.
1-11-5, Kaminagaya, Nishi-ku
Nagoya 451 Japan

Nishimura, J.
Civil Engineering Materials Group,
Research Laboratory
Mitsui Petrochemical Industrial Products, LTD.
No.9 Kawarai-cho, Kuki-City
Saitama 346 Japan

Nitta, Y.
Tohoku Branch
Kiso-Jiban Consultants Co., Ltd.
2-9-23, Gorin, Miyagino-ku
Sendai 983 Japan

Nomaguchi, A.
Construction Project Consultants, Inc.
N8 Matsuda Bldg.
2-1-9 Okubo, Shinjuku-ku
Tokyo 169 Japan

Nonoshita, K.
Daiei Kaihatsu Co., Ltd.
2690 Hiu-machi, Sasebo
Nagasaki 857-11 Japan

Ochiai, H.
Department of Civil Engineering
Kyushu University
6-10-1 Hakozaki, Higashi-ku
Fukuoka 812-81 Japan

Ochiai, T.
Yahagi Construction Co., Ltd.
Civil Engineering Department
19-7 Aoi 3-chome, Higashi-ku
Nagoya 461 Japan

Ogisako, E.
Frontier Technology Dept., Institute of Technology
Shimizu Corporation
4-17, Etchujima 3-chome, Koto-ku
Tokyo 135 Japan

Ohira, T.
Takayama Sougou Industry Co., Ltd.
2-6-12 Minami-tsurusaki,
Ohita 870-01 Japan

Ohkubo, T.
Division of Industrial Science
the Graduate School of Science and Technology
Kobe University
1-1 Rokkodai-cho Nada-ku Kobe
Hyogo 657 Japan

Ohno, S.
Department of Civil Engineering
Kyushu University
6-10-1 Hakozaki, Higashi-ku
Fukuoka 812-81 Japan

Ohshima, H.
Civil Engineering Materials Business Department
Mitsui Petrochemical Industrial Products, Ltd.
Ueno TH Bldg. 8F
39-10 Yushima 3-chome, Bunkyo-ku
Tokyo 113 Japan

Ohta, H.
Department of Civil Engineering
Kanazawa University
2-40-20 Kodatsuno
Kanazawa 920 Japan

Ohtsuka, S.
Nagaoka University of Technology
Department of Civil Engineering
1603-1, Kamitomioka, Nagaoka
Niigata 940-21 Japan

Ohuchi, M.
Kyushu Branch
Chuo Kaihatsu Corporation
6-3-27 Torikai, Johnan-ku
Fukuoka 814-01 Japan

Oikawa, H.
Department of Civil Engineering
Akita University
1-1, Tegatagakuen-machi
Akita 010 Japan

Okabayashi, K.
Department of Civil Engineering
Kochi National College of Technology
200-1 Otsu, Shinkai, Monobe, Nangoku
Kochi 783 Japan

Okamura, Y.
Industrial Textile Technology Department
Teijin Ltd.
3-4-1 Minohara, Ibaraki-shi
Osaka 567 Japan

Okita, A.
Tokyo Road Engineering Ltd.
3-1-3 Yushima, Bunkyo-ku
Tokyo 113 Japan

Okumura, T.
Faculty of Environmental Science & Technology
Okayama University
2-1-1 Tsushima-naka
Okayama 700 Japan

Okuzono, S.
Department of Civil Engineering
Kyushu Sangyo University
3-1 Matsukadai 2-chome, Higashi-ku
Fukuoka 813 Japan

Omine, K.
Department of Civil Engineering
Kyushu University
6-10-1 Hakozaki, Higashi-ku
Fukuoka 812-81 Japan

Oniki, K.
Kajima Technical Research Institute
Kajima Corporation
19-1 Tobitakyu 2-chome, Chofu-shi
Tokyo 182 Japan

Onitsuka, K.
Faculty of Science and Engineering
Saga University
1 Honjo
Saga 840 Japan

Oota, H.
Terre Armee Method Department
Kawasho Corporation
Osaka Fukokuseimei Bldg.
2-4, Komatsubara-cho, Kita-ku
Osaka 530 Japan

Otani, J.
Department of Civil & Environmental Engineering
Kumamoto University
2-39-1 Kurokami
Kumamoto 860 Japan

Otani, Y.
Hirose & Co., Ltd.
Reinforced Earth Division
Toyo Central Bldg.
4-1-13 Toyo, Koto-ku
Tokyo 135 Japan

Ozaki, A.
Civil Engineering Technology Department
Shimizu Corporation
1-29, Akasaka 1-Chome, Chuo-ku
Fukuoka 810 Japan

Ozawa, Y.
Principal Civil Engineering Office
Nikken Sekkei, Ltd.
2-1-2 Koraku, Bunkyo-ku
Tokyo 112 Japan

Pokharel, G.
Yahagi Construction Co., Ltd.
3-19-7 Aoi-cho, Higashi-ku
Nagoya 461 Japan

Porbaha, A.
Geotechnical Centrifuge Facilities
Port and Harbour Research Institute
3-1-1 Nagase
Yokosuka 239 Japan

Pradhan, T.B.S.
Department of Civil Engineering
Yokohama National University
156 Tokiwadai, Hodogaya-ku
Yokohama 240 Japan

Ryokai, K.
Institute of Technology
Shimizu Corporation
4-17, Etchujima 3-chome, Koto-ku
Tokyo 135 Japan

Saitoh, A.
Kawasaki Metal Products Engineering Inc.
Kokusai Kanko Kaikan
8-3, Marunouchi, 1-Chome, Chiyoda-ku
Tokyo 100 Japan

Saitoh, K.
Nakase Geotechnical Institute
Nikken Sekkei
4-11-1 Minamikase, Saiwai-ku
Kawasaki 211 Japan

Sakai, A.
Department of Civil Engineering
Saga University
Honjo-machi 1
Saga 840 Japan

Sakajo, S.
Kiso-Jiban Consultants Co., Ltd.
Soil Engineering Center
1-11-5 Kudan-kita, Chiyoda-ku
Tokyo 102 Japan

Sakamoto, S.
Kyushu Railway Company
3-2-8, Hakataekimae, Hakata-ku
Fukuoka 812 Japan

Sakata, T.
CTI Engineering Co, Ltd.
Highway and Bridge Division
18 Fukuoka BLDG 1-10,
Watanabe-Dori 2-Chome, Chuo-ku
Fukuoka 810 Japan

Sasada, T.
Research Laboratory
Kyushu Electric Power Co., Inc.
2-1-47, Shiobaru, Minami-ku
Fukuoka 815 Japan

Satake, M.
Kozogijutsu center
3-5-7, Hakataekimae, Hakata-ku
Fukuoka 812 Japan

Sato, K.
Department of Civil Engineering
Fukuoka University
8-19-1 Nanakuma Jonan-ku
Fukuoka 814-80 Japan

Sato, K.
Higashi Kanto Road Engineering Co.
Kosumo Park Bldg. 2F
5-7-18 Higashi-Higuresato, Arakawa-ku
Tokyo 116 Japan

Sato, K.
Kyushu Branch, Department of Civil Engineering
Fujita Corporation
Hakata Bldg., 8-36 Cyuogai, Hakata-ku
Fukuoka 812 Japan

Sato, M.
Geotechnical Engineering Department
Institute of Technology
Shimizu Corporation
4-17, Etchujima 3-chome, Koto-ku
Tokyo 135 Japan

Sato, Y.
Nagoya Road Engineer Co.
1-7-33 Sakae Naka-ku
Nagoya 460 Japan

Sekiyama, T.
Nippon Zeon Co., Ltd.
Furukawa Sogo Bldg.
2-6-1 Marunouchi, Chiyoda-ku
Tokyo 100 Japan

Setoyama, K.
Kei Kei Emu Co., Ltd.
3-5-2 Nakamura-Higashi
Miyazaki 880 Japan

Sezaki, M.
Department of Civil Engineering
Miyazaki University
1-1 Gakuen Kibanadai Nishi
Miyazaki 889-21 Japan

Shahu, Jagdish T.
Institute of Lowland Technology
Saga University
Honjo-machi 1
Saga 840 Japan

Shimotani, J.
Products & Market Development Group
Chori Co., Ltd.
4-7 Kawaramachi 2-chome, Chuo-ku
Osaka 541 Japan

Shintani, H.
Civil Engineering Material Division
Mitsubishi Kagaku Sanshi Corporation
Touzan Bldg.
4-4-2, Nihonbashi, Honcho, Chuo-ku
Tokyo 103 Japan

Shiwakoti, D.R.
Department of Civil Engineering
Yokohama National University
156 Tokiwadai, Hodogaya-ku
Yokohama 240 Japan

Shu, K.
Department of Civil Engineering
Miyazaki University
1-1 Gakuen Kibanadai Nishi
Miyazaki 889-21 Japan

Sumita, H.
Civil Engineering Division Group
Penta-Ocean Construction Co. Ltd.
12-11, Kego, 1-chome, Chuo-ku
Fukuoka 810 Japan

Suwa, S.
Geo-Research Institute, Osaka Soil Test Lab.
4-3-2 Itachibori, Nishi-ku
Osaka 550 Japan

Suzuki, H.
Construction Steel Products Dept. Tokyo
Sumitomo Corporation
Josuikai Bldg.
2-1-1 Hitotsubashi, Chiyoda-ku
Tokyo 101 Japan

Suzuki, U.
Kuraray Co., Ltd.
Shin-Nihonbashi Building
3-8-2 Nihonbashi, Chuo-ku
Tokyo 103 Japan

Tada, M.
Disaster Prevention and Snow Section
Research Institute
Japan Highway Public Corporation
1-4-1 Tadao, Machida
Tokyo 194 Japan

Taga, H.
Geo Engineering Division
Fudo Construction Co., Ltd.
2-11-16 Hakataekimae, Hakata-ku
Fukuoka 813 Japan

Tajima, K.
Technical Research Institute
Advanced Technology Division
Mitsubishi Construction Co., Ltd.
1-297 Kitabukuro-cho, Omiya-shi
Saitama 330 Japan

Tajiri, N.
Research & Development Laboratory
Fukken Co., Ltd.
2-10-11, Hikarimachi, Higashi-ku
Hiroshima 732 Japan

Takada, J.
Sales Marketing Department
Kawasaki Metal Products Engineering Inc.
Kokusai Kanko Kaikan
8-3, Marunouchi, 1-Chome, Chiyoda-ku
Tokyo 100 Japan

Takagi, K.
Civil Engineering Division Group
Penta-Ocean Construction Co. Ltd.
12-11, Kego, 1-chome, Chuo-ku
Fukuoka 810 Japan

Takahashi, H.
Technical Division
Nittoc Construction Co., Ltd.
13-18 Akashi-cho, Chuo-ku
Tokyo 104 Japan

Takahashi, Y.
Geo Engineering Division
Fudo Construction Co., Ltd.
4-2-16 Hiranomachi, Chuo-ku
Osaka 541 Japan

Takano, T.
Kyushu Branch
Oriental Consultants, Co., Ltd.
Fujii Bldg., 3-10-24 Hakataekimae, Hakata-ku
Fukuoka 812 Japan

Takasaki, H.
Civil Engineering Division
Shimizu Corporation
1-2-3, Shibaura, Minato-ku
Tokyo 105-07 Japan

Takeuchi, S.
Applied Hydraulic Laboratory
Tetra Co., Ltd.
2-7, Higashi-Nakanuki-machi, Tsuchiura
Ibaraki 300 Japan

Taki, M.
Takamatsu Branch Office
Fukken Co., Ltd.
1-4-40 Nishiki-machi
Takamatsu 760 Japan

Tanabashi, Y.
Civil Engineering Department
Nagasaki University
1-14 Bunkyo-machi
Nagasaki 852 Japan

Tanaka, K.
Kyushu Branch Office
Penta-Ocean Construction Co. Ltd.
12-11, Kego, 1-chome, Chuo-ku
Fukuoka 810 Japan

Tanaka, T.
Marketing & Development Institute
Sekisui Chemical Co., Ltd.
75 Nojiri, Ritto-cho, Kurita-gun
Shiga 520-30 Japan

Taniguchi, S.
Civil Engineering Department
Wakachiku Construction Co., Ltd.
2-23-18 Shimomegro, Meguro-ku
Tokyo153 Japan

Tanoue, Y.
Kiso-Jiban Consultants Co., Ltd.
2-16-7 Hara Sawara-ku
Fukuoka 814 Japan

Tateyama, M.
Railway Technical Research Institute
2-8-38 Hikari-cho, Kokubunji-shi
Tokyo 185 Japan

Tatsuoka, F.
Department of Civil Engineering
The University of Tokyo
7-3-1 Honjo, Bunkyo-ku
Tokyo 113 Japan

Tawa, M.
Sanyu Consultants
1-23 Konya Imamachi
Kumamoto 860 Japan

Tayama, K.
Sakurada Co., Ltd.
Nihon Seimei Hakata-minami 2nd Bldg.
3-1-1, Hakataeki-minami, Hakata-ku
Fukuoka 812 Japan

Tayama, S.
Disaster Prevention and Snow Section
Research Institute
Japan Highway Public Corporation
1-4-1 Tadao, Machida
Tokyo 194 Japan

Terada, T.
Sato Corporation
1-21-32 Shimizu, Minami-ku
Fukuoka 815 Japan

Terashi, M.
Nakase Geotechnical Institute
Nikken Sekkei
4-11-1 Minamikase, Saiwai-ku
Kawasaki 211 Japan

Torii, T.
Geotechnical Engineering Division
Construction Project Consultants Inc.
3-5-25 Utsubo-honmachi, Nishi-ku
Osaka 550 Japan

Toriihara, M.
Technical Research Institute
Civil Engineering Department 1
Obayashi Corporation
640 Shimokiyoto 4-chome, Kiyose-shi
Tokyo 204 Japan

Toyofuku, S.
Kyushu Branch
Obayashi Corporation
Fukuoka Takeda Bldg.
9-12 Simokawabata-machi, Hakata-ku,
Fukuoka 812 Japan

Tsukada, Y.
Construction Engineering Division
Materials and Construction Department
Public Works Research Institute,
Ministry of Construction
Asahi-1, Tsukuba 305 Japan

Tsukamoto, H.
Engineering Planning Section
Construction Project Consultants, Inc.
CPC Bldg., 3-5-25, Utsubohonmachi, Nishi-ku
Osaka 550 Japan

Tsukamoto, S.
Kyushu Branch Consultant Division
Oyo Corporation
2-21-36 Ijiri, Minami-ku
Fukuoka 816 Japan

Tsukamoto, Y.
Science University of Tokyo
Department of Civil Engineering
2641 Yamazaki, Noda City
Chiba 278 Japan

Tsumiya, K.
Geological Survey
Wesco Inc. Tottori Branch Office
270-1 Karo-cho
Tottori 680 Japan

Tsunoda, N.
Civil Engineering Materials Division
Mitsubishi Kagaku Sanshi Corporation
Tozan Bldg., 4-4-2 Nihonbashi, Honcho, Chuo-ku
Tokyo 103 Japan

Tsurukubo, H.
Fukuoka Doro Engineering Co., Ltd.
8-4-2 Watanabedori, Chuo-ku
Fukuoka 810 Japan

Tsutsumi, T.
Earthquake Disaster Prevention Research Center
Public Works Research Institute,
Ministry of Construction
1, Asahi, Tsukuba 305 Japan

Uchimura, T.
Department of Civil Engineering (Tatsuoka Lab.)
The University of Tokyo
7-3-1 Honjo, Bunkyo-ku
Tokyo 113 Japan

Uehara, H.
University of the Ryukyus
Department of Civil Engineering
Senbaru 1 Nishihara-cho
Okinawa 903-01 Japan

Uehara, S.
Institute of Technology & Development
Sumitomo Construction Co., Ltd.
1726 Niragawa, Minamikawachi-machi
Kawachi-gun, Tochigi 329-04 Japan

Uekuma, H.
Kyushu Branch Engineering Department
Oyo Corporation
2-21-36 Ijiri, Minami-ku
Fukuoka 816 Japan

Ueno, M.
Dai Nippon Construction
Research and Development Center
3-56-2 Nangai Higashi-Yamato-Shi
Tokyo 162 Japan

Ueyama, Y.
Kyushu Office
Tenox Corporation
Fukuoka Hiei Bldg., 4-1-17 Tenjin, Cyuo-ku
Fukuoka 810 Japan

Ugai, K.
Civil Engineering Department
Gunma University
Tenjin-1, Kiryu-shi
Gunma 376 Japan

Umebayashi, F.
Okasan Kogyo Co., Ltd.
Tokyo Sumitomo Shibadaimon Bldg.
2-5-5 Shibadaimon, Minato-ku
Tokyo 105 Japan

Umezaki, T.
Department of Civil Engineering
Shinshu University
500 Wakasato
Nagano 380 Japan

Uno, K.
Department of Civil Engineering
Kyushu Kyoritsu University
1-8 Jiyugaoka, Yahatanisi-ku
Kitakyushu 807 Japan

Uzuoka, R.
Engineering & Design Division
Hazama Corporation
2-5-8 Kita-Aoyama, Minato-ku
Tokyo 107 Japan

Wada, H.
Kyushu Office
Raito Kogyo Co., Ltd.
1-28-44 Katakasu, Hakata-ku
Fukuoka 812 Japan

Wakasa, H.
Kyushu Branch
Oyo Corporation
2-21-36 Ijiri, Minami-ku
Fukuoka 816 Japan

Watanabe, K.
Koa Kaihatsu Co., Ltd.
5-3-13 Kotobashi, Sumida-ku
Tokyo 130 Japan

Watanabe, K.
Department of Civil Engineering
Saga University
1 Honjo-machi
Saga 840 Japan

Watanabe, Y.
Kyushu Branch
Kawasaki Geological Engineering Co., Ltd.
1-40 Gionmachi, Hakata-ku
Fukuoka 812 Japan

Watari, Y.
Mizuno Institute of Technology
Miura Bldg., 2-22, Nakajima-cho, Naka-ku
Hiroshima 730 Japan

Weeraya, S.T.
Soil Mechanics Laboratory, Structural Engineering
Department of Civil Engineering,
Hiroshima University
1-4 Kagamiyama 1-chome,
Higasi-Hiroshima 739 Japan

Yahaba, T.
Public Works Materials Group
Tokyo Printing Ink MFG Co., Ltd.
2-7 Tabatashinmachi, Kita-ku
Tokyo 114 Japan

Yaida, O.
Department of Clothing & Textile Science
Kyoto Women's University
35 Imakumano-KitaHiyoshi-cho, Higashiyama-ku
Kyoto 605 Japan

Yakabe, H.
Technical Management Division
Dia Consultants Co., Ltd.
Douwa Kasai Hakata Bldg.
2-9-2 Sumiyoshi, Hakata-ku
Fukuoka 812 Japan

Yamada, H.
Technical Devision
Nittoc Construction Co., Ltd.
8-14-14 Ginza Chuo-ku
Tokyo 104 Japan

Yamada, T.
Material and Construction Method Department
Public Works Research Institute
Ministry of Construction
1, Asahi, Tsukuba 305 Japan

Yamagami, T.
Department of Civil Engineering
The University of Tokushima
2-1 Minamijosanjima-cho
Tokushima 770 Japan

Yamaguchi, E.
Fukken Engineering
Sunhaitu Hirao 3F
9-8 Hirao 2-chome, Chuo-ku
Fukuoka 816 Japan

Yamamoto, A.
Obayashi Corporation
Technical Research Institute
Civil Engineering Department 1
640 Shimokiyoto 4-chome, Kiyose-shi
Tokyo 204 Japan

Yamamoto, K.
Department of Civil & Environmental Engineering
Kumamoto University
2-39-1 Kurokami
Kumamoto 860 Japan

Yamamoto, M.
Fukuoka Branch
Kyowa Engineering Consultants Co., Ltd
Tenjin F Bldg., 3-11-20 Tenjin, Chuo-ku
Fukuoka 810 Japan

Yamamoto, Y.
Kiso-Jiban Consultants Co., Ltd.
2-16-7 Hara, Sawara-ku
Fukuoka 814 Japan

Yamanaka, M.
Department of Civil Engineering
Nagasaki University
1-14, Bunkyo-machi
Nagasaki 852 Japan

Yamanouchi, T.
3-9, Ropponmatsu 4-chome, Chuo-ku
Fukuoka 810 Japan

Yamato, S.
Asahi Chemical Industry Co., Ltd.
Geotechnical Development Dept.
104 Someya, Sakaimachi Sashima-gun,
Ibaraki 306-04 Japan

Yamauchi, T.
Original Engineering Consultants Co., Ltd.
NORITZ Bldg. 4F,
3-1-1 Hakataeki-Higashi, Hakata-ku
Fukuoka 812 Japan

Yamauchi, Y.
Kyushu Laboratory
Geo-Research Institute Osaka
Otemon-daiichihoki Bldg., 3-5-1 Otemon, Chuo-ku
Fukuoka 810 Japan

Yanagisawa, E.
Department of Civil Engineering
Tohoku University
Aoba-ku, Sendai 980-77 Japan

Yang, J.
Department of Civil Engineering
Hachinohe Institute of Technology
88-1 Myo Obiraki, Hachinohe
Aomori 031 Japan

Yano, K.
Ocean Consultant, Japan Co., Ltd.
Kuramochi Bldg. 1
2-3-10 Koutoubashi, Sumida-ku
Tokyo 130 Japan

Yaoita, K.
Okasan Kogyo Co., Ltd.
Tokyo Sumitomo Shibadaimon Bldg.
2-5-5 Shibadaimon, Minato-ku
Tokyo 105 Japan

Yashima, A.
Department of Civil Engineering
Gifu University
1-1 Yanagido, Gifu 501-11 Japan

Yasuda, S.
Department of Civil Engineering
Tokyo Denki University
Hatoyama, Hiki-gun
Saitama 350-03 Japan

Yasufuku, N.
Department of Civil Engineering
Kyushu University
6-10-1 Hakozaki, Higashi-ku
Fukuoka 812-81 Japan

Yasukawa, I.
The development Division of Engineering
Geogyro Corporation
6-4 Onmaeda-cho Nishinanajo, Shimogyo-ku
Kyoto 600 Japan

Yokota, H.
Department of Civil Engineering
Miyazaki University
1-1 Gakuen Kibanadai Nishi
Miyazaki 889-21 Japan

Yokota, Y.
Highway and Bridge Division
CTI Engineering Co, Ltd.
18 Fukuoka Bldg., 2-1-10 Watanabe-Dori,
Chuo-ku, Fukuoka 810 Japan

Yokota, Y.
Technical Department
Maeda Kosen Co., Ltd.
Okinunome, Harue-cho, Sakai-gun
Fukui 919-04 Japan

Yokoya, N.
Department of Geological and Soil Explorations
Fukuyama Consultants Co., Ltd.
1-11-4 Katanoshin-machi, Kokura Kita-ku
Kitakyushu 802 Japan

Yonezawa, T.
Design Division, Head Office
Japan Railway Construction Public Corporation
Sanno Grand Bldg. 6F, 2-14-2 Nagata-cho
Chiyoda-ku, Tokyo 100 Japan

Yosida, N.
Highway and Bridge Division
CTI Engineering Co, Ltd.
18 Fukuoka Bldg., 2-1-10 Watanabe-Dori,
Chuo-ku, Fukuoka 810 Japan

Yoshimura, H.
Tokyo Printing Ink MFG Co., Ltd.
1-28-14 Imazato, Takamatsu,
Kagawa 760 Japan

Yoshimura, S.
Yoshihara Construction
1-21-19 Takasago, Chuo-ku
Fukuoka 810 Japan

Yuda, E.
Civil Engineering Products
Kawasho Corporation
2-4 Komatsubara-cho, Kita-ku
Osaka 530 Japan

Zen, K.
Port and Harbour Research Institute
Ministry of Transport
3-1-1 Nagase, Yokosuka 239 Japan

Zhang, T.
Department of Geotechnical Engineering
Institute of Technology
Shimizu Corporation
4-17, Etchujima 3-chome, Koto-ku
Tokyo 135 Japan

Njiru, J.R.
Building Materials Laboratory
Ministry of Public Works & Housing Materials
P.O. Box 11873, Nairobi Kenya

Ahn, T.B.
Geotechnical Engineering Division
Korea Institute of Construction Technology
104-503, Sungwon Apt, Pyngchondong 898
Donganqu, Anyang Kyungkiodo Korea

Ju, J.W.
Department of Civil Engineering
College of Engineering Sunchon National University
315 Maegok-dong, Sunchon-city
Chonnam, 540-742 Korea

Jung, J.Y.
E&S Engineering Co., Ltd.
3F. Daelim Bldg. Daechi-dong,
Kangnam-ku, Seoul Korea

Kang, I.K.
Vniel Consultant Co., Ltd.
RM301, Tea-Wha Bldg.
903-3 Daechi-Dong, Gangnam-gu,
Seoul Korea

Kim, H.T.
Department of Civil Engineering
Hongik University
72-1, Sangsu-Dong, Mapo-Ku
Seoul Korea

Kim, K.M.
Korea Institute of Construction and Structural Safety
2nd fl. Daelim Bldg. 913-11, Daechi-dong,
Kangnam-ku, Seoul, 135-280 Korea

Kim, J.S.
Daewoo Institute of Construction Technology
San 25, Songjuk-Dong, Jangan-Gu,
Suwon, Kyunggi-Do, 440-210 Korea

Kim, J.Y.
Department of Agricultural Civil Engineering
College of Agriculture
Chonbuk National University
664-14 Duckjin-dong Duckjinku
Chonju-city Chonbuk 560-756 Korea

Kim, S.K.
Department of Civil Engineering
College of Engineering
Dongguk University
Seoul 100-715 Korea

Lee, D.E.
E&S Engineering Co., Ltd.
3F. Daelim Bldg. Daechi-dong,
Kangnam-ku, Seoul Korea

Lee, E.S.
Korea Institute of Construction and Structural Safety
2nd fl. Daelim Bldg. 913-11, Daechi-dong,
Kangnam-ku, Seoul, 135-280 Korea

Shin, E.C.
Department of Civil Engineering
University of Inchon
177 Dowha-Dong, Nam-Gu
Inchon, 402-749 Korea

El Shabrawy, M.
Faculty of Technological Studies.PAAET
P.O.Box 3459 Hawally
Kuwait 32065 Kuwait

Inthavong, S.
Communication Training Center
Km 3 Watt Nak Road, Vientiane, Laos

Faisal, H. Ali
Faculty of Engineering
University of Malaya
59100 Kuala Lumpur Malaysia

Lawson, C.
Royal Ten Cate
11th Floor, Menara Glomac
Kelana Business Centre, 97, Jalan SS 7/2
47301 Petaling Jaya Malaysia

Loke, K. Hooi
Polyfelt Geosynthetics Sdn Bhd
No.2 Jalan SS21/62, Damansara Utama
47400 Petaling Jaya,
Selangor D.R. Malaysia

Mohd, T.A. Wahab
Ranhill Bersekutu Sdn Bhd
Wisma Ranhill, No.2-12Jalan Setiawangsa 10,
Taman Setiawangsa 54200,
Kuala Lumpur Malaysia

Oh., Y. Ching
Nehemiah Reinforced Soil Sdn Bhd
26c, Jalan SS21/62, D'sara Utama, 47400 P.J.
Selangor Darul Ehsan Malaysia

Tack, W. Yee
Royal Ten Cate
11th Floor, Menara Glomac
Kelana Business Centre, 97 Jalan ss7/2,
47301 Petaling Jaya Malaysia

Pannachan, S.L.
APTEC Consultancy (P) Ltd.
Jawalakhel, Kathmandu,
P.O.Box 8975, EPC.338 Nepal

Voskamp, W.
Akzo Nobel Geosynthetics bv
P.O.Box 9300,
NL-6800 SB Arnhem Netherlands

Murillo, C.Y.
School of Civil Engineering
San Agustin National University
Arias Araguez 527 Mariano Melgar
Arequipa, Peru

Klosek, K.
Faculty of Civil Engineering,
The Silesian Technical University of Gliwice,
ul. Akademicka 5
44-100 Gliwice Poland

Cardoso, A.
Faculdade de Engenharia da Universidade do Porto
Departamento de Engenharia Civil-Seccao
de Geotecnia
Rua dos Bragas, 4099 Porto Codex Portugal

Lopes, M.L.
Faculdade de Engenharia da Universidade do Porto
Departamento de Engenharia Civil-Seccao
de Geotecnia
Rua dos Bragas, 4099 Porto Codex Portugal

Pinto, I.
University of Coimbra
Department of Civil Engineering
3049 COIMBRA Codex Portugal

Babin, L.
Ufa State Petroleum Technological University
450062, Kosmonavtov st, 1, Ufa Russia

Kulagin, V.
Ufa State Petroleum Technological University
450062, Kosmonavtov st, 1, Ufa Russia

Shchepin, N.
Ufa State Petroleum Technological University
450062, Kosmonavtov st, 1, Ufa Russia

Spector, Y.
Ufa State Petroleum Technological University
450062, Kosmonavtov st, 1, Ufa Russia

Lim, S.H.
Public Works Department
5 Maxell Road, # 07-00 Tower Block,
MND Complex
Singapore 069110 Singapore

Yogarajah, I
L&M Geotechnical Pte. Ltd.
63 Hillview Avenue
#05-01 Lam Soon Industrial Building
Singapore 669569 Singapore

Guruge, D.C.G.
State Engineering Corporation of Sri Lanka
1/3, 2nd lane, Anura Mawatha,
Kalubowi la, Dehiwala Sri Lanka

Sumanaratne, I.H.D.
Road Construction & Development Co., Ltd.
Station Road, Angulana Sri Lanka

Chang, T.T.D.
Department of Civil Engineering
Chung Yuan University
Chung Li, 32023 Taiwan, R.O.C.

Chen, T.C.
Department of Civil Engineering
National Taiwan University
1 Roosevelt Rd. Sec. 4
Taipei 106 Taiwan, R.O.C.

Huang, C.C.
Department of Civil Engineering
Cheng Kung University
Tainan Taiwan, R.O.C.

Kuo, C.
Measure Instrument Co., Ltd.
3F, No. 216, Sec. 2, Nan Chang Road,
Taipei 100 Taiwan, R.O.C.

Lee, Y.S.
Department of Civil Engineering
National Taiwan University
1 Roosevelt Rd. Sec. 4
Taipei 106 Taiwan, R.O.C.

Lin, C.C.
All-Toffs Industrial Co., Ltd.
4 Alley 136 Lane 39
Sec. 6 Min Taiwan, R.O.C.

Lin, S.S.
Department of Harbor and River Engineering
Taiwan Ocean University
Keelung 20224 Taiwan, R.O.C.

Shen, Y.M.
Ministry of Economy
15 Fu-cho Street
Taipei Taiwan, R.O.C.

Yen, C.W.
Marco Green Ent. Co., Ltd.
7F-3 No.21-1
Bao-ching Str Taiwan, R.O.C.

Bergado, D.
School of Civil Engineering
Asian Institute of Technology
P.O.Box 2754, Bangkok 10501, Thailand

Boontharaksa, P.
Department of Highways
Road Research and Development Center
Sriayudhaya Rd. Rajatheevee
Bangkok Thailand

Mairaing, W.
Civil Engineering Department
Kasetsart University
Bangkok 10900 Thailand

Nontananandh, S.
Civil Engineering Department
Kasetsart University
Bangkok 10900 Thailand

Soralump, S.
School of Civil Engineering
Asian Institute of Technology
P.O.Box 2754, Bangkok 10501 Thailand

Suphakit, W.
Royal Irrigation Dept.
Samsen Road Dusit
Bangkok 10300 Thailand

Taesiri, Y.
Department of Highways
Bangkok Thailand

Wiboon, N.
Royal Irrigation Dept.
Samsen Road Dusit
Bangkok 10300 Thailand

Erken, A.
Civil Engineering Faculty
Istanbul Technical University
Maslak, Istanbul Turkey

Guler, E.
Department of Civil Engineering
Bogazici University
80815 Bebek, Istanbul Turkey

Gurbuz, A.
General Directorate of Electrical Power
Resources Survey and Development Administration
EIE Idaresi Gn. Md. Eskisehir Yolu
7. Km 06520 Ankara Turkey

Austin, R.
Netlon Limited
New Wellington Street
Blackburn, BB2 4PJ UK

Bolton, M.
Department of Engineering
Cambridge University
Trumpington Street, Cambridge
CB2 1PZ UK

Chang, K.T.
Department of Engineering Science
University of Oxford
19 parks road, Oxford, OX1 3PJ UK

Cousens, T.
Department of Civil Engineering
The University of Leeds
Leeds LS2 9JT UK

Greenwood, J.H.
ERA Technology Ltd.
Cleeve Road, Leatherhead
Surrey KT 22 7SA UK

Jones, C.J.F.P.
University of Newcastle upon Tyne
Department of Civil Engineering
Drummond Building, Devonshire Terrace
Newcastle upon Tyne NE1 7RU UK

Kempton, G.
Terram Ltd.
13 Heath Park Drive, Bromley
Kent BR1 2WQ UK

Kenny, M.
Department of Civil Engineering
University of Strathclyde
John Anderson Building
107 Rottenrow, Glasgow G4 0NG UK

Myles, B.
Consultant Civil & Geosynthetic Engineer
Robin Hill, Catbrook, Chestow-Gwent
NP6 6NA UK

Price, D.
Okasan Kogyo
17-B Ellington Road, Muswell Hill,
London N10 3DD UK

Sharma, J.S.
Department of Engineering
Queen Mary & Westfield College
Mile End Road, London E1 4NS UK

Adanur, S.
Textile Engineering Department
Auburn University
Auburn, AL 36849 USA

Bang, S.
Department of Civil and Environmental Engineering
South Dakota School of Mines and Technology
Rapid City, SD, 57701 USA

Bloomfield, R.
The Reinforced Earth Co.
8614 Westwood Center Dr, Suite 1100
Vienna, Va 22182 USA

Dorovdian, M.
Department of Civil and Environmental Engineering
University of California, Los Angeles
CA 90095-1593 USA

Drumm, E.C.
Department of Civil Engineering
University of Tennessee
110 Perlcins Hall Knoxville
TN 37996 USA

Fettig, D.R.
Sixth International Conference on Geosynthetics
Industrial Fabrics Association International
345 Cedar Street, Suite 800
St. Paul, MN 55731 USA

Leshchinsky, D.
University of Delaware
Department of Civil Engineering
Newark, DE 19716 USA

Mahmud, M.B.
Rensselaer Polytechnic Institute
Troy, NY 12180 USA

Nataraj, M.S.
Department of Civil and Environmental Engineering
University of New Orleans-College of Engineering
New Orleans, LA 70148 USA

Stevenson, P.
The Tensor Corpotation
6100 Lake Forest Drive Suite 230
Atlanta GA 30328 USA

Vucetic, M.
Department of Civil and Environmental Engineering
University of California, Los Angeles
CA 90095-1593 USA

White, D.M.
Applied Geotechnology, Inc.
300, 12th Ave. NE, P.O.Box 3885
Bellevue, WA 98005 USA

Wu, J.
Department of Civil Engineering, CU-Denver
University of Colorado
Denver, Co 80217-3364 USA

Zhao, A.
Tenax Corporation
4800 East Monument Street
Baltimore, Maryland 21205 USA

Earth Reinforcement, Ochiai, Yasufuku & Omine (eds) © *1997 Balkema, Rotterdam, ISBN 90 5410 833 9*

Author index